Vulnerability and Adaptation Assessments

An International Handbook

Environmental Science and Technology Library

VOLUME 7

The titles published in this series are listed at the end of this volume.

Vulnerability and Adaptation Assessments

An International Handbook

Version 1.1
April 1996

Edited by

Ron Benioff

Country Studies Management Team,
U.S. Department of Energy,
Washington, DC, U.S.A.

Sandra Guill

Country Studies Management Team,
U.S. Department of Energy,
Washington, DC, U.S.A.

and

Jeffrey Lee[†]

U.S. Environmental Protection Agency,
National Health and Environmental Effects,
Corvallis, OR, U.S.A.

U.S. Country Studies Management Team
1000 Independence Avenue, S.W. (PO-63)
Washington, DC 20585
Phone: 202-426-1628
Fax: 202-426-1540/202-426-1551
Email: CSMT@igc.apc.org

KLUWER ACADEMIC PUBLISHERS
DORDRECHT / BOSTON / LONDON

A C.I.P. Catalogue record for this book is available from the Library of Congress.

ISBN-13: 978-94-010-6621-1   e-ISBN-13: 978-94-009-0303-6
DOI: 10.1007/978-94-009-0303-6

Published by Kluwer Academic Publishers,
P.O. Box 17, 3300 AA Dordrecht, The Netherlands.

Kluwer Academic Publishers incorporates
the publishing programmes of
D. Reidel, Martinus Nijhoff, Dr W. Junk and MTP Press.

Sold and distributed in the U.S.A. and Canada
by Kluwer Academic Publishers,
101 Philip Drive, Norwell, MA 02061, U.S.A.

In all other countries, sold and distributed
by Kluwer Academic Publishers Group,
P.O. Box 322, 3300 AH Dordrecht, The Netherlands.

Printed on acid-free paper

In memory of Dr. Jeffrey Lee
whose leadership and guidance was instrumental
in the preparation of this document

PREFACE

The possible impacts of global climate change on different countries has led to the development and ratification of the Framework Convention on Climate Change (FCCC) and has a strong bearing on the future sustainable development of developing countries and countries with economies in transition. The preparation of analytical methodologies and tools for carrying out assessments of vulnerability and adaptation to climate change is therefore of prime importance to these countries. Such assessments are needed to both fulfill the reporting requirements of the countries under the FCCC as well as to prepare their own climate change adaptation and mitigation plans.

The vulnerability and adaptation assessment guidelines prepared by the U.S. Country Studies Program bring together all the latest knowledge and experience from around the world on both vulnerability analysis as well as adaptation methodologies. It is currently being applied successfully by scientists in over fifty countries from all the regions of the globe. This guidance is being published to share it with the wider scientific community interested in global climate change issues.

This guidance document has two primary purposes:

- To assist countries in making decisions about the scope and methods for their vulnerability and adaptation assessments,
- To provide countries with guidance and step-by-step instructions on each of the basic elements of vulnerability and adaptation assessments.

This document has been designed to be used in tandem with the Intergovernmental Panel on Climate Change (IPCC) draft *Technical Guidelines for Assessing Climate Change Impacts and Adaptation*. It provides a stepwise discussion of the use of selected methods, building on the analytic framework and discussion of alternative methods in the IPCC document. Previous editions of this document have undergone peer review by U.S. and international experts. In addition, previous versions of this document were used at a vulnerability and adaptation assessment training workshop for 56 countries participating in the U.S. Country Studies Program.

These guidelines provide detailed and comprehensive guidance on international vulnerability and adaptation assessment methods and describe a relatively simple approach for each step and stage of analysis. The guidelines have allowed countries to conduct multifaceted assessments tailored to their own specific needs. It is hoped that organizations conducting vulnerability and adaptation studies in different parts of the world can make use of these guidelines as well as improve on them.

The results of these studies supported by the U.S. Country Studies Program will improve global understanding of vulnerability and adaptation issues and contribute to the work of the IPCC and FCCC subsidiary bodies.

Dr. Saleemul Huq
Executive Director
Bangladesh Centre for Advanced Studies

LIST OF CONTRIBUTORS

This document was prepared by a team of experts in the field of analysis of vulnerability and Adaptation to climate change. Direct contact with these experts to address questions is encouraged.

<u>Lead Editors</u>

Ron Benioff

Country Studies Management Team (PO-63)
U.S. Department of Energy
1000 Independence Avenue, S.W.
Washington, DC 20585
phone: 202-426-1635
fax: 202-426-1540
e-mail: rbenioff@igc.apc.org

Sandy Guill

Country Studies Management Team (PO-63)
U.S. Department of Energy
1000 Independence Avenue, S.W.
Washington, DC 20585
phone: 202-426-1635
fax: 202-426-1540
e-mail: sguill@igc.apc.org

Jeffrey Lee

U.S. Environmental Protection Agency
National Health and Environmental Effects
Research Laboratory
200 S.W. 35th Street
Corvallis, Oregon 97333
phone: 503-754-4600
fax: 503-754-4799
e-mail: jeffl@heart.cor.epa.gov

<u>Scenarios, Adaptation</u>

Joel Smith

Hagler Bailly Consulting, Inc.
P.O. Drawer O
Boulder, CO 80306-1906
phone: 303-449-5515 (ext. 318)
fax: 303-443-5684
e-mail: jsmith@habaco.com

General Circulation Models

Roy Jenne

National Center for Atmospheric Research
Scientific Computing Division
P.O. Box 3000
Boulder, CO 80307-3000
phone: 303-497-1215
(Olivia Bortfeld, ext. 1231)
fax: 303-497-1137
e-mail: olivia@ncar.ucar.edu

Agricultural Crop Impacts

Cynthia Rosenzweig

NASA/Columbia University
Goddard Institute for Space Studies
2880 Broadway
New York, NY 10025
phone: 212-678-5591
fax: 212-678-5552
e-mail: cccer@nasagiss.giss.nasa.gov

Gregory Kiker

Department of Agricultural and Biological
Engineering
Cornell University
Riley-Robb Hall
Ithaca, NY 1483-5701
phone: 607-255-2012
(message: 607-255-2481)
fax: 607-255-4080
e-mail: gak2@cornell.edu

Richard Adams

Department of Agricultural and Resource Economics
Oregon State University
Ballard Extension Hall 213
Corvallis, OR 97331-3601
phone: 503-737-1435
fax: 503-737-2563
e-mail: azri@osd.psu.edu

Grassland and Livestock Impacts

Barry Baker

ERI Consulting
P.O. Box 552
Fort Collins, CO 80522-0552
phone: 303-490-8322
fax: 303-493-4496
e-mail: barry@gpsrvl.gpsr.colostate.edu

Forest Impacts

Thomas Smith

Department of Environmental Sciences
University of Virginia
Clark Hall
Charlottesville, VA 22903
phone: 804-924-3107
fax: 804-982-2137
e-mail: tms9a@virginia.edu

Water Resources

Kenneth Strzepek
David Yates

Strzepek & Associates
5343 Aztec Drive
Boulder, CO 80303
phone: 303-492-7317
fax: 303-543-0789
e-mail: strzepek@colorado.edu

Dennis Lettenmaier

Department of Civil Engineering
University of Washington
164 Wilcox Hall, FX-10
Seattle, WA 98195
phone: 206-543-2532
fax: 206-685-3836
e-mail: 52018@uwavm.u.washington.edu

David Tomasko

Environmental Assessment Division
Argonne National Laboratory
9700 South Cass Avenue, EAD/900
Argonne, IL 60439-4832
phone: 708-252-6684
fax: 708-252-3611
e-mail: tomaskod@smtplink.eid.anl.gov

Juan Valdes

Department of Civil Engineering
Texas A&M University
MS 3136
College Station, TX 77843-3136
phone: 409-845-1340
fax: 409-845-6156
e-mail: jvaldes@tamu.edu

Coastal Resources

Stephen Leatherman

Laboratory for Coastal Research
University of Maryland
1113 LeFrak Hall
College Park, MD 20742
phone: 301-405-4059
fax: 301-314-9299
e-mail: s19@umail.umd.edu

Gary Yohe

Department of Economics
Wesleyan University
84 High Street
Portland, CT 06480
phone: 203-347-9411
fax: 203-342-1834
e-mail: gyohe@eagle.wesleyan.edu

Integration and Cross-Sector Issues

Richard Cirillo, ANL Project Manager
Neeloo Bhatti
Guenter Conzelmann
Jayne Dolph

Argonne National Laboratory
9700 South Cass Avenue
Argonne, IL 60439-4832
phone: 708-252-5843
fax: 708-252-6073
e-mail: cirillor@smtplink.dis.anl.gov

Fisheries

Michael Brody

U.S. Environmental Protection Agency
401 M Street, SW
Washington, D.C. 20460
phone: 202-260-2783
fax: 202-260-1935
e-mail: Brody.Michael@epamail.epa.gov

Ihor Hlohowskyj

Environmental Assessment Division
Argonne National Laboratory
9700 S. Cass Avenue
Argonne, Illinois 60439
phone: 708-252-3478
fax: 708-252-6413
e-mail: hlohowsi@smtplink.ead.anl.gov

Robert Lackey

U.S. Environmental Protection Agency
National Health and Environmental Effects
Research Laboratory
200 S.W. 35th Street
Corvallis, Oregon 97333
phone: 541-754-4601
fax: 541-754-4799
e-mail: Lackey@mail.cor.epa.gov

Health

Jonathon Patz

Johns Hopkins School of Public Health
615 N. Wolfe Street, Rm 7041
Baltimore, MD 21205
phone: 202-260-5874
fax: 202-260-6405
email: jpatz@phnet.sph.jhu.edu

Jon Balbus

George Washington University Medical Center
2300 K St., NW
Suite 201
Washington, D.C. 20037
phone: 202-994-2614
fax: 202-994-0011
e-mail: balbus@gwvenus.cmd.gwumc.edu

Wildlife

Adam Markham

World Wildlife Fund
1250 24th Street, NW
Washington, D.C. 20037-1175
phone: 202-293-4800
fax: 202-293-9345
e-mail: markham@R%wwfus@mcimail.com

Jay Malcolm

World Wildlife Fund
1250 24th Street, NW
Washington, D.C. 20037-1175
Phone: 904-378-3019
fax: 904-372-4614
e-mail: jayrmal@nervm.nerdc.ufl.edu

CONTENTS

APPENDICES

TABLES

FIGURES

NOTATION

The following is a list of the acronyms, initialisms, and abbreviations (including chemical symbols and units of measurement) used in this document.

ACRONYMS, INITIALISMS, AND ABBREVIATIONS

ANL	Argonne National Laboratory
AVVA	aerial video-taped vulnerability analysis
CBCPM	Colorado Beef Cattle Production Model
CCC	Canadian Climate Centre
CSMT	Country Studies Management Team
DOE	U.S. Department of Energy
DSSAT	Decision Support System for Agrotechnology Transfer
EPA	U.S. Environmental Protection Agency
GCM	General Circulation Model
GDP	gross domestic product
GFDL	Geophysical Fluid Dynamics Laboratory
GIS	geographic information system
GISS	Goddard Institute for Space Studies
GNP	gross national product
HCOC	high cost, open coast
IBSNAT	International Benchmark Sites Network for Agrotechnology Transfer
ICASA	International Consortium for Application of Systems Approaches
ICZM	integrated coastal zone management
IPPC	Intergovernmental Panel on Climate Change
LCOC	low cost, open coast
NASA	National Aeronautics and Space Administration
NCAR	National Center for Atmospheric Research
NGO	nongovernmental organization
OSU	Ohio State University
PC	personal computer
PET	potential evapotranspiration
RDI	range dependency index
RH	relative humidity
TAMU	Texas A&M University
UKMO	United Kingdom Meteorological Office

USAID   U.S. Agency for International Development
USLE   Universal Soil Loss Equation

VA   vulnerability analysis

CHEMICAL SYMBOLS AND ABBREVIATIONS

CO_2 carbon dioxide

UNITS OF MEASUREMENT

°C   degree(s) Celsius
d   day(s)
g   gram(s)
ha   hectare(s)
km   kilometer(s)
m   meter(s)
mm   millimeter(s)
ppm   parts per million

U.S. COUNTRY STUDIES PROGRAM

GUIDANCE FOR VULNERABILITY
AND ADAPTATION ASSESSMENTS

1 U.S. COUNTRY STUDIES PROGRAM

The *Framework Convention on Climate Change*, which was signed by more than 150 governments worldwide, calls on parties to the Convention to inventory national sources and sinks of greenhouse gases and to develop plans for responding to climate change. To assist developing countries and countries with economies in transition to meet this obligation, the U.S. government committed $35 million to support climate change country studies. The technical and financial assistance is provided through the U.S. Country Studies Program.

The goals of the U.S. Country Studies Program are to:

- Enhance the capabilities of countries and/or regions to inventory their net emissions of greenhouse gases, assess their vulnerabilities to climate change, and evaluate the options available to them to mitigate and adapt to climate change;

- Support countries' efforts to establish a process for developing and implementing national policies and measures to deal with climate change over time; and

- Develop data and information that can be used at the national, regional, and global levels; assess current and future trends in net anthropogenic emissions of greenhouse gases; and further national and international discussions of climate change issues.

Studies supported under this program address various climate change issues of concern to the recipient country, including:

- An inventory of sources and sinks of greenhouse gases,

- An assessment of vulnerabilities to the impacts of climate change and an evaluation of options to adapt to these potential impacts, and

- An evaluation of options to mitigate net emissions of greenhouse gases.

A Country Studies Management Team (CSMT) was formed to manage this program. The team is composed of personnel from the U.S. Environmental Protection Agency (EPA), U.S. Department of Energy (DOE), National Oceanic and Atmospheric Administration (NOAA), U.S. Agency for International Development (USAID), the Department of State, and U.S. Department of Agriculture (USDA). Figure 1.1 depicts the organizational structure of the U.S. Country Studies Program. Fifty-six countries are receiving support. Countries receiving support are identified in Figure 1.2.

R. Benioff et al. (eds.), Vulnerability and Adaptation Assessments, 1-1–1-5.
© 1996 *Kluwer Academic Publishers.*

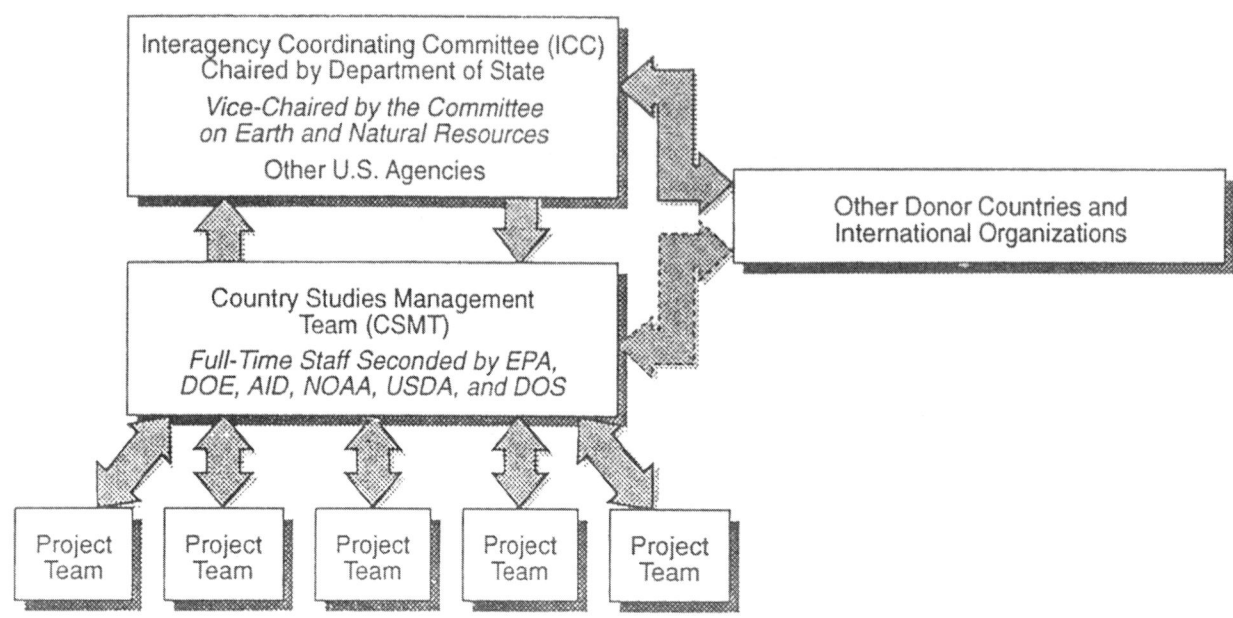

FIGURE 1.1 Structure of the U.S. Country Studies Program

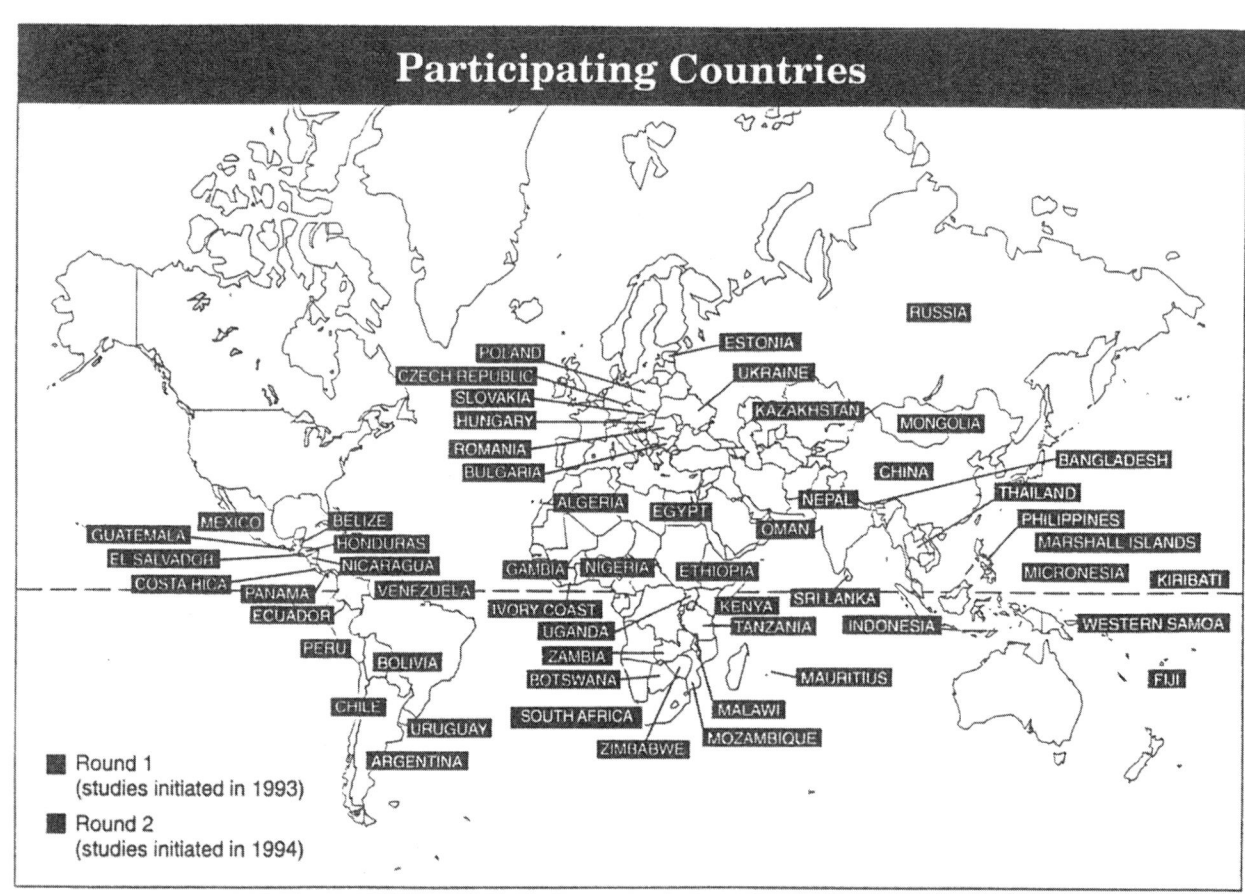

FIGURE 1.2 Countries Participating in the U.S. Country Studies Program

1.1 TECHNICAL ASSISTANCE

The U.S. Country Studies Program (U.S. CSP) is providing technical assistance to countries that are conducting vulnerability and adaptation assessments. The type, scope, and intent of this assistance are defined below.

1.1.1 Objectives of the U.S. CSP Technical Assistance

The U.S. CSP has the following objectives for providing technical support for vulnerability and adaptation assessments:

- Build the technical and institutional capabilities within countries to assess their vulnerability to climate change and evaluate appropriate adaptive responses.

- Provide guidance to the countries on the most effective approaches within resource constraints for conducting vulnerability and adaptation assessments.

- Offer training, analytical tools, and ongoing technical support over the lifetime of the study from experts in the various fields to the researchers in each country.

- Promote exchange of information on vulnerability and adaptation topics among the countries receiving U.S. support and between these countries and other countries and international institutions.

1.1.2 Scope of Assistance

The technical assistance is designed to meet the needs and interests of each of the countries conducting vulnerability and adaptation assessments. Several topics are of common interest to most of these countries, and the technical assistance focuses primarily on the following topics:

- Preparation of climate change and socioeconomic scenarios, including providing General Circulation Model (GCM) data and guidance on its use and interpretation.

- Evaluation of potential biophysical impacts for the most sensitive and important resources in each country, including

    - agricultural crops,

    - grasslands and livestock,

    - forest resources,

    - water resources,

    - coastal resources vulnerable to sea-level rise, and

    - other topics (e.g., health, fisheries, wildlife, etc.), as needed.

- Assessment of the effectiveness of adaptive adjustments to management practices in reducing these impacts.

- Evaluation of economic impacts of climate change.

- Integration of sector-specific results.

- Identification and assessment of adaptation policies designed to implement the adaptive adjustments to reduce vulnerability to climate change.

It was not expected that each country would address all of these topics. Countries were encouraged to address only the topics of greatest concern. In addition, the U.S. CSP is attempting to meet unique or specialized needs that countries may have that go beyond these common topics.

1.1.3 Technical Support Activities

The U.S. CSP has several sources for providing technical assistance in this area. The program arranged for assistance by a team of experts in the field of vulnerability and adaptation assessment. Argonne National Laboratory (ANL) and Hagler Bailly Consulting, Inc. coordinated the work of this team, which drew from various universities, laboratories, and consulting firms. In addition, technical experts from U.S. governmental agencies and nongovernmental organizations complemented the work of this team by assisting countries with specific activities.

The team of experts provided technical assistance through the following activities:

- Prepared and distributed a guidance document that describes a primary recommended approach for conducting vulnerability and adaptation assessments and supplementary approaches to be used if sufficient resources are available. This document and its appendixes provide extensive guidance on the application of several methodologies for conducting vulnerability and adaptation assessments. This document is used in tandem with the IPCC draft *Technical Guidelines for Assessing Climate Change Impacts and Adaptation.*

- Conducted training workshops for countries to provide guidance and "hands-on" training on each major element of vulnerability and adaptation assessment.

- Provided countries with models, data, reports, and other analytical tools and materials necessary to perform their assessments. The analytical tools are described in this guidance document.

- Provided ongoing technical assistance for each country during the full period of their vulnerability and adaptation assessment. The team of experts communicates on a regular basis with researchers in each country and responds to requests for technical advice on methodologies, use of models, interpretation of results, and preparation of reports. This technical support is, of necessity, limited to advice and consultation with country technical staff. The technical experts are not in a position to carry out any of the actual work for a country study.

The U.S. CSP has also sponsored regional workshops to give countries the opportunity to discuss common issues and share their methodologies and results with other countries in their region. One workshop was held on vulnerability and adaptation assessment in Asia and the Pacific, Africa, Latin America, and the former Soviet Union and Eastern Europe and an international conference on adaptation was held in St. Petersburg, Russia. The U.S. CSP cosponsored these workshops with other donors.

2 OVERVIEW OF THE VULNERABILITY AND ADAPTATION ASSESSMENT PROCESS

2.1 ELEMENTS OF VULNERABILITY AND ADAPTATION ASSESSMENTS

An assessment of a country's *vulnerability* to climate change is an evaluation of how changes in climate may affect segments of the natural environment, elements of the national economy, and human health and welfare. Key natural resource sectors that might be susceptible to changes in climate include agricultural crops, livestock, forests, water resources, coastal resources, fisheries, and wildlife. Other sectors potentially affected include human health, energy, infrastructure, and human settlements. A *vulnerability* assessment consists of an analysis of the scope and severity of the potential effects of climate change. For example, a rise in temperature and an increase in rainfall may effectively lower (or raise) the yield of a country's agricultural crops, which, in turn, may reduce (or increase) a country's gross national product (GNP) and its economic well-being.

An assessment of a country's *adaptation* options is an identification and evaluation of changes in technologies, practices, and policies that can be taken to prepare for climate change. For example, agricultural crops may be changed and planting cycles modified to adjust to changes in temperature and rainfall patterns.

2.2 VULNERABILITY AND ADAPTATION ASSESSMENT PROCESS

The process for conducting vulnerability and adaptation assessments is depicted in Figure 2.1. The first step is to define the scope of the problem and assessment process. This step, described in Section 3, includes defining terminology, identifying the issues to be addressed, and selecting the sectors to be studied, together with the geographic study area and time frame for the assessment. The second step involves defining and describing the scenarios underlying the assessment. (Section 4 provides guidance on how to select and define scenarios for analysis).

In the third step, the biophysical impacts are determined from examining the effect of the baseline, climate, and environmental scenarios on each sector: agricultural crops, grasslands and livestock, forests, water resources, coastline resources, and other resources. An evaluation of alternative practices and technologies for adapting to these impacts, including economic considerations, is part of this step. Section 5 provides guidance on the primary approach, resource requirements, and issues related to each sector. In the fourth and fifth steps, the results of each impact sector are integrated (Section 6) and adaptation policies and programs are analyzed (Section 7). In the final step, results are documented and presented (Section 8).

R. Benioff et al. (eds.), Vulnerability and Adaptation Assessments, 2-1–2-3.
© 1996 *Kluwer Academic Publishers.*

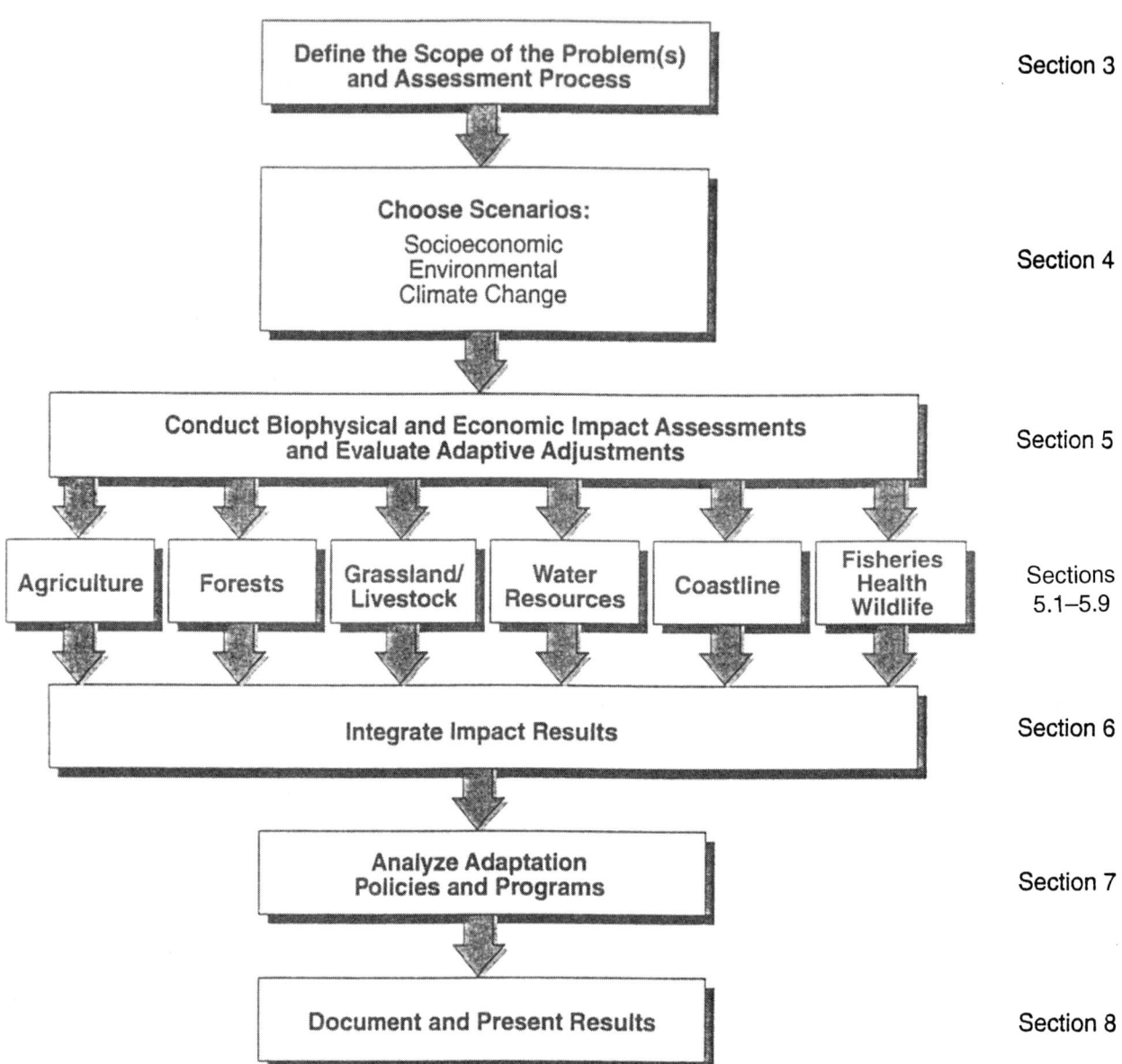

FIGURE 2.1 The Climate Change Vulnerability and Adaptation Assessment Process

2.2.1 Purpose and Use of this Guidance Document

This guidance document, including its appendixes, provides guidance for countries in the conduct of vulnerability and adaptation assessments in eight selected sectors: (1) agriculture (crops), (2) grasslands and livestock, (3) forests, (4) water resources, (5) coastal resources, (6) health, (7) fisheries, and (8) wildlife. These sectors were chosen because they have the widest interest to most countries and are most susceptible to climate change. It is not expected that every country will address each of these topics. Countries are encouraged to address only those topics of greatest concern and interest.

This guidance document describes, for each sector, a *primary approach* for conducting a vulnerability and adaptation assessment. The primary approach represents the most practical and widely used techniques for carrying out the analysis. This approach is based on prior experience with this type of study.

In describing the primary approach, this document presents a structured and systematic process for its implementation, including identification of data requirements, available analytical tools (e.g., computer programs), and advantages and disadvantages. All countries conducting vulnerability and adaptation assessments should review the primary approach presented here for vulnerability and adaptation assessments addressing these eight sectors. **Where appropriate, countries are encouraged to use their own models or approaches, which may provide for more thorough or accurate assessments.**

The primary approach is not, in general, the most sophisticated or detailed analysis possible. Instead, it is meant to be a reasonable method that can be used within the resource limitations of the overall program.

In addition to the primary approach, some *supplementary approaches* are also presented. This information may involve the use of more comprehensive data, more sophisticated analytical tools, or more thorough evaluations. Countries should feel free to use these supplementary approaches if they can do so within the time and resource constraints of the program.

2.2.2 Organization of this Guidance Document

Section 4 discusses the selection of scenarios to be used in the assessments, including climate change scenarios and non-climate-related scenarios (e.g., population growth and economic development). This section also discusses how to use the results of a GCM.

Section 5 provides information on the conduct of biophysical impact assessments in each of the eight sectors included in this document. This section summarizes the primary and supplementary approaches. The structure of the subsections dealing with each of the sectors is basically the same.

Section 6 describes the procedures for integrating the results of the vulnerability and adaptation assessments of each of the individual sectors into an overall evaluation. Section 7 describes procedures for evaluating policies and programs to adapt to climate change. Finally, Section 8 gives a typical outline of how the results of the vulnerability and adaptation analyses can be presented.

The appendixes provided at the end of the document give more detailed, technical information on conducting analyses in each sector.

3 DEFINING THE SCOPE OF AN ASSESSMENT

The first step in conducting a vulnerability and adaptation assessment is to define the scope of the proposed assessment, including the problems of interest and the assessment process to be used (Figure 2.1). The objective of this scoping effort is to focus on important questions and issues and to use limited resources efficiently. The IPCC has outlined a general approach for scoping and problem definition (IPCC 1992). The IPCC outline and Carter and de Rozan (1993) make it possible to describe the steps that can be taken to determine the proper scope of an assessment. The steps are as follows:

- Identify assessment goals,

- Define sectors to be studied,

- Select the study region,

- Select the time frame,

- Determine data needs,

- Develop the context for assessment, and

- Develop a schedule.

3.1 IDENTIFY ASSESSMENT GOALS

The overall goals of a climate change vulnerability and adaptation assessment, as stated by the IPCC, are to (1) evaluate how climate affects human activities and natural systems and estimate uncertainties of these effects; (2) evaluate sensitivities, thresholds, and vulnerabilities of natural systems to likely scenarios of climate; and (3) identify and evaluate the possible adaptation and policy options. These goals define broad objectives for an assessment.

More detailed, country-specific goals also need to be identified. The following issues need to be considered:

- *Identify who will use the results of the assessment.* For example, decision makers could use the work to evaluate policies for adapting to climate change. The scientific community could use the results to develop research plans. Results could also provide information for international cooperative efforts in climate change. Each of these goals could require different studies and analyses.

- *Determine what information should be generated from the assessment.* It is important to consider what type of information would be the most useful to those who will use the assessment. This information will help to guide the type of analysis and the level of detail.

- *Determine what level of detail is necessary for the assessment.* The type of results desired from the assessment should be determined. If the goal is to develop a first-order estimate of the

R. Benioff et al. (eds.), Vulnerability and Adaptation Assessments, 3-1–3-6.
© 1996 *Kluwer Academic Publishers.*

magnitude of climate change impacts and to develop plans for more detailed studies, a relatively simple analysis process can be used. If the work is a follow-up study to other work, more detailed work needs to be planned, and a more rigorous analysis should be done.

It is important to identify the country's goals early in the process to ensure efficient allocation of limited resources.

3.2 DEFINE SECTORS TO BE STUDIED

The second step in the scoping process is to identify sectors in the country that are most vulnerable to climate change and will have the greatest impact on the population and the economy. Table 3.1 lists some of the sectors that could be impacted by climate change.

When attempting to determine whether a country should include a particular sector in a vulnerability and adaptation assessment, the following questions can be asked:

- *Is the sector a major element of the national system?* Examples include agriculture as a major (or minor) component of GNP, extensive (or limited) low-lying coastline areas, and extensive (or limited) natural forests.

- *Do any current conditions indicate that the sector is especially sensitive to changes in climate?* Examples include agricultural crops that are sensitive to changes in rainfall, coastal areas that flood regularly, and water supplies that are currently marginal or sporadic.

- *Would a disruption in the sector because of climate change have a major impact on human populations or economic activity?* Examples include large populations living in potential inundation areas and significant food shortages from crop losses.

- *Is there a foreseeable benefit to taking some action to deal with climate change impacts on the sector in the short or medium time horizon?* Examples include the need to begin work on managing water resources and on planning for seawalls.

TABLE 3.1 Sectors Impacted by Climate Change

Sector	Potential Impacts
Agricultural crops	Changes in crop yields Shifts in relative productivity and production
Grasslands and livestock	Changes in yields Shifts in relative productivity and production
Forests	Changes in species range of growth Changes in forest production Migration of vegetation
Water resources	Changes in supplies Changes in drought and floods Changes in water quality and hydropower
Coastal resources	Inundation Saltwater intrusion Increased flooding
Human health	Shifts in range of infectious diseases Changes in heat stress and cold weather afflictions
Industry and energy	Changes in industrial product demand Changes in energy demand Changes in hydropower
Human settlements	Changes in drinking water availability Changes in transportation (road, water) Changes in housing requirements (heating, cooling) Changes in population density and land use
Marine ecosystems	Loss of habitat Migration of new species Invasion of new species

Sources: U.S. Office of Technology Assessment (1993) and IPCC (1994).

If these questions can be answered affirmatively for a sector, the issue should be considered for inclusion in a vulnerability and adaptation assessment. It would also be useful to rank the sectors in terms of importance to the vulnerability and adaptation issue to allow work to focus on high-priority sectors.

3.3 SELECT THE STUDY REGION

The selection of the study region is guided by the goals of the assessment and by sectors chosen for study. The following candidate study regions were selected on the basis of IPCC (1992):

- *Administrative units (e.g., district, town, province, state, and nation).* These units are useful for dealing with social and economic issues. They are also convenient because boundaries are clearly defined.

- *Geographic or physiographic units (e.g., river basin, plain, mountain range, and lake regions).* These units focus the assessment on areas closely related from a physical and biological perspective. The same type of impacts can be seen in the entire unit.

- *Ecological zone (e.g., wetland, forest, moorland, and savannah).* These units are useful for investigating impacts to a specific biological community.

- *Climatic zone (e.g., desert, monsoon zone, and rain shadow area).* These areas are likely to have similar climate change patterns.

- *Sensitive regions (e.g., tree lines, ecotones, coastal zones, ecological niches, and marginal communities).* These regions are likely to be especially sensitive to climate change. The first indications of impacts may show up in these areas because of their sensitivity.

The study region can be selected by careful consideration of available data, including both physical information and socioeconomic data.

3.4 SELECT THE TIME FRAME

The time frame used for vulnerability and adaptation assessments is generally long (20-100 years) because many of the impacts are not expected to become evident or significant in the short term. However, it is also important to consider short-term issues (5-10 years). Consideration must be given to adaptation strategies that need to be implemented in the relatively short term in order to be effective in dealing with long-term problems. For example, if an analysis shows that sea-level rise will inundate large areas, and it is decided to construct seawalls to deal with the problem, the planning and implementation of these measures must be started early. Section 4 discusses time-frame issues in detail.

3.5 DETERMINE DATA NEEDS

All of the determinations of scope addressed so far must be tempered with consideration of data availability. If sufficient data of reasonable quality are not available to use as a basis for an assessment, it is not necessary to carry out sophisticated quantitative analyses.

The data requirements for carrying out assessments in each of the sectors covered by this document are summarized in Section 5 and expanded upon in the appendixes. Countries are urged to review these data requirements and to determine the availability of the necessary information before committing to a vulnerability assessment in any of the sectors. Furthermore, the analytical methodology to be used in a vulnerability assessment in any sector should be selected on the basis of whether the quality of the available data justifies the use of the approach. For example, the agricultural crop computer simulation models should not be used if sufficient information on soil classification, historical weather data, and current plant populations is not known. The alternative nonsimulation approaches might be more appropriate in areas with serious data deficiencies.

3.6 DEVELOP THE CONTEXT FOR ASSESSMENT

The next step in the scoping process is to develop the broader context of the assessment. Included in this activity is identifying other work (either completed or in progress) that might support the assessment, developing a coordinated and well-managed project, and determining the extent to which the results will be integrated with other work, once completed. These factors help establish the role that the assessment will play in the national decision-making process.

Of special importance is identifying "stakeholders" in the outcome of the assessment. Some members of the community are either directly or indirectly involved in the assessment process. Included in this group are policy makers, climate researchers, government officials, educational leaders, nongovernment organizations, and the general public. Each of these groups may have a different interest in the outcome of the work. To the extent practical, the interests of stakeholders should be considered when designing the scope of the assessment.

One important consideration in addressing the interests of the decision makers and stakeholders is the manner in which (1) their input to the assessment will be obtained and (2) the results of the assessment will be conveyed to them. Of necessity, these issues will need to be addressed in different ways in each country. In some instances, a country may wish to convene a formal advisory committee with representatives from each of the interested groups. This committee can be briefed on the type of assessment to be done, the methodologies to be used, and the expected results. The committee can meet regularly during the assessment to review intermediate results and to provide feedback to the analysts. An alternative would be to approach each stakeholder group separately to solicit input to the assessment and obtain reaction to the results. Whatever technique is used, it is important that the analysis receive exposure to a wide range of stakeholder and decision-making groups. The development of a broad base of input will enhance the ability of a country to implement the recommendations resulting from the assessment.

3.7 DEVELOP A SCHEDULE

Table 3.2 gives a sample schedule for completing the assessment. This schedule should be considered as a general guide and should be modified to suit the specifications required by individual countries.

TABLE 3.2 Sample Schedule for a Vulnerability and Adaptation Assessment

Months from Project Start	Activity
0-2	Define scope of problem and assessment process and methods
2-4	Select and develop scenarios (climate, nonclimate)
2-12	Conduct vulnerability and adaptation assessment for each sector
12-13	Present workshop (in country) on preliminary vulnerability results and discuss integration and adaptation options
13-15	Integrate results across sectors and analyze the adaptation policy, including presenting an adaptation workshop (in country)
16-18	Draft vulnerability and adaptation report; present results for review
18-19	Present workshop (in country) on results
19-20	Finalize results and report

3.8 REFERENCES FOR SECTION 3

Carter, T.R., and M. Blantran de Rozan, 1993, *Workbook Methods for Problem Definition and Scoping in Climate Impact Assessment*, in UNEP-Canada Workshop on Impacts and Adaptation to Climate Variability and Change, Toronto, Canada, November 29 to December 3, 1993.

IPCC, 1994, *Technical Guidelines for Assessing Climate Change Impacts and Adaptations*, Intergovernmental Panel on Climate Change (draft).

IPCC, 1992, *Preliminary Guidelines for Assessing Impacts of Climate Change,* Intergovernmental Panel on Climate Change, Environmental Change Unit, United Kingdom.

U.S. Office of Technology Assessment, 1993, *Preparing for an Uncertain Climate, Summary*, Washington, D.C., Sept.

4 DEFINING AND APPLYING SCENARIOS AND USING GCM DATA FOR VULNERABILITY ASSESSMENTS

As shown in Figure 2.1, step 2 of the vulnerability and adaptation analysis is the selection of scenarios to be used in assessing the effects of climate change. Selecting scenarios is one of the most important steps in assessing a country's vulnerability to climate change because it affects the magnitude and the direction of results. An "extreme" scenario can yield extreme results, whereas a "mild" scenario can yield mild results. Thus, selection of scenarios should be a balanced, carefully planned decision.

Predicting the effect of climate change would be easier if the relationship between an increase in greenhouse gases and changes in climate was well defined, particularly at the regional level. Although an increase in greenhouse gas emissions is likely to raise global temperatures, precipitation, and sea level, no reliable predictions can be made about regional changes in climate. In addition, it is uncertain how climate changes will vary (Houghton et al. 1990, 1992). Furthermore, future socioeconomic and environmental conditions are unknown.

Climate change scenarios can be used to assess a country's (or a region's) vulnerability to climate change. Such scenarios can be used to:

- Determine whether a sector such as agriculture is potentially vulnerable to climate change,

- Identify thresholds at which impacts become negative or severe, and

- Identify relative vulnerability among sectors in the same regions or among similar sectors in different regions.

This section examines the options for developing scenarios and the advantages and disadvantages of those options. The discussion addresses the issue of how far into the future climate change vulnerability should be assessed. "Nonclimate," or baseline, scenarios that estimate how the world would change without climate change are also discussed. Baseline scenarios are important tools for assessing vulnerability because baseline changes can affect a sector's sensitivity to climate change. This section also presents options for developing climate change scenarios and recommendations for their use. Details about GCMs are presented in Section 4.4.

R. Benioff et al. (eds.), Vulnerability and Adaptation Assessments, 4-1–4-22.
© 1996 *Kluwer Academic Publishers.*

4.1 TIME FRAME

Studies of the effects of climate change usually focus on climate changes caused by a doubling of CO_2 levels (referred to as $2 \times CO_2$) in the atmosphere (e.g., Smith and Tirpak 1989; Tegart et al. 1990).[1] These studies use estimates of regional climate change from GCMs, which generally compare an effective $2 \times CO_2$ with single levels of CO_2 (referred to as $1 \times CO_2$).[2] The IPCC projected that, if current emission trends remain unchanged, global temperatures will increase by 0.3 °C per decade, and $2 \times CO_2$ will cause an increase in temperature of 2.5 °C. At that rate of warming, temperatures in 2075 should be about 2.5 °C warmer than present-day (1994) temperatures.

A country study analysis needs to assess both long-term (e.g., 80 years from now) and near-term impacts. Policy makers need information on the potential effects of climate changes in the next few decades because they are currently developing plans that will affect water resources, coastal areas, nature reserves, and other areas. Policy makers need to know how near-term climate changes could affect these decisions.

This section develops climate change scenarios through the next century. Although this study focuses on the year 2075, both near- and long-term scenarios are also examined.

4.2 BASELINE SCENARIOS

A brief discussion of the process for developing baseline scenarios is presented here. A more detailed description is given in Appendix A.

Baseline changes cannot be accurately predicted, so the goal of this study is more modest. *Baseline scenarios are developed to help identify how changes in baseline conditions affect the sensitivity of sectors to climate change.* In other words, baseline scenarios are used to identify whether changes in population, income, or technology make a sector such as agriculture more or less sensitive to climate change.

To help identify the degree of sensitivity to changes in baseline conditions, at least two baseline scenarios are tested because baseline scenarios are always controversial. In addition, testing includes a scenario based on no changes in current conditions. It is useful to test against current conditions because people may find it difficult to understand changes in baseline conditions and then changes in climate imposed on them. Comparing the baseline scenario with the "no-change" scenario can help identify the baseline variables (e.g., population growth) that have the most important effect on climate sensitivity. Identifying such sensitivities can help in designing policies for adaptation to climate change.

---

[1] Cline (1992) argues that examining impacts only under $2 \times CO_2$ does not go far enough into the future to fully assess the potential effects from greenhouse gas emissions. He argues that atmospheric CO_2 levels could effectively increase eight times over the next several centuries. We do not discourage researchers from examining climate change scenarios beyond $2 \times CO_2$. However, the difficulty in doing so is that it is not clear that the world will continue to use fossil fuels well beyond the early part of the 21st century. Even if we were sure about those emissions, few GCMs have been run for conditions beyond $2 \times CO_2$.

[2] Most GCMs use from 300 to 330 ppm of CO_2 for $1 \times CO_2$. These levels were nearly reached at mid-century. The "effective doubling" of CO_2 means that the combined contribution to atmospheric warming of all greenhouse gases is equivalent to doubling CO_2 levels alone.

Baseline scenarios include estimates of climate, socioeconomic, and environmental conditions. The baseline climate scenario is for no change in climate. Socioeconomic scenarios consider population, income, and productivity (including technology) levels. Environmental scenarios consider pollution levels, land use, Desertification, and ecosystem health.

4.2.1 Baseline Climate Scenario

Thirty years of current climate data are used in developing the baseline scenario. A 30-year period is considered long enough to have a good representation of wet, dry, warm, or cool periods. Selecting a recent 30-year period is preferred because it not only represents current climate but also, in most cases, has the most accurate data.

The 30-year baseline climate data can be based on one of two periods: 1951-1980 or 1961-1990. The advantage of the latter is that it is more recent; the disadvantage is that the 1980s recorded some of the warmest years in average global temperature (Jones et al. 1990). It is not desirable to include a warming trend in the baseline, particularly one that *could* have been caused by increased greenhouse gas concentrations (Carter et al. 1994). For that reason, 1951-1980 should be used as the baseline climate period.

Researchers can gather observed data on climate variables needed for analysis for the 1951-1980 period. In some cases, observations for the entire 30 years are not available. If data on some variables are not available, researchers can use weather generators to derive estimates for those variables (see Appendix C on climate data; also see the discussion of the Decision Support System for Agrotechnology Transfer [DSSAT] model in Section 5.1.4.2).

4.2.2 Baseline Socioeconomic Scenarios

Baseline socioeconomic scenarios can be developed for variables that affect the sensitivity of a sector to climate change. Once a list of important variables is developed, scenarios can be designed to account for at least changes in population levels, national and per-capita income, and productivity. In general, projections of economic activity beyond 10 to 20 years are unreliable. Long-term economic projections should be used with caution (i.e., only to understand climate sensitivities). Estimates for these variables should be internally consistent; that is, changes in population will probably affect national and, perhaps, per-capita income.

4.2.2.1 Population Scenarios

Population is probably the easiest variable to use to develop scenarios. The World Bank (1992) and the United Nations (1989a) have published country-by-country population estimates through 2100.

4.2.2.2 Income Scenarios

Both The World Bank (1993) and the United Nations (1990) have published estimates of regional or national changes in income. However, these estimates only project incomes for 10 to 25 years into the future. A scenario could assume that the rate of growth in the last estimated period would continue throughout the century, or a regional estimate of income growth developed by Pepper et al. (1992) for the IPCC could be used.[3] While these estimates are not given for individual countries, they are provided through 2100. Pepper et al. also estimated for population and per-capita income. The advantage of using this source is that estimates for the variables are internally consistent.

4.2.2.3 Income Per-Capita Scenarios

Income per capita is an indicator of standard of living. It is also an indicator of productivity, although gross domestic product (GDP) per worker is a better measure. Income per capita can be calculated by dividing national income by population.

4.2.3 Baseline Environmental Conditions

Like socioeconomic conditions, changes in environmental conditions can also significantly affect the sensitivity of sectors to climate change. For example, increased water pollution could make fish and other species that inhabit water bodies more vulnerable to climate change. Increased fragmentation of habitat could put species with small populations at greater risk of extinction from climate change. Desertification could reduce arable land and degrade water quality. To examine changes in these potentially important environmental conditions, baseline scenarios for pollution levels, land use, Desertification, and ecosystem health should be developed.

Few estimates of changes in environmental conditions, particularly for individual countries have been published. The United Nations (1989b, 1990), the World Resources Institute (1992, 1993), and Meadows et al. (1992) have published information on current environmental conditions. As with the estimates for changes in productivity, scenarios should be developed based on extrapolation of past trends and guided by expert judgment.

4.2.3.1 Pollution Level Scenarios

Baseline pollution scenarios only need to be developed for water or air pollution levels because other types of pollution are not likely to be relevant for climate change vulnerability analyses. Pollution-level scenarios should consider the effects of increased population and income, as well as improved technology and enforcement of air and water quality standards. In addition, these scenarios can estimate baseline pollution control costs because effects of climate change may be expressed as changes in pollution control costs.

4.2.3.2 Land Use Scenarios

Increases in population lead to an increase in the use of land for living space and other purposes. Baseline scenarios for land use must consider the following issues:

- The degree to which urbanization will increase;

- The degree to which agriculture or other activities, such as mining, will expand;

---

[3] Houghton et al. (1992) report global economic growth increasing 2.3% per year from 1990 to 2100 under Scenario "A."

- The degree to which deforestation trends will continue and changes in habitat fragmentation will occur.

The Food and Agricultural Organization (1993) has published projections of land use for agriculture.

4.2.3.3 Desertification

The UNEP (1992) and Dregne and Chou (1992) have published information on current amounts of Desertification. Neither source projects future Desertification. Because Desertification is a factor of climate change and land use, baseline Desertification scenarios should only consider effects unrelated to climate. Desertification scenarios should be consistent with population, agriculture, and land-use scenarios.

4.2.3.4 Ecosystem Baselines

All of the environmental factors discussed above affect ecosystem health. Increases in pollution, land development, and Desertification could threaten the diversity of habitats and species. Baseline scenarios should also consider other factors such as excess harvesting of species (e.g., fish).

4.2.4 Baseline Institutional Scenarios

Changes in institutions and laws can significantly affect the sensitivity of many sectors to climate change. Furthermore, these changes could be more important than other changes. However, because it is difficult to predict changes in institutions, baseline scenarios should assume that present-day institutions and laws do not change. This assumption allows researchers to examine the ability of current institutions to respond to climate change. Changes in institutions should be analyzed as part of the examination of adaptation options.

4.2.5 Other Considerations for Baseline Scenarios

Developing baseline scenarios for socioeconomic and environmental conditions is challenging. Many researchers spend lifetimes investigating issues related to changing socioeconomic or environmental conditions. Developing baseline scenarios to assess vulnerability to climate change should not become an end in itself. Rather, researchers should either use available estimates or make simple assumptions about how conditions may change in the future. If resources permit, two baseline scenarios should be derived: a "pessimistic" scenario (e.g., high population, low income, low productivity, high pollution, high land use, and high Desertification) and an "optimistic" scenario (e.g., high income and low pollution).

4.3 CLIMATE CHANGE SCENARIOS

A brief discussion of the process for developing climate change scenarios is presented here. A more detailed description is given in Appendix B.

Climate change scenarios are not constructed to help predict the future. They are designed to help identify the sensitivity of sectors to climate change. Climate change scenarios can help identify the potential direction of effects (e.g., does runoff increase or decrease) and the potential magnitude of impacts (e.g., by how much does runoff change). Applying climate change scenarios can also help to identify the relative sensitivity of a sector to changes in different meteorological variables. For example, is forest growth more sensitive to changes in temperature or precipitation?

To be useful for assessing the potential vulnerabilities to climate change, the climate change scenarios must meet the following conditions:

1. The scenarios *must* be consistent with predictions of climate change (e.g., 1.5 to 4.5 °C for $2 \times CO_2$). Vulnerability assessments determine sensitivity to climate changes that result from increased greenhouse gas emissions. A scenario that is inconsistent with this basic premise would not help identify vulnerabilities.

2. The scenarios *must* be physically plausible; that is, they cannot violate basic laws of physics. It is not plausible, for example, to assume that a country as vast as Russia or Brazil would have a uniform change in temperature, but no change in precipitation. However, such a scenario is plausible on a smaller scale.

3. The scenarios *must* estimate a sufficient number of variables on a spatial and temporal scale needed for vulnerability assessment (Smith and Tirpak 1989; Viner and Hulme 1992). Many impacts models need scenario data for meteorological variables, such as temperature, precipitation, solar radiation, humidity, and winds. In addition, daily information may be needed for many locations.

4. Climate change scenarios *should*, to a reasonable extent, reflect potential regional climate changes. For example, a set of scenarios that examines only a relatively large or small amount of warming will not help to identify the full range of sensitivities to climate change.

Climate change scenarios should *not* be interpreted as predictions of future climate change (Viner and Hulme 1992; Robock et al. 1993).

4.3.1 Options for Climate Change Scenarios

Three basic options are available for creating climate change scenarios: (1) GCMs, (2) analogue (historical) warm periods, and (3) incremental changes in climate variables (e.g., +2 °C, +4 °C; ±10% precipitation). The advantages and disadvantages of each option are discussed below, and Section 4.3.2 recommends which scenarios should be used in the vulnerability assessments.

4.3.1.1 GCM Scenarios

General circulation models are mathematical representations of atmosphere, ocean, and land surface processes based on the laws of physics. Such models consider a wide range of physical processes that characterize the climate system and have been used to examine the impact of increased greenhouse gas concentrations on climate (Gates et al. 1990). These models estimate changes for dozens of meteorological variables in regional climate in grid boxes that are typically 3° or 4° up to 10° (latitude and longitude). (Section 4.4 discusses GCMs in more detail.)

Two types of GCM output can be useful for vulnerability assessments. The first is $1 \times CO_2$ and $2 \times CO_2$ output. Almost all GCMs are run for current climate ($1 \times CO_2$) and $2 \times CO_2$ conditions. The difference between these estimates is a scenario of how climate will change with an effective doubling of CO_2 concentrations. The second type of GCM output is output from transient scenarios. Transient scenarios are run by assuming a steady increase in greenhouse gas concentrations and examining how climate could change over time. Few of these transient runs are available to the vulnerability assessment community.

Many vulnerability assessment studies have used GCMs as the basis for creating scenarios (e.g., Parry et al. 1988; Smith and Tirpak 1989; Strzepek and Smith forthcoming). These studies combined average monthly changes between $2 \times CO_2$ and $1 \times CO_2$ from GCMs with 30-year climate normal data sets. Using the observed climate data provides spatial and temporal variability, although it assumes that these data do not change from current conditions.

Creating transient scenarios is more difficult. Smith and Tirpak (1989) created a transient scenario by repeating the 30-year climate normal period and combining that data set with average changes from each decade of a GCM transient (Appendix B). Another option for transient scenarios is to select certain decadal averages from the GCMs and combine them with the 30-year climate data set. This option gives more information about potential climate effects at a particular point in the future.

The major advantage of using GCMs as the basis for creating climate change scenarios is that they are the only tool that estimates changes in climate due to increased greenhouse gases for a large number of climate variables in a physically consistent manner. The GCMs estimate changes in a host of meteorological variables that are consistent with each other within a region and around the world, and thus they fully meet conditions 1 and 2, and partially satisfy 3.

A major disadvantage of using GCMs is that, although they accurately represent global climate, their estimates of current regional climate are often inaccurate. In many regions, GCMs significantly underestimate or overestimate current temperatures and precipitation (Grotch and MacCracken 1991; Kalkstein 1991). Another disadvantage is that GCMs do not produce output on a geographic and temporal scale fine enough for vulnerability assessments (condition 3). General circulation models estimate uniform climate changes in grid boxes several hundred kilometers across, and although they estimate climate on a daily or even twice daily basis, results are generally reported as monthly averages. This problem can be overcome by combining GCM output with the baseline climate record. An additional disadvantage is that a single GCM or even several GCMs may not represent the range of potential climate change in a region (condition 4).

4.3.1.2 Analogue Scenarios

The second option for creating climate change scenarios is to base them on historical warm periods. Some periods in the past have been warmer than the present. For example, in the United States, the 1930s were warmer than current conditions, particularly in the Midwest. In the mid-Holocene period, which was about 6,000 to 9,000 years ago, global temperatures were approximately 1 °C warmer than they are today (Webb 1992).

A major study recently conducted used the 1930s as an analogue for warming in the 2030s in the northern Midwestern United States (Rosenberg and Crossen 1991). In addition, some researchers used information on warm periods from paleohistorical periods (e.g., Vinnikov and Lemeshko 1987).

An advantage of using warm periods in the historical record is that climate change data are available on a daily and local scale, which is finer than that provided by GCMs (condition 3). Thus, scenarios can be created without combining data sets. The advantage of using paleoclimate data instead of historical data is that the temperature differences compared to current climate, on average, are greater. The disadvantage is that paleoclimate data are only available for seasonal changes in temperature and precipitation.

The major disadvantage of using analogue climates for climate change scenarios is that the changes in historical climates were probably not caused by increased concentrations of greenhouse gas. More likely, these changes in paleoclimate were caused by changes in the earth's rotation around the sun. The reasons for the warming in the 1930s are uncertain. Thus, these scenarios are not consistent with human-induced increases in greenhouse gas concentrations, particularly for $2 \times CO_2$ levels (condition 1). Furthermore, analogue scenarios tend to be at the low end or even below the range of potential climate warming, thus violating condition 4.

4.3.1.3 Incremental Scenarios

The final option for creating climate scenarios is to use incremental changes in such meteorological variables as temperature and precipitation. For example, temperature changes of +2 °C and +4 °C can be combined with no change and ±10 and 20% changes in precipitation (e.g., Nash and Gleick 1993; Poiani and Johnson 1993). These incremental changes are usually combined with an observed climate database to yield an altered 30-year record of daily climate with spatial variability.

A major reason for using incremental scenarios is that they capture a wide range of potential climate changes (condition 4). In addition, because individual variables are changed independently of each other, incremental scenarios also help identify the relative sensitivities of sectors to changes in specific meteorological variables. A disadvantage of using incremental scenarios is that they may not be physically plausible (condition 2), particularly if uniform changes are applied over a very large area. Although it is not necessary to limit their application to two variables, most studies that use incremental scenarios have only examined changes in temperature alone or in temperature and precipitation. Few have examined changes in other climate variables. A possible problem with examining only a few variables is that potentially important sensitivities to changes in other variables may be missed.[4]

4.3.2 Primary Approach for Creating Climate Scenarios

The primary approach for creating climate scenarios should be based on output from GCMs and incremental scenarios. The GCMs are the best source of information about how climate may change as a result of increased concentrations of greenhouse gas. However, GCM scenarios produce unreliable regional climate change estimates. Even using several GCMs probably will not yield a range of scenarios broad enough to cover potential climate changes in a region.

---

[4] For example, if runoff is being estimated and only the change in temperature is examined, potential evapotranspiration would increase 6% for every degree increase in temperature (Waggoner 1990). However, changes in vapor pressure, winds, and solar radiation, combined with higher CO_2 levels, can partially or completely offset the effect of temperature increase on evapotranspiration (Rosenberg et al. 1990).

Therefore, incremental scenarios for temperature and precipitation should be used to complement GCMs. Incremental scenarios provide a wider range of sensitivities than do GCMs. They also help determine relative sensitivities to changes in temperature and precipitation.

In addition, analogue scenarios can also be used; however, they may be less useful than GCMs or incremental scenarios because they represent relatively small climate changes in relation to present conditions. Paleoclimate scenarios can also be produced in only a few areas. However, analogue scenarios combined with GCMs and incremental scenarios will give a richer appreciation of climate change sensitivities.

4.3.2.1 Selection of GCM Scenarios

At least three GCMs should be used for creating $2 \times CO_2$ regional climate change scenarios. Using one GCM scenario can create the impression of a prediction, and using two GCMs shows only minor variance among scenarios. As discussed in Section 4.4, at least five $2 \times CO_2$ GCM runs are available at the U.S. National Center for Atmospheric Research (NCAR). The following steps should be taken to select three GCMs for creating climate scenarios:

- Obtain $1 \times CO_2$ output from all GCMs at NCAR.

- Compare the regional $1 \times CO_2$ output with observed climate data. Use a spatially averaged data set such as the RAND data set for current climate (Section 4.4.3). This procedure is preferable to selecting observed data sets from single sites.

- Select the three (or more) GCMs that best reflect current climate.[5] Models that give a better estimate of current (observed) conditions will also give better estimates of changed conditions. There is no guarantee, however, that this observation is true.

Once GCMs are selected, regional scenarios can be created by combining average monthly output obtained by taking the difference between equilibrium $2 \times CO_2$ and $1 \times CO_2$ runs with observed climate data from the 30-year baseline climate period; adding the change in temperature ($2 \times CO_2 - 1 \times CO_2$) to the observed temperatures; and multiplying ratio changes ($2 \times CO_2/1 \times CO_2$) in other variables by the observed values of the other variables.[6] In general, using a linear interpolation between the four nearest grid points is recommended. Without interpolation, sudden changes in climate could occur in impacts at the boundaries of GCM grid boxes. If only one site is being studied, it is acceptable to simply use the climate data from the nearest grid point. The technique for interpolation is described in Appendix B.

---

[5] If the GCMs do not satisfactorily estimate current seasonal climate patterns, researchers could use only annual average changes from the models (Wescoat and Leichenko 1992). If a single GCM grid box does a relatively poor job (compared to other regions) of representing current climate, researchers could average output from several grid boxes. Either adjustment, however, results in a loss of information from GCMs about potential seasonal or spatial variations.

[6] In cases in which the $1 \times CO_2$ estimates of precipitation and other variables from a GCM are close to zero, the scenario could be based on the difference between $2 \times CO_2$ and $1 \times CO_2$, rather than the ratio between the two. Using the difference avoids using very high ratios that are mainly due to low values in the denominator. In applying the differences to baseline climate, users should ensure that the total of the daily changes in the variable equals the monthly difference from the GCM.

The application of GCMs is limited because it assumes that daily and interannual variability do not change from the baseline climate. That is, the same day-to-day pattern of temperature and precipitation within a month is maintained. Only the absolute amount of temperature and precipitation changes. Year-to-year changes also stay unchanged. In addition, variability between sites or stations within grid boxes basically remains unchanged (use of the interpolation technique will affect spatial variability). Appendix B describes some sophisticated techniques for developing scenarios with temporal and spatial variability. For example, Mearns et al. (1992) developed a technique to modify interannual variability of monthly temperature and precipitation from a historic record. Wilks (1992) demonstrated how to use a weather generator to change daily variability. In addition, Wigley et al. (1990) developed a statistical technique for examining spatial variance for temperature and precipitation within GCM grid boxes. Other techniques, such as use of nested meso-scale models, may also be available (Giorgi 1990). Use of simple methods, as discussed below, are recommended for ease of application.

4.3.2.2 Application of Incremental Scenarios

Incremental scenarios should be applied for the combinations of temperature and precipitation changes in Table 4.3.1. This broad range of temperature and precipitation changes captures the range of potential climate change in a region. Incremental changes in temperature and precipitation should be combined with the baseline climate data to create incremental scenarios. Researchers are encouraged to examine incremental changes in other climate variables that may significantly affect a sector. In addition, researchers may wish to test the sensitivity of different seasonal patterns of change (e.g., relatively more warming and precipitation in the winter).

4.3.2.3 Development of Transient GCM Scenarios

It may be important to policy makers to understand not only the potential climate change impacts in the latter half of the next century but also the potential effects within a few decades. Thus, transient scenarios can also help identify potential near-term effects.

The NCAR has made the Geophysical Fluid Dynamics Laboratory (GFDL) transient available to all participants in the U.S. Country Studies Program. In addition, the transient run at the Max-Planck Institute in Hamburg, Germany, will be made available when it is released.

TABLE 4.3.1 Incremental Climate Change Scenarios

Temperature/ Precipitation (%)	+2 °C	+4 °C	+6 °C
-20	X	X	X
-10	X	X	X
0	X	X	X
+10	X	X	X
+20	X	X	X

Either of the procedures discussed in Section 4.3.1.1 can create transient scenarios. If feasible, the first procedure should be used. This procedure is a continuous transient scenario from the present until the end of the transient simulation. The procedures for creating this scenario are discussed in detail in Appendix B. Readers may also wish to consult Carter et al. (1994) for the IPCC approach to creating transient scenarios.

4.3.3 Other Climate Change Scenarios

Scenarios must be developed for two other variables: the effects of CO_2 fertilization and sea-level rise.

4.3.3.1 Atmospheric CO_2-Level Scenarios

Higher levels of CO_2 enhance plant growth and increase efficiency of water use (Bazaaz and Fajer 1992). These higher levels can significantly affect crop yields and grassland and forest growth. Under some scenarios and for some crops, consideration of CO_2 effects can result in estimates of increased crop yields rather than decreased yields (e.g., Rosenzweig 1990; Easterling et al. 1993).

The use of CO_2 fertilization is discussed in detail in Section 5.1. The sensitivity of a sector to CO_2 fertilization should be tested by using scenarios with and without higher CO_2 levels. If resources do not permit both experiments, only scenarios including higher CO_2 levels should be used. The levels of atmospheric CO_2 recommended for use in climate change scenarios are 355 parts per million (ppm) for 1990, 440 ppm for 2030, and 580 ppm for 2075 (Wigley 1993a,b). Linear interpolations may be made between these data points to derive CO_2 values in other years.

4.3.3.2 Sea-Level Rise Scenarios

Sea-level rise is a highly probable result of global warming (Mitchell et al. 1990). The latest estimates for sea-level rise are for a 0.15- to 0.9-m rise by 2100, and the most likely estimate is about 0.5 m (Wigley and Raper 1992).

The sensitivity of coastal resources to current sea-level rise (0.2 m by 2100) and to accelerated rates of sea-level rise of 0.5 and 1.0 m by 2100 should be tested. These scenarios reflect the most likely range of sea-level rise by 2100. Researchers may wish to use even higher rates of sea-level rise, such as 2.0 m by 2100, to further understand sensitivities.

4.3.4 Summary of Primary Approach to Developing Scenarios

Table 4.3.2 summarizes the primary approach used for developing baseline and climate change scenarios recommended for vulnerability assessment.

TABLE 4.3.2 Primary Approach to Developing Baseline and Climate Change Scenarios

Scenario	Description	Source
Baseline		
Climate	1951-1980	
Socioeconomic	Population Income Productivity	United Nations (1990) World Bank (1993)
Environmental	Pollution Land use Desertification Ecosystem health	UNEP (1992) World Resources Institute (1992) Dregne and Chou (1992)
Climate Change		
$2 \times CO_2$ GCMs	At least three GCMs	NCAR
Incremental	Combinations of +2, +4, +6 °C and ±10 and 20% and no change in precipitation	
Transient	One to two GCM-based transient scenarios	NCAR
CO_2 levels	355 ppm in 1990; 440 ppm in 2030; and 580 ppm in 2075	Wigley 1993b
Sea-level rise	About 0.2, 0.5, and 1.0 m by 2100	Wigley and Raper (1992)

4.3.5 References for Sections 4-4.3

Bazzaz, F.A., and E.D. Fajer, 1992, "Plant Life in a CO_2-Rich World," *Scientific American* 266(1): 68-74.

Carter, T.R., M.L. Parry, S. Nishioka, and H. Harasawa, 1994, *Technical Guidelines for Assessing Climate Change Impacts and Adaptations*, Intergovernmental Panel on Climate Change, Geneva, Switzerland.

Cline, W.R., 1992, *The Economics of Global Warming*, Institute for International Economics, Washington, D.C.

Dregne, H.E., and N.-T. Chou, 1992, "Global Desertification Dimensions and Costs," in H.E. Dregne (ed.), *Degradation and Restoration of Arid Lands*, Texas Technical University, Lubbock, Texas.

Easterling, W.J., et al., 1993, "Paper 2. Agricultural Impacts of and Responses to Climate Change in the Missouri-Iowa-Nebraska-Kansas (MINK) Region," *Climatic Change* 24:23-61.

Food and Agricultural Organization, 1993, *Agriculture: Towards 2000*, The United Nations, Rome, Italy.

Gates, W.L., P.R. Rowntree, and Q.-C. Zeng, 1990, "Validation of Climate Models," in J.T. Houghton, G.J. Jenkins, and J.J. Ephraums (eds.), 1990, *Climate Change: The IPCC Scientific Assessment*, Cambridge University Press, New York, N.Y.

Giorgi, F., 1990, "Simulation of Regional Climate Using a Limited Area Model Nested in a General Circulation Model," *Journal of Climate* 3:941-963.

Grotch, S.L., and M.C. MacCracken, 1991, "The Use of General Circulation Models to Predict Regional Climatic Change," *Journal of Climate* 4:286-303.

Houghton, J.T., B.A. Callander, and S.K. Varney, 1992, *Climate Change 1992—The Supplementary Report to the IPCC Scientific Assessment*, WMO/UNEP Intergovernmental Panel on Climate Change, Cambridge University Press, Cambridge, England.

Houghton, J.T., G.J. Jenkins, and J.J. Ephraums (eds.), 1990, *Climate Change: The IPCC Scientific Assessment*, Cambridge University Press, New York, N.Y.

Jones, P.D., T.M.L. Wigley, and P.B. Wright, 1990, "Global and Hemispheric Temperature Anomalies," in T.A. Boden, P. Kanciruk, and M.P. Farrell (eds.), *Trends '90*, ORNL/CDIAC-36, Carbon Dioxide Information Analysis Center, Oak Ridge National Laboratory, Oak Ridge, Tenn.

Kalkstein, L.S. (ed.), 1991, *Global Comparisons of Selected GCM Control Runs and Observed Climate Data*, 21P-2002, U.S. Environmental Protection Agency, Washington, D.C.

Meadows, D.H., D.L. Meadows, and J. Randers, 1992, *Beyond the Limits: Confronting Global Collapse, Envisioning a Sustainable Future*, Chelsea Green Publishing Company, Post Mills, Vt.

Mearns, L.O., et al., 1992, "Effect of Changes in Interannual Climatic Variability on CERES-Wheat Yields: Sensitivity and $2 \times CO_2$ General Circulation Model Studies," *Agricultural and Forest Meteorology* 62:159-189.

Mitchell, J.F.B., S. Manabe, T. Tokioka, and V. Meleshko, 1990, "Equilibrium Change," in J.T. Houghton, G.J. Jenkins, and J.J. Ephraums (eds.), *Climate Change: The IPCC Scientific Assessment*, Cambridge University Press, New York, N.Y.

Nash, L.L., and P.H. Gleick, 1993, *The Colorado River Basin and Climate Change*. U.S. Environmental Protection Agency, EPA-230-R-93-009, Washington, D.C.

Parry, M.L., T.R. Carter, and N.T. Konijn (eds.), 1988, *The Impacts of Climatic Variations on Agriculture: Volume 1—Assessments in Cool Temperate and Cold Regions and Volume 2 — Assessments in Semi-Arid Regions*, Kluwer Academic Publishers, Norwell, Mass.

Pepper, W., et al., 1992, *Emission Scenarios for the IPCC. An Update: Assumptions, Methodology, and Results*, ICF, Inc., Fairfax, Va.

Poiani, K.A., and W.C. Johnson, 1993, "Potential Effects of Climate Change on a Semi-Permanent Prairie Wetland," *Climatic Change* 24:213-232.

Robock, A., et al., 1993, "Use of General Circulation Model Output in the Creation of Climate Change Scenarios for Impact Analysis," *Climatic Change* 23:293-335.

Rosenberg, N.J., and P.R. Crossen, 1991, *Processes for Identifying Regional Influences of and Responses to Increasing Atmospheric CO$_2$ and Climate Change—The MINK Project: An Overview*, DOE/RL/01830T-H5, prepared for the U.S. Department of Energy, Resources for the Future, Washington, D.C.

Rosenberg, N.J., B.A. Kimball, P. Martin, and C.F. Cooper, 1990, "From Climate and CO$_2$ Enrichment to Evapotranspiration," in P.E. Waggoner (ed.), *Climate Change and U.S. Water Resources*, John Wiley & Sons, New York, N.Y.

Rosenzweig, C., 1990, "Crop Response to Climate Change in the Southern Great Plains: A Simulation Study," *Professional Geographer* 42:20-37.

Smith, J.B., and D. Tirpak (eds.), 1989, *The Potential Effects of Global Climate Change on the United States*, EPA-230-05-89-050, U.S. Environmental Protection Agency, Washington, D.C.

Strzepek, K.M., and J.B. Smith (eds.) (forthcoming), *As Climate Changes: The Potential International Impacts of Climate Change*, Cambridge University Press, New York, N.Y.

Tegart, W.J. McG., G.W. Sheldon, and D.C. Griffiths, 1990, *Climate Change—The IPCC Impacts Assessment*, WMO/UNEP Intergovernmental Panel on Climate Change, Australian Government Publishing Service, Canberra, Australia.

United Nations, 1990, *Overall Socioeconomic Perspective of the World Economy to the Year 2000*, New York, N.Y.

United Nations, 1989a, *World Population Prospects 1988*, New York, N.Y.

United Nations, 1989b, *The State of the World Environment 1989*, U.N. Environment Programme, New York, N.Y.

UNEP, 1992, *World Atlas of Desertification*, United Nations Environment Programme, Edward Arnold, Hodder, & Stoughton, Kent, U.K.

Viner, D., and M. Hulme, 1992, *Climate Change Scenarios for Impact Studies in the UK*, Climatic Research Unit, University of East Anglia, Norwich, U.K.

Vinnikov, K.Y., and N.A. Lemeshko, 1987, "Soil Moisture Content and Runoff in the USSR Territory with Global Warming," *Journal of Meteorology and Hydrology* Vol. 12.

Waggoner, P.E., (ed.), 1990, *Climate Change and U.S. Water Resources*, John Wiley & Sons, New York, N.Y.

Webb, T., III, 1992, "Past Changes in Vegetation and Climate: Lessons for the Future," in R.L. Peters and T.E. Lovejoy (eds.), *Global Warming and Biological Diversity*, Yale University, New Haven, Conn.

Wescoat, J., and R. Leichenko, 1992, *Complex Indus River Basin Management in a Changing Global Climate: The Indus River Basin in Pakistan, A National Assessment*, CADSWES, University of Colorado, Collaborative Paper U.S., Boulder, Colo.

Wigley, T.M.L., 1993a, "Balancing the Carbon Budget. Implications for Projections of Future Carbon Dioxide Concentration Changes," *Tellus* 45B:1-17.

Wigley, T.M.L., 1993b, personal communication (National Center for Atmospheric Research).

Wigley, T.M.L., and S.C.B. Raper, 1992, "Implications for Climate and Sea Level of Revised Press. IPCC Emissions Scenarios," *Nature* 357: 293-324.

Wigley, T.M.L., et al., 1990, "Obtaining Sub-Grid-Scale Information from Coarse-Resolution General Circulation Model Output," *Journal of Geophysical Research* 95:1943-1953.

Wilks, D.S., 1992, "Adapting Stochastic Weather Generation Algorithms for Climate Change Studies," *Climatic Change* 22:67-84.

World Bank, 1993, *Global Economic Prospects in the Developing Countries,* Washington, D.C.

World Bank, 1992, *World Population Projections, 1992-93*, World Bank, Washington, D.C.

World Resources Institute, 1993, *The 1993 Information Please Environmental Almanac*, Houghton Mifflin Company, New York, N.Y.

World Resources Institute, 1992, *World Resources 1992-93*, Oxford University Press, New York, N.Y.

4.4 GCM MODEL AND OBSERVED CLIMATE DATA

Section 4.3 discussed the selection and application of climate change scenarios. This section provides information about the data from climate models.

The National Center for Atmospheric Research has established a data bank of selected output from climate models to support studies of the economic and social impacts of climate change. The data bank has the variables (e.g., temperature, precipitation, and radiation for crop growth) most needed for assessment studies. Data are available for both $1 \times CO_2$ and $2 \times CO_2$ runs and for transient model runs that gradually increase the trace gases in the atmosphere. The data are packaged for ready use for a variety of assessment studies.

To prepare the data for assessment studies, NCAR extracted data from the models' global archives and put them into one simple format with appropriate documentation. Data are available on floppy disks, networks, and tapes. In addition to the data contained in the model simulation described here, data are available on tape at NCAR from other model simulations.

4.4.1 General Description of Climate Models

A climate model is a sophisticated set of computer programs that solve the equations that describe, for example, how winds blow and how pressure and temperature change in the atmosphere. The purpose of a climate model is to describe how the climate differs when a major change (such as $2 \times CO_2$) is put into the model. The models simulate radiation and clouds; they evaporate water from the soil and use excess precipitation to build up soil moisture. When it snows, the model changes the reflectance of the surface. These low-resolution models cannot directly handle the details of a thunderstorm, but the statistical effects of such storms over a broad region are included.

At least a simple ocean and its sea ice must also be included in any climate model. The effects of ocean heat storage and sea ice are too significant to be ignored.

The reader might ask why one should believe a climate model when a weather forecast model (run at better resolution) does not have good skill beyond a few days. Climate models are still useful for the long term because they do not have to forecast the exact timing and shape of each weather system as a forecast model does. They only need to give the correct statistics about these events. Climate models should be able to tell whether the main belt of westerly storms would shift in latitude or intensity if more CO_2 were added to the atmosphere. If the model correctly handles clouds, surface evaporation, and the radiative effects of gases, for example, it should then be able to provide insight about the earth's climate. If something in the earth's system (or in the model) changes, such as the amount of CO_2 or the land surface properties, the model should be able to give information about how the earth's climate will respond.

The climate models are sophisticated and useful, but they are still simple compared with the complexity of the real climate system. Assessments of climate change use the outputs from several climate models so that the differences among the models give some feeling for their reliability. The models still require substantial development.

Climate models have had relatively low resolution because of the computer time needed to run these models. The nine-layer Goddard Institute for Space Sciences (GISS) model run in 1982 had a horizontal resolution of about 900 km. Various models run during 1984-1988 usually had about nine vertical levels and a resolution of 500 to 550 km. In 1989-1990, some models provided high-resolution data (350-450 km) (Table 4.4.1).

4.4.2 Climate Models for Use in Assessment Studies

This section describes the main models available for use in assessment studies. Appendix C provides more detailed information. The more recent model runs typically have better physics and higher resolution. Model runs that have been made available by the Country Studies Program to date include the 1982 GISS run, the 1989 GFDL, R-30 run, the 1989 CCC run, and the 1991 GFDL transient run. Table 4.4.1 gives some characteristics for these and other model runs.

4.4.2.1 Model Runs at the Goddard Institute for Space Sciences—GISS Model

Three climate model experiments developed at GISS are considered here, including one transient model. The first model run was completed in 1982 and is still relatively competitive (Hansen et al. 1983). It has a diurnal cycle, with low horizontal resolution (7.83° × 10.0°) and nine vertical levels. The GISS group originated the Q-flux procedure to improve simulation of the ocean and sea ice. Ocean water temperatures and ice cover are computed on the basis of an hour-by-hour energy exchange with the atmosphere, the specified pseudo-ocean heat transports (Q flux), and the ocean mixed-layer heat capacity. The model uses a simple, so-called slab ocean, not more than 65 m deep for the $1 \times CO_2$ and $2 \times CO_2$ runs, with some variation of mixed depth over the seasonal cycle; the depth is shallower in mid-latitudes in summer than it is in winter. The model sensitivity is 4.2 °C (i.e., the global surface temperature warms by 4.2 °C in the model for the $2 \times CO_2$ run compared with the $1 \times CO_2$ run).

The second model is a transient simulation completed in 1985. The transient case is similar to the $1 \times CO_2$ and $2 \times CO_2$ cases, but it has a deeper mixed depth, when appropriate.

The third model is a new version of the original $1 \times CO_2$ and $2 \times CO_2$ run. It has a better resolution (3.9° × 5.0°), and a Q-flux slab ocean. A preliminary 70-year run with only nine vertical levels was completed about February 1994 (control run and 1% gas increases). After 70 years, the temperature increase was 1.5 °C. The output was similar to that from other main models, but this model has river runoff into the ocean, and the sea level is calculated. After 70 years, the sea-level rise was 11 cm; about 7 cm of this increase was due to thermal expansion. The model uses a more advanced ocean model than has been used in some runs. Computation time is two months on an IBM 590 workstation to calculate 50 years or simulation (one-third of this time is used for the ocean).

TABLE 4.4.1 Characteristics of Selected Climate Models[a]

Climate Model	When[b] Calculated	Model Resolution (latitude × longitude)	Vertical Levels	Diurnal Cycle	Base 1XCO$_2$ (ppm)	ΔT for 2XCO$_2$ (°C)	Increase in Global Precipitation (%)	Data Use Code[c]
1XCO$_2$, 2XCO$_2$								
GISS	1982	7.83° × 10.0°	9	Yes	315	4.2	11.0	1, 0
GISS	Nov. 1994	3.9° × 5.0°	12-18	Yes	315	TBD[d]	TBD	0, 2
OSU	1984-1985	4.00° × 5.0°	2	No	326	2.84	7.8	1, 0
NCAR, T42	June 1994	2.81° × 2.81°	18	Yes	300	TBD	TBD	0, 3
GFDL R-15[e]	1984-1985	4.44° × 7.50°	9	No	300	4.0	8.7	1, 0
GFDL[e]	Feb. 1988	4.44° × 7.50°	9	No	300	4.0	8.3	1, 0
(Better ocean)								
GFDL R-30[e]	May 1989	2.22° × 3.75	9	No	300	4.0	8.3	0, 1
GFDL R-30[e], V2	March 1994	2.22° × 3.75	14	No	300	3.2	TBD	0, 2
U.K.	June 1986	5.00° × 7.50°	11	Yes	320	5.2	15.0	1, 0
U.K. (high-resolution)	Nov. 1989	2.50° × 3.75°	11	Yes	320	3.5	9.0	0, 1
Canada (CCC), T32	Nov. 1989	3.75° × 3.75°	10	Yes	330	3.5	3.8	1, 1
German	1990	5.63° × 5.63°	19	--[f]	--	2.6	--	0, 1
Transient								
U.K.	1994	TBD	TBD	TBD	TBD	TBD	--	1, 0
GISS	1984-1985	7.83° × 10.0°	9	Yes	315[g]	(e4.2)[h]	--	0, 2
German (dynamic ocean)	1990	5.63° × 5.63°	19	Yes	--	(2.6)	--	0, 2
GFDL (dynamic ocean)	1991	4.44° × 7.50°	9	No	--	(e4.0)	--	0, 1

a All models are global in extent, have a smoothed topography that varies between models, and have an annual cycle. All models (except the transients) give data for the present climate (1XCO$_2$) and double CO$_2$ climate (2XCO$_2$).

b The last date (or only date) given is the time the model run was completed.

c The first digit tells whether the model run has been available and used in previous assessment studies (0 = no, 1 = yes). The second digit can be 0 (not intended for use in CSP), 1 (for use, available now), 2 (for use, available somewhat later), and 3 (probably too late to use).

d TBD = to be determined.

e This spectral model (R-15) has 15 waves around the equator; the R-30 has 30 waves. The NCAR model (T42) has 42 waves. Most of the other models are grid-point models with resolution as given.

f Hyphens denote no data available.

g This amount was the observed amount of CO$_2$ in 1958.

h The transient runs do not really have a sensitivity. This number (e 4.2) shows the temperature sensitivity when the model or its near equivalent is used in a 1XCO$_2$ or 2XCO$_2$ run.

The GISS is developing new versions of the model that have new physics. One version with half of the improvements will also be run at a resolution of 3.9° × 5.0° and with nine vertical levels. The run will be completed about August 1994 and released to NCAR in November 1994. For the atmosphere, it is 33% slower than that of the present ocean-plus-atmosphere version. The version with all of the new physics will be completed about November 1994, with the data released to NCAR around March 1995. It will have 12 to 18 levels (with a top of about 3 mbar) and a 3.9° × 5.0° resolution. None of these newer model runs is available from NCAR at this time.

4.4.2.2 Model Runs at the National Center for Atmospheric Research—NCAR Model

The NCAR has developed a community climate model called "CCM2," which has a diurnal cycle with a sophisticated land surface. A $1 \times CO_2$ and $2 \times CO_2$ run was expected to start in January 1994 and be completed in June 1994. This run has 18 vertical levels and a slab ocean with a variable depth and uses a Q-flux procedure. The resolution is 2.81° × 2.81° (312 km); a global field has 128 × 64 points. Data was released in October 1994.

4.4.2.3 Model Runs at the Geophysical Fluid Dynamics Laboratory—GFDL Model

The GFDL has four model runs. The first version of the model uses a slab ocean 68 m deep and has a resolution of 4.44° × 7.50° (R-15, a spectral model with 15 waves) (Manabe and Wetherald 1987). This run was completed in 1985 and did not have ocean currents or a pseudo-heat-flux to improve the water temperatures for the present climate. Another model run at GFDL was completed in February 1988. The model is about the same as before, except that it includes a Q-flux procedure to reproduce the sea surface temperature and the sea ice extent of the present climate. The GFDL calls this model the Q-flux model. The sensitivity of both model runs is 4.0 °C (global change from $1 \times CO_2$ to $2 \times CO_2$).

A third run was conducted in May 1989 with largely the same model but with a better resolution, i.e., 2.22° × 3.75° (R-30, a spectral model with 30 waves). It still does not have a diurnal cycle. These simulations start with the results from the R-15 runs and process 10 more years of simulations (for both $1 \times CO_2$ and $2 \times CO_2$). With the better resolution, the model precipitation is more realistic when compared with that of the present climate. The soil moisture is also more realistic, probably because the precipitation is better. The sensitivity of the R-30 run is still 4.0 °C. This run gives unrealistic changes in precipitation from one grid point to the next, and these will be smoothed out for the users.

The GFDL is conducting a new run with 14 vertical levels. The resolution is R-30, and it has a Q-flux slab ocean, as before. A diurnal cycle is not included. With changes in high clouds and ice albedo, the temperature sensitivity may be somewhat below the previous 4.0 °C.

The GFDL 100-year transient run was completed in 1990. The atmosphere has a latitude-longitude resolution of 4.4° × 7.5° (R-15). The ocean is the same resolution in latitude, but it is better in longitude (3.75°), which helps obtain a better simulation of the western boundary currents. There are 9 vertical levels in the atmosphere and 12 in the ocean. The bottom ocean level is 5 km deep. This dynamic ocean has currents and is better than the slab oceans used in earlier models. The atmospheric model is essentially the same as the GFDL models that have a sensitivity of 4.0 °C for $2 \times CO_2$. The earliest run was made in 1989 and used in the IPCC studies. The time step in the ocean caused some minor problems. This run is a 1990 rerun with a somewhat shorter time step in the ocean model.

4.4.2.4 Model Runs at the Canadian Climate Centre—Canadian Climate Model

The Canadian Climate Centre (CCC) has developed a model and has run it for $1 \times CO_2$ and $2 \times CO_2$ conditions (Boer et al. 1991). The model has a higher resolution ($3.75° \times 3.75°$) than many climate simulations, with 10 vertical levels. It has a 50-m-deep slab ocean and uses a Q-flux procedure. The model has a typical relative humidity scheme to develop clouds and a diurnal cycle. The climate sensitivity is lower than many models: it shows a temperature increase of 3.5 °C and a 3.8% increase in precipitation when CO_2 doubles (Table 4.4.1). Data were released for general use in May 1991.

4.4.2.5 Model Runs at the U.K. Meteorological Office

The U.K. Meteorological Office (UKMO) has two model runs. The first $1 \times CO_2$ and $2 \times CO_2$ run was completed about June 1986 (Wilson and Mitchell 1987). This grid-point model has a resolution of $5.00° \times 7.50°$ and 11 vertical levels (on sigma surfaces), plus the surface, and a diurnal cycle. It has a slab ocean that is 50 m thick, similar to other models. The Q-flux procedure correctly simulates the present ocean water temperature and sea ice. The model uses a spin-up period of about 20 years and then collects 15 years of data. The model climate warms by 5.2 °C for $2 \times CO_2$. Most other models show only about 4 °C. The higher sensitivity results because this model has more layers, which permits better development of the important high clouds. In general, a warming of this extent is believed to be too high.

The UKMO completed a second run in November 1989, with double the resolution ($2.50° \times 3.75°$). This run included water content of clouds, penetrating cumulus clouds, more realistic cirrus dissipation, but not variable optical properties. The model still has a diurnal cycle and a Q-flux slab ocean. The sensitivity for $2 \times CO_2$ is a global temperature increase of 3.5 °C and a 9% increase in precipitation. The UKMO has just released a new transient model run to NCAR.

4.4.2.6 Model Runs at the Max-Planck Institute in Hamburg—Hamburg Climate Model

Three transient runs and one $1 \times CO_2$ and $2 \times CO_2$ run are available from the Max-Planck Institute in Hamburg, Germany. The model has 19 vertical levels in the atmosphere and a coupled dynamic ocean, not just a slab ocean. A global array has 64×32 points at a resolution of $5.63° \times 5.63°$. It has one control run for 100 years and two transient runs (A and D) with differing greenhouse forcing. Run A has gases for "business as usual," whereas run D has a slow increase of greenhouse gases. In the global average, the surface atmospheric temperature rises by 2.6 °C in scenario A and by 0.6 °C in scenario D for the 100-year period. The NCAR received year-monthly output data from the three long transient runs. Many variables are at the surface (e.g., temperature, precipitation, radiation, heat fluxes, and others) and data for 15 vertical levels at 1,000, 950, and 900 through 30 mbar. The data are on 15 cartridge tapes (about 200 Mbyte each). The data from this run have just been released to NCAR.

Runs for $1 \times CO_2$ and $2 \times CO_2$ were also conducted. The model sensitivity for $2 \times CO_2$ was also 2.6 °C. The data from this run are soon to be released to NCAR.

4.4.3 Climatological Data to Compare with Climate Models

The NCAR has global climatological grids of sea-level pressure; surface air temperature and moisture; and upper-air temperature, moisture, and winds. The surface temperature and dew point and the upper-air climate are from Taljaard et al. (1969) and Crutcher and Meserve (1970). These tapes at NCAR also have many upper-air fields for the world. The climatological precipitation used for comparisons has been published in the RAND set (Schutz and Gates 1971), where the precipitation was taken from Möller (1951). Other sources of precipitation grids will also be used in future comparisons. None of the climatologies of precipitation over the ocean is reliable, and enough is now known that they could be improved. Data sets are derived from satellite measurements that give the albedo of the earth (the amount of short wave energy from the sun reflected back to space) and the outgoing infrared radiation. The NCAR also has several data sets with global elevation and land use data.

4.4.4 Data Available for the Country Studies Program Vulnerability Assessment Studies

The data most used by researchers who study crops, forests, and rivers are long-period mean model data (about 10 years) for each month for the present climate ($1 \times CO_2$) and for the $2 \times CO_2$ climate. In the transient runs, means for each decade are usually used. These data are available for the Country Studies Program (Table 4.4.1).

The Country Studies Program assessment studies uses output from several climate models. The temperature change when changing from $1 \times CO_2$ to $2 \times CO_2$ is shown in Table 4.4.1. This change is also called "model sensitivity." The data-use code in the table has two digits (e.g., 0, 2); see footnote c in the table for an explanation.

The NCAR has prepared the data most needed into a simple common format. For most models, NCAR has only long-term monthly averages of the most important 7 to 20 variables. The primary variables include temperature, precipitation, surface incident solar radiation, surface-air moisture, total cloud cover, surface average wind speed (if available), evaporation, and sea-level pressure. The NCAR also has data for individual months for some models, and sometimes for all 170 fields in the model. Several years of daily model data are also present from one model.

All of the climate model data at NCAR are on many tapes; however, the most important global monthly mean data for assessment studies are on one tape (132 Mbyte). The most important three or four variables (e.g., temperature and precipitation) for two to four models total just a few megabytes and can often be provided on floppy disks.

The user can use the access software to extract a table of all grid points from one global field within a specified latitude-longitude region. Another software tool allows a user to obtain data in another table with data for each of the 12 months, for $1 \times CO_2$ and $2 \times CO_2$, at one grid point. Appendix C has more detailed information about how to use this software and about the model data.

4.4.5 References for Sections 4.4-4.5

Boer, G.J., N. McFarlane, and M. Lazare, 1991, "Greenhouse Gas-Induced Climatic Change Simulated with the CCC Second-Generation GCM," *Journal of Climate* (accepted for publication).

Crutcher, H.L., and J.M. Meserve, 1970, *Selected Level Heights, Temperatures and Dew Points for the Northern Hemisphere*, NAVAIR 50-1C-52 (revised), Chief of Naval Operations, Washington, D.C.

Hansen, J., G. Russell, D. Rind, P. Stone, A. Lacis, S. Lebedeff, R. Ruedy, and L. Travis, 1983, "Efficient Three-Dimensional Global Models for Climate Studies: Models I and II," *Monthly Weather Review* III:609-662, April.

Manabe, S., and R.T. Wetherald, 1987, "Large-Scale Changes in Soil Wetness Induced by an Increase in Carbon Dioxide," *Atmos. Sci.* 44:1211-1235, April.

Möller, F., 1951, "Vierteljahreskarten des Niederschlags für die ganze Erde," *Petermanns Geographische Mitteilungen*, Justus Perthes, Gotha., pp. 1-7.

Schutz, C., and W.L. Gates, 1971, *Global Climate Data for Surface, 800 millibars, 400 millibars*, R-915-ARPA, Rand Corporation, Jan.

Taljaard, J.J., H. van Loon, H.L. Crutcher, and R.L. Jenne, 1969, *Climate of the Upper Air: Southern Hemisphere. 1: Temperatures, Dewpoints, and Heights at Selected Pressure Levels*, NAVAIR 50-1C-55, Chief of Naval Operations, Washington, D.C., p. 135.

Wilson, C.A., and J.F.B. Mitchell, 1987, "A Doubled CO_2 Climate Sensitivity Experiment with a Global Climate Model Including a Simple Ocean," *GR*, 92:D11-11/20 13, 315-13,343.

5 CONDUCTING BIOPHYSICAL IMPACT AND TECHNICAL ADAPTATION ASSESSMENTS

As shown on the vulnerability and adaptation assessment flow diagram (Figure 2.1), after the scenarios have been selected, the next step in the process is to conduct the biophysical impact and technical adaptation assessments. These assessments focus on the specific biological and physical effects that climate change may have on various sectors of the economy and the natural environment. This section describes procedures for conducting assessments on five sectors:

- Crops,

- Grasslands and livestock,

- Forests,

- Water resources, and

- Coastal resources.

Other sectors can also be affected by climate change. Countries should feel free to expand on this list in the country study as time and resources allow.

In general, the biophysical assessment contains two parts. The first is an impact assessment of the effects of climate change in each sector using the scenarios described in Section 4. This part is designed to develop insight into the direct effects of climate change (e.g., changes in crop yields or water availability). Either an empirical-statistical or a simulation approach is used to develop projections of the biological and physical effects.

The second part of the assessment recalculates those effects, including consideration of steps taken to adapt to the changes (e.g., changing crop planting cycles or water use patterns). This recalculation of effects is termed "technical adaptation" because it presumes that the steps taken to adapt do not have any major impact on the scope and extent of the climate change. (Steps to effect the extent of climate change by, for example, reducing CO_2 emission, are considered "mitigation" strategies and are covered elsewhere.)

It is important to include both parts of the biophysical assessment in the country study. They will indicate the effectiveness of various actions to reduce a country's vulnerability to potentially damaging effects of climate change.

The process of extending the analysis of biological and physical effects into economic and social effects and of combining the results from the different sectors into an integrated analysis is discussed in Section 6.

Each of the following sections, which are structured in a similar fashion for each of the sectors, describes a primary approach for conducting the biophysical impact and technical adaptation assessment. Each describes the methodology, specifies the analytic procedures that can be used, identifies available computer models, gives data requirements, and shows the results that can be generated. Each section also indicates the limitations of the approach.

R. Benioff et al. (eds.), Vulnerability and Adaptation Assessments, 5-1-5-133.
© 1996 *Kluwer Academic Publishers.*

The primary approach should be considered as the basic recommendation of experts in the field who have previously carried out these types of analyses. However, it should not be considered as mandatory for completion of the country study. Countries are free to substitute other approaches (including the supplementary approaches described for each sector) as their needs dictate and in consultation with the Country Studies Management Team.

5.1 CROP IMPACT AND ADAPTATION ASSESSMENTS

Simulating the effects of climate change on agricultural production at a national level requires a coordinated effort that integrates data, software, and expertise from various disciplines and institutions. Researchers gather data on weather, soil, and management practices for crops of national importance—both commercial and subsistence—for selected sites and regions. It is useful to include sites representative of both major production areas and regions that may be vulnerable to a changing climate regime. Models that simulate crop growth and climate change scenarios can then be used to estimate changes in yields. By using a common set of climate change scenarios and compatible models, agricultural impact researchers can develop a set of results that can be compared with results from both their own and other regions.

Because of the uncertainties associated with predicting climate change, researchers commonly use climate scenarios to estimate how climate affects a system (in this case, agricultural production). Scenarios derived from GCMs and arbitrary sensitive tests (e.g., +2 °C and +4°C temperature changes, ±10% precipitation change) are recommended to estimate potential future change in yield and other agronomically important variables. (Section 4 discusses climate change scenarios in detail.)

The yield changes that result from national studies can be used in several types of assessments: (1) studies with agricultural economics models to estimate changes in production and trade, (2) studies of the magnitudes and nature of food deficits in regions currently vulnerable or possibly vulnerable to the predicted climate change, and (3) analyses of adaptive responses to climate change.

Economic assessments can provide information on potential gains and losses across space and time as well as benefits and costs to society. Economic modeling can estimate the potential effects of climate change on the comparative advantage of food-producing regions, food prices, and patterns of food production and trade.

Vulnerability to climate change can be assessed in terms of national food balance, agroecological potential, and vulnerable socioeconomic groups. Case studies of particularly vulnerable regions or groups are useful for addressing potential food deficits and the severity of these deficits.

Adaptive response analysis is recommended to evaluate changes in agricultural practices (e.g., irrigation, planting dates, varieties, and species) or programs (e.g., crop breeding and genetic development, and food stocks) that would minimize adverse climatic impacts. To the extent possible, regional and national agricultural program managers in each country should be informed of the agricultural climate change impact studies and asked to suggest adaptive policies or programs.

5.1.1 Primary Approach

The primary approach for analyzing agricultural impacts consists of a preliminary screening technique and simulation techniques. These techniques are described in the following sections.

5.1.1.1 Preliminary Screening Technique

Preliminary screening begins with an assessment of vulnerability to climate change. The assessment is based on expert judgment obtained by consulting in-country agricultural specialists. Such an assessment can provide a qualitative (or simple quantitative) analysis of potential impacts and adaptations. One objective is to identify areas in need of more detailed analyses. In some cases, where models cannot be used and/or data are not available for further analysis, such a vulnerability assessment may substitute for modeling results.

The preliminary vulnerability assessment for agriculture could include the following:

- Identification of the crops, geographical regions, and rural populations most likely to be vulnerable;

- Description of the vulnerable crops, regions, and groups as well as the reasons for their vulnerability;

- Analysis of analogous regions (e.g., warmer regions of the country) under current climate regimes to help identify implications for future response to climate change;

- Projections of the expected magnitude of the impacts, expressed qualitatively (e.g., positive, negative, or no impact on crop yield); and

- Examination of potential adaptation measures, with explicit study of their expected effectiveness and costs.

A number of methods are available for obtaining expert input to the preliminary screening analysis. These methods include surveys, workshops, and reports from key agencies. Both agronomists and agricultural economists can supply useful information on the biophysical and socioeconomic impacts, respectively. Also useful are summaries of previous research on potential climate change impacts and adaptations in the country or region.

Results of the preliminary agricultural vulnerability assessment can be in the form of survey analyses, workshop reports, and government agency reports. Key results can be summarized in a table. (Table 5.1.1 gives an example for crops.)

TABLE 5.1.1 Example for Crops

System	Characteristics Sensitive to Climate Change				Degree of Vulnerability	Adaptation Alternatives	
	Yield	Irrigation Needs	Suscepti- bility to Disease	Etc.		Type	Effective- ness, Cost
Grain crops Wheat Corn etc.							
Other crops Fruits Cassava etc.							

5.1.1.2 Simulation Techniques

The IBSNAT-ICASA[7] network allows evaluation and dissemination of appropriate crop models and climate change impact methodologies so that country study participants can use specific weather, soil, and crop information to simulate the effects of climate change, thus identifying potential agricultural vulnerabilities. This approach was selected because the crop models (1) have been validated in a wide range of environments around the world; (2) are easy to use; (3) have consistent inputs and outputs, which facilitate comparison of results; and (4) have a decision support system that allows comprehensive testing of climate change scenarios, increasing levels of CO_2, and farm-level adaptations. Each country's agricultural scientist is encouraged to follow a set of simulation tasks. Methodology will be disseminated for extending the crop model results from individual sites to regional and national estimates of change. The linkages among resource economists, which are needed to evaluate economic adaptations to projected climate changes, will be identified.

The IBSNAT-ICASA models are made up of parameterizations of important physiological processes responsible for plant growth and development, evapotranspiration, and partitioning of photosynthate to produce economic yield. The functions help predict the growth of crops, as influenced by the major factors that affect yields, i.e., genetics and climate (e.g., daily solar radiation; maximum and minimum temperatures; and precipitation, soils, and management practices). The models include a soil moisture balance submodel, which allows them to predict both rainfed and irrigated crop yields. The cereal models simulate the effects of nitrogen fertilizer on crop growth. Changes in soil moisture can be compared with those calculated with water resource models (Section 5.4).

---

[7]The International Benchmark Sites Network for Agrotechnology Transfer (IBSNAT) was a 10-year project of USAID. The International Consortium for Application of Systems Approaches to Agriculture (ICASA) is an independent follow-on to IBSNAT.

The Decision Support System for Agrotechnology Transfer (DSSAT) is a comprehensive software system that integrates crop growth models with crop, weather, and soil data and with various application programs (IBSNAT 1989). The DSSAT can be the major tool for integrating the databases and crop models in the climate change studies. This software was developed by the IBSNAT project to assist scientists studying crop growth, development, and yield responses to various soil, weather, and management conditions. It is a user-friendly system in which a "shell" program resides in computer memory and provides users with a range of functions via "pop-up" menus on the screen. The system's three main components—crop models, databases, and crop data—are referenced by site. Users can enter site-specific experimental data for crop model validation or for sensitivity analysis. Utility programs assist users in data entry, graphic display, and linkage to crop models contained in the system. Appendix D contains a more detailed description and technical procedures for the IBSNAT-ICASA model approach.

5.1.2 Description of the Methodology

Tasks for the agricultural scientists conducting the country study include:

- Specifying the soils and crop management inputs necessary to run the IBSNAT-ICASA crop models at the selected sites.

- Defining the geographic boundaries of the major production regions of the country and estimating the current production of major (both commercial and subsistence) crops in those regions.

- Providing observed climate data for representative sites within these regions.

- Specifying the soils and crop management inputs necessary to run the crop models at the selected sites.

- Validating the crop models with experimental data from field trials.

- Running the crop models with the baseline-observed data and climate change scenarios, with and without simulations of the direct effects of CO_2 on crop growth, irrigated production (if applicable), additional sensitivity tests, and adaptation responses (planting date, appropriate varieties, or species).

- Reporting modeled yield (T/ha) changes and other results, that is, changes in season length (d), growing season precipitation (mm), growing season evapotranspiration (mm), and water used for irrigation (mm) arising from the climate change scenarios and other simulations.

- Identifying and evaluating alterations in agricultural practices that would lessen any adverse consequences of climate change.

- Collaborating with agricultural and resource economists to project the economic consequences of crop modeling results.

- • Writing a report that describes the agricultural system modeled; the methods and results of the crop modeling work, including adaptation responses, the economic aspects of the study, and the implications of the projected climate; and yield changes on agriculture in the country.

A flow diagram illustrating the above tasks is given in Figure 5.1.1.

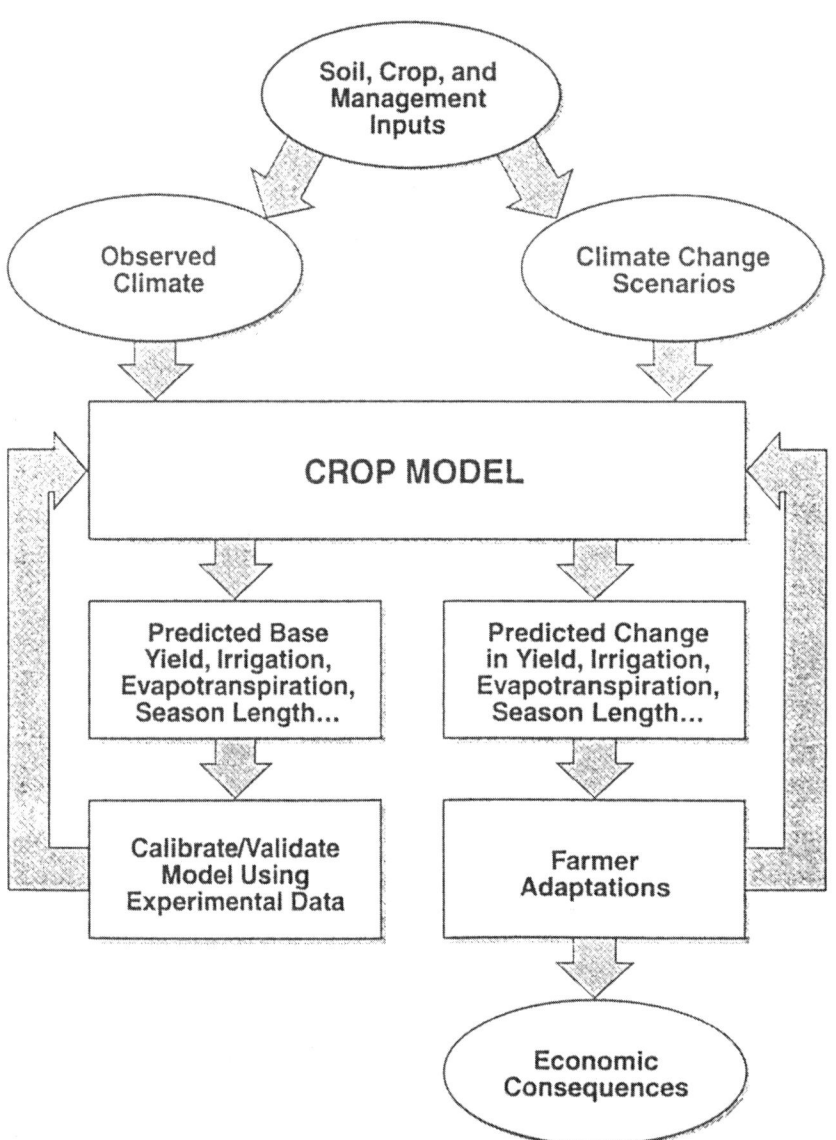

FIGURE 5.1.1 Flow Diagram for the Crop Model

5.1.3 Scope of the Analysis

5.1.3.1 Spatial Dimensions in Model Validation

Crop modelers should choose the most productive agricultural regions in each country. These regions are defined as similar geographic areas of land, characterized by particular patterns of soil, climate, water resources, land use, and type of farming. Studying two or more regions in each country is useful for examining regional differences in response to projected climate changes, provided that weather and crop data are available to validate and run the crop models.

Representative sites for which the needed data are available should be selected for each major agricultural region. If possible, the sites should represent low, medium, and high productive capacity within the region.

5.1.3.2 Methodological Design for Baseline and Climate Change Scenarios

Scenarios for the simulation studies can be devised by changing an observed baseline set of climatological data (e.g., daily minimum and maximum temperature, daily precipitation, and daily incident solar radiation for the IBSNAT/ICASA models) to prescribed anomalies, some arbitrary and some derived from climate models (GCMs). Section 4 describes the climate change scenarios available for use and the suggested order of use in the country studies.

The GCM results needed to create the climate change scenarios for the IBSNAT-ICASA models are changes in minimum and maximum temperature, precipitation, and incident solar radiation. Other crop models may also require changes in dewpoint temperature, humidity, and wind speed. Thresholds of vulnerability can be determined by simulating a set of climate change scenarios with the crop models and analyzing the results. Changes in climate variability, such as changes in the frequencies of drought and storms, can significantly affect crop yields; such types of changes should be tested, where possible.

5.1.3.3 Methodological Design for Simulations

Each agricultural scientist is to run the IBSNAT-ICASA crop model(s) for the baseline and climate change scenarios at each selected site, depending on availability of time and resources.

5.1.3.3.1 Simulations with the Direct Effects of CO_2. Crop growth and water use benefit from increased levels of CO_2 (Cure and Acock 1986). The CERES-Wheat, CERES-Maize, SOYGRO, and other IBSNAT-ICASA crop models have been modified to simulate the changes in photosynthesis and evapotranspiration caused by increases in atmospheric CO_2 (Peart et al. 1989). These models can produce results for climate change and the direct effects of CO_2 on crop growth and water use. Section 4.3.3.1 discusses the appropriate levels of CO_2 for use in crop simulations for the equilibrium and transient climate change scenarios.

5.1.3.3.2 Irrigated Production. If irrigated production is widespread under the current climate, or if the climate change scenarios suggest drying for an agricultural region, both dry land and irrigated production should be simulated. This scenario aids in giving a critical estimate of how the demand for water from agriculture may change as climate changes.

5.1.4 Economic Analyses

The previous sections discuss some of the assumptions, procedures, and challenges in forecasting the magnitude and geographic distribution of crop yield changes that result from climate change. Once measured, it is useful to translate such yield responses into economic or other measures of human welfare; these economic measures are one component of the information that policy makers need to understand how climate change affects humans.

A complete assessment of the economic consequences requires three types of information: (1) the differential changes that climate change causes in production and consumption opportunities (e.g., crop yields); (2) the responses of input and output market prices to these changes; and (3) the input and output changes that affected individuals make to minimize losses or maximize gains from changes in production and consumption opportunities and in the prices of these opportunities. The screening techniques and crop modeling studies described above are the primary sources of information for the first requirement.

Evaluation of the latter two requirements represents the economics portion of the benefits assessment. If an environmental change causes substantial changes in outputs, price and quality changes can occur, which, in turn, lead to further market-induced output changes. Moreover, even if prices remain constant, accurate indications of output changes are needed in those cases when individuals can alter production practices and the types of outputs produced in response. Thus, information on the economic consequences of environmental change for agriculture is accurate only if the reciprocal relations between physical and biological changes and the responses (adaptation) of individuals and institutions are explicitly recognized.

5.1.4.1 Assessment Framework

Probable economic consequences of climate change on agriculture can be identified and perhaps ordered in a general framework. As a starting point, consider the following general findings on the economic impacts of environmental stress on agriculture:

- Increasing environmental stress causes an increase in economic losses at a faster rate.

- Growers may gain from yield losses because of environmental stress, due to price increases, to a certain point.

- Consumer losses are a substantial portion of the total loss from environmental stress.

- Economic losses (in terms of percentage changes) are less than the underlying biophysical yield changes because producers and consumers adjust activities.

- Environmental stress affects both productivity and demand for inputs.

- Environmental stress has differential effects on the comparative advantage of regions or countries.

- Climate change alters international trade flows with attendant gainers and losers.

Quantitative analysis can be achieved by constructing simplified economic databases based on in-country data. Resource economists can provide technical support. The availability of land and labor must be taken into account in these analyses.

5.1.4.2 Evaluation of Economic Impacts

The country studies approach uses existing or "off-the-shelf" plant biophysical simulation models. Because of their flexibility, such simulation models are readily adapted to the specific climate, edaphic, and agronomic conditions of each country. No similar off-the-shelf economic models are available for easy transference to other countries. A few countries in the Country Studies Program (e.g., Egypt) have economic models unique to their agricultural sector. In such cases, the output from the plant science modeling efforts should be integrated with these existing economic models. The Basic Linked System (Fischer et al. 1988) family of models recently used by Rosenzweig and Parry (1994) offers economic modeling capability for other selected countries. For most countries, however, the use of more general "rules-of-thumb" or simple economic constructs will be required if economic effects and vulnerabilities are to be a part of the overall assessment.

Researchers in such countries should first construct a simple checklist of available economic and agricultural information for the country. The availability and quality of such information can help determine the existence and possible magnitude of economic vulnerability in the agricultural sector. This information falls into three general categories: production, consumption, and policy.

Production information suggests the vulnerability of agriculture to climate changes or other environmental stress. This type of information includes (1) the number of alternative crops, (2) the number of production techniques (e.g., irrigation), (3) the nature and extent of resource use in agricultural production, and (4) the cost of production information.

The first three categories of production information suggest the substitution possibilities in the agricultural sector. Specifically, the greater the number of crop alternatives and production techniques in a country, the greater the likelihood of adaptation possibilities. Conversely, for countries with few crops and few production techniques, the greater the potential vulnerability. Cost of production information can help define the costs of such adaptation as well as the conditions under which supplies of each commodity are likely to increase (decrease) as prices of commodities increase (decrease).

Information dealing with the consumption (demand) aspects of agriculture is also useful in understanding possible economic consequences and vulnerabilities. Important economic information includes the role of each crop in the country's overall food consumption; the percent of crops consumed domestically versus that which is exported; and the price movements of commodities.

The extent to which a crop contributes to domestic consumption has two potentially important effects. For example, a crop produced primarily for domestic consumption (i.e. a staple) is likely to be important to the well-being of both producers (in terms of their own consumption) and consumers. Reduction in the supply of such crops implies some increased vulnerability in terms of diet. A crop produced largely for export is typically a major source of foreign exchange earnings for the country. Reduction in exports can reduce the economic well-being of both producers and others in the economy who depend on those export earnings (e.g., beneficiaries of government programs funded by the export earnings). As with production, countries with more crop alternatives in terms of both domestic and export use are less likely to be vulnerable to climate change.

These consumption and production effects are reflected in markets through the movements in prices for each commodity. Information on domestic price movements can reflect the degree to which crops are staples (highly inelastic demands). Prices in export markets reflect international conditions (unless the country is the major supplier of a commodity, in which case the changes in prices reflect both international consumption patterns as well as domestic production). The nature of price and consumption information can thus signal the degree of vulnerability in a particular country.

Information on government policies toward the agricultural sector can indicate the extent to which economic adaptations are encouraged. Specifically, government intervention in agriculture (or other economic sectors) typically distorts economic processes, resulting in less efficient resource allocation decisions than would be achieved in the "free market" situation. Government intervention is common in both developed and developing economies, although the goals of the intervention may differ. Most developed countries are attempting to reduce government intervention/distortions in agriculture, in part because of the direct cost of such intervention to national treasuries. Government policies tend to build rigidities into the agricultural sector by protecting producers from events that would be expected to guide producers toward alternative uses of their resources. Removal of some government policies is thus expected to facilitate more rapid adjustment to environmental change. Conversely, increased government intervention implies less flexibility and perhaps greater vulnerability to climatic stress. Hence, some understanding of the present and future government involvement in the agricultural sector can be useful in forecasting potential vulnerabilities.

5.1.5 Adaptation Techniques

Many possible adaptation options at the farm level are appropriate for responding to climate change. These adaptation options include altered planting dates; change to a crop more adaptable to the new climate; change to a crop more suitable to the new climate; change in the crop rotation pattern; application of fertilizer; application of additional irrigation water, if an irrigation system is in place; and installation of an irrigation system, if one is not in place. These adaptation options can be tested by using the IBSNAT/ICASA crop growth models, which allow comparison of climate change scenarios with and without the given adaptation. The DSSAT model includes a routine for analyzing changes in risks to farmers based on the different adaptations to climate change; risk is an important consideration for farmer response to changing conditions. The crop models can also be used to explore the genetic resources needed to adapt to extreme climate change. The adaptation techniques listed in Table 5.1.2 should be evaluated with the crop models.

Potential adaptations should also be evaluated with respect to their economic consequences. For example, installation of irrigation systems may be costly and require regional planning and management. Furthermore, water for such systems may not be available under changed climate conditions. This possibility should be examined in cooperation with the researchers studying the effects of potential climate change on water resources. Costs and availability of additional fertilizer supplies should also be evaluated.

Economic analyses also provide insight into adaptations available to both producers and consumers of agricultural commodities (Table 5.1.2). Such analyses are particularly useful in projecting possible regional shifts in resource allocation and commodity production as well as shifts in consumer patterns. Both

TABLE 5.1.2 Adaptation Measures Recommended for Evaluation

Biophysical (with crop model)
  Changes in planting date
  Changes in cultivar
  Application of irrigation
  Changes in levels of fertilization
  Changes in crop

Economic
  Substitution possibilities for other crops
  Availability and costs of alternative production techniques
  Alteration of level of government intervention in agriculture

---

biophysical and socioeconomic analyses are necessary to obtain a full picture of potential adaptations to climate change.

5.1.6 Data Requirements

The following data are required for running the IBSNAT-ICASA crop models. Written documentation of these inputs is highly recommended.

5.1.6.1 Soils

The soil classification for each site is needed and will be assigned a generic soil type (e.g., shallow, medium, or deep silty clay; silt loam; sandy loam; or sand), with associated IBSNAT-ICASA soil characteristics for model runs at each site. If a particular soil at a site differs from these generic soils, appropriate specific soil characteristics should be used.

5.1.6.2 Climate

Validation, baseline simulations, and creation of climate change scenarios require 30 years (or as long a record as possible) of daily weather data for corresponding periods at each site, including minimum and maximum temperature as well as precipitation. Also needed are daily solar radiation, if available, daily sunshine duration, or computer-simulated daily solar radiation for each site. A weather generator is embedded in the IBSNAT-ICASA decision support system and can be used to generate daily solar radiation and other climate variables.

5.1.6.3 Other Required Inputs

Latitude and longitude of site, the variety commonly sown, plant population, row spacing, sowing depth, planting date, and crop data for validation are also required.

5.1.6.4 Crops

The crop modelers simulate the most important export crop(s) and/or the most important staple food crop(s) for their country. Among the most important crops for world food trade and subsistence are wheat, maize, soybeans, and rice. However, other crops, such as sorghum, millet, barley, peanuts, beans, potatoes, and cassava, are also of interest.

5.1.6.5 Calibration and Validation

The IBSNAT-ICASA crop models should be calibrated to regional experimental field data according to the procedures described in the DSSAT. If necessary, IBSNAT-ICASA scientists can assist in the calibration and validation process by visiting the site. Researchers should have all weather and experimental data on hand before the visit occurs.

5.1.7 Results Generated

Modeled values of the following variables are tabulated for the baseline case and for each climate change scenario at each site: yield (T/ha), season length (d), growing season precipitation (mm), growing season evapotranspiration (mm), and water used for irrigation (mm). Simulation results also include these quantities for (1) runs with and without the direct effects of CO_2, (2) rainfed and irrigated runs, and (3) the adaptive responses selected by each researcher.

Crop model results from the climate change simulations can be compared with crop model results from simulations with baseline climate. This comparison allows the user to determine the directions and potential magnitudes of change in the key agronomic variables (e.g., yield and water used for irrigation). Analyzing the crop model simulation results elucidates the roles of direct physiological effects of increasing levels of CO_2, rainfed and irrigated production systems, and adaptations to climate change such as changes in variety, planting date, and fertilization. Data may be presented graphically by means of the DSSAT software system.

Simple spreadsheet analyses of current national economic supply-and-demand data and relationships, focusing on the agricultural sector, can be developed. These analyses can then be modified according to the crop modeling and national yield change estimation techniques to provide an initial assessment of the economic consequences of the biophysical crop responses. To the extent possible, results from the economic analyses should include regional changes in supply and demand, based on producer-level economic modeling, national supply-and-demand relationships, and potential changes in such measures as consumer welfare.

5.1.8 Advantages and Disadvantages

5.1.8.1 Advantages of the Methodology

One of the primary advantages in using the IBSNAT-ICASA crop models is that a worldwide network of scientists conducts both model development and validation research. As a result, the models have been applied to many different climates, cultivars, and growing techniques. In addition, the structure of the input and output parameters allows examination of specific changes in farming practices or adaptive measures at a more detailed level than is possible in other more generalized crop models.

Including economic analysis in the country studies allows researchers to estimate the responses of input and output market prices to the biophysical changes predicted by the crop models. Economic analysis also allows identification of the adaptations that producers, consumers, and resource owners can make to minimize losses or maximize gains from changes in opportunities and prices. Economic analysis can provide information on gains and losses across space and time as well as benefit and cost to society. This combination of crop and economic models gives greater insights into the social dimensions of food production and potential climate change effects.

5.1.8.2 Disadvantages of the Methodology

One limitation of the IBSNAT-ICASA crop models is their demand for detailed input parameters. The information for specifying these parameters is sometimes hard to find or unavailable. However, consultation with IBSNAT-ICASA model specialists can usually resolve the problem. This type of consultation is also beneficial for users who find that a specific model assumption may not be applicable for their environment or circumstances. While modeling of particular sites is useful for detailed agronomic analysis, the extension of the results to regions requires simplifying assumptions. The economic analyses of crop model results should be augmented by other types of in-country data, particularly in regard to groups vulnerable to the risk of hunger, in formulating social responses to the effects of climate change.

5.1.9 Supplementary Approaches

Interaction with other study groups is one option for researchers who wish to extend the scope of their crop model studies. For example, crop model output could be combined with results from analyses of water resources to study potential changes in irrigation and possibly competing water demands in agricultural production areas. Coastal scientists could use crop model data to study areas with potential loss to sea-level rise and saltwater intrusion. Many potential collaborations are possible.

Numerous crop models are available for use besides the IBSNAT-ICASA models and for other crops not included in the IBSNAT-ICASA decision support system (Appendix D lists other crop models). Other modeling methodology is also available for vulnerability assessment. One such method is the potential production approach of Doorenbos and Kassam (1979), which uses temperature, length of growing season, and incident solar radiation and crop coefficients to predict maximum crop yields. To follow this method, site data on monthly temperature and cloud cover, and crop planting time and length of growing season, are used along with crop coefficients in the equations developed by Doorenbos and Kassam (1979). This approach is more generalized than that incorporated into dynamic process crop growth models, such as the IBSNAT-ICASA models, but it can provide initial estimates of the sensitivity of other crops, including vegetables and fruits, to climate and CO_2 changes.

5.1.10 References for Section 5.1

Cure, J.D., and B. Acock, 1986, "Crop Responses to Carbon Dioxide Doubling: A Literature Survey," *Agricultural and Forest Meteorology* 38:127-145.

Doorenbos, J., and A.H. Kassam, 1979, *Yield Response to Water*, Food and Agriculture Organization of the United Nations, Rome, Italy.

Fischer, G., K. Frohberg, M.A. Keyzer, and K.S. Parikh, 1988, *Linked National Models: A Tool for International Food Policy Analysis*, Kluwer, Dordrecht, the Netherlands.

IBSNAT, 1989, *Decision Support System for Agrotechnology Transfer Version 2.1 (DSSAT V2.1)*, Department of Agronomy and Soil Science, College of Tropical Agriculture and Human Resources, University of Hawaii, Honolulu, Hawaii.

Peart, R.M., J.W. Jones, R.B. Curry, K. Boote, and L.H. Allen, Jr., 1989, "Impact of Climate Change on Crop Yield in the Southeastern USA," in J.B. Smith and D.A. Tirpak (eds.), *The Potential Effects of Global Climate Change on the United States*, U.S. Environmental Protection Agency, Washington, D.C.

Rosenzweig, C., and M.L. Parry, 1994, "Potential Impact of Climate Change on World Food Supply," *Nature* 367:133-138.

5.2 GRASSLAND/LIVESTOCK IMPACT AND ADAPTATION ASSESSMENTS

Sustaining grasslands is critical for maintaining diverse outputs, such as forage for both wild and domestic herbivores; seed sources for agricultural, reclamation, or landscaping purposes; water quality and quantity; open space; threatened and endangered plants and animals; recreational use; plant and animal diversity; human community stability; and scenic quality (Joyce 1989). Understanding the potential impact of extreme climatic events, as predicted by the GCMs, on grassland ecosystems is essential for both the ecological sustainability and the economical feasibility of grassland-based livestock production.

Several grassland ecosystem models have been written since the creation of the International Biological Program biome ELM model (Innis 1978). Detailed reviews of these models can be found in Hanson et al. (1985) and Agren et al. (1991). Currently, three simulation models provide the level of detail required for considering the effects of climate change on grassland ecosystems. The GEM (Hunt et al. 1991) and SPUR2 (Hanson et al. 1992) models are short-time step models (1 day or less), while the CENTURY model (Parton et al. 1987) is a long-time step model (1 to 12 months). Long-time step models typically simulate the long-term dynamics (10 to 500 years) of ecosystems. Therefore, the data used to drive these models are much more general than models that have finer temporal resolution. Short-time step models generally simulate processes that occur over 2 to 50 years and require driving variables that occur on a daily or weekly basis.

Although the GEM and CENTURY models are well suited for examining the feedbacks of climate change on grassland biogeochemical processes (Schimel et al. 1990; Burke et al. 1991; Hunt et al. 1991; Schimel et al. 1991), neither model was constructed to be a grassland/livestock management model. Both the GEM and CENTURY models have an aboveground herbivore component; however, the effect on grassland quantity and quality on the herbivore either is not simulated or is very simplified. These models are discussed in detail in Section 5.2.9.

The SPUR2 model is used here to simulate the effects of climatic change on grassland/livestock production. SPUR2 has been used to simulate the effects of climate change on grassland ecosystem processes and cattle production on U.S. rangelands (Baker 1991; Baker et al. 1993; Hanson et al. 1993). The mechanistic process-oriented structure of SPUR2 makes it well suited for examining the interactions between management decisions and climatic influences on short-term ecological processes and evaluating possible adaptive management strategies. Also, SPUR2 is the only grassland model available at the temporal and spatial resolutions required for this type of analysis, which incorporates a process-oriented livestock model.

5.2.1 Primary Approach

The primary approach for analyzing grassland/livestock impacts consists of a preliminary screening technique and simulation techniques. These techniques are described in the following sections.

5.2.1.1 Preliminary Screening Technique

The purpose of the preliminary screening technique is to identify those areas where more detailed analysis may be needed and to develop some initial quantitative assessments of the area's vulnerability. Vulnerable regions or sectors are identified. These regions or sectors may be defined by either political or ecological boundaries.

Political boundaries can be used to delineate areas that are economically or sociologically dependent on livestock production from grasslands or savannas. The degree of dependence can be determined by deriving a dependency index. Baker et al. (1993) derived a range dependency index (RDI) to locate areas in the United States that were economically dependent on rangelands. The RDI is defined as the percentage of a county's income derived from unfed beef cattle sales. Clearly, areas economically dependent on grasslands for livestock production are the most vulnerable to changes in climatic conditions.

Ecological boundaries can be used to define areas that are ecologically at risk from changes in climate. These areas could be defined by their susceptibility to degradation, loss of biodiversity, desertification, and so on. Baker and Hanson (1993) developed a socioecological risk index to define areas in the western United States that have marginal rangeland resources and that were susceptible to degradation. The purpose of the index was to locate areas where degradation of rangeland resources is most likely to occur and where degradation would have economically significant societal impacts.

To examine the effects of possible climatic changes in vulnerable areas, scenarios could be constructed from analogous regions. Analogous regions are regions with present-day climates similar to the predicted future climate. The expected impact of such a change could be evaluated qualitatively. Also, present-day management strategies used in the analogous region could indicate the possible adaptive strategies likely to be required for the study region. Therefore, it would be possible to examine the expected effectiveness and costs of adaptation.

Ideally, these techniques should be used in conjunction with a geographic information system (GIS). However, hand-drawn maps work equally well to demonstrate the spatial location of vulnerable areas. If the creation of hand-drawn maps is not feasible, the same data can be presented in tabular form (Table 5.2.1).

TABLE 5.2.1 Preliminary Screening Techniques for Grassland/Livestock Assessment

Region Identification	Type of Vulnerability or Susceptibility	Impact of Future Climate	Use of Analogous Regions — Adaptive Alternative	
			Type	Effectiveness, Cost
Political Boundaries	Economic Dependence, Sociologic Dependence, etc.			
Ecological Boundaries	Loss of Diversity, Susceptibility to Degradation, etc.			

5.2.1.2 Simulation Techniques

The SPUR2 model—a newer and enhanced version of SPUR (Wight and Skiles 1987)—is a general grassland ecosystem model that simulates the cycling of carbon and nitrogen through several compartments, including standing green, standing dead, live roots, dead roots, seeds, litter, and soil organic matter. It also simulates competition between plant species and the impact of grazing on vegetation. The SPUR2 model has been modified to simulate the direct effects of CO_2 on plant production. The model is driven by daily inputs of precipitation, maximum and minimum temperatures, solar radiation, and wind run. These variables are derived either from existing weather records or from use of a stochastic weather generator. The soils/hydrology component calculates upland surface runoff volumes, peak flow, snow melt, upland sediment yield, and channel stream flow and sediment yield. Soil-water tensions, used to control various aspects of plant growth, are generated by using a soil-water balance equation. The Soil Conservation Service curve number procedure computes surface runoff, and soil loss is computed by the modified universal soil loss equation. The snow melt routine uses an empirical relationship between air temperature and energy flux of the snow pack.

5.2.1.2.1 Cattle Model. The Colorado Beef Cattle Production Model (CBCPM) has been incorporated into SPUR2. The CBCPM is a second-generation beef-cattle production model that was a modification of the Texas A&M University (TAMU) Beef Simulation Model (Sanders and Cartwright 1979). The CBCPM is a herdwide, life-cycle simulation model and operates at the level of the individual animal. The biological routines of CBCPM simulate animal growth, fertility, pregnancy, calving, death, and demand for nutrients. Currently, 14 genetic traits related to growth, milk, fertility, body composition, and survival can be studied. The user can define the size and age distribution of the cow herd, the breeding season, the calving season, and the weaning date as well as specify a particular culling strategy. Intake of grazed forage is calculated by FORAGE, a deterministic model that interfaces with CBCPM and SPUR2 (Baker et al. 1992). The model is driven by weight from the animal growth curve, the animal's demand for grazed forage, and the quantity and quality of forage available for each time step of the simulation. FORAGE determines the intake of grazed forage by simulating the rate of intake and grazing time of each animal in the time step.

5.2.1.2.2 Steer Model. The basic structure of the steer model in SPUR2 was adapted from the TAMU Beef Simulation Model (Sanders and Cartwright 1979). The grazing season is defined by Julian dates of turnout to pasture and removal from the pasture. The initial physical and physiological status of the steers is inferred from their age and weight at turnout. Supplemental feed can be offered between input Julian dates. Except for the calculation of forage intake, the SPUR2 steer model is identical to the model in SPUR. Forage intake and diet selection in SPUR2 are calculated by the FORAGE interface.

5.2.2 Description of the Methodology

The methodology for implementing the grassland/livestock assessment is divided into nine steps (Figure 5.2.1). Step 1 determines the areas of the country in which the simulation sites will be located. This preliminary screening procedure is discussed in Section 5.2.1.1. Once the regions have been identified, simulation sites can be selected.

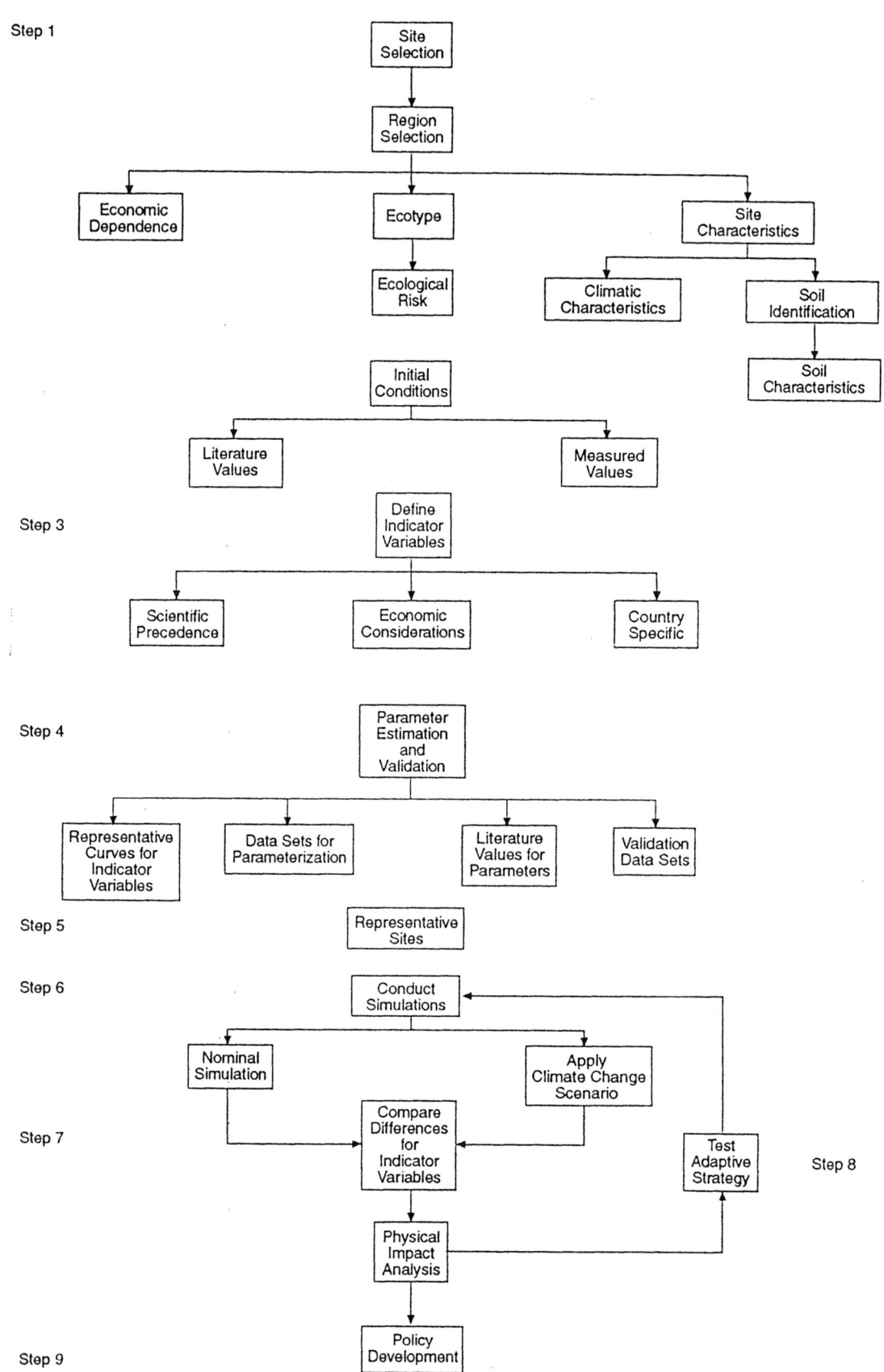

FIGURE 5.2.1 Flow Diagram for CSP Grassland/Livestock Impact and Assessment Methodology

Sites within the region are determined in part by the availability and completeness of data needed for the simulations. Data needed for site characterization include weather data for the length of the simulation and soil and vegetation data. Input data requirements are described in more detail below. The characteristicsite data should be generalized so that the site describes the region. However, if the region is large or extremely heterogeneous, several sites may be needed.

Representative soils can be determined by identifying the soil type that occurs most frequently within the region. Once the soil has been identified, distributions for soil properties can be calculated and the mode used to describe the physical properties for the representative soil.

Step 2 determines the initial conditions for the simulations. Initial conditions are needed for the plant, soil, and livestock models. These data can be obtained from the scientific literature, government publications, or measured values.

Management strategies for the simulations should be determined to establish a baseline. To compare future predictions of change, it is essential to define present-day practices. A single management strategy should be used for all simulation sites in the initial simulations to minimize the interactions between climate change and management. This strategy allows for comparisons between sites and simulation scenarios.

Step 3 defines indicator variables. Indicator variables are model-derived state or intermediate variables used to test the hypotheses under examination. These variables are determined from scientific precedence, economic importance, or objectives of the simulation study. A partial list includes peak standing crop, water use efficiency, soil organic matter, weaning weights, and forage intake.

Step 4 is parameter estimation and model validation. Quantitative data are needed to develop the appropriate parameter sets for the models. Data should include representative curves for the indicator variables and raw data. Types of curves needed include monthly standing crop, forage crude protein or nitrogen, animal weight gains, animal intake and digestibility of diet, yearly peak standing crop, weight and age at weaning or slaughter, and average number of offspring born per female. Additional parameter estimates for model parameterization can be found in the scientific literature.

Additional data sets will be needed for model validation. This process involves comparing model output with real world observations. The length of the data set will be determined in part by how much data are available. The validation processes should be conducted on representative simulation sites. The number of validation sites depends on how many distinct geographic regions are being simulated.

Step 5 selects representative simulation sites. Representative sites are used for validating and evaluating the within-year effect of climatic change on selected indicator variables.

Step 6 conducts the simulation experiment. The number of simulations per site depends on how many GCM climate change scenarios are chosen. A control or "nominal" simulation is required for every simulation site. This simulation uses an unaltered version of the historical weather file.

For the GCM scenario simulations, the nearest grid point to a simulation site is chosen. The historic weather data are to be adjusted for $2 \times CO_2$ by the recommended GCM adjustment statistic for monthly average temperature, precipitation, and solar radiation. As stated in Section 5.1.3.2, at least three $2 \times CO_2$

equilibrium GCM scenarios should be simulated to more adequately describe the sensitivity of the system to climatic perturbations.

Step 7 compares the differences between the "nominal" and climate change scenario runs. Multiple comparisons, such as Scheffe's multiple comparison procedure, should be conducted to test main effects (year and scenario) on all indicator variables.

Step 8 tests adaptive strategies. Adaptive management strategies to be tested depend on the current and accepted management strategies within a country and the type of livestock being studied. Adaptation techniques are discussed more fully in Section 5.2.5.

Step 9 develops policies to mitigate the negative impacts or to take advantage of possible positive impacts of climatic change. Policy development should be combined with a complete socioeconomic analysis.

A more complete description of SPUR2 and livestock models to be provided by the Country Studies Program, computer requirements, step-by-step procedures, data requirements, and analyses are provided in Appendix E.

5.2.3 Scope of the Analysis

Much of the scope of the analysis has already been discussed. The spatial dimension of the analysis depends on how many relatively homogenous areas can be determined within the geographical regions of study. A GIS would be useful for this study in determining both simulation sites and demonstrating the spatial component of the impacts (Burke et al. 1991; Baker et al. 1993).

This type of analysis should be no more than 30 simulated years for two reasons. The time scale at which the ecological effects of climate change on a grassland ecosystem may occur is on the order of decades to centuries. However, human influences, such as livestock production, suppression of fire, and so on, accelerate changes in ecosystems and the feedback to the environment within months or years (Ojima et al. 1991).

To assume that agricultural practices will remain unchanged during the simulation period is unrealistic. In fact, agriculturalists constantly modify their management practices to adapt to changing climatic and economic conditions. Therefore, the usefulness of trying to predict the interactions between climatic change and livestock management decisions is diminished for longer simulations.

5.2.4 Economic Analysis

Economic analyses can be conducted to examine the effects of climatic change on the local or producer level, at a more aggregated level to examine the effects on secondary supply and demand within a region or country. These analyses can be extrapolated further to explore the effect of world supply and demand of meat and animal products produced from grasslands. The following discussion and Figure 5.2.2 outline a simplified method for examining some of the economic consequences of climate-induced alterations in livestock production.

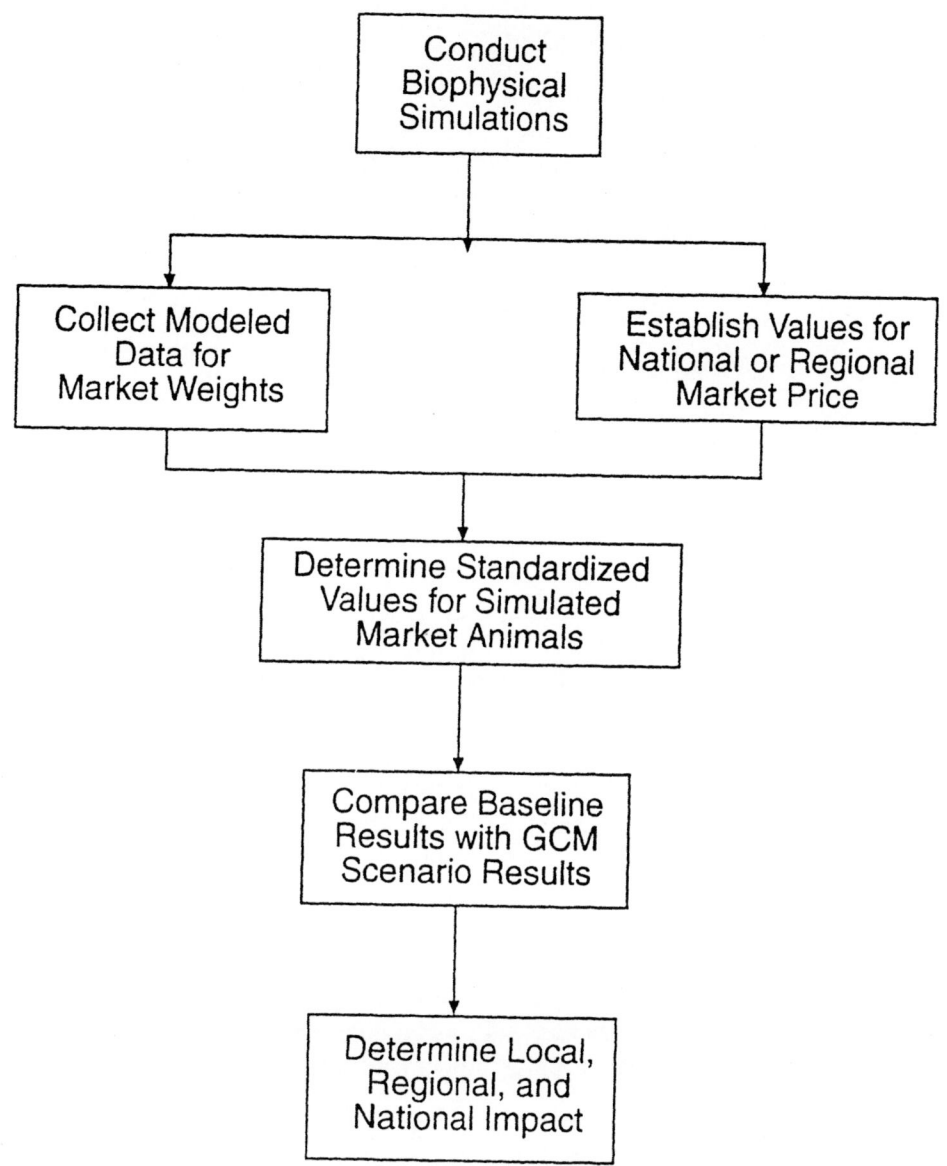

FIGURE 5.2.2 Flow Diagram for the Grassland/Livestock Economic Analysis

Several model-generated indicator variables can be used to conduct an economic analysis of the effects of climate change on livestock production. However, four variables would have utility for most analyses: the total weight of males sold, the total weight of females sold, the total weight of culled animals sold, and the total weight supplemental feed used. (Culled animals are animals that are sold because they are too old, have health problems, or are no longer considered to be necessary to maintain herd size.)

Market prices should be established on a per kilogram ($ kg^{-1}) live weight basis for the three classes of animals for all scenarios. Multiply market price times the total weight for each class and standardize by dividing by the number of hectares being simulated ($ ha^{-1}). Summing the values for each site and comparing results from the climate change simulations to the nominal run give an estimated reduction in the marketed value of range land production. Results can be presented in tabular form (Table 5.2.2).

TABLE 5.2.2 Marketed Output and Values

Site and Item	Males	Females	Culled	Total
Site name				
Nominal run (kg/ha)				
Value, nominal run ($/ha)				
Climate change scenario[a] (kg/ha)				
Climate change scenario ($/ha)				
Change in value (%)				

[a] Repeat for as many climate change scenarios as are being simulated.

In some production systems, supplemental feed represents as much as 16% or more of the cash costs for cow-calf production systems. Therefore, changes in the amount of supplemental feed consumed may be an important component in some analyses. Other country-specific economic indicator variables may be needed in the analysis.

The figures calculated in Table 5.2.2 represent the potential impact of climate change on private returns for ranchers in each area. These figures could be aggregated to qualitatively estimate economic impacts of the potential climatic change on regional and national livestock production. Further analyses on the effect of regional, national, and global supply-and-demand structure should be conducted.

5.2.5 Adaptation Techniques

The response of agriculturalists to climatic change can be described in terms of adjustments in management strategies and tactics or as adaptations. Adjustments refer to the short-term modifications that agriculturalists make in response to year-to-year fluctuations in climate. These changes are consciously applied to management practices to offset environmental risk (Warrick et al. 1986). Some of these changes include altering the timing and location of grazing, changing the number of animals grazing a specific area or pasture, changing the migration routes for nomadic pastoralists, and changing the use patterns of marginal lands. Adaptations to climatic change may take several years to several generations and may not be consciously recognized as having any relationship to changes in climatic conditions or environmental fluctuations (Warrick et al. 1986). Some experts view adaptation to climatic change as a result of an aggregation of short-term responses to environmental changes or risk. Adaptations include a spatial shifting of grazing resource use, a shifting of the genotype of grazing animal, and even an alteration of the species of animals used for grazing.

As stated in Section 5.2.2, adaptive or adjustment strategies to be tested depend on the country's current and accepted livestock management practices. However, some of the management strategies to be tested include (1) alternative grazing systems, (2) changes in stocking rates, (3) changes in the timing of the grazing period, (4) changes in the genotype of the livestock, and (5) other country-dependent alternative strategies. These options can be tested by using the SPUR2 model. Comparison of climate change scenarios with and without the given change in management practices can be made against the nominal climate scenario. More specific details describing what parameters to change so as to conduct the various adaptation options are given in Appendix E.

Potential alternative management strategies should be evaluated with respect to their short- and long-term economic and ecological consequences. For example, changes in precipitation patterns and temperature may increase plant production. Economically, a producer may be enticed to increase the number of animals grazing the area. However, this tactic may not be an ecologically sustainable management practice. On the basis of changes in soil moisture and temperature regimes, nutrient uptake may exceed nutrient return, thereby establishing a ecologically destabilizing situation

5.2.6 Data Requirements

To conduct the simulations, 30 years of continuous historical weather data are needed for each simulation site. Weather data requirements include daily measurements for maximum and minimum temperature (°C), precipitation (mm), wind run (km d^{-1}), and solar radiation (langleys).

Recommendations for constructing the GCM climate change scenarios are outlined in Section 4.3. Specific data needed from the GCM models include average monthly temperature, precipitation, and solar radiation for both the $1 \times CO_2$ and $2 \times CO_2$ equilibrium runs.

Data for the hydrology model include soil name and texture; slope; percentage of sand, clay, and silt; organic matter; parameters for universal soil loss equations; soil evaporation; bulk density; and water holding capacity at -1/3 and -15 bar. These data can be obtained from empirical measurements, from scientific literature, or from documentation provided with the model once the soil texture has been identified.

Plant model initial conditions include biomass estimates for green shoots, live roots, propagules, standing dead, dead roots, litter, and soil organic matter as well as an estimate of the amount of nitrogen ($g \times m^{-2}$) in green shoots, live roots, propagules, standing dead, dead roots, litter, soil organic nitrogen, and soil inorganic nitrogen. These data may need to be collected by empirical means if data sets for a simulation site do not exist.

Data for the animal model include animal weight, estimate of genetic potential for gain, milk production, birth weight, weaning weight, yearling weight, mature weight, estimates of lifetime growth curve weights, reproductive capacity, and forage intake. These data may be found in scientific literature for specific breeds or from experimental data. Other animal management data include birthing dates, castration dates, weaning dates, breeding dates, slaughter or sale dates, supplemental forage type and when fed, grazing systems, and herd structure.

5.2.7 Results Generated

To capture the effect of changes in biotic processes due to climatic influences, output data for the indicator variables should be collected daily, weekly, or monthly (Ojima et al. 1991). Data to be analyzed include monthly averages for peak standing crop (g m^{-2}), carbon-to-nitrogen ratio, soil organic matter (g m^{-2}), precipitation (cm), potential evapotranspiration (mm), transpiration (mm), water use efficiency, intake of grazed forage (kg head^{-1}), digestibility of diet, forage-to-supplement ratio, body condition scores, milk production (kg head^{-1}), and weight at market (kg head^{-1}). Other specific country data may be needed.

Peak standing crop, transpiration, water use efficiency, potential evapotranspiration (PET), and precipitation are used to examine the potential effects of climatic change on plant and plant-soil moisture interactions. Soil organic matter is used to monitor the status of below-ground nutrient sources. The carbon-to-nitrogen ratio of the aboveground biomass indicates the change in plant tissue quality.

Diet digestibility, intake of grazed forage, forage-to-supplement ratio, and body condition scores are used to evaluate the effect of climatic change on feed intake. Milk production is used to monitor the potential effects of the change in calf performance. Market weights of the animals are used in the economic analysis to determine the economic impact of climatic change.

As stated, the multiple comparison tests of the data for the year and scenario should be conducted by means of the SAS Institute (SAS 1990) or another equivalent statistical program for each of the indicator variables. Data generated from the representative sites should also be analyzed for within-year trends over the simulation period. These data will be used to demonstrate the effect of within-year timing of climatic change events such as earlier or later precipitation or temperature increases. These data can be used as baseline data for the adaptation and mitigation analyses.

A GIS would be useful for demonstrating the spatial component of the impacts (Burke et al. 1991; Baker et al. 1993). If a GIS is available, output data could be interpolated using kriging, inverse distance weighting, or other appropriate techniques to demonstrate geographical or regional trends.

5.2.8 Advantages and Disadvantages

Several advantages result from this approach. Biophysiological simulation models mechanistically simulate ecological and physiological processes; therefore, they are useful for integrating the nonlinear effects of climate change. Because these models are process driven, they can be applied to many different environments. The models can also test the sensitivity and stability of the system to a range of changes in climatic conditions.

The major disadvantages for using a biophysical simulation approach for this problem are that complete data sets for parameterizing the model rarely exist. Also, the models used in this approach are point models, which require making simplifying assumptions when results are aggregated to the regional level. The most limiting assumptions of this analysis are that management practices will remain constant over the simulated period and that the region of aggregation is homogenous. In reality, management strategies, microclimate, vegetation structure, and soil types may be quite heterogeneous. Therefore, some care must be exercised when describing discrete regions.

5.2.9 Supplementary Approaches

Two other grassland ecosystem models have been used to examine the potential effects of climatic change on grassland production. The CENTURY model was developed to simulate monthly biogeochemical cycling of carbon, nitrogen, phosphorus, and sulfur in both natural grassland and managed agro-ecosystems (Parton et al. 1987, 1988). Driving variables and major parameters for the model include surface-soil physical properties, monthly precipitation and temperature, plant nitrogen and lignin contents, and land use. The model's simplistic design allows it to be used to predict regional trends in key ecosystem processes, such as carbon and nitrogen fluxes, including primary production and the fate of soil carbon and nitrogen across a wide range of climates and soil types. However, the model is mechanistic enough to explore regional responses to climate change (Schimel et al. 1990, 1991; Burke et al. 1991).

The objectives of the GEM model are to predict seasonal and year-to-year biomass dynamics of primary producers, microbes, and soil fauna and nitrogen availability in grasslands, and the effects of CO_2 level and climate change on these dynamics (Hunt et al. 1991). Information needed to run this model includes daily precipitation; weekly mean, maximum, and minimum air temperature; wind speed; relative humidity; and monthly soil temperature. The model's structure allows for investigation of the effects of climate change and elevated CO_2 on feedbacks through the trophic structure of the grassland ecosystem. GEM has been used to predict ecosystem level effects of climate change in a short grass steppe (Hunt et al. 1991).

Other types of models available include decision support systems such as the Australian models (McKeon et al. 1988; Howden et al. 1991; McKeon and Howden 1991). These models have been used in Australia to examine the impact of climate change on pastoral production of beef cattle.

Additional data that would be useful for this analysis include vegetation maps, soils maps, and land use maps. Remote sensed satellite data could be used to establish baseline conditions. The cost of acquiring the data would not warrant the use of the data, however, if they were not already available.

5.2.10 References for Section 5.2

Agren, G.I., et al., 1991, "State-of-the-Art of Models of Production-Decomposition Linkages in Conifer and Grassland Ecosystems," *Ecological Applications* 1:118-138.

Baker, B.B., 1991, *A National Analysis of the Potential Effects of Climate Change on Rangeland Ecosystems*, Ph.D. Dissertation, Colorado State University, Fort Collins, Colo.

Baker, B.B., and J.D. Hanson, 1993, "Simulating the Effects of Climate Change on Beef Cattle Production: A Methodology for Determining Simulation Sites," in *Proc. 46th Annual Meetings, Society Range Management* 46:9-10.

Baker, B.B., et al., 1993, "The Potential Effects of Climate Change on Ecosystem Processes and Cattle Production on U.S. Rangelands," *Climatic Change* 25:97-117.

Baker, B.B., et al. 1992, "FORAGE: A Simulation Model of Grazing Behavior for Beef Cattle," *Ecological Modeling* 60:257-279.

Burke, I.C., et al., 1991, "Regional Analysis of the Central Great Plains," *Bioscience* 41:685-692.

Hanson, J.D., et al., 1993, "Comparison of the Effects of Different Climate Change Scenarios on Rangeland Livestock Production," *Agricultural Systems* 41:487-502.

Hanson, J.D., et al., 1992, *SPUR2 Documentation and User's Guide*, U.S. Department of Agriculture, ARS, Great Plains Systems Research Technical Report-1, Fort Collins, Colo.

Hanson, J.D., et al., 1985, "Plant Production of Grassland Ecosystems: A Comparison of Modeling Approaches," *Ecological Modeling* 29:131-144.

Howden, S.M., et al., 1991, "Managing Pastures in Northern Australia to Minimize Greenhouse Gas Emissions: Adaptation of an Existing Simulation Model," in *Proc. Ninth Biennial Conference on Modeling and Simulation,* Simulation Society of Australia, Greenmount, Australia, Dec.

Hunt, H.W., et al., 1991, "Simulation Model for the Effects of Climate Change on Temperature Grassland Ecosystems," *Ecological Modeling* 53:205-246.

Innis, G.S. (ed.), 1978, *Grassland Simulation Model,* Springer, New York, N.Y.

Joyce, L.A., 1989, *An Analysis of the Range Forage Situation in the United States: 1989-2040*, RM-180, U.S. Department of Agriculture, Forest Service, Rocky Mountain Forest and Range Experiment Station, Ft. Collins, Colo.

McKeon, G.M., and S.M. Howden, 1991, "Adapting Northern Australian Grazing Systems to Climate Change," *Climate Change Newsletter* 3:5-8.

McKeon, G.M., et al., 1988, "The Effect of Climate Change on Crop and Pastoral Production in Queensland," in G.I. Pearman (ed.), *Greenhouse: Planning for Climate Change*, CISRO, Melbourne, Australia, pp. 546-563.

Ojima, D.S., et al., 1991, "Critical Issues for Understanding Global Change Effects on Terrestrial Ecosystems," *Ecological Applications* 1:316-325.

Parton, W.J., et al., 1988, "Dynamics of C, N, P, and S in Grassland Soils: A Model," *Biogeochemistry* 5:109-131.

Parton, W.J., et al., 1987, "Analysis of Factors Controlling Soil Organic Matter Levels in Great Plains Grasslands," *Soil Science Society of American Journal* 51:1173-1179.

Sanders, J.O., and T.C. Cartwright, 1979, "A General Cattle Production Systems Model. I. Description of the Model," *Agricultural Systems* 4:217-227.

SAS Institute, Inc., 1990, *SAS/STAT User's Guide, Version 6, Fourth Ed.*, Vol. 2, Cary, N.C.

Schimel, D.S., et al., 1991, "Terrestrial Biogeochemical Cycles: Global Interactions with the Atmosphere and Hydrology," *Tellus* 43AB:188-203.

Schimel, D.S., et al., 1990, "Grassland Biogeochemistry: Links to Atmospheric Processes," *Climatic Change* 17:13-25.

Warrick, R.A., et al., 1986, "CO_2, Climate Change and Agriculture," in B. Bolin, B.R. Doos, J. Jager, and R. Warrick (eds.), *The Greenhouse Effect, Climate Change, and Ecosystems,* John Wiley & Sons, Chichester, U.K., pp. 393-473.

Wight, J.R., and J.W. Skiles (eds.), 1987, *SPUR: Simulation of Production and Utilization of Rangelands, Documentation and User Guide*, ARS 63, U.S. Department of Agriculture, Agricultural Research Service.

5.3 FOREST IMPACT AND ADAPTATION ASSESSMENTS

The methods used to evaluate the potential impacts of climate change on forest ecosystems depend on the questions being asked. Determining how forests respond to changes in environmental conditions (such as climate) involves the study of patterns and processes that range from near instantaneous responses at the leaf level to broad-scale changes in the distribution of species over a time scale of decades to centuries. Models that simulate how basic plant processes respond to features of the environment are being used to address a range of questions across these spatial and temporal scales.

5.3.1 Primary Approach

5.3.1.1 Preliminary Screening Technique

Although the remainder of Section 5.3 addresses the application of modeling (simulation) techniques to assess the potential impacts of a climate change on forest ecosystems, preliminary screening can be conducted by using nonsimulation approaches. This screening can be used either as a first step to identify areas and/or species of greatest vulnerability or as a substitute for more quantitative analyses where insufficient data are available for model development and application.

The initial step in any screening procedure is to identify factors within the scenario (e.g., climate change) that may influence forest ecosystems. In the context of global climate change, this procedure would involve a direct analysis of features of the climate that influence forest distribution and productivity. A number of climate indices are related to basic plant processes and vegetation distribution (Appendix F). Changes in the length of the growing season and thermal indices such as growing-degree days (Appendix F) or absolute minimum temperature directly influence species distribution and forest productivity. In addition to thermal indices, changes in patterns of precipitation directly influence plant productivity.

An important part of the preliminary screening process involves identifying areas critical to forest resources, such as conservation or production forestry areas. Furthermore, within these areas, critical species or groups of species can be identified. By examining the climate changes and other environmental features predicted by the global change scenarios, local experts can evaluate the susceptibility of the identified areas (or species) to the predicted changes in environmental conditions. This approach can be limited to qualitative assessments (e.g., high impact, low impact), but it provides a critical step in identifying areas in need of further quantitative analyses, such as those outlined in Section 5.3.1.2. One approach is to examine analogous regions. By identifying areas that have corresponding climate conditions to those predicted for a location under the climate change scenario, the patterns of productivity and vegetation structure for the analog site can be used as an indicator of the patterns that might be expected under the new climate patterns.

Once the projections of the expected impacts have been established, potential adaptations can be evaluated for effectiveness and cost.

These techniques can be presented in the form of maps that illustrate the location of vulnerable areas, or they can be summarized in tabular form.

5.3.1.2 Simulation Techniques

The primary approach for evaluating the potential impacts of climate change on forest ecosystems uses two simulation models: the Holdridge life zone classification model and the forest gap model. This section outlines a framework for applying these models, which explicitly relates vegetation patterns to environmental conditions at two different temporal and spatial scales. In addition, this section discusses how these models can be used to examine the effects of climate change on forest ecosystems. Each model examines the response of vegetation to climate conditions; however, the models differ in the way in which they describe vegetation and examine processes.

The Holdridge model relates the current spatial distribution of vegetation to features of the climate system. This model is suitable for examining (1) broad-scale patterns of vegetation as they relate to climate and (2) the influence of climate changes on the suitability of a region to support different vegetation/forest types. However, this approach does not address vegetation processes *per se* and as such cannot be used to predict the temporal dynamics of species composition and stand productivity, features of importance in evaluating the potential impacts of climate change on forest resources and conservation.

The forest gap model is an individual-based model of forest dynamics that simulates the response of basic plant processes to environmental conditions. The model can examine patterns of composition and productivity; however, it is site-specific and requires detailed information on the attributes of species and on site-specific factors.

Each of these models addresses a different feature of vegetation pattern. The Holdridge model provides a regional mapping system for interpreting spatial changes in climate patterns throughout the country or region. The forest gap model evaluates the temporal dynamics of a given forested site in response to climate changes. Together, the two models provide an integrated approach to evaluating regional impacts (Figure 5.3.1). This integrated approach is outlined in the following sections. Appendix F provides a more detailed description and step-by-step procedures for using the two models.

5.3.2 Description of the Methodology

5.3.2.1 Holdridge Life Zone Classification Model

5.3.2.1.1 Model Description. One of the simplest approaches to relating vegetation pattern to climate change is climate-vegetation classification. By assuming that the broad-scale patterns of vegetation (e.g., biomass) are at equilibrium with present climate conditions, the distribution of major vegetation/plant types can be correlated with biologically important features of the climate. Holdridge (1947) has developed a system of vegetation-climate classification. This system relates vegetation to present climate patterns at a regional-to-global scale.

The Holdridge model (Figure 5.3.2) is a climate classification scheme that relates the distribution of major ecosystem complexes to the climatic variables of biotemperature, mean annual precipitation, and the ratio of PET to precipitation (Holdridge 1967). The life zones are depicted by a series of hexagons in a triangular coordinate system. Two climate variables—biotemperature and annual precipitation—determine the classification. Biotemperature is a temperature sum over a year with the unit temperature values (i.e., average daily, weekly, or monthly temperatures) used in computing the index set to 0 °C if values are less than or equal to 0 °C.

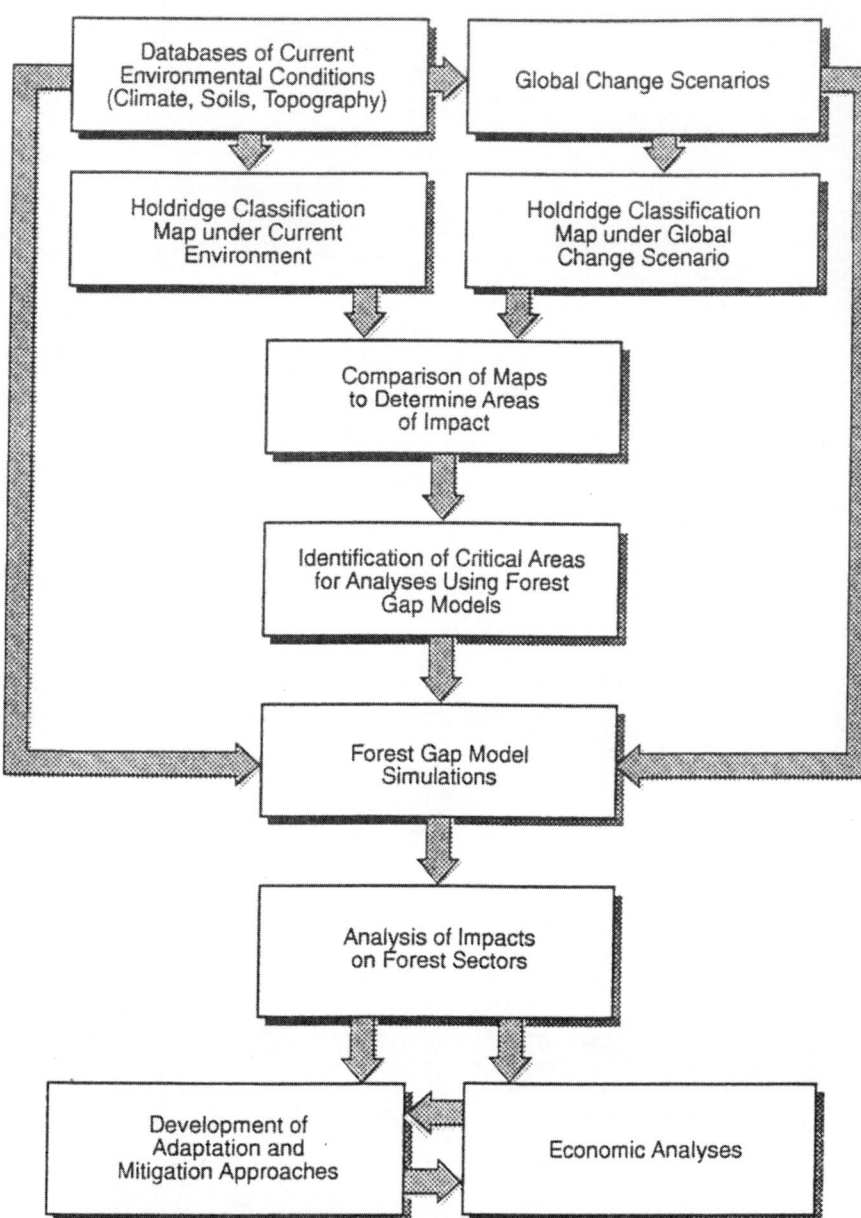

**FIGURE 5.3.1 Flow Diagram Outlining the Steps Involved in
Implementing the Forest Impacts Assessment**

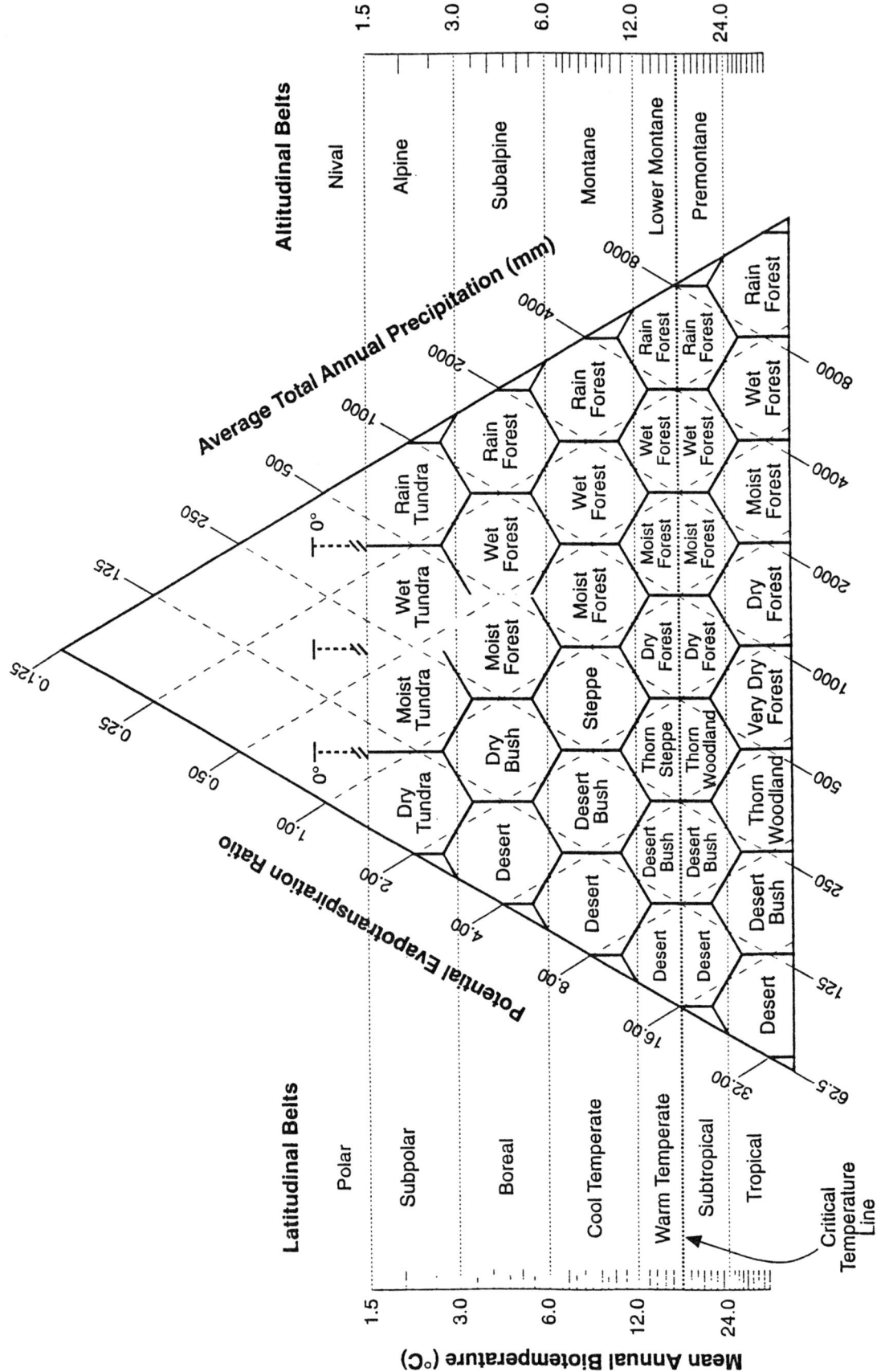

FIGURE 5.3.2 Holdridge Life Zone Classification Model (Source: Holdridge 1967)

Identical axes for average annual precipitation form two sides of an equilateral triangle. The PET ratio forms the third side, and an axis for mean annual biotemperature is oriented perpendicular to its base. By striking equal intervals on these logarithmic axes, hexagons are formed that designate the Holdridge life zones.

The potential evapotranspiration ratio is the quotient of PET and average annual precipitation. On the basis of data from several ecosystem types, Holdridge (1959) assumes that PET is proportional to biotemperature (constant of proportionality of 58.93). The PET ratio in the Holdridge diagram therefore depends on the two primary variables—annual precipitation and biotemperature.

One additional division in the Holdridge model is based on the occurrence of killing frost. This division occurs along a critical temperature line that divides hexagons between 12 and 24 °C into warm temperate and subtropical zones. The complete Holdridge classification at this level includes 37 life zones.

For regional analyses, a more detailed form of the classification exists, which further subdivides the 37 life zones. This higher resolution description of vegetation within the broad classes shown in Figure 5.3.2 allows for a more detailed description of vegetation. This description can be related to species distribution and patterns of productivity within a region.

5.3.2.1.2 Implementation. Implementation of the Holdridge model for a region requires only data on annual precipitation and biotemperature for a grid network based on latitude and longitude. Biotemperature is calculated from mean temperature values at either a daily, weekly, or monthly resolution. The classification can be applied at any spatial resolution, but for regional analyses of impacts, the highest possible resolution should be used. The values of mean annual precipitation and biotemperature are then used to classify each grid cell to determine the potential land cover based solely on climate. The resulting database of potential land cover (life zones) can be mapped, providing a base map of the country.

The map based on the Holdridge model represents the potential distribution of vegetation based solely on climate. This map should be compared with existing maps of vegetation for the region to incorporate other features of the environment that may influence vegetation pattern by modifying the two primary climate variables (biotemperature and annual precipitation). Factors such as variation in topography (e.g., slope and aspect) and soils can modify the moisture and temperature conditions from those characterized by the statistics for a given grid cell.

5.3.2.1.3 Incorporation of Scenarios. The changes in climate patterns predicted from the GCMs discussed in Section 4 can be easily incorporated into the modeling framework. The changes in monthly temperature and precipitation predicted by the GCM are used to modify the data on current patterns of annual precipitation and biotemperature used to generate the initial map of land cover for the country or region. By using the new values of biotemperature and annual precipitation for each grid cell, the Holdridge model generates a new map of predicted vegetation cover based on the changed climate conditions. This new map can be compared with the original map used on current climate patterns to examine potential shifts in vegetation distribution under the changed climate patterns (Figures 5.3.3a and 5.3.3b).

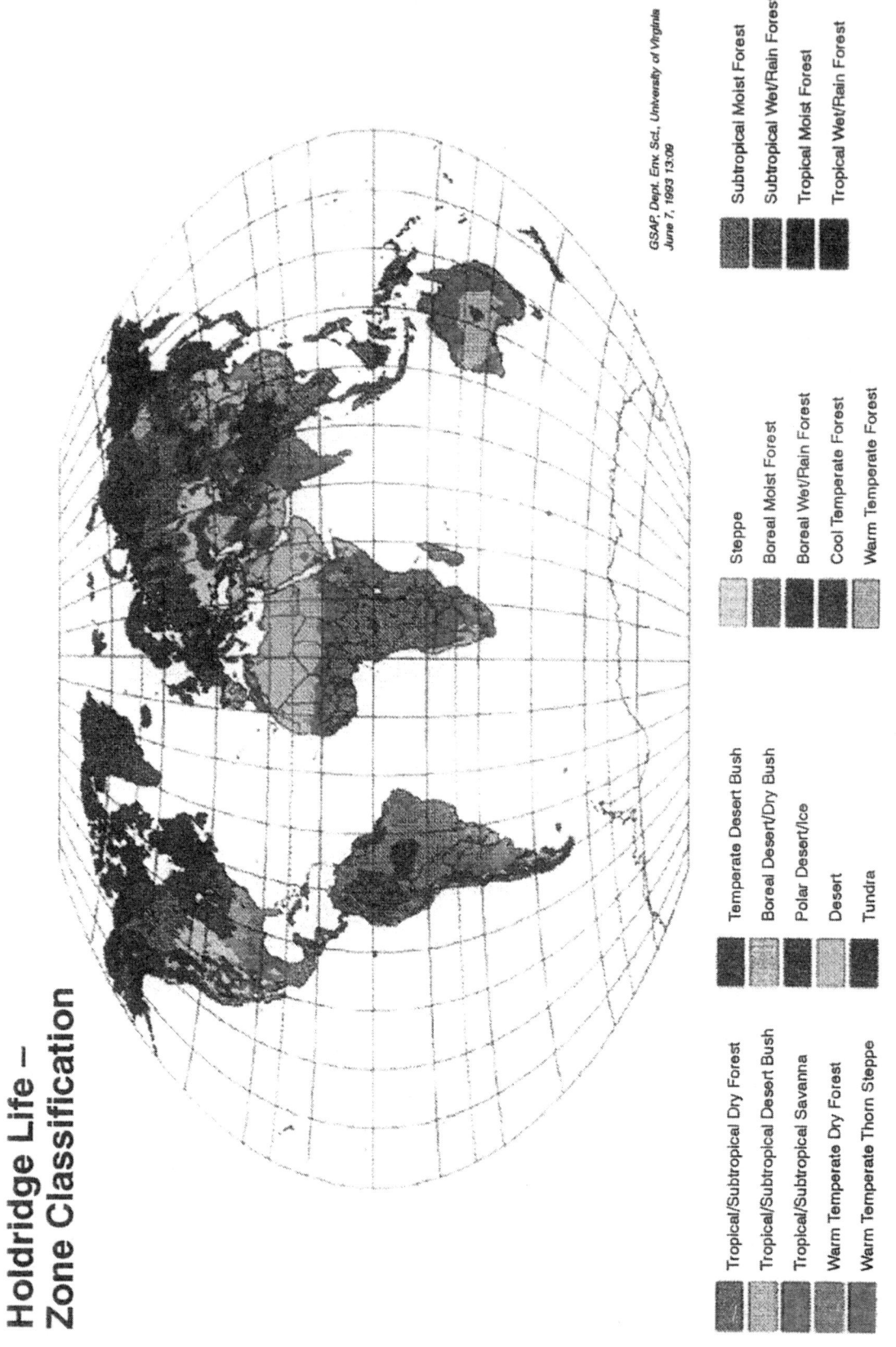

FIGURE 5.3.3a Global Map of Holdridge Life Zones under Current Climate Scenario

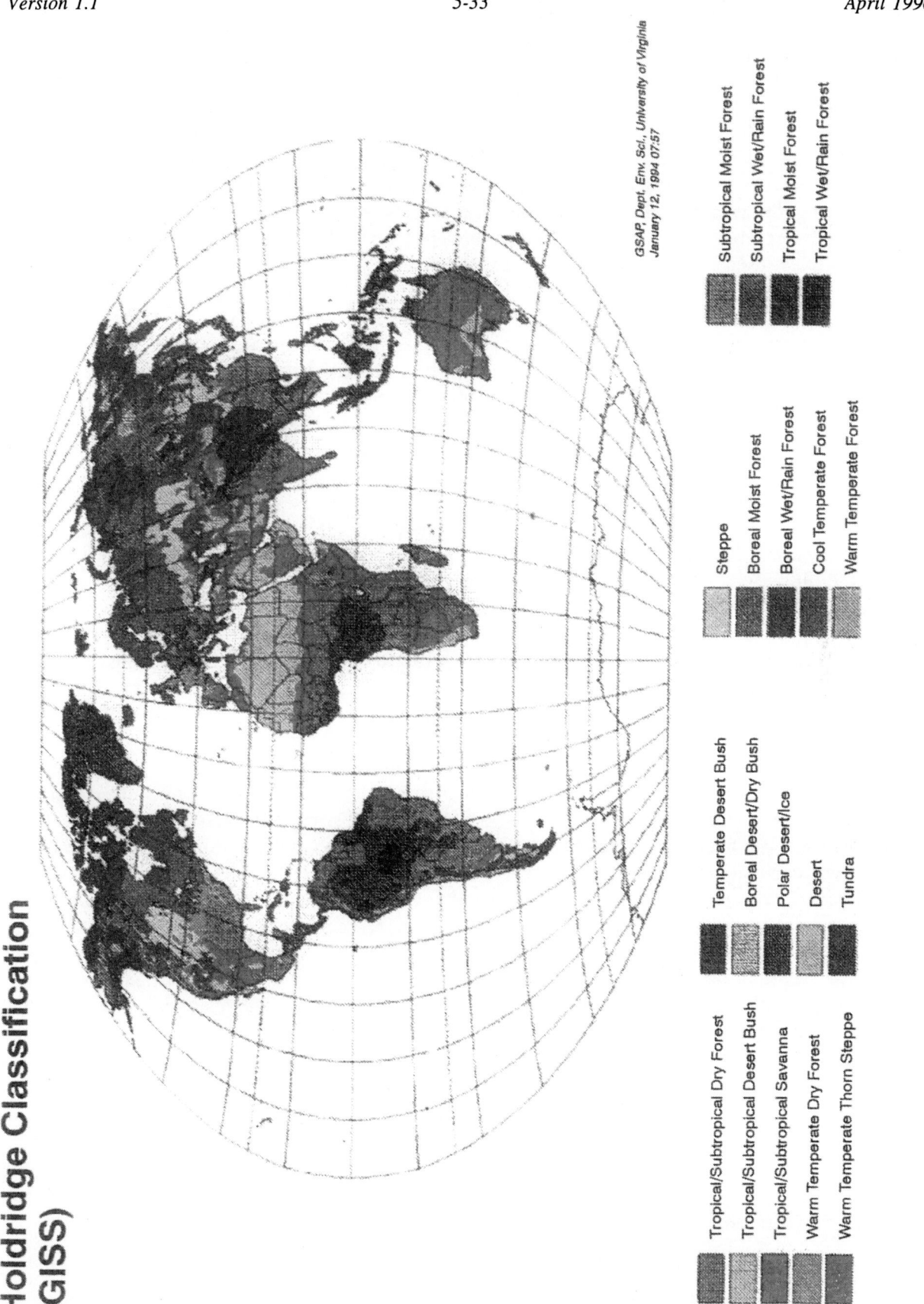

FIGURE 5.3.3b Global Map of Holdridge Life Zones under Climate Change Scenario Based on the GISS General Circulation Model

5.3.2.2 Forest Gap Model

5.3.2.2.1 Model Description. The forest gap model evaluates changes in the species composition and productivity of specific forested sites. Forest gap models simulate the establishment, growth, and mortality of individual trees on a forest stand (<1 ha) on an annual time step. Because the model simulates the response of individual trees on a forest plot, it can predict changes in species composition, forest structure (e.g., size-class distributions), and productivity. In addition, the model incorporates forest management practices (e.g., selective cutting and stand thinning), which allows for evaluation of adaptive strategies.

The potential growth of each tree is estimated from species-specific optimal growth curves (Figure 5.3.4). These growth curves are derived from either field data or estimated from simple silvicultural data on maximum tree size and longevity (Botkin et al. 1972; Shugart 1984). The environmental conditions on the plot (e.g., temperature and available moisture, nutrients, and light) modify the optimal growth for an individual tree. The species responses to these environmental factors are defined through either procedures that relate species distributions to regional climate patterns or through basic silvicultural data (for examples, see Shugart 1984). In many cases, species are categorized into groups that have a common response to a single environmental factor (Figure 5.3.5). For example, growth response to available light is often categorized into two classes: shade tolerant and shade intolerant. Light, soil fertility, and soil moisture are ranked tolerances, whereas temperature is a function of minimum and maximum growing-degree days based on the geographic distribution of the species.

Forest gap models have been developed and applied to a variety of forest ecosystems around the world, ranging from tropical to boreal forest. Locations and references to existing forest gap models are presented in Figure 5.3.6.

5.3.2.2.2 Implementation. The forest gap model can simulate species composition and structure on an annual basis (Figure 5.3.7). The model can either simulate forest growth on bare soil, or it can be initialized for a given forest by defining the numbers, species, and sizes of trees to be simulated. Given the environmental conditions for a forested site, the model simulates a number of forest plots. The results are then averaged to provide a statistical description of the forest. Larger forested areas can be simulated by establishing a network of sites that represent the environmental variation in the area (e.g., topography and soils) that can influence species distribution and dominance, or gradients of productivity.

Validation of the model for a given region is an important step in implementing the model for a new region. Validation generally involves comparison of the simulated patterns of forest composition and structure to some independently collected (not used for model parameterization) set of field data describing the forest (e.g., age or size class structure, relative species abundances, and biomass). Validation procedures are discussed in detail in Shugart (1984).

5.3.2.2.3 Incorporation of Scenarios. Because forest gap models simulate the temporal dynamics of forests in response to environmental conditions, a transient climate change scenario should

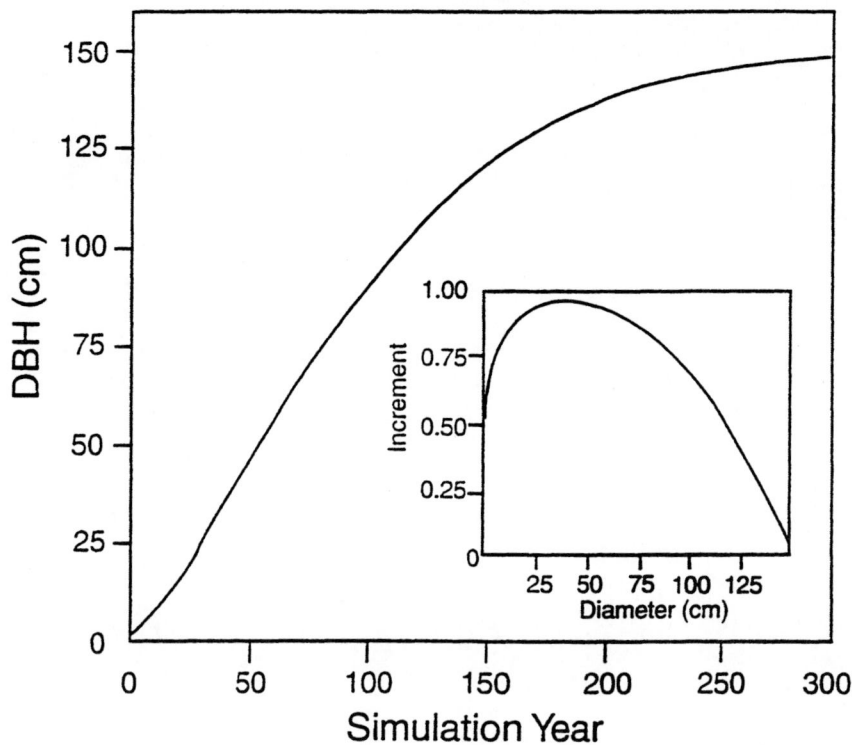

FIGURE 5.3.4 Diameter Growth as a Function of Tree Age and (inset) Diameter Increment as a Function of Current Diameter as Simulated by Forest Gap Models

be incorporated into the modeling framework. Section 4 discusses the construction of transient climate change scenarios. The annual changes in monthly temperature and precipitation from the transient scenario are used to modify the environmental conditions on the forest stand, and the response of individual trees to those changes is simulated.

5.3.2.3 Integrating Models. Together, the Holdridge and the forest gap models provide an integrated view of impacts on forested ecosystems within a region. The maps generated using the Holdridge model define the broad-scale patterns of vegetation change within the region. These maps can identify areas of potential forest decline or areas that are currently not forested but that could potentially support forest under the changed climate conditions. Specific areas of interest (e.g., forest production areas or nature reserves) can be identified and sites established so that the forest gap models can be applied.

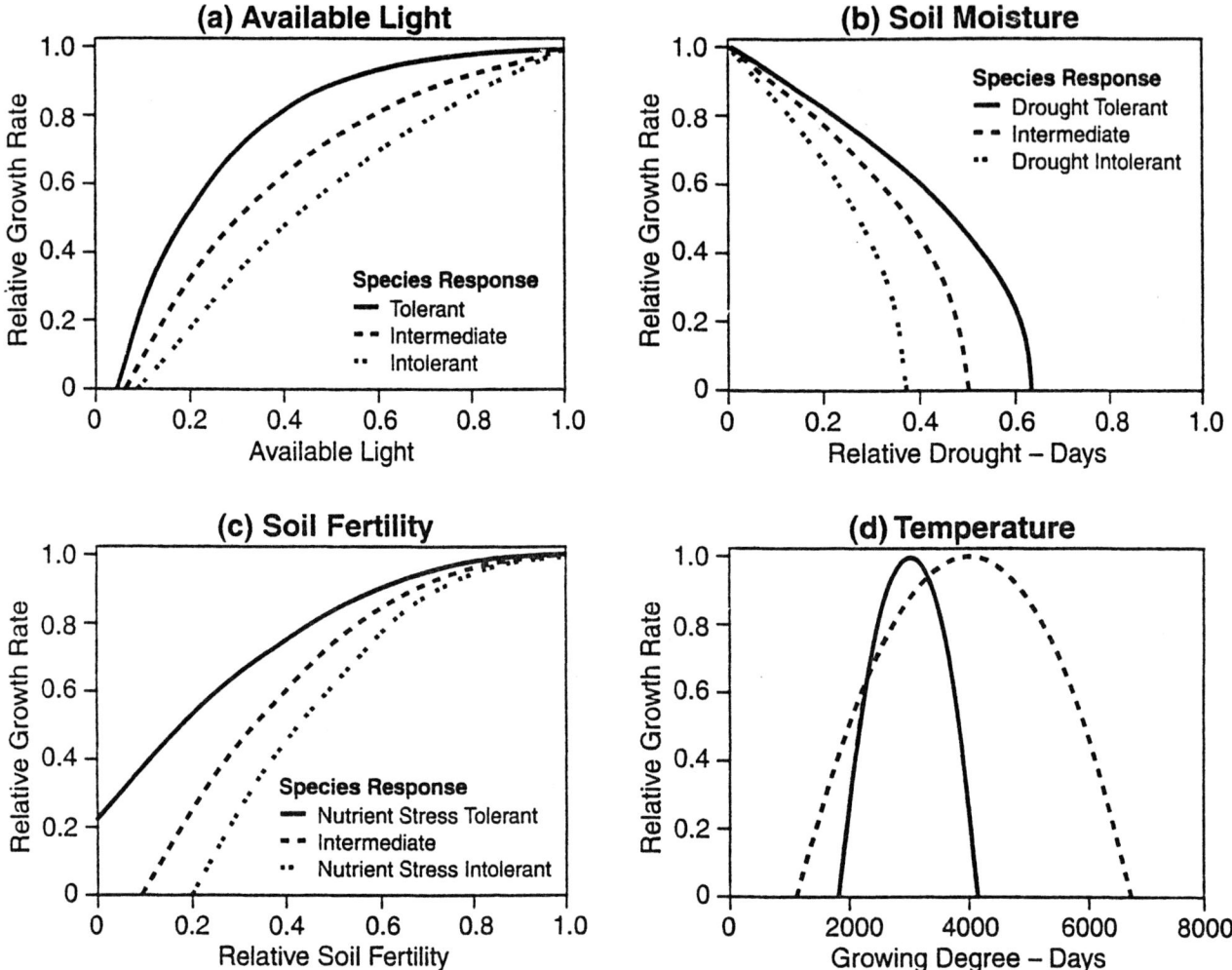

FIGURE 5.3.5 Environmental Response Functions Implemented by Forest Gap Models

The two model approaches can be integrated in two ways for the following purposes:

- To examine the general agreement between predicted patterns of changes in vegetation cover from the Holdridge model with changes in species composition predicted from the forest gap model and

- To examine the rates of expected changes for the predicted changes in vegetation cover from the Holdridge model.

Initially, sites are selected in areas where the Holdridge model predicts shifts in vegetation cover (e.g., forest to grassland or change in forest type) under the scenario. By applying the gap models to these sites, the model predictions can be tested for general agreement. For example, if the Holdridge model predicts a shift from forest to grassland, the forest gap model should predict an increase in mortality and decline in productivity for the site. This integrated analysis also estimates the temporal dynamics associated with the predicted shift in vegetation cover. These estimates are particularly important in developing adaptation strategies and assessing economic impacts.

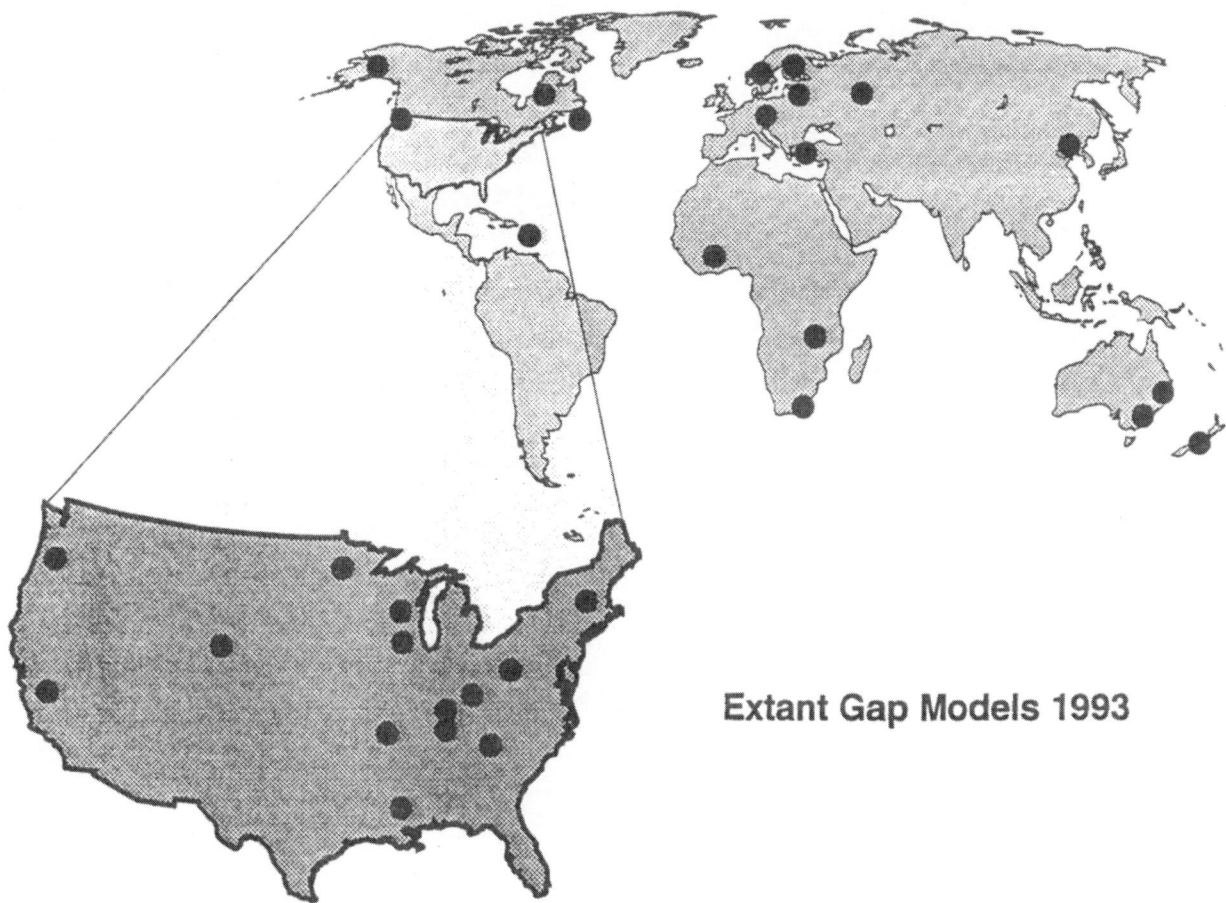

Extant Gap Models 1993

FIGURE 5.3.6 Geographic Locations of Existing Forest Gap Models

5.3.3 Scope of the Analysis

The Holdridge model can be applied at any spatial resolution, but analyses of regional impacts require the highest possible resolution. Global analyses have been undertaken at a spatial resolution of 0.5° × 0.5° (latitude and longitude); however, this rather coarse spatial resolution may prove inadequate for regional analyses of vulnerability and development of adaptation/mitigation strategies. Ultimately, the appropriate spatial scale for vegetation mapping and impacts assessment depends on the size of the region, the availability of climate (i.e., numbers and locations of meteorological stations) and topography data, and the specific areas of interest to the investigators.

Likewise, the application of forest gap models to specific forested sites depends on the availability of basic silvicultural and site data. Investigators may focus on commercially important species or areas of particular importance for production forestry or conservation. Combining the two modeling approaches allows for a systematic analysis of the impacts of climate change. Because the forest gap models have limited application to large areas, a sampling strategy should be developed to define sites at which to apply the forest gap models. Two criteria should be used. First, areas of particular importance (e.g., forestry and conservation) should be mapped and sites established for model simulations. Second, the maps of vegetation change under the scenarios generated by the Holdridge model should define areas of vulnerability. These areas may be currently forested areas that are predicted to decline under the changed climate conditions (i.e., GCM scenarios).

Southern Appalachian Deciduous Forest

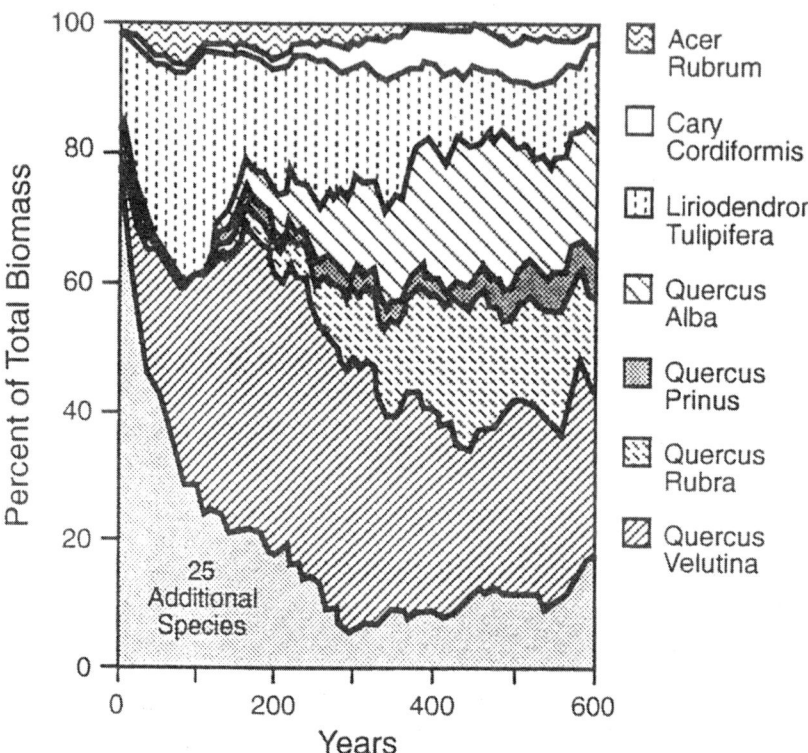

FIGURE 5.3.7 Annual Changes in Species Composition for a Southern Appalachian Deciduous Forest in Eastern North America as Simulated by the Forest Gap Model (Source: Shugart and West 1977)

The analysis of impacts should include multiple scenarios. Given the large degree of uncertainty in the regional results of GCMs, the greatest value in evaluating the potential impacts of the various climate change scenarios lies in examining the sensitivity of the forested areas within a region. Sensitivities examined include changes in the regional climate system, particularly increases in temperature and possible associated drying.

5.3.4 Economic Analysis

An economic analysis of the impacts of climate change on forest resources requires information on both the impacts on supply and subsequent impacts on pricing of timber and other forest products. The two modeling approaches outlined for impacts assessment can provide information on the first requirement (changes in supply). The two models provide different levels of information for assessing changes in the production of forest resources. Simple analyses can be performed by overlapping areas of forest impact from the Holdridge model with areas of current forest production. This procedure can identify areas currently used for production forestry, but are predicted to change to some other type of land cover that is incompatible with timber production (e.g., shrubland, grassland). This approach, together with the screening techniques discussed above, can provide a coarse resolution qualitative assessment of impacts on the supply of forest resources.

In contrast, forest gap models can perform more detailed analyses. These models can directly evaluate changes in productivity and timber production for specific areas under the climate change scenarios. They can predict changes in timber production by species and therefore provide detailed input of changes in supply for economic models.

The second requirement for an economic analysis is to evaluate the impacts of changing supply on pricing of timber and other forest products. Models must consider regional patterns of forest product utilization (e.g., fuel wood, saw timber, pulp) and foreign markets. In the latter case, the analyses must view how changes in supply in other regions influence market pricing. Even where prices are constant, changes in supply are needed to evaluate the economic feasibility of adaptation strategies (e.g., changing land use, changes in management, or harvest practices).

5.3.5 Adaptation Techniques

Both models can be used to develop adaptation strategies. The land cover maps derived from the Holdridge model identify areas suitable for future forest production, but currently used for other purposes (e.g., agriculture). Identifying possible changes in the suitability of land use allows for planning future land use policies. Such policies could offset impacts due to possible declining productivity in current forest management areas under the changed climate conditions.

The forest gap models directly evaluate specific forest management strategies (e.g., changes in rotation time or cutting practices). In addition, they evaluate the response of individual species to the climate change scenario, which allows for development of management plans that favor a given species (e.g., selective cutting) that may be more resistant to the predicted climate changes. Another option is to use the model to evaluate the introduction of species from other areas more suitable to the new environmental conditions.

5.3.6 Data Requirements

5.3.6.1 Holdridge Model

The Holdridge model requires data on annual precipitation and biotemperature. Biotemperature can be calculated from mean daily, weekly, or monthly data for a given site. The PET ratio is calculated as a direct function of biotemperature and therefore requires no additional data. Climate data used to calculate these two variables should represent long-term averages for the site (e.g., 30 years). Global databases of mean monthly temperature and precipitation at a spatial resolution of $0.5° \times 0.5°$ are available; however, primary data from meteorological stations within the country are preferred. Interpolation of climate data to provide continuous coverage depends on regional climate and topographic patterns. Exact procedures will be developed on a case-by-case basis.

5.3.6.2 Forest Gap Model

An example of the data required to describe species growth and environmental response is shown in Table 5.3.2. In addition to data on species growth and response to environmental conditions, the model requires data that describe the environmental conditions of the forest stand. Environmental data include monthly average temperature and precipitation for the area, soils (e.g., texture and water holding capacity), and topography (e.g., slope and aspect).

TABLE 5.3.2 Basic Parameters Used to Define Species Attributes in the FORET Forest Gap Model[a]

Species	Maximum Observed Height (m)	Maximum Observed Diameter[b] D_{bh} (m)	Maximum Observed Age (yr)	Shade Tolerance[c]	Drought[d]	Growing-Degree Day[e]	
						D_{min}	D_{max}
Liriodendron tulipifera	35	1.5	300	2	0.160	2,300	5,993
Acer rubrum	30	1.0	150	2	0.230	1,260	6,600
Quercus alba	35	1.0	400	1	0.330	1,721	5,537
Quercus falcata	35	1.0	400	1	0.423	2,660	5,993

[a] FORET simulates the dynamics of a deciduous hardwood forest in southeast North America. The table does not represent a complete species list.

[b] Maximum observed diameter at breast height.

[c] 1 = shade tolerant, 2 = shade intolerant (Figure 5.3.5).

[d] Proportion of growing season that species can tolerate soil moisture below or at wilting point (Figure 5.5.5).

[e] D_{min} and D_{max} denote minimum and maximum values, respectively, of growing-degree days associated with the northern and southern range limits for the species (response function defined as parabola (Figure 5.3.5).

Source: Shugart and West 1977.

5.3.7 Results Generated

5.3.7.1 Holdridge Model

The maps and associated databases generated from the Holdridge model for current climate and the GCM-based climate change scenarios can be used to calculate changes in land area associated with different categories of vegetation cover (e.g., grassland and forest).

The analysis is valuable for comparing the predicted spatial changes in potential vegetation cover with the actual patterns of land use for the country or region.

As stated, the resulting vegetation map from the Holdridge model represents potential vegetation cover based on the correlation between a region's vegetation distribution and climate. To assess the potential impacts on forest resources, areas of forest cover must be located within the mapping system. An example of such a database (digitized maps) locating forest conservation areas for Costa Rica is presented in Figure 5.3.8. By combining data that identify specific land areas of concern with the maps generated by the Holdridge model, investigators can evaluate specific changes in land cover under the GCM scenarios. The predicted changes in vegetation cover for these specific areas of interest can help to interpret the potential impacts of changing vegetation cover on such activities as forest production, conservation of flora and fauna, or suitability for various other land use activities.

Although the Holdridge model does not predict species distributions directly, data on species composition or distributions of commercially valuable species can be related to the forest categories used in the classification. This approach of combining species-level data with the vegetation classes, as defined

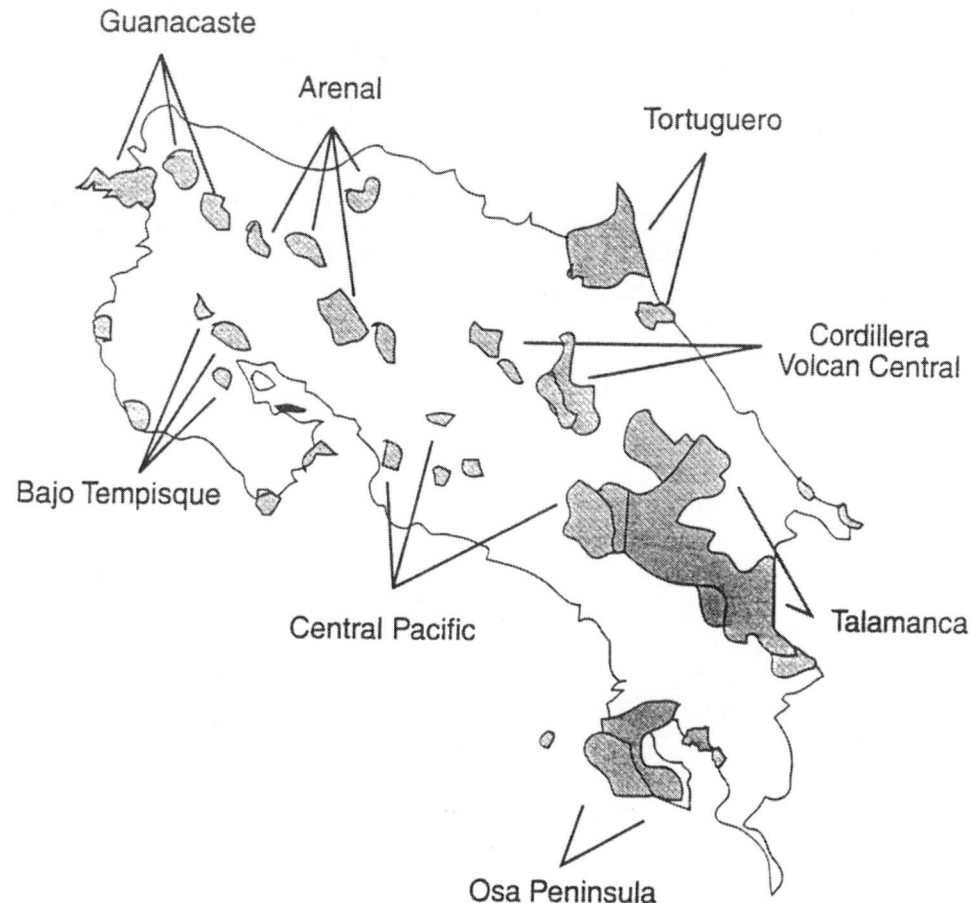

FIGURE 5.3.8 Digitized Map Showing the Location of Forest Conservation Regions for the Country of Costa Rica

by the Holdridge model, can greatly assist in interpreting potential impacts that result from changes in vegetation cover.

5.3.7.2 Forest Gap Model

The forest gap model can be used to examine the response of forested systems to climate changes, both in the reconstruction of prehistoric Quaternary forests (Solomon et al. 1980, 1981; Solomon and Shugart 1984; Solomon and Webb 1985; Bonan and Hayden 1990; Bonan et al. 1990), as well as to project possible consequences of future climate change (Solomon et al. 1984; Solomon 1986; Pastor and Post 1988; Urban and Shugart 1989; Bonan et al. 1990; Overpeck et al. 1990; Smith et al. 1992a).

Unlike the Holdridge model, the forest gap model directly examines the potential impacts of a climate change on forest composition and productivity (Figure 5.3.8). The model examines the time scales associated with forest response to climate change. In addition, the model incorporates adaptive strategies such as planting new species or introducing specific forest management practices.

Output from the forest gap model includes the stem diameter, height, and species of each individual tree on the forest stand. These data can be used to calculate biomass, basal area, size class distribution, and net primary productivity on a species or stand level.

One disadvantage of this model is its fine spatial resolution, which limits the model's ability to extrapolate results at a regional scale. This limitation is due to spatial variation in the environmental variables, which influence the response of the various species included in the analysis.

5.3.7.3 Model Integration

Figure 5.3.9 combines the Holdridge model and the forest gap model approaches for the North American Boreal Zone. The map of North America shows predicted changes in the distribution of boreal forest under the Oregon State University (OSU) climate change scenario. Areas are categorized as follows:

- Stable (areas of current boreal forest that remain in boreal forest under the scenario),

- Boreal or nonforest zone (areas currently in boreal forest that are predicted to change to grassland under the scenario), and

- Boreal or other forest zone (areas currently in boreal forest that are predicted to change to cool temperate forest).

The diagrams showing changes in biomass with time result from simulations made by a forest gap model for sites currently dominated by black spruce (*Picea mariana*). Simulations include both control (current climate) and climate change scenarios. The climate change scenarios used for the simulations were transient estimates based on $2 \times CO_2$ equilibrium simulations from various GCMs. Section 4 describes the methodology for constructing transient scenarios. Values of biomass at year 0 of the x axis represents current stand structure (i.e., initial conditions). From year 0 to year 75 of the simulations, the climate changes from currently observed patterns (i.e., control) to those predicted under the $2 \times CO_2$ scenario. From year 75 on, the climate remains constant.

In this example, the Holdridge model provides the broad-scale patterns of change within the boreal zone. The forest gap model then estimates rates change and quantitative data on changes in forest biomass and productivity for key species such as black spruce.

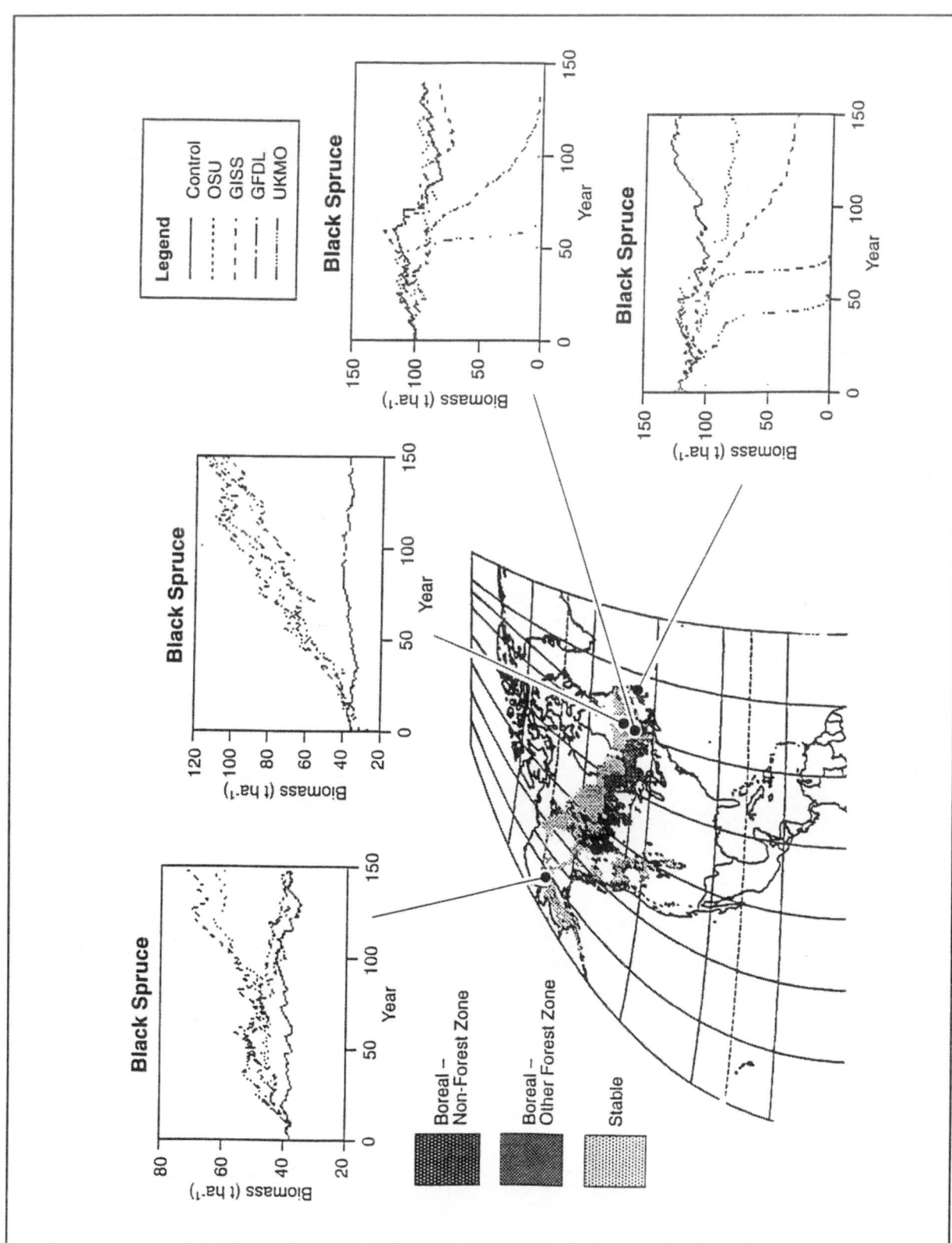

FIGURE 5.3.9 Map of America Showing Areas of Forest Change in the Boreal Zone under the OSU Climate Change Scenario and the Results of Forest Gap Model Simulations of Changes in Biomass for a Series of Sites Dominated by Black Spruce

5.3.8 Advantages and Disadvantages

5.3.8.1 Holdridge Model

This approach is essentially an equilibrium solution to a dynamic process of vegetation change. In many ways, it is analogous to the equilibrium solutions of the GCMs for climate patterns under conditions of increased CO_2. The approach has many implicit assumptions about the stability of vegetation associations or types under changed climate conditions, and the ability of vegetation to track (migrate with) spatial changes in climate patterns. However, the approach provides an important tool for exploring potential sensitivities of different regions and vegetation types to changing climatic conditions. Applied at a regional scale, this model can identify potentially vulnerable areas and thus assist in locating sites for more intensive investigation.

5.3.8.2 Forest Gap Model

In contrast to the Holdridge model, the forest gap model is a high-resolution approach in that it can predict species composition, vegetation structure, and associated productivity and standing biomass through time. However, the predicted patterns of forest response can not easily be extrapolated to different sites in the region because of (1) variation in the environmental conditions of the site (e.g., climate or soils) and (2) changes in species composition of the forest. The limited ability to extrapolate the results limits the use of this model in providing a complete regional assessment. The application of the forest gap model to provide total coverage over broad regions would be virtually impossible because of both computational and data limitations. As an alternative, sampling approaches could provide large-scale coverage over broad environmental gradients (Solomon 1986; Bonan 1990a,b; Smith et al. 1992b).

This model has an additional limitation. If the environmental conditions change enough that the species currently inhabiting a forest site can no longer survive, the rates of immigration of other species currently not present on the site (suitable to the new climate conditions) must be estimated. This limitation exists only if long-term simulations of climate change are undertaken.

5.3.9 Supplementary Approaches

The two modeling approaches presented in this section as a primary approach for assessing climate change impacts on forested ecosystems represent only a small subset of the models and approaches that can examine the response of vegetation to changing environmental conditions. The Holdridge model is one of a number of climate-vegetation classifications applied at the regional and global scales. The Box (1981) and BIOME (Prentice et al. 1992) models are similar to the Holdridge model in that they relate the large-scale distribution of vegetation or plant types to climate patterns. However, these models require additional climate variables and calculations of evapotranspiration. Of the models that simulate the dynamics of stand structure and species composition, the forest gap model most easily adapts to new sites because parameters can generally be derived from simple silvicultural data and information on species distributions.

In addition to the climate-vegetation classification and forest stand models, more physiologically detailed models are available to examine the implications of both climate change and the direct effects of increasing concentrations of atmospheric CO_2 on forests. These models (e.g., Running and Coughlan 1988; Bonan 1989) generally simulate the processes of photosynthesis and transpiration of forest canopies.

Although this approach is suitable for examining patterns of CO_2 and water flux on both an intra- and an interannual basis, typically these models require that vegetation cover be defined as an input. As such, they are of limited value in examining the potential for changes in forest structure and composition that may occur under changing climate conditions. Recent work has focused on likening these models of canopy processes with individual-based forest gap models (Smith et al. 1992c) to examine the potential direct effects of rising CO_2 concentrations on patterns of water use efficiency and net primary productivity under changed climate conditions; however, data requirements for parameterization limit the ease with which this approach can be applied to a variety of forested ecosystems.

5.3.10 References for Section 5.3

Bonan, G.B., 1990a, "Carbon and Nitrogen Cycling in North American Boreal Forests. I. Litter Quality and Soil Thermal Effects in Interior Alaska," *Biogeochemistry* 10:1-28.

Bonan, G.B., 1990b, "Carbon and Nitrogen Cycling in North American Boreal Forests. II. Biogeographic Patterns," *Canadian Journal of Forest Research* 20:1077-1088.

Bonan, G.B., 1989, "A Computer Model of the Solar Radiation, Soil Moisture, and Soil Thermal Regions in Boreal Forests," *Ecological Modeling* 45:275-306.

Bonan, G.B., H.H. Shugart, and D.L. Urban, 1990, "The Sensitivity of Some High-Latitude Boreal Forests to Climatic Parameters," *Climatic Change* 16:9-29.

Bonan, G.B., and B.P. Hayden, 1990, "Using a Forest Stand Simulation Model to Examine the Ecological and Climatic Significance of the Late-Quaternary Pine-Spruce Pollen Zone in Eastern Virginia," *U.S.A. Quaternary Research* 33:204-218.

Botkin, D.B., J.F. Janak, and J.R. Wallis, 1972, "Some Ecological Consequences of a Computer Model of Forest Growth," *Journal of Ecology* 60:849-873.

Box, E.O., 1981, *Macroclimate and Plant Forms: An Introduction to Predictive Modeling in Phytogeography*, Junk, Hague, the Netherlands.

Holdridge, L.R., 1947, "Determination of World Plant Formations from Simple Climate Data," *Science* 105(2727):367-368.

Holdridge, L.R., 1959, "Simple Method for Determining Potential Evapotranspiration from Temperature Data," *Science* 130:572.

Holdridge, L.R., 1967, *Life Zone Ecology*, Tropical Science Center, San Jose, Costa Rica.

Overpeck, J.T., D. Rind, and R. Goldberg, 1990, "Climate-Induced Changes in Forest Disturbance and Vegetation," *Nature* 343:51-53.

Pastor, J., and W.M. Post, 1988, "Response of Northern Forests to CO_2-Induced Climate Change," *Nature* 334:55-58.

Prentice, I.C., W. Cramer, S.P. Harrison, R. Leemans, R.A. Monserud, and A.M. Solomon, 1992, "A Global Biome Model Based on Plant Physiology and Dominance, Soil Properties and Climate," *Journal of Biogeography* 19:117-134.

Running, S.W., and J.C. Coughlan, 1988, "A General Model of Forest Ecosystem Processes for Regional Applications. I. Hydrological Balance, Canopy Gas Exchange and Primary Production Processes," *Ecological Modeling* 42:125-154.

Shugart, H.H., 1984, *A Theory of Forest Dynamics*, Springer-Verlag, New York, N.Y.

Shugart, H.H., and West, 1977, "Development of an Appalachian Deciduous Forest Succession Model and Its Application to Assessment of the Impact of the Chestnut Blight," *Journal of Environmental Management* 5:161-179.

Smith, T.M., H.H. Shugart, G.B. Bonan, and J.B. Smith, 1992a, "Modeling the Potential Response of Vegetation to Global Climate Change," *Advances in Ecological Research* 22:13-113.

Smith, T.M., R. Leemans, and H.H. Shugart, 1992b, "Sensitivity of Terrestrial Carbon Storage to CO_2-Induced Climate Change: Comparison of Four Scenarios Based on General Circulation Models," *Climatic Change*.

Smith, T.M., J.F. Weishample, H.H. Shugart, and G.B. Bonan, 1992c, "The Response of Terrestrial Carbon Storage to Climate Change: Modeling Carbon Dynamics at Varying Temporal and Spatial Scales," *Water, Air and Soil Pollution* 64:307-326.

Solomon, A.M., 1986, "Transient Responses of Forests to CO_2-Induced Climate Change: Simulation Modeling Experiments in Eastern North America," *Oecologia* 68:567-569.

Solomon, A.M., and H.H. Shugart, 1984, "Integrating Forest-Stand Simulations with Paleoecological Records to Examine Long-Term Forest Dynamics," in G.I. Agren (ed.), *State and Change of Forest Ecosystems: Indicators in Current Research*, Report No. 13, pp. 333-357, Swedish University of Agricultural Science, Uppsala, Sweden.

Solomon, A.M., and T. Webb, III, 1985, "Computer-Aided Reconstruction of Late-Quaternary Landscape Dynamics, " *Ann. Rev. Ecological Systems* 16:63-84.

Solomon, A.M., H.R. Delcourt, D.C. West, and T.J. Blasings, 1980, "Testing a Simulation Model for Reconstruction of Prehistoric Forest-Stand Dynamics," *Quat. Res.* 14:275-293.

Solomon, A.M., M.L. Tharp, D.C. West, G.E. Taylor, J.M. Webb, and J.L. Trimble, 1984, *Response of Unmanaged Forests to CO_2-Induced Climate Change: Available Information, Initial Tests and Data Requirements*, Technical Report TR009, U.S. Department of Energy Carbon Dioxide Research Division, Washington, D.C.

Solomon, A.M., D.C. West, and J.A. Solomon, 1981, "Simulating the Role of Climate Change and Species Immigration in Forest Succession," in D.C. West, H.H. Shugart, and D.B. Botkins (eds.), *Forest Succession*, pp. 154-177, Springer-Verlag, New York, N.Y.

Urban, D.L., and H.H. Shugart, 1989, "Forest Response to Climate Change: A Simulation Study for Southeastern Forests," in J. Smith and D. Tirpak (eds.), *The Potential Effects of Global Climate Change on the United States*, EPA-230-05-89-054, pp. 3-1 to 3-45, U.S. Environmental Protection Agency, Washington, D.C.

5.4 WATER RESOURCE IMPACT AND ADAPTATION ASSESSMENTS

Water resources management is the task of transforming natural hydrologic resources into a managed resource to be applied by society to a variety of uses. For the past decade, attention has focused on assessing the impacts of climate change on the world's water resources, specifically on the hydrologic impacts or the supply side. However, the impact on demand (the water uses themselves) can be significant and often greater than the impacts on supply. Very few studies of demand impacts and alternative water management strategies have been conducted. Only when supply and demand are jointly examined can a nation's vulnerability be assessed. Once the vulnerability is defined, alternative management strategies for adaptation to climate change impacts can be examined.

5.4.1 Primary Approach

The primary approach to be followed in assessing a country's water resources vulnerability to climate change is a comprehensive assessment of the full water resource system, including supply, demand, and management.

Supply is analyzed in two stages. Stage 1 assesses river runoff impacts by means of one of the methods recommended below. Stage 2 assesses the effect of the impacted river runoff on the management of the water resource system and the resulting water supply.

Demand is analyzed on a very aggregated national basis. Because water demand is significantly tied to the socioeconomic driving forces, an important part of demand impact assessment depends on the base scenario projection of population, agricultural and industrial production, and energy demand. Water resource management is to be assessed by using systems analytic methodologies that include the impacts on supply and demand as well as potential adaptation measures.

5.4.1.1 Preliminary Screening Technique

In many cases, the effort required to conduct a comprehensive assessment of a country's water resources vulnerability to climate change will be beyond the resources available. In these cases, a preliminary screening of the four major areas of water resources vulnerability should be carried out to identify critical areas. In addition, some countries may lack sufficient data to perform the preliminary simulation; however, a simpler screening technique may be possible. Appropriate screening methodologies are outlined in Table 5.4.1.

While water resources assessment is primarily cast in a quantitative framework, qualitative assessment of water resources is possible. However, this assessment requires water resource professionals with extensive knowledge of a nation's or region's water resources. These professionals may be able to assess qualitatively the impacts that climate change will have on water supply, demand, and system performance as well as propose adaptation options.

TABLE 5.4.1 Appropriate Screening Methodologies for Major Areas of Water Resources Vulnerability

Impacts on hydrologic resources—supply

    Data analytic methods:

- Aridity index

- Climate/runoff elasticity approach

Impacts on water uses—demand

- Linear regression techniques

- Analogous country approach

Impacts on demand/supply balance—vulnerability

- Application of the Shuval index using the screening methods above

- A Delphi methodology using local experts

Alternative management strategies—adaptation

- A workshop and roundtable meeting of water resource, agricultural, and economic experts from the country and experts from countries at the developmental level at which the country expects to be in base year 2075

- A Delphi methodology using local experts

5.4.1.2 Simulation Techniques

The simulation approach is an analysis of the joint impact of population growth, economic development, and climate change on regional water resources. Thus, a base scenario of water supply and demand without climate change will be developed. The approach will have a four-step procedure that investigates:

- Impact on hydrologic resources—supply,

- Impact on water uses—demand,

- Impact on supply-and-demand balance—vulnerability, and

- Alternative management strategies—adaptation.

5.4.1.2.1 Supply. Each country will be divided into its river basin units. These units represent entire river basins or the national portion of an international river basin. For each basin, two monthly water balance models will be developed. The first, a simplified model, is a mean monthly runoff estimate based on inputs of mean monthly temperature and precipitation values. The mean monthly values are obtained from climate databases.

The second, an extended model, is a modification of the mean monthly approach to a time series simulation model. The time series of monthly temperature and precipitation values is input to the model with a time series of monthly runoff as output. The models allow for GCM scenario-based changes in precipitation and temperature to be input directly to the models and the resulting changes in runoff to be output. For countries downstream in international river basins, modeling of the upstream portion of the river basin is necessary.

5.4.1.2.2 Demand. Within each basin, estimates of water demand without climate change based on estimates of population and economic growth to the year 2075 are made on the basis of Kulshreshtha (1993). Climate-change-induced changes in water demand driven directly (e.g., agriculture) or indirectly (e.g., coastlines) are then estimated.

5.4.1.2.3 Vulnerability. A supply-and-demand balance for each river basin unit is undertaken, and a measure of vulnerability is assigned. Vulnerability is defined by using Shuval's water deficit criterion (1987). National vulnerability is assessed via a gross national supply-and-demand balance and by a restrictive basin method.

5.4.1.2.4 Adaptation. National level adaptations are examined via additional national vulnerability assessments that incorporate the possibility of interbasin transfers of water and shifts in population and economic growth. One or more key or representative river basins are selected to examine alternative management strategies for adaptation to climate change impacts at a basin level. This assessment includes alternative operation of the existing water resource infrastructure as well as the addition of capital investments such as reservoirs or canal linings. National implications drawn from insights gained from the representative basins study will be developed.

5.4.2 Scope of the Analysis

As noted, the scope of the analysis is on a river basin scale. However, because of the varying sizes of the countries under study, the variability in the hydrometeorological data network, and the variability in size of individual river basins, the spatial scale of the analysis will vary. Most river basins are composed of a number of distinct subbasins or tributaries where runoff records may exist. Thus, the runoff modeling effort takes place at the scale dictated by the measured runoff stations and might even be different within countries because important basins may possess detailed records, while other basins are not even gauged. As a result, in some countries, small basins are modeled on the order of hundreds of square kilometers, while elsewhere in the country or in other countries, basins are modeled on the order of hundreds of thousands of square kilometers. Where local data are not available and the world gridded climatic data are used, data are available at a $0.5° \times 0.5°$ resolution. This defines the coarsest scale of the climatic parameters. A final reason for varying spatial scale is limited time and funds. Larger river basin units may be selected even if data exist for smaller scale analysis.

The temporal scale is monthly for some analyses and annual for others. Research has shown that, although storm event time scales are most appropriate for hydrologic modeling, such fine resolution is not warranted, given the uncertainties in the climate change scenarios. Section 5.4.9 discusses alternative approaches. Water balances using annual average parameters may underestimate runoff because of the seasonal distribution of precipitation and PET. The monthly time scale appears to be the "optimal time step" for climate change assessments because it models the seasonal distribution and is large enough to avoid computational and data problems. For water supply, annual parameters are estimated by summing monthly values. For water demand, some uses (e.g., navigation, municipal, and industrial) can be estimated annually, but others (e.g., environmental, hydropower, and irrigation) need to reflect the seasonal distribution. The latter require monthly estimates, with annual estimates, which are a summation of the monthly estimates.

The scope of the assessment methodology presented in Section 5.4.3 is designed to take a comprehensive view of climate change on water resources. The scope is not limited to water resources supply but includes water demand and, most important, water resources management. Impacts on hydrologic resources are reported in terms of water quantity and seasonal distribution. Water demand impacts are estimated both directly from climate parameters as well as indirectly via impacts from other biophysical impacts. The river basin as the scope of the vulnerability assessment is an important feature of the analysis because one of the most important factors in water resource assessment is the spatial distribution of water supply and demand. It is necessary to include water resources management in the analysis because the goal of water resources management is to redistribute water supply both spatially and temporally to meet water demand. Any assessment of water resource vulnerability and adaptation must include management because management is the vehicle for adaptation.

5.4.3 Description of the Methodology

The framework for assessing the impacts of climate change on national water resources and analyzing possible adaptation assessments is outlined in Figure 5.4.1. This framework is made up of four distinct components: supply, demand, vulnerability, and adaptations. Each of the components of the assessment could be carried out via a number of different methodologies. This section describes the methodologies chosen for this particular countries studies activity.

Many methodologies are available; however, few are appropriate for assessing the impacts of climate change on a nation's water resources. The rationale for selecting the methodologies was driven by five criteria: (1) methods applicable in data-rich or data-sparse conditions; (2) methods where the modeling effort does not overshadow the analysis of and insights into potential climate change impacts; (3) simple, easily understood methods that allow for a comprehensive analysis within the time and resource constraints; (4) methods successfully used in previous climate change impact studies; and (5) methods with a high probability of successful implementation in most countries.

In addition, because the goal of the study is the assessment of vulnerability and adaptation, the methodologies have to be balanced to ensure that each component is allocated the time and effort appropriate to achieve its goals at the same scope and scale as the other components. One goal was to avoid spending 90% of the resources on hydrologic modeling on a daily time step and kilometer length scale and then linking these results with a coarse- scale analysis of the other three components. The components do not require equal time and resource efforts, but are allocated the resources necessary to ensure a consistent analysis in terms of spatial and temporal scale.

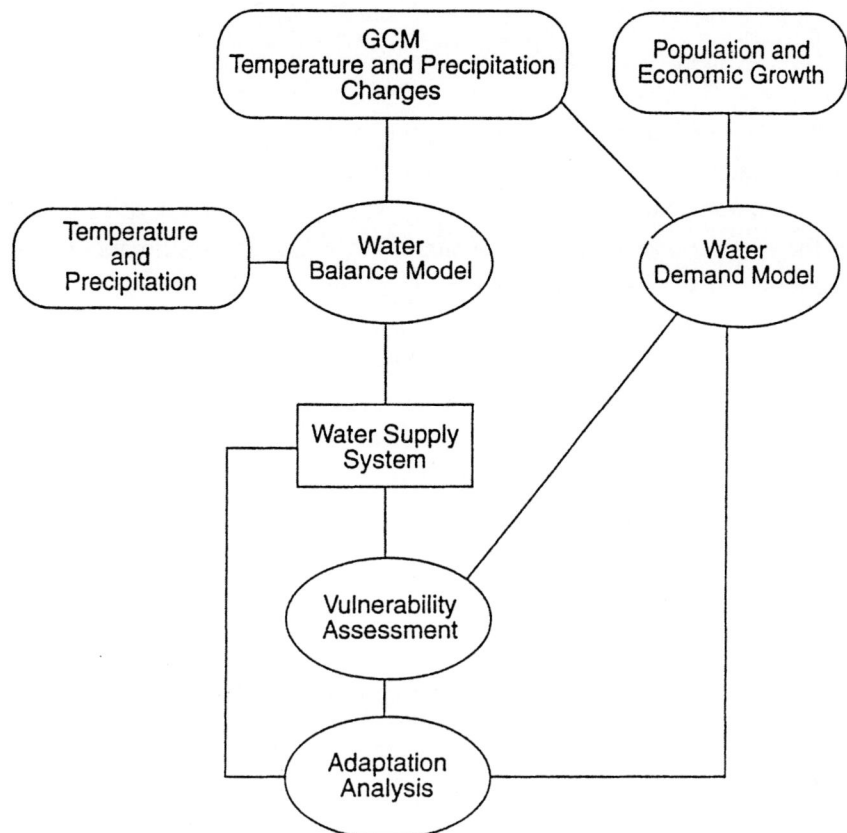

FIGURE 5.4.1 Framework for Assessing Water Resources Vulnerability

5.4.3.1 Impacts on Hydrologic Resources—Supply

A monthly, spatially lumped, one-dimensional water balance model was selected to model the hydroclimatic cycle. The model illustrated in Figure 5.4.2 is based on the work of Kaczmarek (1993), and various forms of it have been applied to model hydrologic impacts of climate change in a number of international studies across a variety of climatic zones. To apply the model, a watershed is selected that has one or more stream flow measuring stations with "sufficiently" long records. The watershed area upstream of the stream flow stations is delineated, and weighed average values of precipitation and temperature are estimated. The procedures for determining the weighted average values are described in Appendix G.

In the model, PET can be estimated by using a variety of methodologies based solely on average monthly temperature and daily sunlight duration. Precipitation is divided into interception, percentage snow, direct surface runoff, and infiltration. A snow water equivalent account is kept that includes accumulation, storage, and melt. The melt is modeled as direct surface runoff. The soil moisture zone is modeled as a nonlinear reservoir with discharge to the mouth of the watershed and no storage over the year or losses. There is within year storage.

The model is calibrated for the "base" or current climate by estimating four model parameters to get the best fit (annual flow and seasonal distribution) of the model to mean monthly stream flow when mean monthly precipitation and temperature are used as input. The model is then used to estimate the impact of climate change on the mean monthly and annual runoff. The impact to climate change is analyzed by inputting to the model-modified mean monthly temperature and precipitation values. The modified values are obtained by using GCM-generated monthly differences in temperature and precipitation between $2 \times CO_2$ and $1 \times CO_2$. After selecting the most appropriate GCM grid cell for the watershed, new input values are calculated. That is, the base temperatures are added to changes in temperatures, and the precipitation ratio is multiplied by the observed base. The model is run holding all parameters constant, and runoff estimates are generated. These new runoff values represent the estimate of the impact of the associated GCM scenario on water supply, both in terms of total annual supply and seasonal distribution.

The extended model is calibrated over a portion of the time series and validated over the remaining portion.

The GCM-based mean monthly differences of temperature and precipitation are combined with the time series to produce new GCM-modified time series. Details of GCM-modified time series are presented in the technical appendixes.

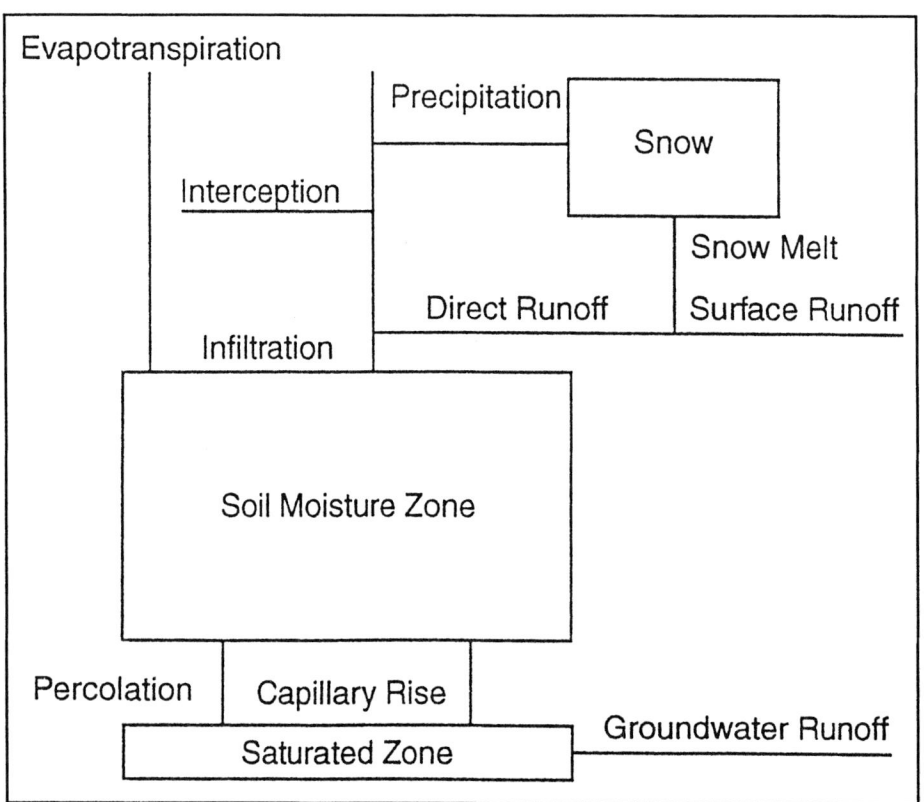

FIGURE 5.4.2 Schematic of One-Dimensional Soil Column Model

While the monthly water balance approach is the primary approach used by countries that do not have other hydrologic modeling resources, countries that use other models for assessing climate change are encouraged to use these models in parallel with, or in place of, the water balance approach presented here.

5.4.3.2 Impacts on Water Uses—Demand

Three key elements dominate estimating the impact of climate change on water demand: (1) the establishment of the "base" scenario, (2) the assumptions about technological improvements in water use sectors, and (3) the degree of adaptation possible on top of the projected base technological improvements.

The methodology chosen for this component is a simple model that highlights the underlying assumptions rather than buries them in a complex model. The methodology is as follows. Four aggregate water use classes are selected for the demand assessment: agricultural, domestic, industrial, and energy-related. For each river basin unit, current water use and associated economic activity data are collected. Per unit water use coefficients are calculated (e.g., for domestic use - water per capita). Growth rates for each economic activity in the basin are then provided, and forecasts of economic activity are given for the base year 2075. An alternative approach is to have specified national level projections of base year economic activity and disaggregate these national estimates to river basin estimates based on existing ratios or some other specified method.

To estimate water demand in the base year 2075, multiply the forecasts of economic activities by a water use coefficient. However, historically, water use coefficients seldom remain constant. As income rises, domestic water use rises. Until it hits a threshold, industrial water use can actually observe reductions in per unit water use because of strict environmental standards. Therefore, careful estimations of unit water use coefficients are required for many years in the future. The dynamics of water use coefficients are very sensitive to the point on the development curve where each river basin is now and where it is projected to be in the base year. In addition, these coefficients have varied by hydroclimatic zone. A series of functions of water use versus some measure of development for different climate zones will be provided for the four economic activities.

Figure 5.4.3 shows the complex nature of estimating the impact of climate change on water demand. In static analysis, the forecasts of economic activity for each basin do not change; only the water use coefficient responds. A dynamic analysis gives feedback of the cost of adjusting to climate change, and the economic activities change as a result. Dynamic analysis is beyond the scope of this study, so impacts of climate change on water use are determined solely by changes in the water coefficients. The change for agriculture comes from the agriculture biophysical assessment (Section 5.1 and Appendix D); analysts estimate others on the basis of local conditions. These changes are caused by the change in climate, such as more domestic water use for lawn watering or more cooling water for thermal electric generating stations.

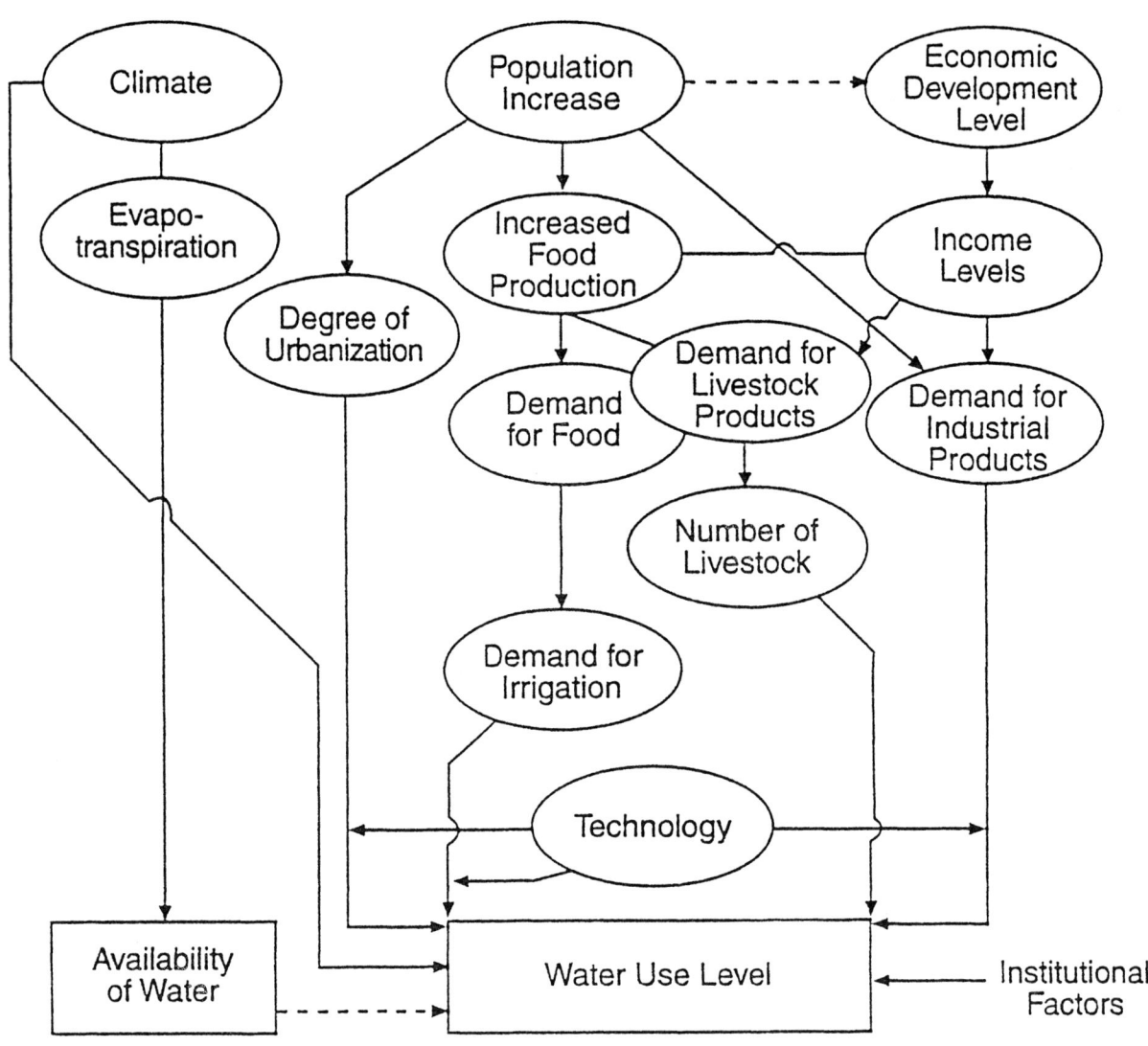

FIGURE 5.4.3 Interrelationship of Water Demand (Source: Kulshreshtra 1993)

As part of the adaptation component described below, it is necessary to estimate changes in the water use coefficient because of adjustments to climate changes. In most cases, these adjustments result in either additional costs or loss of utility on the part of society. An example is the reduction in per-capita water use as homeowners replace lawns with low water use grasses or xeriscape, drastically reducing water use. Another example is when an industry changes to total water recycling to save water. These adaptations and the estimated cost must be clearly presented and highlighted in the adaptation analysis.

5.4.3.3 Impacts on Supply-and-Demand Balance—Vulnerability

After water supply and demand have been estimated for each river basin unit, vulnerability can be assessed. Basin supply and demand are entered into a spatial database tool (e.g., a GIS), and this tool calculates the water surplus or deficit. By using Shuval's vulnerability index (Figure 5.4.4), the vulnerability of each basin is then classified. Each basin is assigned a yield index. The yield index is a number between 0 and 1 that is multiplied times the natural water supply to reflect the amount of water that can be currently supplied to meet water demand. This index is based on the monthly distribution of water supply, the amount of installed reservoir storage capacity in the basin, and a measure of water distribution infrastructure. A new vulnerability classification using the effective water supply (yield index times natural water supply) is carried out. Finally, linkage between river basin units based on actual and reasonable potential interbasin water transfer is identified. The model then develops an additional vulnerability classification after first minimizing the total national water deficit by providing for the transfer of water from surplus basins to deficit basins. These analyses are carried out for the base and GCM scenarios.

5.4.3.4 Alternative Management Strategies—Adaptation

At the national level, adaptation consists of supply-and-demand management. Supply management, such as more capital investment in reservoirs and infrastructure, results in increasing the yield index and allowing for more interbasin water transfers. Demand management means reductions in water demand via investment in new water-saving technologies and changed use practices. As outlined in Figure 5.4.1, the vulnerability classification is carried out for the base and GCM scenarios and again with the changes as a result of the adaptations.

To conduct a national-level adaptation study, the country team will gather multidisciplinary experts, together with local and international knowledge and data, to suggest possible future adaptation scenarios. The scenario generation process usually takes place at one- or two-day workshops, where all experts are present to brainstorm "realistic" future adaptation options based on local constraints and projected changes.

Because many aspects of both supply-and-demand management cannot be accurately represented by the yield index or basinwide averages of water demand, one or more representative basin studies should be undertaken. The basin study takes a closer look at adaptation via supply-and-demand management of the detailed elements of the water resource system of the basin. This analysis requires the use of a hydroeconomic river basin simulation model and/or river basin optimization model. With these tools, changes in the operation of the current infrastructure as well as the addition of new elements (e.g., reservoirs) can be examined.

Various river basin simulation models are available. These models range from simple, comprehensive models such as MITSIM to complex, comprehensive models such as HEC-5. The appropriate choice of model depends on the level of basin development and the availability of data and resources. The time series of runoff output from the models is the driving function of the simulation models. These models require physical, engineering, hydrologic, and economic data. Although the models provide great insights at a basin or regional level, they require a focused effort of time and resources to gather the data, learn the model, perform the run, and analyze the results.

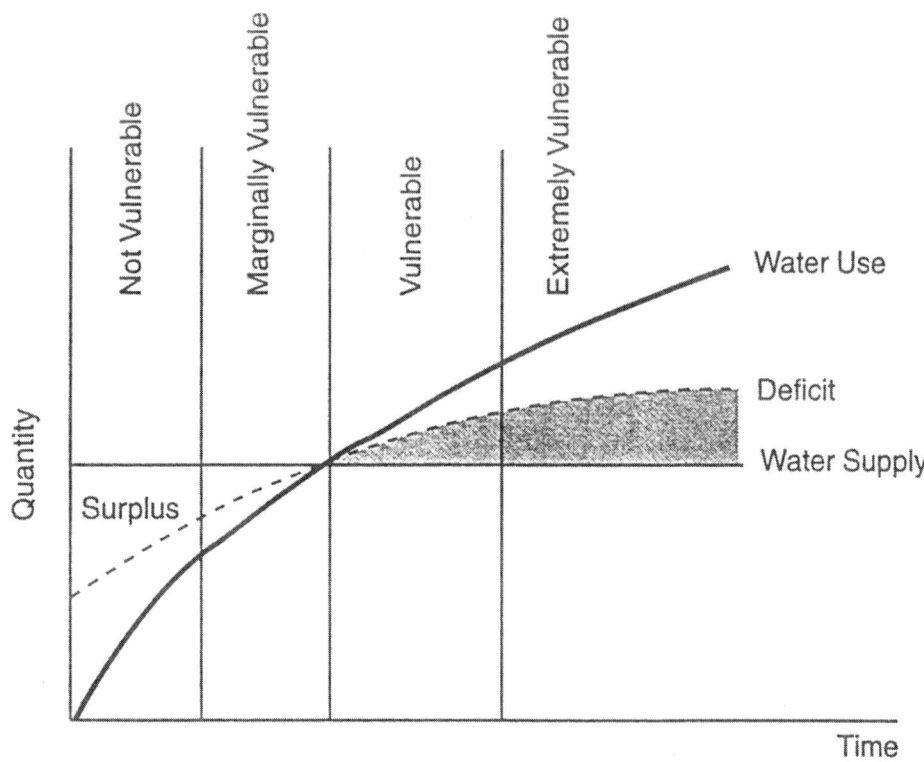

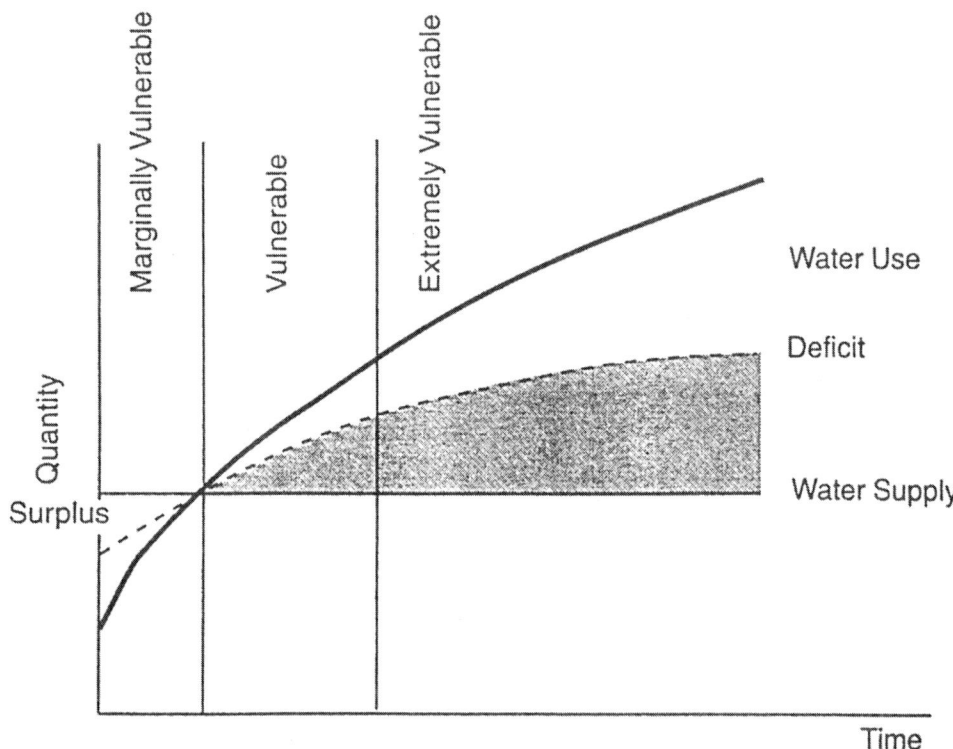

FIGURE 5.4.4 The Impact of Climate Change on Vulnerability

5.4.4 Economic Analysis

Economic impacts of climate change are hard to identify on a national level, but are easily addressed on regional levels. However, other effects such as direct and indirect or economywide effects are not easy to address. Three levels, or a scale of economic impact assessment, are proposed.

In the case of static direct sectoral impact, water use in each sector (agricultural, domestic, industrial, and energy-related) would be estimated on the basis of the water demand component. An economic value (GDP) of each of these activities is estimated for the base year 2075. Increases or decreases in the water resource supply are equally allocated to each sector, and the resulting changes in each sector are measured. This approach is very crude, but simple.

Economic models of each sector can be developed with water as direct input to the sectoral production. This method would allow for substitution of water by other factors, which would more accurately represent economic behavior. This approach has been used in a U.S. EPA study (Strzepek and Smith 1994).

The previous two approaches attempt to measure the direct economic impacts of water. To measure the indirect economywide impacts, intermediate activities must be modeled. These activities do not directly use water, but use inputs from sectors that directly use water, such as cotton in the textile industry, hydropower in the aluminum industry, or navigation in the transport sector. Economywide input-output models are one approach; however, Berck et al. (1991) have used a general equilibrium model with water as a factor input to find economywide impacts of water resource availability for California. Strzepek and Smith (1994) have done the same for Egypt.

5.4.5 Adaptation Techniques

Three basic adaptations are possible in the water resources sector: increase water supply, reduce water demand, and manage supply and demand differently. Examples of possible adaptations follow:

- *Increase water supply.* Modify basin vegetation, construct reservoirs, drain swamps, reduce evaporation, develop groundwater resources, and use interbasin transfers.

- *Reduce water demand.* Decrease the activities that require water, decrease the amount of water demand for each unit of economic activity, modify behaviors to use less water, reuse water, and recycle water.

- *Manage supply and demand differently.* Modify cropping patterns so that water use better matches hydropower releases or natural flow patterns. Change recreational water use to match hydropower or agricultural needs. Change hydropower from a peaking to a baseload energy provider.

The impacts of adaptations 1 and 2 mitigate the impacts of climate change and can be directly measured by using the models described previously. Primarily, changes in model parameters and, in some cases, changes in model results are fed to other models. However, the management adaptations are much harder to identify directly at a national level and can actually only be measured at a river basin level. River basin management models are available and can be easily used to analyze the management options proposed.

5.4.6 Data Requirements

The assessment methodology requires substantial reliable data to lend credence to the results being generated. The data required are a mixture of hydrometeorological and socioeconomic data. Much of the data can be found in published sources, while other estimates such as, technology and development in the base year 2075, require the educated guesses of experts. Listed below in general terms are the data needs for each of the assessment components followed by sources for these data.

5.4.6.1 Impacts on Hydrologic Resources—Supply

- Monthly river runoff values for each river basin to be studied:

  - Local hydro-met service or Global Runoff Data Centre Koblenz.

- Spatially averaged mean monthly values of temperature and precipitation covering the watershed of each river basin unit modeled for the time period that coincides with the time period from which the monthly stream flow values were calculated:

  - Local hydro-met service or IIASA Global Climate Database.

- Total and elevation area values for each river basin unit to be modeled:

  - Local survey/mapping service or IIASA Global Climate Database.

- GCM-generated spatially averaged estimates of the changes in mean monthly temperature and precipitation due to $2 \times CO_2$:

  - Provided by the Country Studies Program team (Section 4).

Issues associated with providing spatially averaged values and generating mean monthly values are discussed in detail in Appendix G.

5.4.6.2 Impacts on Water Uses—Demand

- Current water use for agricultural, domestic, industrial, and energy-related sectors for each river basin unit: local hydro-met service.

- Current economic activity indicators for agricultural, domestic, industrial, and energy-related sectors for each river basin unit: local economic/information service or United Nations/World Bank database.

- Estimates of economic activity level for base year 2075 for agricultural, domestic, industrial, and energy-related sectors for each river basin unit or annual growth rates: local economic/information service, United Nations/World Bank database, and local and Country Studies Program technical experts.

- Estimates of technological changes in water use reflected in modified water use coefficients: local economic/information service, United Nations/World Bank database, and local and Country Studies Program technical experts.

- Impacts of climate change scenarios on water use coefficients: other country studies biophysical assessments and local and Country Studies Program technical experts.

5.4.6.3 Impacts on Supply-and-Demand Balance—Vulnerability

- Estimates of current and future water resource infrastructure development, especially reservoir storage for each river basin unit: local and Country Studies Program technical experts.

- Estimates of current and potential interbasin transfers for each river basin unit: local and Country Studies Program technical experts.

5.4.6.4 Alternative Management Strategies—Adaptation

- Estimates of adaptations and costs in water use and management at a coarse national and river basin unit scale: local and Country Studies Program technical experts.

Also required are detailed hydrologic, engineering, economic, and institutional project data for current and future river basin components of the representative basin analysis. Reservoir storage for each river basin unit is also needed. Issues associated with obtaining detailed project level data are discussed in Appendix G.

5.4.7 Results Generated

The goal of this approach is to provide a comprehensive assessment of the vulnerability and the adaptability of the water resource system to climatic change, as defined by supplied GCM scenarios. Therefore, the two primary results to be presented are (1) a measure of the national water resource vulnerability using the Shuval index presented in Figure 5.4.4 and (2) the change in that index as a result of adaptation and a measure of the costs of those adaptations in monetary or utility terms. These indicators are obtained by analyzing the impacts of climate change on water supply, measured in physical quantities of monthly stream flow, and on water demand, measured in water use per economic activity.

Insights can be gained and conclusions drawn by analyzing the physical impacts on water supply and demand, but the approach focuses on providing insights for decision and policy makers and researchers and scholars as to the vulnerability of the managed water resource system to potential climate change.

5.4.8 Advantages and Disadvantages

The approach presented here has three main advantages. First, the techniques can be used with easily obtainable data. Second, the techniques are simple yet accurate, allowing the users to gain insight into the climate change impacts. Third, the techniques have been tried and tested for assessing climate change impacts on water resources for a range of hydroclimate regions.

The overall approach is limited because it is national and comprehensive in scope. This approach does not give detailed analyses of microlevel processes, which may be where most of the impacts of climate change occur. In terms of methodology, the mean monthly water balance model does not provide information on floods, droughts, or other submonthly hydrologic processes. In addition, groundwater impacts are not explicitly addressed in terms of impacts on recharge or increased demand for the resource as an alternative to surface resource. Glacial melt is not explicitly modeled in this approach. The temporal and spatial scale ignore many important hydrologic processes.

The demand analysis ignores many industrial sectors that may have local importance and does not provide for dynamic estimates of future economic activities. The vulnerability index has inherent in it all the failings of a single indicator that does not capture the nuances of local conditions. In summary, the major limitation of the approach is the coarse spatial and temporal scale at which the assessment is carried out.

5.4.9 Supplementary Approaches

Two different choices are available for addressing alternatives to the approach outlined above. The first is the choice of approach. A comprehensive assessment of the vulnerability of the "water resource system," which includes supply, demand, and management or an alternative assessment, focuses on hydrologic or water supply vulnerability. The current approach provides for a comprehensive but less detailed assessment of each component, while a hydrologic assessment would examine the more detailed biophysical processes and their relationship to climate change impacts on hydrology.

The second is the choice of methodologies for carrying out each of the component analyses. The alternative methodologies are described in detail in the appendixes. The key issues driving the applicability or appropriateness of methodologies are the temporal and spatial scale and the extent of data. Many modeling methodologies exist, but they are only as good as the data input. The methodologies presented above are believed to best meet the criteria given in Section 5.4.3.

5.4.10 References for Section 5.4

Berck, P., S. Robinson, and G. Goldman, 1991, "The Use of Computable General Equilibrium Models to Assess Water Policies," in A. Dinar and D. Zilberman (eds.), *The Economics and Management of Water and Drainage in Agriculture*, Kluwer Academic Publishers, Hingham, Mass.

Kaczmarek, Z., 1993, "Water Balance Model for Climate Impact Analysis," *ACTA Geophysica Polonica* 41(4):1-1.

Kulshreshtra, S.N., 1993, *World Water Resources and Regional Vulnerability: Impact of Future Changes*, RR-93-10, International Institute for Applied Systems Analysis, Laxenburg, Austria.

Shuval, H.I., 1987, "The Development of Water Reuse in Israel," *Ambio* 16:186-192.

Strzepek, K.M., and J.B. Smith, 1994, *As Climate Changes: International Impact and Implications*, Cambridge University Park (to be published).

5.5 COASTAL IMPACT AND ADAPTATION ASSESSMENT

An accelerated rise in sea level is one of the more certain responses to global warming and presents a major challenge to humankind. Rising sea level causes erosion, submergence, salinization, higher water tables, and a greater risk of impacts from flooding and storms (National Research Council 1987). However, a rise in sea level only manifests itself over long time periods (i.e., decades to centuries).

The most recent and comprehensive analysis of sea-level rise suggests that in the last 100 years (1880-1980), the eustatic rate of sea-level rise was 1.8 mm/yr (Douglas 1991). The relative or actual sea-level rise has often been much higher for many coastal areas because of the additional problem of land subsidence. Thus, without any change in present trends, significant problems exist in the coastal zone. Best estimates are that by 2100, the global rise will range from 0.33 to 1.10 m, with the most likely rise being 0.66 m (IPCC 1990). Thus, while the magnitude of future sea-level change is uncertain, the consensus is that the sea will rise in response to global warming. Simply knowing the direction of change greatly assists planning appropriate response strategies to such change.

Coastal scientists are increasingly being requested to assess the physical and societal impacts of sea-level rise and to investigate appropriate response strategies. Such assessments are difficult in many developing countries because they lack physical, demographic, and economic data, particularly appropriate topographic information, which are needed for the first (physical) phase of the analysis.

To overcome these difficulties, a new rapid and low-cost reconnaissance technique called "aerial videotape-assisted vulnerability analysis" (AVVA) has been developed (Leatherman et al. 1994). This technique has also been termed "aerial videomapping." Combining a video record of the coastline with ground-truth information characterizes the coastal topography and, with the use of an appropriate land loss model, allows estimates of the physical impact for different sea-level rise scenarios. Land loss estimates based on AVVA raise important issues such as the appropriate seaward limit of the beach profile. Response options also involve issues such as the long-term cost of seawalls. Therefore, realistic low and high estimates have been developed.

Subsequent to developing the AVVA approach, the IPCC (1992) formalized a seven-step procedure for vulnerability analysis (VA) to accelerated sea-level rise, termed the "common methodology" (Figure 5.5.1). The common methodology is a comprehensive, stepwise approach for assessing vulnerability to accelerated sea-level rise; it also assesses physical, ecological, and socioeconomic impacts in coastal zones. The common methodology uses scenarios for global changes, national development, and local response options. The AVVA method addresses the same problems and is one approach toward vulnerability assessment in the context of the common methodology. Specifically, AVVA supports step 2 of the common methodology, which is the delineation of case study and the collection of natural system data. These physical data, combined

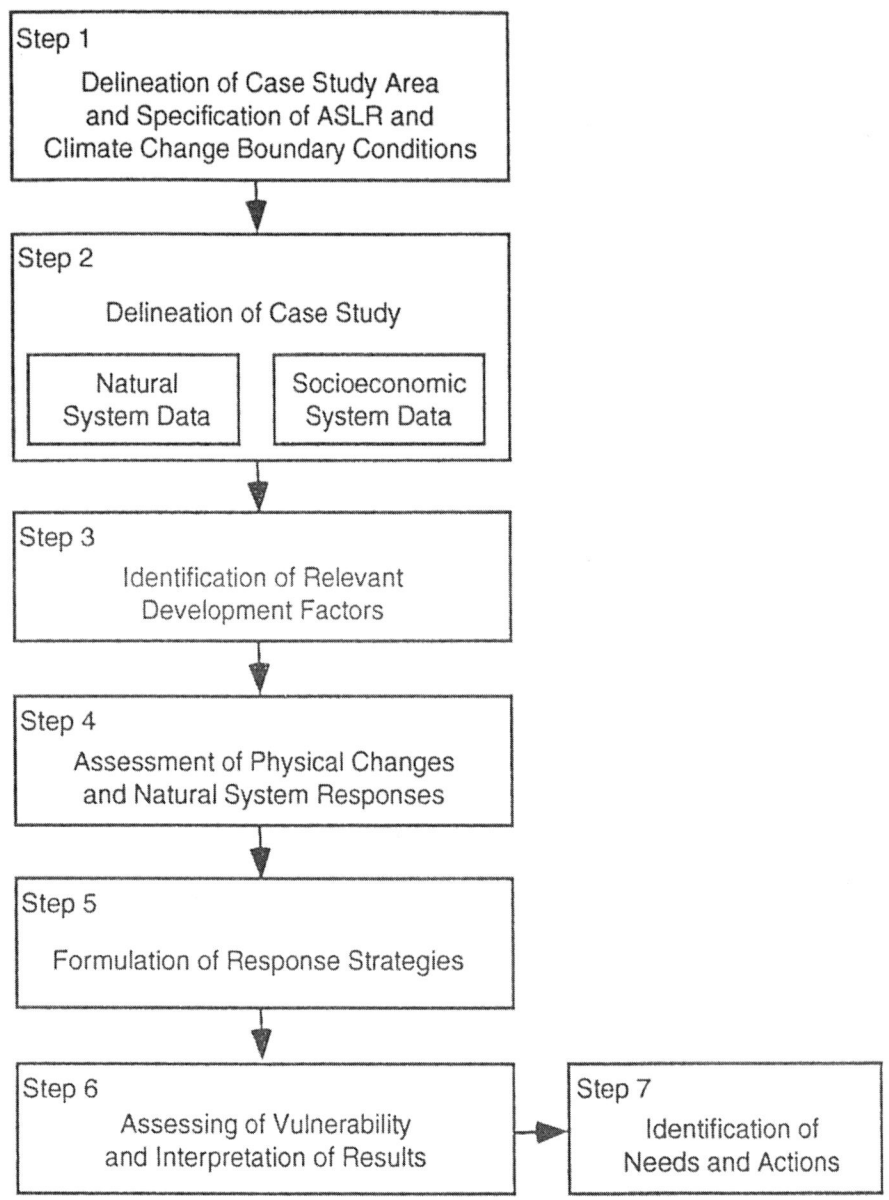

FIGURE 5.5.1 IPCC Common Methodology

with relevant socioeconomic information, promote identification of relevant development factors and, combined with land loss estimates, provide a quantitative assessment of physical changes and the vulnerability of the natural system (steps 3-6 of the common methodology). Thus, the IPCC common methodology (with AVVA) plus economic impact and adaptation analyses constitute the full primary approach to coastal impact and adaptation assessment (Figure 5.5.2).

The ability to conduct a VA depends critically on the availability of a range of physical, demographic, and economic data. In developed countries such as The Netherlands, sophisticated analyses have been possible, including developing new physical and modeling concepts (Peerbolte et al. 1991). A long observational database on coastal evolution was an important prerequisite for the successful completion of this analysis. However, in the

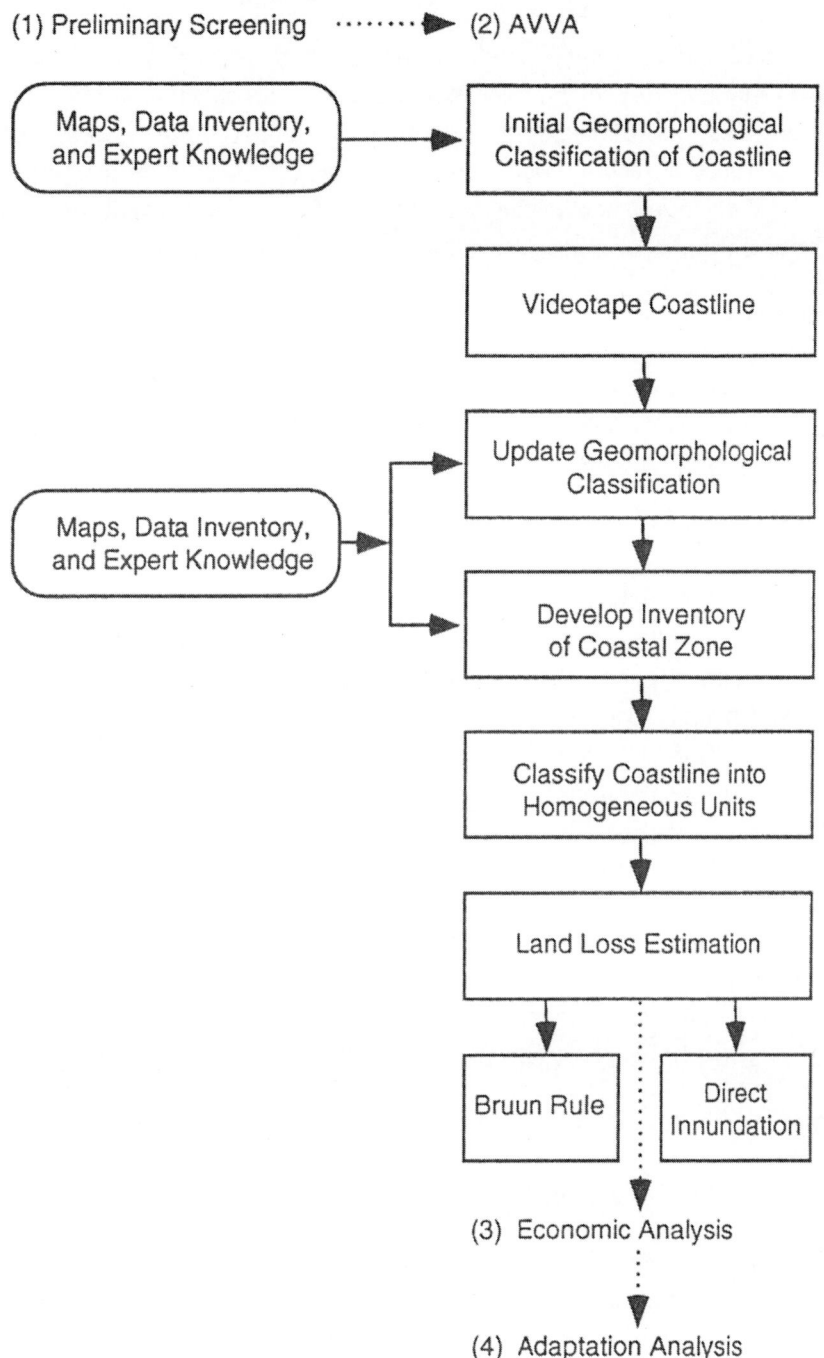

FIGURE 5.5.2 Primary Approach

developing world, the necessary data often do not exist or are found in inappropriate forms for VA, and many analyses are essentially qualitative. Clearly, more quantitative VA needs to be provided for such developing countries. Furthermore, these VAs should be completed in a reasonably short period (less than a few years). The approach described in this section is consistent with the IPCC common methodology and provides quantitative results for VA. Assessment of the vulnerability of coastal areas to climate change is one of the first steps toward better coastal zone management planning.

5.5.1 Primary Approach

Figure 5.5.2 illustrates the primary approach to be used in assessing the impacts of climate change on coastal areas. The primary approach consists of preliminary screening, AVVA, and vulnerability and economic analyses. These steps are discussed in more detail in Section 5.5.3.

5.5.2 Scope of the Analysis

Aerial videomapping can be applied to any coastal nation or part thereof. It is most appropriate for nondeltaic environments, where the shoreline is relatively straight and well defined. However, this approach can also be used in coastal marshy and estuarine areas. It may be advisable to videotape sandy beaches during the high-energy season, when the beaches are naturally narrow, to more clearly define the present vulnerability of coastal development. For countries with long and undeveloped coastlines (e.g., Indonesia), it is not cost-effective to videotape the entire coastline; overflights should be confined to areas of human habitation and areas of most importance and suspected high vulnerability to sea-level rise. For instance, the southern coast of Argentina is largely unpopulated, and therefore only the pockets of development were included in our surveys and national assessment.

Aerial videomapping can be undertaken from any small, single-engine plane, such as the Cessna 172, which is ubiquitous worldwide. Larger, more stable platforms can be used, but the low airspeeds (70-90 miles per hour) and very low flying elevations (30-50 m) severely limit their overall utility.

5.5.3 Description of the Methodology

5.5.3.1 Preliminary Screening Technique

The purpose of preliminary screening is to identify those areas within a country where detailed vulnerability and adaptation analyses may be needed and to develop some initial qualitative (or simple quantitative) assessments of a region's vulnerability to sea-level rise. Preliminary screening involves a general assessment of a country's vulnerability to climate change. The assessment may be based on expert judgment obtained by consulting in-country coastal specialists and/or by compiling information from maps, workshops, surveys, and agency reports. Also useful are summaries of previous research in the country on potential climate change impacts and adaptations. The findings from the preliminary screening should identify those regions within the country where more detailed analysis using the primary approach is required.

5.5.3.2 Aerial Video-Taped Vulnerability Analysis

One of the major problems with studying the impacts of sea-level rise in developing countries is the lack of detailed information on coastal geomorphology and the existing pattern and scale of development (IPCC 1990). Fundamentally, information on coastal elevations is lacking. For instance, the best topographic maps of Senegal have 5-m contours, but these maps cover local areas only (near the capital city, Dakar). Most maps have a 40-m contour interval, which is of limited value when analyzing land loss for any reasonable sea-level rise scenario. Other developing countries have similar problems. Furthermore, new topographic surveys using aerial photography and satellite imagery are prohibitively expensive and time-consuming, and, in the case of satellite imagery, a 10-m contour is the highest accuracy possible (Theodossiou and Dowman 1990). However, any available satellite data or archival photography should be used because they are a valuable source of supplementary information.

Aerial videotape-assisted vulnerability analysis was developed as a new, rapid, low-cost reconnaissance technique to overcome these difficulties. Focusing primarily on the land loss impacts of sea-level rise (erosion and inundation), AVVA involves (1) obliquely videotaping the coastline of low elevation from a small airplane, (2) making limited ground-truth measurements, and (3) performing archival research. Combining the video record with the ground-truth information characterizes the coastal topography and, using an appropriate land loss model, allows estimates of the physical impact of different sea-level rise scenarios. A feasibility study on the Chesapeake Bay demonstrated that the method is unbiased and suitable for reconnaissance surveys (Leatherman et al. 1994). A pilot study of Senegal demonstrated the utility of the method at a country scale (Nicholls and Leatherman 1994). Other national studies have been completed for Uruguay, Venezuela, and Argentina. These results give (1) national assessments of vulnerability to sea-level rise and (2) allow the attention of future studies to be directed to those areas most "at risk" from sea-level rise.

The field work procedure is as follows:

- Divide the coast into a working geomorphic classification by using published information and expert knowledge and obtain the best maps.

- Videotape the coastline at low and (sometimes) high altitudes of about 70/100 m and 300 m, respectively, including a recorded commentary and accurate positioning.

- View the videotapes and update the geomorphic classification, as appropriate.

- Visit as many representative coastal types as possible and collect topographic data and information concerning land and property values, agriculture, future plans for development, and so on.

The first stage of the analysis is to use the videotape record to develop an inventory of the coastal zone, including coastal geomorphology, coastal land use and development, and estimates of coastal elevation. Thus, the coastline can be classified into sections with similar characteristics, which can be considered homogeneous units. Subsequent analysis, such as estimates of land loss, uses procedures described in the following sections. The video record is still important to the analysis for the following reasons: (1) infrastructure losses are estimated directly from the video record, and any available maps, by overlaying the predicted recession and estimating the number of buildings that would be destroyed; and (2) the lengths of coastline requiring protection are measured from the video record.

Two important problems can emerge in preparing these assessments: quantifying land loss and estimating the cost of response measures. The Bruun Rule can be used to estimate land loss due to beach erosion on sandy coasts. However, uncertainties require construction of realistic high and low erosion estimates. The cost of response is also difficult to assess because of a lack of experience with the range of options. Again, realistic high and low estimates can be developed to overcome this difficulty.

5.5.3.3 Vulnerability Analysis: Estimation of Land Loss

Land loss is estimated by two different methods, depending on the shore type. For sandy beaches and erodible bluffs along open coast, the Bruun Rule is applied. For coastal wetlands and other nonoceanic coastal lowlands, the direct inundation or "drowning" concept is applied.

The sensitivity of coastal resources to accelerated rates of sea-level rise of 0.5 and 1.0 m by 2100 will be determined in accordance with IPCC (1990) guidance and the scenarios of Section 4. The 1-m scenario is considered the benchmark for long-term planning and coastal engineering design. Researchers may wish to use even higher rates of sea-level rise, such as 2.0 m, by 2100, to further understand sensitivities.

5.5.3.3.1 Erosion. The Bruun Rule is used to calculate erosion due to sea-level rise (Bruun 1962). The following form is used (Hands 1983):

$$R = GS[L/(B + h)] \, ,$$

where

R = shoreline recession due to a sea-level rise S;

G = overfill ratio;

S = sea-level rise (projected);

L = active profile width from the dune to the depth of closure;

B = dune or erodible bluff height (note $H = h + B$ in Figure 5.5.3); and

h = depth of closure.

These parameters are estimated from field data and available charts. The value of G is assumed to be unity (all eroded material is sand), except where material is not erodible (hard rock), so these recession estimates are minimum values. The depth of closure h (and hence the active profile width L) is the variable most difficult to estimate (Leatherman 1991a). In particular, it depends on time scale: the longer the period of interest, the larger the depth of closure. This fact is important to the analysis because if all things are equal, a larger value of h implies more beach erosion (and higher beach nourishment costs) (Figure 5.5.3). No accepted methods are available for estimating depth of closure over the long time scales of interest. Therefore, for calculational purposes, low and high estimates of closure depth are constructed.

5.5.3.3.2 Inundation. For wetlands and other coastal lowlands, a direct inundation or "drowning" concept is applied. Inundation is most significant in deltas and in wetlands around estuaries. For these low-lying areas, the video record is insufficient to define an estimate of elevation and must be integrated with available maps, plus expert judgment. This technique does not undervalue the video record because it (1) checks the validity of the maps and (2) helps define the present extent of wetlands and mangroves.

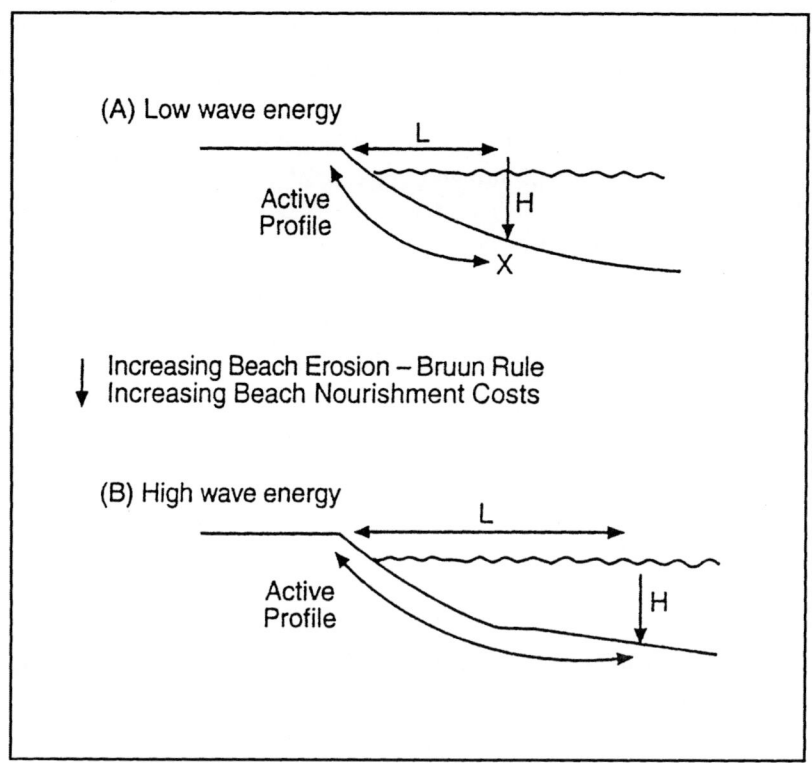

FIGURE 5.5.3 The Importance of Active Profile Width

Where possible, 1-m contours (above high water), plus the area of existing wetlands or mangroves, are estimated. Sedimentation in the inundated area is considered as best as possible. Coastal marshes and mangroves can accrete vertically in response to slow rates of sea-level rise and are only inundated above a certain threshold value of sea-level rise (Gehrels and Leatherman 1989; Ellison and Stoddart 1991). Deltas may or may not be receiving sediment from the river catchment. In most cases, the availability of sediment is poorly understood, and wetland evolution due to relatively small amounts of sea-level rise (up to 0.5 m) is difficult to assess. Often the best estimate of land loss is a linear interpolation of land loss from the higher scenarios (e.g., 2 m) of sea-level rise.

5.5.4 Economic Analysis

The ultimate goal of any economic analysis of future sea-level rise is to compute time trajectories of true economic cost in the context of future social, cultural, institutional, and economic development within threatened coastal zones. It will, therefore, eventually become necessary to project time series of the future values of all of the important economic activities that integrally depend on the physical characteristics of coastal zones and to judge their inherent vulnerability to sea-level rise along specific scenarios. These time series will have to incorporate any socioeconomic adaptation that might be expected in response to projected vulnerability, and they will have to be cast against the costs of more proactive protective strategies to determine which activities should be preserved, to some extent, and which should be abandoned.

Questions of timing are paramount in analyzing protection decisions, and answers depend on both the quality of information about costs, values, and future physical impacts and the degree to which the affected populations believe that decisions to protect will be carried through. Time series of true economic cost can be produced by carefully considering how these factors will move into the future. The result should, in particular, be a series of statistics that includes the cost of future economic activities that are abandoned to rising seas (net of the savings that accrue due to adaptation in anticipation of abandonment, including the cost of that adaptation) and the costs associated with protecting other, more favored activities.

Methods are being designed to complete all of these tasks, starting with developed coastal zones in the United States where advanced real estate markets work well to incorporate natural threats (like earthquakes and/or flooding) into the economic depreciation of property and where large stores of investment capital are available to underwrite deserving protection projects (Yohe et al. 1994). The general methodology that frames this initial work will be extended to accommodate other social, cultural, and economic structures. However, it is not necessary to wait for that extension to begin analyses in countries that do not exactly fit the underlying advanced market economy context of the United States model (Yohe 1993). Any analysis that eventually turns to computing true economic cost will be well informed and efficiently directed by preliminary investigations of economic vulnerability, which produce a series of statistics computed in terms of the current value of current economic activities and assets where methods like AVVA are highlighted.

Appendix H records the details of a systematic procedure designed to focus on economic assets. Table 5.5.1 outlines the general procedure for estimating economic vulnerability. Subsequent work may be required to capture the current values of activities that depend on threatened coastal zones—dependence that should be clearly defined either in terms of the services and support provided to economic activity by the physical characteristics of coastal zones or by virtue of the physical location of an asset or activity within their boundaries. To a first-order approximation, though, the values of economic activity should be embodied in the values of assets that the preliminary procedure is designed to capture.

TABLE 5.5.1 Economic Analysis Steps

Step	Description
1	Define a working list of "coastal units"
2	Specify sea-level rise scenarios
3	Calibrate baseline sea-level rise
4	Compute total sea-level rise
5	Determine the economic value of land loss
6	Determine the economic value of threatened structure
7	Compute the total vulnerability
8	Estimate the cost of protection
9	Determine the aggregate vulnerability/protection

Armed with raw vulnerability data, researchers can move forward by formalizing links between potentially significant sources of change in economic value and changes in the physical character of coastal zones that might be associated with sea-level rise. Tracking scenarios of these physical changes into the future along specific sea-level rise trajectories is then all that is required to correlate rising oceans to time series of economic vulnerability (e.g., Yohe 1990, 1991).

Time series of economic vulnerability can support an extremely informative first- order review of response options. They reflect the potential cumulative economic vulnerability of coastal zones when people decide to do nothing—protecting nothing in an indiscriminate retreat from the sea. Vulnerability series can also support preliminary evaluations of various alternative strategies. Continuing only current protection policies, protecting all economically developed and/or exploited properties and resources, and protecting the entire coastline are three obvious candidates on a potentially long list of possible options. Comparisons of discounted cumulative vulnerability with the discounted cost of protection for any option can, more specifically, support a preliminary sorting of coastal zones into three categories:

- Zones for which the need for protection seems clear because the discounted stream of vulnerability is so much greater than the discounted stream of costs along every potential sea-level trajectory;

- Zones for which the case for abandonment seems equally clear because the discounted stream of vulnerability is so much smaller than the discounted stream of costs along every potential sea-level trajectory; and

- Zones for which the protection decision is not clear because the relative size of the discounted streams of costs and vulnerability depends critically upon the choice of discount rates, sea-level trajectories, and other parameters.

While careful, cost-based analysis would certainly be required to confirm that assessments based on vulnerability estimates accurately place one zone or another in either of the first two categories, estimating true economic cost statistics that can be used in lieu of economic vulnerabilities in the planning protection strategies will pay the greatest dividends if they are first developed for and applied to zones that initially fall into the last category.

5.5.5 Adaptation Techniques

The next step in the vulnerability analysis is response planning and coastal zone management. While vulnerability analysis quantifies the serious consequences of accelerated sea-level rise, such analyses do not indicate the optimum response option.

For the vulnerability analysis, four response options are considered:

- *No protection.*

- *Present protection.* Existing protection is maintained.

- *Developed areas protection.* Medium to highly developed areas are protected (i.e., cities, tourist beaches, factories). Using the video record, medium-to-high development is defined as areas with more than 15% of the ground covered in structures. For tourist areas, beach nourishment is assumed. Elsewhere, seawalls are used.

- *Total protection.* All coastal areas with a population greater than 10 people/km^2 are protected (IPCC 1990). Seawalls are only used to protect developed areas because of the high cost of such structures.

5.5.5.1 Nourishment and Seawall Designs

The protection options should be developed as follows: for seawalls and groins, the cost of purchasing, transporting, and placing rock, plus the cost of design (10%) and maintenance (20%) determine the total cost.

5.5.5.2 Beach Nourishment

The volume of beach fill is calculated by raising the entire active profile by the sea-level scenario; however, as with beach erosion, the active width is difficult to determine (Hallermeier 1981; Leatherman 1991b). In this case, it is logical to use the low- and high-active widths defined for the Bruun Rule as low- and high-cost designs, respectively.

An additional cost on long open beaches is longshore loss of sand out of the nourished area (Dean 1983). To stop such losses, large terminal groins are proposed at the ends of each nourishment project, which, in effect, converts the open beach into a pocket beach. Costs do not rise uniformly as sea level rises because the groins are a large, but declining proportion of the total cost as the sea-level rise scenario is raised. While this procedure significantly raises the costs of beach nourishment, the resulting costs are more realistic than simply considering the cost of a single beach fill (Titus et al. 1991).

5.5.5.3 Seawalls

To calculate the cost of erecting a seawall, three simple seawall designs were developed. The type of design used depends on the wave environment. These designs are a low- and high-cost seawall for open (or wave-exposed) coasts and a seawall for sheltered coasts. They all use 1:2 slopes and a 2-m berm.

The low-cost, open-coast (LCOC) and high-cost, open-coast (HCOC) designs are necessary because it is uncertain how a beach in front of the seawall will respond to a rise in sea level. In the best case (lowest cost), the beach would not erode at all; in the worst case (highest cost), the beach would be completely lost. Actual behavior is expected to fall somewhere in between these extremes, with total beach loss being more likely with higher sea-level rise scenarios. The LCOC design assumes no beach erosion and only considers the effects of a more severe nearshore wave climate. The HCOC design assumes total beach loss and the consequent need to prevent undermining of the wall. The size and cost of the HCOC design increase with wave climate. Thus, the LCOC and HCOC designs reflect the two cost extremes. The sheltered coast design is lower than the LCOC design because it would have limited problems with a less severe nearshore wave climate. In general, existing seawalls are uncommon in the countries being considered, and the problems of costing upgrades of existing structures are relatively minor.

A dynamic analysis of response planning for different sea-level scenarios involves considerable methodological development and must consider the following:

- Timing of the response,

- Uncertainty concerning future sea levels, and

- Other changes to the coastal zone, such as increased development as populations expand.

5.5.6 Data Requirements

The following data sets should be collected to facilitate the sea-level rise impact analysis using the AVVA approach:

- Topographic maps, preferably with at least 3-m contour intervals;

- Aerial photographs;

- Satellite imagery (not required, but useful for geomorphic interpretation, if available);

- Bathymetric maps, particularly of the nearshore areas (less than 10 m of water);

- Tide gauge data to determine long-term sea-level rise and storm tides and frequency;

- Wave data, either from gauges or shore observations over the longest period of record possible;

- Historical data, including flood maps, photos taken during floods, and storm impact data; and

- Historical shoreline change data to calculate erosion rates where accurate and available.

- Sediment impoundment through construction of dams on rivers, seawalls and jetties (or proposed projects) that could adversely affect the sediment budget of a beach (National Research Council 1990).

5.5.7 Results Generated

The AVVA approach permits the following determinations:

- *Index of terrain and relief changes.* A cross-sectional view along the coastline enables a subjective estimate of the relative topography of the coastline.

- *Types of coastal environments.* The size and location of wetlands, sandy beaches and dunes, hard rocky coastlines, and so on, have been identified. This information is useful for determining the dominant land loss mechanism (erosion or inundation).

- *Land use practices.* Coastal land use such as agricultural plots and aquacultural facilities have been identified.

- *Infrastructure.* The coastal infrastructure has been recorded.

- *Population.* Quantitative indicators of the population living in the coastal zone have been identified.

5.5.8 Advantages and Disadvantages

The AVVA is a rapid and low-cost method to conduct reconnaissance vulnerability analyses at large scales. It provides a focus on which areas can be most usefully studied with more conventional and expensive techniques and hence maximizes the results from a limited national budget for such an analysis. The flexibility of the method allows rapid mobilization, for example, after the passage of a hurricane.

The videotape record provides much useful information on land use, infrastructure, and "brings the coast to your desk." Therefore, it may be useful even if high-quality topographic data are available. Because AVVA has been constructed as a series of modules whereby individual elements of the overall procedure can be improved, the remaining elements of the approach stay robust. Therefore, as understanding improves, vulnerability assessment can be easily and rapidly repeated to generate new and improved estimates of the impacts of sea-level rise.

Many possible improvements to the procedures described are possible and would help to more precisely define impacts and appropriate responses to sea-level rise. In particular, improved understanding of beach erosion would be useful as well as the costs of beach nourishment and seawalls. In this regard, better wave data are essential. In addition, a more explicit description of the time element of the problem is required, including factors such as economic growth and development.

Aerial video techniques are also useful for more detailed studies when high-quality topographic data are available. However, other problems remain, such as the appropriate seaward limit of the active beach profile. These problems are independent of the data source used. Procedures have been developed to provide a realistic range of values for these estimates. Further work is required on these problems, including establishing observational databases for the world's coastlines, in particular a meaningful description of wave climate.

The most important limitation of the AVVA approach is that, unlike traditional procedures involving aerial photographs, parallax is not used to estimate elevations. Instead, researchers use the video record to classify the coastal landforms and subjectively interpolate elevations between occasional ground-truth data (either reliable spot heights or surveyed transects). Clearly, this procedure would not provide an accurate contour for anyone who needed to know whether a particular parcel of land would be inundated. However, validation experiments in Chesapeake Bay, Maryland, have demonstrated that this method is unbiased and reasonably accurate when estimating land loss at the large scale.

5.5.9 Supplementary Approaches

For projecting future erosion rates in response to sea-level rise, it is important to have historical shoreline change data. However, this projection depends on the availability of high-quality and large-scale historical maps and aerial photographs. If such data exist, the metric mapping GIS program can calculate erosion rates and at least verify and calibrate the Bruun Rule computations. Trend analysis can also predict future erosion rates on the basis of the historical trend. While the primary approach focuses on physical impacts, this initial work provides an opportunity to look at the broader needs of more extensive, supplementary approaches. The end goal is "integrated coastal zone management" (ICZM). Considerable work is needed, however, to design ICZM approaches based on the site specifications of individual countries.

Several stylized methodologies have been applied to coastal zone management, and these methods could be generalized. The approaches should support the beginning of an applicable integrated approach to issues of timing, adaptation, and uncertainty that could form the basis of a procedures handbook. Ultimately, the key to assessing the cost of sea-level rise, and thus, to judging the efficacy of adaptation under any social metric, is to produce dynamic portraits of the social and economic activities that would be hurt or helped by rising sea levels and by any adaptation that might be considered. This general structure will not change, and so future requirements can be anticipated now as necessary methodologies are being refined.

Efforts to collect data should reflect future needs so that the collection of supplementary data is contemporaneous with the data required for the primary approach. It is important that continuity exists in georeferencing and temporal boundary conditions. These additional data are primarily social-scientific and include (1) anticipated changes in population and population density; (2) anticipated future development, value appreciation or depreciation, and descriptions of private, social, and political decision-making rules and/or norms; (3) values and uses of (slightly) inland buildings and real estate; (4) alternative uses of coastal zone property gleaned from down (up) coast and resources; and (5) correlations of coastal zone and inland systems with other economic activity.

5.5.10 References for Section 5.5

Bruun, P., 1962, "Sea-Level Rise as a Cause of Shore Erosion," *American Society of Civil Engineers Proceedings. Journal of Waterways & Harbors Division* 88:117-30; 1983, "Review of Conditions for Uses of the Bruun Rule of Erosion," *Coastal Engineering* 7:77-89.

Dean, R.G., 1983, "Principles of Beach Nourishment," in P.D. Komar (ed.), *Handbook of Coastal Processes and Erosion*, CRC Press, Boca Raton, Fla.

Douglas, B.C., 1991, "Global Sea-Level Rise," *Journal of Geophysical Research* 96:6981-6992.

Ellison, J.C., and D.R. Staddart, 1991, "Mongrare Ecosystem Collapse during Predicted Sea-Level Rise: Holocene Analogous and Implications," *Journal of Coastal Research* 7:151-166.

Gehrels, R., and S.P. Leatherman, 1989, *Sea-Level Rise: Animator and Terminator of Coastal Marshes*, Vance Bibliography, Monticello, Ill.

Hallermeier, R.J., 1981, "A Profile Zonation for Seasonal Sand Beaches from Wave Climate," *Coastal Engineering* 4:253-77.

Hands, E.B., 1983, "The Great Lakes as a Test Model for Profile Responses to Sea Level Changes," in P.D. Komar (ed.), *Handbook of Coastal Processes and Erosion*, CRC Press, Boca Raton, Fla.

IPCC, 1990, *Strategies for Adaption to Sea-Level Rise*, Report of the Coastal Zone Management Subgroup, Intergovernmental Panel on Climate Change, Response Strategies Working Group, Rijkswaterstatt, the Netherlands.

IPCC, 1992, *Global Climate Change and the Rising Challenge of the Sea*, Report of the Coastal Zone Management Subgroup, Intergovernmental Panel on Climate Change, Response Strategies Working Group, Rijkswaterstatt, the Netherlands.

Leatherman, S.P., 1991a, "Coasts and Beaches," in G.A. Kiersch (ed.), *The Heritage of Engineering Geology: The First Hundred Years,* Geological Society of America, Boulder, Colo., Centennial Special Issue, Vol. 3.

Leatherman, S.P., 1991b, "Modeling Shore Response to Sea-Level Rise in Sedimentary Coasts," *Progress in Physical Geography* 14:447-464.

Leatherman, S.P., R.J. Nicholls, and K.C. Dennis, 1994, "Approaches to Vulnerability Analysis for Accelerated Sea-Level Rise," *Journal of Coastal Research* Special Issue No. 14 (in press).

National Research Council, 1987, *Responding to Changes in Sea-Level: Engineering Implications,* National Academy Press, Washington, D.C.

National Research Council, 1990, *Managing Coastal Erosion,* National Academic Press, Washington D.C.

Nicholls, R.J., and S.P. Leatherman, 1994, "Sea-Level Rise," in *As Climate Changes: Potential Impacts and Implications,* Cambridge University Press (in press).

Peerbolte, E.B., J.G. de Ronde, L.P.M. de Vrees, B. Mann, and G. Baarse, 1991, *Impact of Sea-Level Rise on Society: A Case Study of the Netherlands,* Delft Hydraulics.

Theodossiou, E.I., and I.J. Dowman, 1990, "Heighting Accuracy of Spot," *Photogrammetric Engineering and Remote Sensing* 56:1643-1649.

Titus, J.G., R.A. Park, S.P. Leatherman, J.R. Weggel, M.S. Green, P.W. Mausel, S. Brown, C. Gaunt, M. Trehan, and G. Yohe, 1991, "Greenhouse Effect and Sea-Level Rise: Potential Loss of Land and the Cost of Holding Back the Sea," *Coastal Management* 19:39-58.

Yohe, G., 1990, "The Cost of Not Holding Back the Sea—Toward a National Sample of Economic Vulnerability," *Coastal Management* 18:403-431.

Yohe, G., 1991, "Uncertainty, Climate Change and the Economic Value of Information: An Economic Methodology for Evaluating the Timing and Relative Efficacy of Alternative Responses to Climate Change with Application to Protecting Developed Property from Greenhouse Induced Sea Level Rise," *Policy Sciences* 24:245-269.

Yohe, G., P. Marshall, J. Neumann, and H. Ameden, "The Economic Cost of Greenhouse Induced Sea Level Rise: National Estimates for the United States," Electric Power Research Institute (forthcoming 1995).

Yohe, G., et al., 1994, "Assessing the Economic Cost of Greenhouse Induced Sea Level Rise: Methods and Application in Support of a National Survey," *Journal of Environmental Economics and Management* (forthcoming 1995).

5.6 SOIL EROSION IMPACT AND ADAPTATION ASSESSMENTS

Maintaining the soil resource is an essential step in sustaining the productivity of agricultural and grassland systems. Removal of soil by wind or water can reduce productivity of these lands, sometimes to a significant degree. Soil erosion can also cause secondary environmental effects, such as increased dust in the air or increased transport of sediments to rivers and lakes. Increased dust (particulate matter) can have an adverse effect on human health. Increased sediment transport can cause many adverse effects such as reducing fisheries, increasing flooding, hampering transportation by boat, and decreasing hydroelectric power.

Changes in climate and CO_2 can change the rates of soil erosion by wind or water. Soil erosion can be directly affected by changes in the quantity and quality of erosive forces; i.e., the amount and intensity of precipitation (erosion by water), and wind speed and direction (erosion by wind) on an event basis. Changes in climate and CO_2 can indirectly affect erosion through effects on the degree and timing of crop cover, and the production and decay of residue. Changes in soil water content, as affected by changes in the ratio of precipitation to evapotranspiration, can also influence erosion. Generally, water erosion increases (Wischmeier and Smith, 1978) and wind erosion decreases (Woodruff and Siddoway, 1965) as soil becomes wetter. Water erosion and wind erosion both tend to be dominated by extreme events which might occur only rarely.

The most important factor influencing soil erosion on agricultural and grasslands is management by the people using the land. Practices that disturb the soil the least, leave the greatest amount of plant residue on the soil surface, and minimize the time without crop cover tend to cause the least amount of erosion. Management practices such as terracing might be used specifically to decrease soil erosion. Thus, an assessment of the vulnerability of these lands to soil erosion under changed climate must take into account management practices, erosive forces of wind and water, and changes in crop cover and residue. It must also account for typical and extreme events.

This project uses a widely used model which integrates the necessary features, the Erosion/Productivity Impact Calculator (EPIC). EPIC was developed by the U.S. Department of Agriculture to analyze the relationship between soil erosion and agricultural productivity. It uses a daily time-step to simulate crop growth, soil erosion by wind and water, nutrient cycling (including soil carbon), and hydrology. The model is capable of simulating non-point source pollution of surface and groundwaters by agricultural pesticides and nutrients. It allows for detailed specification of management practices. Detailed descriptions of EPIC modules, validation, and input requirements are contained in the model documentation and users' manual (Sharpley and Williams, 1990a,b), and in the draft User's Guide.

5.6.1 Primary Approach

The primary approach consists of (1) a preliminary screening technique to identify regions and systems of concern, and (2) application of simulation techniques to assess vulnerability for changed climate, CO_2, and management regimes.

5.6.1.1 Preliminary Screening Technique

The purpose of the preliminary screening is to identify and prioritize regions and systems for simulation analysis. To the greatest extent possible, the screening for erosion assessment should build on the screenings for crop and grassland assessments. Wherever possible, common sites and datasets should be used.

Three categories should be considered high-priority for erosion assessment:

- Major crop or grassland systems. It's likely that these will be identified and analyzed as part of the crop and grassland assessments, and that much of the data needed for the erosion simulations will be gathered for these assessments. Because of the economic importance of these systems, they should be considered for erosion assessment.

- Currently impacted systems. Systems for which erosion is currently a problem should be considered, especially systems where a moderate problem might become a severe problem. These can be identified by consulting with erosion experts, or, more coarsely, by using databases such as the Global Assessment of Soil Degradation (GLASOD) database (Oldeman et al., 1990).

- Systems vulnerable to vegetation change. Systems that are vulnerable to changes in vegetation might also be vulnerable to increased erosion if a decrease in vegetation cover is likely. For example, a grassland might become more desert-like. Changes in crop management might also affect erosion. For example, an area that shifted from continuous wheat to wheat-fallow might be more vulnerable to erosion because of increased periods of uncovered soil.

5.6.1.2 Simulation Techniques

EPIC consists of several linked sub-models. Most of the sub-models used in EPIC were adapted from earlier, stand-alone models. Many of these, and some of the input data, are very similar to those in the SPUR2 grassland model.

- *Vegetation:* A generic model, parameterized for specific crops (Table 5.6.1), is used to simulate crop growth. Capture of photosynthetically active radiation (PAR) is modeled as a function of date, latitude, cloudiness, and leaf area index. Daily conversion of PAR into biomass, and allocation into root, above-ground, and yield components, is constrained by plant development and by water, nutrient, and temperature stress. The crop model has been modified to incorporate the direct effects of CO_2 and water deficit on net primary production (NPP) and water use (Stockle et al., 1992).

- *Water erosion:* Soil erosion on an event basis is simulated by the Universal Soil Loss Equation (USLE; Wischmeier and Smith, 1978), Modified Universal Soil Loss Equation (MUSLE; Williams, 1975), or by the Onstad-Foster modification of USLE (Onstad and Foster, 1975). All three equations used by EPIC calculate soil erosion rate as the product of factors for rainfall or runoff energy (R), percent slope (S), slope length (L), soil erodibility (K), cover (C), and erosion control practice (P). For USLE, R is determined by rainfall variables, which provides an estimate of erosion that includes soil moved only a short distance within a field. USLE requires use of a sediment delivery ratio to estimate net erosion off-site (Robinson, 1979). For MUSLE, R is based on runoff rather than rainfall, and was designed to more closely represent net erosion (Williams, 1975). S, L, and P are specified by the user. C and R are calculated by EPIC from the simulated crop and residue cover, and from the simulated daily weather, respectively. The K factor is based on the characteristics of the soil surface layer. Over time the K factor changes as surface layers are eroded away and mixed with deeper soil layers.

TABLE 5.6.1 Crops currently parameterized for EPIC

Soybeans	Peanuts
Corn (Maize)	Rice
Grain Sorghum	Potatoes
Winter Wheat	Winter Peas
Spring Wheat	Lentils
Durham Wheat	Sorghum Hay
Barley	Alfalfa
Oats	Range
Sunflowers	Summer Pasture
Stripper Cotton	Winter Pasture
Picker Cotton	Pine Trees

- *Wind erosion:* Wind erosion is modeled by a variation of the Wind Erosion Equation (WEQ; Woodruff and Siddoway, 1965) modified to simulate daily soil losses (Cole et al., 1983). Daily soil loss depends on field length, width, and orientation; daily vegetation and residue cover; ridge roughness; precipitation and evapotranspiration; and daily wind energy.

- *Soil dynamics:* EPIC simulates soil dynamics in each of ten layers, derived from the two to four horizons listed in the input data. Organic matter decomposition and release of nutrients are driven by temperature, moisture, C:P or C:N ratio, and residue condition in each layer as simulated daily by EPIC. Two major soil C pools, corresponding to fresh residue and soil humus, are defined. Fresh residue is further divided into carbohydrate-like, cellulose-like, and lignin-like pools, depending on the stage of decay. The main mechanisms for introduction of fresh residue into a layer are incorporation of surface residue through tillage, and root growth and death within the layer. Soil humus is sub-divided into stable and active pools, initially based on how long the soil has been cultivated.

- *Weather:* EPIC uses a stochastic model to generate daily weather from monthly climatic parameters. Theoretically, a 100 year simulation should capture rare as well as typical events. This is important for erosion, which is mainly driven by relatively rare but large events.

5.6.2 Description of the Methodology

The methodology consists of six steps. The first step is the preliminary screening, described above. This screening will identify and prioritize the regions and management systems that will be simulated. More than one system might be used to grow a specific crop in a given region. For example, wheat might be grown continuously, or in a wheat-fallow rotation. Corn might be grown using conventional tillage or reduced tillage. Each way of growing the crop is a separate system.

The second step is to develop input files for representative sites. The ideal situation would be to use actual sites for which data are available, preferably the same sites used for the crop and grassland assessments. An alternative is to define theoretical sites based on a typical soil, management system and climate for the region. An important part of Step 2 is developing management schedules describing what operations are used in growing the crops, and when they are used. In some situations, it might be desirable to also model management operations which are recommended by agronomists, but which might not have been adopted by farmers.

In Step 3, EPIC is run for each site for 100 years. The reason for using a long simulation time is to include rare events of large magnitude, as well as typical events. If default values are used for the input data, the model should be run for somewhat longer (for example, 102 years). The purpose of the extra years is to allow the model to stabilize; the results from these extra years at the beginning of the run should not be used.

Step 4 repeats the simulations for future climate and CO_2 conditions. The future conditions used for erosion assessment should be chosen from among those used for crop and grassland assessments; however, it may not be necessary to use all the climate/CO_2 scenarios. The management schedules should be changed to include likely or necessary adaptations indicated by the crop and grassland assessments.

Step 5 is the vulnerability assessment. Regions or systems would be considered vulnerable if soil erosion causes reduced productivity or unacceptable secondary environmental impacts such as increased duststorms or transport of sediment to lakes or rivers. If the results from Step 4 do not indicate that soil erosion will be a significant problem in the future, no further analysis is needed. Otherwise, it is necessary to proceed to the adaptation assessment.

Step 6, the adaptation assessment, should focus on those regions and systems which Step 5 identified as vulnerable to unacceptable rates of soil erosion. Management schedules should be modified to include practices specifically designed to reduce erosion. These might range from relatively simple changes, such as reducing tillage to maintain crop residue on the soil surface, to major changes in cropping systems. These changes should be judged in terms of their practicality and likelihood of being implemented, as well as the decreases in erosion indicated by the simulation results using the revised management schedules.

5.6.3 Defining the Scope of the Analysis

The spatial scope will be defined when the regions and systems of interest are identified by the initial screening. The temporal scope is the time of the $2 \times CO_2$ climate. The long time for simulations (100 years) is needed to include the potential effects of rare, large events. Multi-generational simulations (i.e., several hundred years) of selected soils are used to characterize productivity over the useful life of the soil.

5.6.4 Economic Analysis

The economic effects of decreases in yield associated with soil erosion should be analyzed using the same procedures used in the crop and grassland analyses. The costs of implementing management practices to ameliorate erosion should be estimated. The monetary and non-monetary costs of the secondary effects of erosion (e.g., sediment transport to lakes and rivers, duststorms) should be considered, if possible.

Multiple generation impacts should also be assessed on major soil groups to identify soil productivity relationships as the upper soil layers are eroded away. This analysis will identify the point at which the rate of productivity loss increases at an increasing rate. This can be used as an indicator of a critical cumulative soil loss.

Climate changes may change the shape of this multi-generation productivity curve and shift the critical point. Historically, some civilizations have been dramatically impacted as these critical soil loss levels were reached and exceeded (Benson et al., 1989).

Productivity losses such as those illustrated in Figure 5.6.1 may only be identified after hundreds of years of erosion. Each point in Figure 5.6.1 on the yield-cumulative erosion curve represents 100 years of erosion. Note that although there is little yield change in the first 600 years, we have moved to a point that any further erosion will cause considerable reduction in productivity. This rapid loss begins after the first 300 mm of soil have been eroded away. There is a 10 to 20 percent yield loss with the next 300 mm of soil eroded and nearly a 30 percent loss with the next 300 mm of soil loss.

Many countries around the world have had land in cultivation for hundreds or even thousands of years and may be approaching or be beyond the critical point described above. This multi-generation analysis illustrates that societies may gain more from a small amount of preventive conservation before this threshold is reached than from considerable soil reclamation.

5.6.5 Adaptation Techniques

There are a great number of practices which could be adopted to decrease erosion by wind or water. For example, increasing the amount of crop residue on the soil surface or decreasing the amount of time without vegetation cover can make soil more resistant to erosion. Terracing and other manipulations can reduce the erosive force of runoff. In some cases, it might be necessary to retire the land from cropping, perhaps by establishing forest or pasture. For grasslands, it might be necessary to reduce grazing intensity.

5.6.6 Data Requirements

At first glance, EPIC seems to require a very large amount of input data. The minimum data set the user needs to supply is, however, much smaller because (1) standard data sets have been developed, and (2) the program can estimate many parameters. Detailed descriptions of the input data are given in the User's Guide.

- *Weather:* The stochastic weather generator used by EPIC requires monthly parameters describing temperature, precipitation, and wind. The basic data set needed for each site is a 30 year continuous record of daily maximum and minimum temperature, precipitation amount, and solar radiation; a program supplied with EPIC generates the required input from these data. For wind, monthly mean speed at a 10 m height and frequency of direction are needed. EPIC weather data sets have been developed for many locations.

- *Crops:* EPIC has been parameterized for the crops listed in Table 5.6.1. These data should be adequate for most cases.

- *Soil:* EPIC can accept up to 20 parameters for 10 soil layers. However, only 6 parameters are required: depth, percent sand, percent silt, pH, percent organic carbon, and percent calcium carbonate. Specifying data for 2-3 layers should be adequate for most applications. EPIC soil data sets have been developed for many soils.

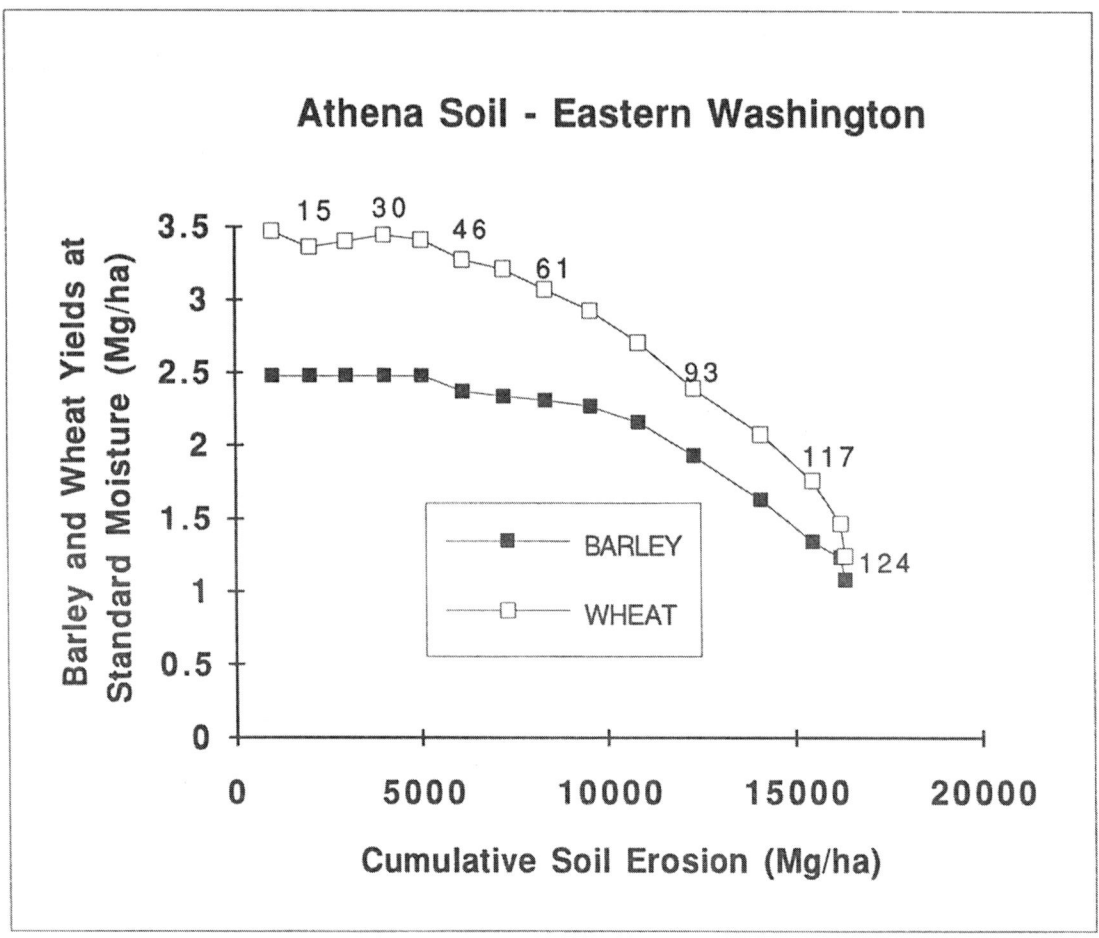

Figure 5.6.1. The cumulative effects of soil erosion by wind and water on yield of wheat and barley on the Athena soil of eastern Washington, USA, as simulated by EPIC. Numbers next to data points are depth of soil that has been eroded (cm).

5.6.7 Results Generated

The focus of this assessment is on the amount of soil removed by erosion, and the possible effects on yield. Erosion rates should be captured monthly and annually, and yield captured annually. It is also useful to output values of soil variables at the end of the simulation. Multi-generation simulations focus on results of successive 50 or 100 year periods.

The effect of changes in climate, CO_2, or management can be characterized by calculating the means, standard deviations, and standard errors from the 100 annual values of water, wind, and total erosion. It is also possible to get insights into the effects on extreme events. If 100-year simulations are used, then the largest annual value represents a 100-year event, the fifth largest a 20-year event, and the tenth largest a 10-year event. The effects on extreme monthly erosion can be characterized by selecting the largest monthly value for each year, and using these values in place of the annual values.

The effects of soil erosion on yield can be characterized in several ways. One way is test for trends in yield by regressing yield versus year. A second approach is to compare the average yields for the first and last ten years. Either of these can detect large changes in yield associated with erosion. However, both approaches could give false results if the average weather simulated by the stochastic weather generator changes with time. The preferred approach is to repeat a simulation using the values of the final soil variables as initial conditions, and comparing yields for the first ten years. This procedure forces the weather to be the same, so that any differences in yield can unambiguously be attributed to changes in soil conditions. In addition, multiple successive 100-year periods can be examined to plot the potential soil productivity curve under current and changed climate.

5.6.8 Advantages and Disadvantages

This approach has the advantage of integrating the direct effects of changes in climate on wind and precipitation with the indirect effects of changes in crop and residue cover. It incorporates the effects of rare, large events as well as typical events. It provides an analysis of specific changes in management.

The disadvantage is that, because a large amount of data is required, it may be necessary to rely on default or standard data sets. This should be adequate for comparative analyses.

5.6.9 Supplementary Approaches

The USLE, MUSLE, and Revised USLE (Renard and Ferreira, 1993) can be used independently of EPIC to estimate the direct effects of climate change on water erosion (Phillips et al. 1993), and WEQ can be used for wind erosion. It would be difficult, however, to integrate vegetation and residue effects with this approach, or to investigate several alternative management schedules. These equations could be useful for preliminary screening of a few regions and systems.

Two new models are being developed which simulate erosion and vegetation: the Water Erosion Prediction Project (WEEP; Laflen et al. 1991) and Wind Erosion Prediction System (WEPS; Hagen 1991) models. These models are still experimental, and do not have well-developed support systems and datasets.

5.6.10 References for Section 5.6

Benson, V.W., O.W. Rice, P.T. Dyke, J.R. Williams, and C.A. Jones. 1989. *Conservation impacts on crop productivity for the life of a soil.* J. Soil and Water Conservation 44:600-604.

Cole, G.W., L. Lyles and L.J. Hagen. 1983. *A simulation model of daily wind erosion soil loss.* Trans., ASAE 26:1,758-1,765.

Hagen, L.J. 1991. *A wind erosion prediction system to meet user needs.* J. Soil and Water Conservation 46:106-111.

Laflen, J.M., L.J. Lane, and G.R. Foster. 1991. WEEP: *A new generation of erosion prediction technology.* J. Soil and Water Conservation 46:34-38.

Lee, J. J., D. L. Phillips and R. Liu. 1993. *The effect of trends in tillage practices on erosion and carbon content of soils in the US corn belt.* Water, Air, and Soil Pollution 70:389-401.

Oldeman, L.R., R.T.A. Hakkeling and W.G. Sombroek. 1990. *World Map of the Status of Human-Induced Soil Degradation: An Explanatory Note.* International Soil Reference and Information Center, Wageningen, Netherlands; and United Nations Environment Programme, Nairobi, Kenya.

Onstad, C.A. and G.R. Foster. 1975. *Erosion modeling on a watershed.* Trans. ASAE 18:288-292.

Phillips, D.L., D. White, and C.B. Johnson. 1993. *Implications of climate change scenarios for soil erosion potential in the USA.* Land Degradation & Rehabilitation 4:61-72.

Renard, K.G., and V.A. Ferreira. 1993. *RUSLE: Model description and database sensitivity.* J. Env. Quality 22:458-466

Robinson, A.R.: 1979, *Sediment Yield as a Function of Upstream Erosion, in:* Peterson, A.E. and Swan, J.B. (eds), Universal Soil Loss Equation: Past, Present, and Future. Soil Sci. Soc. Am., Madison, WI, pp. 7-16.

Sharpley, A.N. and Williams, J.R. (eds.). 1990a. *EPIC—Erosion/Productivity Impact Calculator: 1.* Model Documentation, US Department of Agriculture Technical Bulletin No. 1768, Washington, DC.

Sharpley, A.N. and Williams, J.R. (eds): 1990b, *EPIC–Erosion/Productivity Impact Calculator: 2.* User Manual, US Department of Agriculture Technical Bulletin No. 1768, Washington, DC.

Stockle, C.O., J.R. Williams, N.J. Rosenberg and C.A. Jones. 1992. *A method for estimating the direct and climatic effects of rising carbon dioxide on growth and yield of crops: Part I - Modification of the EPIC model for climate change analysis.* Agricultural Systems 38: 225-238.

Williams, J.R.. 1975. *Sediment yield prediction with Universal Equation using runoff energy factor. In: Present and Prospective Technology for Predicting Sediment Yields and Sources.* USDA Agric. Res. Service ARS-S-40, pp. 244-252.

Wischmeier, W.H. and Smith, D.D. *1978, Predicting Rainfall Erosion Losses, a Guide to Conservation Planning,* US Department of Agriculture Agricultural Handbook No. 537, Washington, DC..

Woodruff, N.P. and F.H. Siddoway. 1965. *A wind erosion equation.* Soil Sci. Soc. Am. Proc. 29:602-608.

5.7   HUMAN HEALTH VULNERABILITY ASSESSMENT

Understanding of the relations between climate, ecosystems, and human health is in its infancy. As a result, the methodology to conduct human health impact assessments for global climate change is in its earliest stages of development. Unlike classical health risk assessment for toxic exposures, which uses data from animal experimentation and historical human exposures to extrapolate and predict the consequences of a new exposure scenario, health vulnerability assessment for climate change has no existing experimental data for extrapolation, and historical human exposures have not been in the range of predicted climate changes. Therefore, we describe a multi-tiered approach to impact assessment. The first stages use expert judgment and relatively simple data collection to make broad inferences about likely consequences of climate change for human health in a given region or country. The latter stages use more sophisticated analytic and modeling tools to attempt more quantitative and integrative analysis. While such integration of multiple factors is an essential ultimate goal of predictive modeling, our poor present understanding of many of the interactions increases the possibility that such integration will result in large errors. Until integrated models have been fully validated for present conditions, their use as accurate predictors of future conditions is limited (IPCC, 1994).

Much of our present knowledge of climate-health interactions come from historical analogue studies, in which historical extremes of climate variation (such as El Nino/ENSO events) are examined for health consequences. While such studies are critical for advancing our understanding of climate-health interactions, as well as for validating integrated models, historical analogue studies are not listed as a method for country studies because of the time needed to assemble and analyze the data. Any existing studies for a given country should be reviewed as part of the vulnerability assessment, however.

5.7.1   Primary Approach

The initial step in vulnerability assessment is the baseline assessment survey (Figure 5.7.1). The purpose of this survey is to comprehensively review the country's current public health problems from the standpoint of sensitivity to climate change.   All areas of public health, including nutrition, sanitation/environmental health, infectious diseases, and principle causes of morbidity and mortality should be reviewed to avoid an early focus on one area that may not have the greatest public health significance.

It is important that a multi-disciplinary team perform the initial survey. Government and academic public health professionals, environmental engineers, sociologists, economists, and agricultural experts all have a role to play in performing the baseline assessment.

Because human health is ultimately dependent on the integrity of natural and managed ecologic systems and human social systems, changes in global climate are likely to affect human health through alterations in other sectors of the country studies project, such as agriculture, water resources, coastal resources, fisheries, grasslands and livestock. Methods for addressing how these sectors may be affected by climate change are described elsewhere in this guidance document. These methods, however, do not address how human health may be affected by changes in these sectors. For example, they do not describe how potential reductions in crop production may affect nutrition of the population and consequently, health. As these other sectors collectively represent much of the foundation of public health, their potential impact is large. To the extent that vulnerability assessments are available, the results for these sectors should be included in baseline assessments.

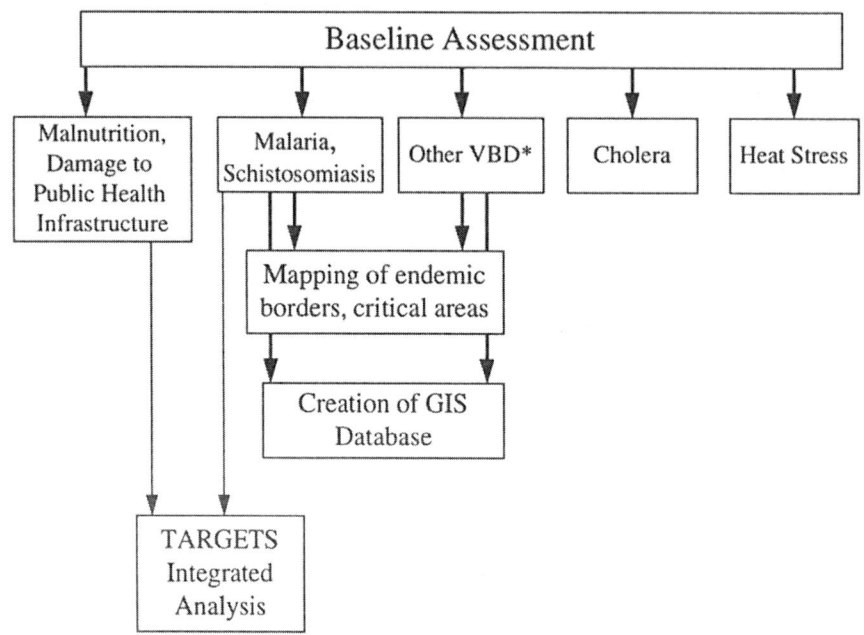

FIGURE 5.7.1. Baseline Assessment

The baseline assessment survey is expected to give a broad overview of health problems within a country that may be exacerbated by climate change, as well as to identify those areas and populations that are likely to be most affected by climate change. Thus, a matrix consisting of the most serious current health problems arranged by their anticipated sensitivity to climate change should be created to help prioritize regions or problems for further vulnerability assessment. The marginal zones around areas endemic for vector-borne diseases should be delineated. Maps and geographic information systems (see below) should be developed to identify sensitive regions and subpopulations.

The main limitations of the baseline assessment survey are that 1) linkages between predictive climate models and health data are not fully developed, so that there remains uncertainty in the extent and even direction of changes in disease systems or human conditions as a result of climate change; and 2) no attempt is made to formally integrate the different health impacts and population changes to be able to make more accurate predictions. The survey provides qualitative information on sensitive regions and subpopulations upon which to base further, more detailed analysis.

5.7.2 Description of the Methodology

5.7.2.1 Simple Mapping Studies

Useful information can be obtained by comparing plots of disease incidence or prevalence, geographic features which form boundaries, demographic data, current climate data, and climate predictions.

Because vector-borne diseases tend to be limited in range by geographic and climatic features, this mapping exercise may demonstrate potential future changes in the ranges of infectious diseases in relation to population centers.

A number of studies have created maps of isolated features mentioned in 3.2.1, such as seasonal minimum and maximum temperatures in Zimbabwe (Unungai), current land use in Honduras (Almendares et al, 1994), and altitude in Zimbabwe (Taylor and Mutambu, 1986). Such maps can then be compared with maps of vector populations and disease incidence to make inferences about critical areas of susceptibility.

The result of the mapping exercise should be the identification and characterization of geographic areas that are most susceptible to climate-related changes in health, primarily because they lie on the other side of a geoclimatic border from an area of high disease activity. The mapping exercise can be used to direct further field work in assessing a given disease's dependence on climate, locate areas in need of surveillance and monitoring facilities, or locate areas in need of health care or other interventional facilities. In general, mapping will be most appropriate for analysing vector-borne diseases.

The main limitation of mapping is that it does not provide a quantitative estimate of the impacts of climate change on human diseases. Also, without analytic capability, mapping is unable to provide an integrated assessment of the various factors influencing human health, although factors relevant for a given region can be identified through this process. Lastly, as with all analyses of climate change impacts, the availability of appropriate data is likely to be limited.

5.7.2.2 GIS and Remote Sensing

In addition to the creation of analog maps, a geographic method of organizing and storing data on disease incidence, vector populations, demographics, and climate with linkage to specific geographic locations will facilitate both vulnerability assessments and future research into climate disease linkages. A number of Geographic Information Systems (GIS) software programs provide this capability. The use of GIS systems for predicting disease incidence is still under development, but especially in combination with satellite remote sensing, there appears to be promise in using ecosystem parameters such as vegetation types to help predict where disease outbreaks might occur. It should be stressed that GIS systems will not provide higher quality outputs than the quality of the data which is entered into them, therefore attention needs to be paid on the proper collection and organization of the data before computerized analysis is initiated.

At present, there have been few applications of GIS to disease prediction. One study in California used satellite images and GIS technologies to predict population levels of anopheles mosquito in rice fields, although there was no ability to further correlate these results with subsequent human disease (Wood, et al., 1991). The WHO has supported a demonstration project using GIS technology to organize a database of schistosomiasis in Egypt (Yoon, 1993). While this project primarily used GIS for database organization and detection of spatial clustering, the existence of data organized in this fashion could facilitate future modeling or statistical analysis to examine the impacts of climate change. A third study of African trypanosomiasis used remote sensing to identify vegetation types which, combined with air humidity measures, were then used to predict tsetse fly survival rate (Rogers and Randolph, 1991).

The main utility of GIS is to organize and allow the superimposition of geographically-based data. Thus, the first expected product would be the creation of such a database for each country, with separate layers for geopolitical boundaries, demographics, nutritional status, and individual disease incidence/prevalence. Additional products would be the use of the database to demonstrate climate-disease linkages or in conjunction with an empirical model to use projected changes in climate to predict future patterns of disease.

Because of the ability of GIS to superimpose multiple layers of data, it is essential that the resolution of scale of the various layers be as similar as possible, if not identical. In addition, the data have to be formatted compatibly in terms of spatial organization. Such detailed and compatible data often do not exist, and the costs of developing such a database can be very high (Glass et al., 1993). Other specific limitations include the cost of the software (although less expensive products are available) and the need to train technicians to use the software. Lastly, most applications of GIS and remote sensing are not able to dynamically model the interactions between climate change, vector populations, and susceptible human populations.

5.7.2.3 Integrated Modeling

Because of the complexity of ecosystems and the variety of social interactions involved in mediating the effects of climate change on human diseases, and because predicted changes in climate lie outside the range of recorded experience, traditional methods of epidemiology are not likely to predict climate-related changes in human disease accurately, especially vector-borne diseases (McMichael, 1995). Integrated mathematical modeling may be one method of addressing this problem. To date, the models which have been developed are still in the process of validation, but they can be used in conjunction with the other methods of vulnerability assessment to investigate the range of possible outcomes and the potential impacts of various socioeconomic factors, such as population migration, malnutrition, and health resource investment.

A model chosen frequently by country studies participants is the TARGETS (Tool to Assess Regional and Global Environmental and health Targets for Sustainability) model, developed by the National Institute of Public Health and Environmental Protection (RIVM) of the Netherlands (Rotmans et al., 1994). TARGETS uses a systems approach to interrelate social, economic, and environmental forces in predicting the behavior of human diseases or other human behaviors. Climate data are generated by Global Circulation Models (GCM's) linked to the Integrated Model to Assess the Greenhouse Effect (IMAGE) (Rotmans 1990). Discrete but interrelated components of the model include modules for land and water use, energy resource development, population growth, and toxic chemical release. The use of a systems approach which includes inputs of societal factors allows exploration of the relative impacts of different levels of population growth, resource allocation, and energy use, among other human factors. These factors can be dynamically modeled or input as fixed projections based on available data sources such as the FAO. For country studies, scaled-down disease-specific modules for malaria and schistosomiasis have been created which dynamically model population growth, nutrition and other diseases (i.e., measles, diarrhea) but use fixed projections for other areas such as land use, water resources, etc.

While it is useful to integrate the many determinants of human health, the quantitative relations between these factors are not known with great precision, and iterative repetitions of predictive equations will result in large degrees of uncertainty. Integrated models are highly dependent on initial conditions, so it is essential to perform a number of model runs with varying initial conditions to determine how sensitive the outcomes are to specific initial parameters and assumptions.

5.7.3 Defining the Scope of the Analysis

Because of the complexity of climate and health interactions, it is preferable for the baseline assessment to be comprehensive, encompassing both a wide range of potentially affected diseases and a wide geographic area. The results of the baseline assessment may then be used to narrow the scope of further analysis to critical areas and diseases.

5.7.4 Economic Analysis

Unlike other sectors whose products are inherently economic in nature, such as agriculture and water resources, human health is historically difficult to quantify and associate with economic value. Nonetheless, there are techniques to try to describe human disease in terms of its economic impact. Changes in disease incidence for different diseases can be compared by estimating lost productivity due to the disease. This requires knowing the average age of onset of the disease, case fatality rates, average extent and duration of disability due to the disease. These data are likely to be very difficult to obtain, and the levels and types of studies from which the data come need to be examined and found comparable (Aron and Davis, 1993). Should these data be obtainable, an estimate of healthy years of life lost can be made. The further valuation of years of life lost in economic terms is quite problematic. Regional differences in wages, employment, and even gender roles will complicate the comparison of disease impacts.

5.7.5 Adaptation Techniques

Just as the differing conditions among countries necessitate different approaches to health sensitivity and impact assessments, so too will the approaches to adaptation differ by region. For many developing countries, the health problems that are likely to be exacerbated by climate change are significant current problems. Thus, adaptive strategies developed in anticipation of future climate conditions may have substantial utility for the present situation. Many of the adaptive strategies discussed below are not specific to climate change and, in fact, should not be viewed in isolation or out of context of the more generalized problem of global environmental degradation and compromised public health infrastructure in much of the developing world (Patz, 1995).

Ongoing monitoring, both of human diseases and critical ecosystem indicators, will be essential to the timely institution of interventions as disease systems change. Because of the inertia of large ecosystems, and the fact that changes in human diseases due to climate factors generally represent the end result of ecosystem changes, substantial ecosystem changes will have occurred by the time an increase in disease incidence is detected, and intervention will be far more difficult. Therefore, surveillance should include intermediate indicators such as vector populations and vegetation changes as well as actual disease incidence. A strategy for global monitoring of health effects of climate change has been proposed involving remote sensing and extensive telecommunications networks of environmental and health professionals (Haines et al, 1994). Such an effort is strongly needed on a global scale, but smaller efforts on a regional scale, targeted at the critical geographic areas identified in the sensitivity analysis, will be important for regional adaptation strategies as well.

Water treatment facilities and shelter, already in short supply in many areas, may be further threatened by severe storms and sea level rise. For these threatened areas, investment in expanded facilities may have substantial current benefit, and attention paid to safe location of the facilities with respect to sea level rise and extreme weather events will be of use in adapting to future conditions. Consideration should also be given to improving efficiency of existing water systems as well as reducing demand for water where possible. Involvement of local communities in planning and developing water systems is essential (World Bank, 1992). This is another area in which present day investment will have public benefits with or without impacts of climate change.

Human behavior has a considerable influence on disease incidence. Some behaviors, such as the storage of open water containers or the improper disposal of human wastes, create favorable environmental conditions for disease-causing agents to reproduce. Other behaviors, such as the type of clothing worn and the filtering of drinking water, affect exposure to disease-causing agents. Public education efforts will be needed, both to inform about the causes of disease and human impacts on disease, as well as to instruct on ways to minimize the health impacts of climate change. The need may be greatest in the critical areas where experience with disease is limited but the risk of the spread of disease is high.

Educating diverse groups of people in a way that does not conflict or negate present belief systems can be quite difficult. Experience with public education efforts in Tanzania on malaria has shown that educational methods need to be adapted to the local ethnic belief systems. Without the education and involvement of local communities, regional adaptation efforts will not succeed (Schiff, C personal communication).

5.7.5.1 Impact-specific Adaptation Strategies

In addition to the general adaptive strategies discussed above, a variety of impact-specific options are available. In general, these are measures which have been employed for present-day problems. The list of options below is not intended to be complete, but rather to initiate discussion and evaluation of a variety of options that will be decided upon by local and regional health experts and policy makers. Furthermore, it should be emphasized that many of these measures will only be temporary in their effects; this list should not be viewed as an alternative to addressing the root causes of global warming through policy initiatives.

For weather extremes and sea-level rise:

- Maintaining disaster preparedness programs, including tools for local public health facilities to conduct rapid health needs assessments.

- Engineering measures such as strengthening sea-walls or requiring building contractors to follow hurricane standards in coastal areas.

- Adopting landuse planning to minimize erosion, flash-flooding, precarious residential placements.

- Siting intakes for water facilities far enough upstream to tolerate saline intrusion from storm surges and sea level rise.

For vector-borne diseases:

- Installing window screens in areas endemic to insect-borne diseases.

- Expanding coverage of vaccination programs in critical areas aimed at infectious diseases that are likely to increase with climate change, e.g., yellow fever (unfortunately, no vaccines exist as yet for some sensitive diseases such as malaria and dengue).

- Public education to encourage elimination of manmade breeding sites (i.e., small water containers for aedes mosquitoes).

- Education campaigns to sensitize health care givers in geographically vulnerable regions.

- Release of sterilized male insects to reduce reproductive capacity of vector populations.

- Promoting the use of pyrethroid impregnated mosquito bed-nets.

For water-borne diseases:

- Possibly creating early warning systems based on algal blooms to predict cholera.

- Public education on sources of infection.

- Distribution of low-technology water filtration systems (i.e., nylon mesh, cloths).

For heat-related illness:

- Designing buildings to be more heat resistant (insulation, air flow).

- Planting trees within cities to reduce the urban heat-island effect.

- Establishing new weather watch/warning systems that focus on health-related adverse conditions, such as oppressive air masses.

- Creating public education campaigns regarding precautions to take during heat waves.

- Work schedules that avoid peak daytime temperatures for outdoor laborers.

For agricultural stresses:

- Producing climate-resistant transgenic plants (genetically engineered).

- Reducing the proportion of monocultural farming for better crop resistance to pests.

- Promoting land reforms that would favor environmentally sound land usage.

When considered in total, these adaptive measures will offer varying amounts of protection to human health. Many diseases may not be amenable to preventive actions (Haines and Parry, 1993). Problems may occur at many stages: inability to provide sufficient resources for engineering options, development of pesticide and/or drug resistance in disease agents and vectors, and lack of local public support. Critical evaluation of implemented strategies will be needed to inform a dynamic process of adaptation.

5.7.6 Data Requirements

For the baseline assessment:

- Principal causes of mortality by age group.

- Principal infectious diseases with incidence/prevalence data arranged by age group and geographic distribution.

- Populations living at low elevations in coastal zones, populations with poor or marginal drinking water quality, arranged by geographic distribution.

- Prevalence of malnutrition, arranged by age group and geographic distribution.

- General demographic data on income, education, geographically arranged, with population growth predictions.

- Any available GCM predictions for the country or regions within the country

- Any existing historical analogue studies based in a given country should be reviewed and included in the baseline assessment.

For mapping/GIS studies:

- Current climate data, including minimum and maximum seasonal temperatures, precipitation levels, average monthly temperatures, by smallest geographical and political area possible.

- Predicted changes in above climate data from available GCM models. If such data are lacking, arbitrary changes encompassing the range of possible changes can be substituted.

- For infectious diseases, it may be useful to map vegetation types to assess suitability and continuity of habitats for vector species. This data may be obtainable from other sectors (grasslands, agriculture, coastal resources, etc.).

- Also, for infectious disease, information on surface water irrigation and other means of water distribution is useful.

For GIS and remote sensing:

- Geographic data arranged in a suitable format (raster or vector).

- Data from baseline assessment arranged in the same GIS format.

- Geographic (including land-use), ocean (including temperature and plankton populations) and vegetation data may be obtained from satellite images, allowing relatively easy updating for future analyses.

For TARGETS:

- Population and food intake data.

- GNP and governmental health investment levels.

- Climate data; baseline (preferably 1951–1980) available from local weather stations; and projections, either from GCM models or using a range of arbitrary changes.

- Baseline disease incidence/prevalence data.

5.7.7 Results Generated

The result of the mapping exercise should be the identification and characterization of geographic areas that are most susceptible to climate-related changes in health, either because they lie on the other side of a geoclimatic or ecologic border from an area of high disease activity, or because they contain populations with increased susceptibility (e.g., due to age, poverty, malnutrition, etc.). The mapping exercise can be used to direct further field work in assessing a given disease's dependence on climate, locate areas in need of surveillance and monitoring facilities, or locate areas in need of health care or other interventional facilities.

The main utility of GIS is to organize and allow the superimposition of geographically-based data. Thus, the first expected product from GIS-based efforts would be the creation of such a database for each country, with separate layers for geopolitical boundaries, demographics, nutritional status, and individual disease incidence/prevalence. Additional products would be the use of the database to demonstrate climate-disease linkages or in conjunction with an empirical model to use projected changes in climate to predict future patterns of disease. As new developments in GIS analysis allow more extensive analysis of disease and climate data, it can be anticipated that present efforts to develop a GIS database will facilitate future assessments.

5.7.8 Advantages and Disadvantages

Because the assessment of human health vulnerability to climate change has not been widely performed, it is not possible to compare the approach in this section to other approaches. This section has provided a range of techniques from simple data collection and mapping to computer-intensive mathematical modeling. There are a several advantages to combining a number of different approaches, as compared to, for example, simply using available mathematical models: first, mathematical models do not exist for a number of the anticipated impacts of climate change on health, thus proper prioritizing can not be performed. Second, a qualitative baseline approach may help with the interpretation of the results of mathematical models. Because integrated models are still in the early stages of development, their results should be used to inform a comprehensive assessment rather than to stand alone as a forecast.

Disadvantages and limitations of individual methods are discussed below.

The main limitation of mapping is that it does not provide a quantitative estimate of the impacts of climate change on human diseases. Also, without analytic capability, mapping is unable to provide an integrated assessment of the various factors influencing human health, although factors relevant for a given region can be identified through this process. Lastly, as with all analyses of climate change impacts, the availability of appropriate data is likely to be limited.

One limitation of GIS systems is the need for the different layers of data to be highly compatible in terms of scale and format. Such detailed and compatible data often do not exist, and the costs of developing such a database can be very high (Glass et al. 1993). Other specific limitations can include the cost of the software (although less expensive products are available) and the need to train technicians to use the software. Lastly, most applications of GIS and remote sensing are not able to dynamically model the interactions between climate change, vector populations, and susceptible human populations. This capability is likely to be developed in the near future, however.

There are a number of limitations of the TARGETS and other integrated models. Because of the model's dependence on initial conditions, the baseline data must be authentic and verified. The aggregation of assumptions and variability inherent in integrated statistical models can lead to such a confidence interval for outcomes that the results become less useful. This is one reason that fixed projections are used for certain socio-economic data instead of dynamic models. An additional problem is that the resolution of the global circulation models is still too crude to allow accurate climate predictions on a local basis. TARGETS uses a grid of 0.5° latitude by 0.5° longitude, a scale more refined than the 2° by 4° scale of most GCMs, but one that may still be too large to take into account some local climate-altering geographical features such as mountain ranges or lakes. Thus, there is a need to couple the outputs of an integrated predictive model, such as TARGETS, with an evaluation of local geoclimatic factors influencing potential disease spread. Alternatively, a range of different climate projections for a specific region can be used to explore the range of possible outcomes. Lastly, the TARGETS model has been developed on a global scale and modified to be applicable to the smaller scale necessary for country studies. The development of such a global model assumes that the disease systems being modeled exhibit certain universal relations. As there may be important differences in habitats and relations with local species among different regions which can not be included in TARGETS, this assumption of universality may not hold for all areas. For these reasons, the integrated models, and TARGETS specifically, should be viewed as providing insight into the interactions of various societal and natural systems and giving suggestions of trends in the behavior of human disease systems, rather than as providing quantitative predictions of actual disease incidence or impact.

5.7.9 Supplementary Approaches

Although other integrated mathematical models have been developed (i.e., Matsuoka and Kai, 1995), the multi-tiered approach used in this chapter encompasses most of the techniques that have been used in human health assessments. There are no other known approaches to supplement the methods already mentioned.

5.7.10 References

Almendares, J., Sierra, M., Anderson, P.K., Epstein, P.R., (1994) *Critical regions, a profile of Honduras.* in Health and Climate Change, The Lancet, London. 1994.

Aron, J.L., Davis, P., (1993) *A comparative review of the economic impact of selected infectious diseases in Africa.* Johns Hopkins University, Baltimore.

Glass, G.E., Aron, J.L., Ellis, J.H., Yoon, S.S., (1993) *Applications of GIS Technology to Disease Control.* Johns Hopkins University, Baltimore.

Haines, A., Epstein, P.R., McMichael, A.J., et al. (1994) Global Health Watch: monitoring the impacts of environmental change. Lancet 342:1464-1469.

Intergovernmental Panel on Climate Change (IPCC) (1990) *Scientific Assessment of Climate Change: Report to IPCC from Working Group II.* World Meteorological Association and UN Environment Programme. Geneva & Nairobi.

IPCC (1994) *Technical Guidelines for Assessing Climate Change Impacts and Adaptations.* Department of Geography, University College London and Center for Global Environmental Research, National Institute for Environmental Studies, Tsukuba, Japan.

Matsuoka, Y., Kai, K., (1995) An estimate of climatic change effects on malaria. J Global Environ Eng. 1:43-57

McMichael, A.J., (1995) *Conceptual and Methodological Challenges in Predicting the Health Impacts of Climate Change.* Medicine and War 11: 195-201

Parry, M.L., Rosenzweig, C. (1993) *Food Supply and the Risk of Hunger* Lancet 342:1345-7

Patz, J.A., (1995) *Health Adaptation to Climate Change.* Climate Research Nov 1995.

Rogers, D.J., Randolph, S.E., (1991) *Mortality rates and population density of tsetse flies correlated with satellite imagery.* Nature 351: 739-741.

Rotmans, J. (1990) IMAGE: an Integrated Model to Assess the Greenhouse Effect. 1990; Kluwer Academic Publishers, Dordrecht/Boston/London, The Netherlands.

Rotmans, et al. (1994) Global change and sustainable development- a modeling perspective for the next decade. Rijksinstituut voor volksgezondheid en milieuhygiene (RIVM) report no. 461502004 June 1994. Bilthoven, the Netherlands.

Taylor, P. and Mutambu, S.L., (1986) *A review of the malaria situation in Zimbabwe with special reference to the period 1972-1981.* Transactions of the Royal Society of Tropical Medicine and Hygiene 80:12-19.

Wood, B., Washino, R., et al. (1991) *Distinguishing high and low anopheline-producing rice fields using remote sensing and GIS technologies.* Preventive Veterinary Medicine 11: 277-288

World Bank (1992). *World Development Report 1992: Development and the Environment.* Oxford University Press, New York, USA.

Yoon, S., (1993) *The value of geographic information systems with regard to schistosomiasis control*, in (de Lepper, MLC, Scholten HJ and Stern RM, Eds.) The Added Value of Geographical Information Systems in Public and Environmental Health. Kluwer Academic Publishers, Dordrecht and World Health Organization Regional Office for Europe, Copenhagen.

5.8 FISHERIES VULNERABILITY AND ADAPTATION ASSESSMENT

Several of the participating countries in the Country Studies Program identified the need for the development of vulnerability assessment guidance for fisheries resources. Only existing methods and tools were evaluated for use in assessing potential climate change impacts on freshwater and coastal marine fisheries resources. Programmatic constraints, as well as time and resource limitations, precluded the development of new analytical tools or the use of tools which require acquisition of new data or field studies.

The variety of methods identified in this section reflect the variety of climate factors that can affect fisheries resources, the variety of aquatic habitats that may have to be evaluated, the availability or lack of existing data and technical capabilities, and the lack of a single methodology or tool that can be used for all fisheries resources and evaluate multiple climate variables. In addition, the use of multiple methods will produce a more robust vulnerability assessment than an approach using only a single method.

This section, and Appendix J, identifies the rationale that was used to identify, screen, and select the methods, and for each method identifies the assessment objectives, data needs, and output. This report also identifies the environmental and anthropogenic factors that were not considered in the selection of the assessment methods and discusses why assessment methods were not developed for these factors.

5.8.1 Primary Approach

An evaluation of the available literature identified three general approaches that have been used to evaluate potential impacts to fisheries resources from climate change: 1) predicting changes in the availability of *thermal habitat* by evaluating changes in the thermal structure of lakes and streams; 2) predicting effects of temperature changes on fish physiological processes, particularly growth and feeding, using bioenergetics models; and 3) predicting impacts of changes in *physical habitat* features (i.e., water temperatures, flow rates, and water levels) to important life history stages such as migration periods and spawning times. These general approaches have been identified by some researchers as the basic framework for evaluating climate change impacts on fisheries resources (Regier and Meisner 1990; Shuter and Meisner 1992).

Studies that have evaluated potential impacts of climate change on fisheries in temperate and northern latitudes have primarily followed the first two approaches identified above, and have used methods focusing primarily on the thermal aspects of climate change (see Transactions of the American Fisheries Society 1990, Volume 119). Seasonal water availability, which affects lake levels, stage and flow in rivers, extent and timing of floodplain inundation, and sea level, plays an important role in maintaining the life history of tropical fish species (Welcomme 1976, 1985; Lowe-McConnell 1987). While the magnitude of temperature changes in tropical areas of the world is expected to be relatively minor in comparison to the magnitude of change in more temperate and northern latitudes, water availability in tropical regions is expected to be affected to a greater extent by potential climate changes (Meisner and Shuter 1992). Because of the importance of water availability and the importance of seasonal flooding to the life history of many fish species (Lowe-McConnell 1987; Welcomme 1976, 1985), changes in water availability due to changes in precipitation will probably affect some tropical fisheries resources more than will changes in temperature (Meisner and Shuter 1992).

Because of the nature of the output of the general circulation models and the environmental factors directly affecting fish, the vulnerability assessments must link predicted climate change parameters to changes in aquatic habitats. For example, GCM outputs include predictions of changes in air temperatures and precipitation magnitude, duration, and distribution, but do not provide direct information on water quality or hydrology parameters that affect fisheries resources. Thus, before the fisheries vulnerability assessment can be performed, appropriate methods must be employed to translate the predicted atmospheric climate changes to changes in lake and sea levels, stream flow, and water temperatures (Figure 5.8.1) (Meisner et al. 1987; Christie and Regier 1988; Kennedy 1990). Only when these linkages are made can the response of fisheries resources to climate-induced changes in aquatic habitats in specific ecological and biological parameters (such as growth rates, reproductive success, mortality, and distribution) be identified and evaluated (Figure 5.8.2). In particular, the assessment of impacts will rely very strongly on the results of the Water Resource and Coastal Resource vulnerability assessments (see Section 5.4), and staff assessing the fisheries resources will have to interact with technical staff from both resource areas when developing hydrographs, estimating sea and lake levels, and estimating floodplain inundation under different climate change scenarios.

5.8.2 Scope of the Analysis

The major aquatic habitats present in countries participating in the Country Studies Program include, but are not limited to 1) lakes such as Lakes Malawi and Victoria, 2) man-made reservoirs, 3) large river and floodplain systems, and 4) coastal habitats such as estuaries, mangrove swamps, and deltas. Each habitat may have to be evaluated separately for responses to climate change. In addition to this diversity of habitats, a variety of ecologically and/or economically important species occupy these habitats and may have to be considered with regards to potential climate change impacts.

Because of this great diversity in habitats and species as well as the diversity in the habitats and species that any single participating country may have to consider, no single assessment approach is suitable for use by all countries or for all aquatic habitats.

5.8.3 Description of the Methodology

5.8.3.1 Preliminary Screening Technique

The purpose of the preliminary screening is to identify and prioritize those fisheries within a country for which detailed vulnerability and adaptation analyses might be needed. In many cases, expert judgement can be applied to the fisheries to rank their vulnerability to potential changes in precipitation, temperature, or sea level rise as low, moderate, or high. All fisheries ranked as highly vulnerable should be considered for detailed analyses. Fisheries with that have high economic or ecological importance and are ranked as moderately vulnerable should also be considered.

5.8.3.2 Summary of Methods

The methods described in Appendix J, and summarized in Tables 5.8.1, 5.8.2, and 5.8.3, apply to lacustrine, riverine, and coastal fisheries, including penaeid shrimp.

The methods for evaluating climate change impacts on lacustrine fisheries (Table 5.8.1) include: 1) comparing growth and mortality under historic and predicted temperature regimes; 2) estimating fish yield under historic and predicted temperature regimes; and 3) estimating habitat availability under historic and predicted lake levels and surface water area.

The assessment of potential climate change impacts on fisheries resources is most problematic for riverine-based fisheries. The methods (Table 5.8.2) are focused to assess the impacts of temperature, dissolved oxygen, and precipitation on fish growth, life history, habitat suitability and yield.

**General Circulation Model
Predictions**

Altered air temp, precipitation,
solar radiation, wind speed

|

V

**Identify coastal marine and freshwater conditions vulnerable
to climatic changes**

- Increased water temperatures
- Reduced DO levels
- Altered lake and sea levels
- Changed river hydrographs

|

V

Identify and Collect Appropriate Data

- Historic limnological and climatological data
- Historic flow (hydrograph) data
- Morphometric/bathymetric data
- Topographic information (maps, aerial photography)

|

V

Implement Assessment Approaches

- Predict surface water temp from air temp
- Predict DO levels
- Develop thermal and DO profiles
- Predict sea and lake levels and river hydrographs
- Quantify physical changes in habitats due to changes
in sea and lake levels using topographic information
- Quantify areal extent of floodplain inundation

|

V

Predict Changes in Marine and Freshwater Conditions

Figure 5.8.1. Conceptual framework for identifying potential changes in marine and freshwater habitats under different climate change scenarios.

**General Circulation Model
Predictions**

|
V

Predicted Changes in Marine and Freshwater Conditions

|
V

**Identify Habitat Parameters Vulnerable to
Predicted Changes in Marine and Freshwater Conditions**

- Change in habitat availability (area, seasonality)
- Change in habitat quality (temperature, DO)

|
V

**Identify and Collect Appropriate Biological and Fisheries Data
and Develop Habitat and Catch/Yield Models**

- Physiological parameters (thermal niche and tolerance, DO requirements)
- Habitat requirements (flow, substrate, depth) and other life history information
- Individual process rates (growth and mortality rates)
- Historic fish yield or catch estimates
- Develop empirical models relating fish yield or catch
with habitat or climate factors
- Develop habitat suitability index models
- Develop relationships between population abundance and thermal habitat
- Develop temperature-process relationships and predict process responses

|
V

Implement Assessment Approaches

- Predict fish yields from empirical models and GCM climate predictions
- Predict fisheries response to changes in habitat availability
- Predict fisheries response to changes in habitat quality
- Evaluate changes in growth using bioenergetics model and
predicted changes in temperature

|
V

Predict Impacts to Fisheries Resources

Figure 5.8.2. Conceptual framework for identifying potential impacts to fisheries resources resulting from different climate change scenarios.

Table 5.8.1. Suggested approaches for evaluating potential impacts of climate change to lacustrine fisheries resources.

Climate variable to be evaluated	Method approach	Vulnerability assessment and method output	Climate-environment linkage	Linkage to other sector assess-ments	Historic physical data needs	Fisheries and Biological data needs
Temperature	Species-specific comparison of growth and feeding rates. Rates estimated using bioenergetics model. *(Appendix J.2.1)*	Vulnerability assessed on differences in estimated growth and feeding rates for target species under different environmental temperatures.	Link air temperature to water temperature using regression analysis.	Climate scenario sector	Air and water temperatures.	Species-specific ingestion, respiration, and excretion rates.
Temperature	Species-specific comparison of natural mortality. Mortality estimated with empirical model using length, weight, and air temperature. *(Appendix J.2.2)*	Vulnerability assessed on differences in estimated growth and feeding rates for target species under different environmental temperatures.	Same as above	Climate scenario sector	Air and water temperatures.	Species-specific asymptotic length and weight.
Temperature	Comparison of estimated maximum sustainable yield. Yield estimated with empirical models using air temperature and morphoedaphic index of lakes. *(Appendix J.2.3)*	Vulnerability assessed on differences in estimated maximum sustainable yield under different mean annual air temperatures.	Air temperature used directly.	Climate scenario sector	Mean annual air temperatures, morphoedaphic indices for lakes of concern, estimated fishing intensity.	No biological data needed.
Precipitation	Comparison of estimated annual fish catch. Catch estimated with empirical model using lake surface area under historic and predicted precipitation. *(Appendix J.2.4)*	Vulnerability assessed on differences among estimated annual catch under historic and predicted precipitation regimes.	Link precipitation to surface runoff inflow to lake surface area using water resources output and mapping techniques.	Climate scenario sector for precipitation scenarios, water resources sector for surface runoff predictions.	Precipitation, surface water runoff, and surface water area.	Annual fish yield
Precipitation	Qualitative estimates of impacts from estimates of habitat availability under historic and predicted lake levels. *(Appendix J.2.5)*	Vulnerability assessed on differences in estimated area of known important habitats under historic and predicted lake levels.	Same as above.	Same as above.	Precipitation, surface water runoff, and lake levels.	Location of known important fisheries habitat, such as spawning and nursery grounds.

Table 5.8.2. Suggested approaches for evaluating potential impacts of climate change to riverine fisheries resources.

Climate variable to be evaluated	Method approach	Vulnerability assessment and method output	Climate-environment linkage	Linkage to other sector assessments	Historic physical data needs	Fisheries and Biological data needs
Temperature	Comparison of historic and predicted annual productivity of river on basis of river width, fish community, fish food production, and mean annual water temperature. *(Appendix J.3.1)*	Vulnerability assessed on differences in estimated annual productivity for a river under historic and predicted mean environmental temperatures.	Link air temperature to water temperature using regression analysis.	Climate scenario sector	Air and water temperatures, stream width, acidity or alkalinity of the river.	Annual productivity of river, type of fish population present, and fish food production.
Temperature	Species-specific comparison of growth and feeding rates. Rates estimated using bioenergetics model. *(Appendix J.3.2)*	Vulnerability assessed on differences in estimated growth and feeding rates for target species under different environmental temperatures.	Same as above	Climate scenario sector	Air and water temperatures.	Species-specific ingestion, respiration, and excretion rates
Temperature	Comparison of river segments under historic and predicted climates where water temperature is outside the tolerance of individual species of concern. *(Appendix J.3.3)*	Vulnerability assessed on differences in location and extent of river reaches where water temperature exceeds species-specific tolerance levels.	Link air temperature to water temperature using regression analysis.	Climate scenario sector	Air and water temperatures for river reaches of concern.	Species-specific temperature preferences, and maximum temperature tolerances.
Precipitation	Comparison of historic and predicted flow regimes in relation to timing of known important life history events such as spawning and nursery periods. *(Appendix J.3.4)*	Vulnerability assessed on differences in hydrographs under historic and predicted precipitation, and relationship of hydrograph to life history events.	Link precipitation to stream flow using Water Resources output.	Climate and Water Resources Sectors.	Stream flow, flood duration, timing, and magnitude.	Timing of important species-specific life history activities, such as spawning and nursery periods.
Precipitation	Predict annual catch with regression model using predicted areal extent of floodplain inundation *(Appendix J.3.5)*	Vulnerability assessed on differences among estimated annual catch under historic an predicted flood conditions.	Link precipitation to stream flow and floodplain inundation using Water Resources output and mapping techniques.	Same as above.	Stream flow and areal extent of floodplain inundation.	Historic annual catch for river reaches of concern.
Precipitation and Temperature	Develop species-specific habitat suitability models that incorporate water temperature and stream flow. *(Appendix J.3.6)*	Vulnerability assessed on differences in estimated habitat suitability under historic and predicted climate scenarios.	Link precipitation to stream flow and air temperature to water temperature as above.	Same as above.	Stream flow and water temperatures for important fisheries areas.	Species-specific habitat preferences, including water temperature and stream flow.

Table 5.8.3. Suggested approaches for evaluating potential impacts of climate change to coastal marine fisheries resources.

Climate variable to be evaluated	Method approach	Vulnerability assessment and method output	Climate-environment linkage	Linkage to other sector assess-ments	Historic physical data needs	Fisheries and Biological data needs
Precipitation	Estimate total annual shrimp catch using regression analyses to develop empirical models relating catch to previous years total annual rainfall. *(Appendix J.4.1)*	Vulnerability assessed on differences in estimated annual shrimp catch under historic and predicted total annual rainfalls.	GCM-predicted precipitation values used directly.	Climate scenario sector.	Total annual rainfall.	Historic total annual shrimp catch for each shrimp fishery area of concern.
Temperature	Estimate stabilized commercial shrimp yield from mean annual air temperatures using regression model developed from empirical data. *(Appendix J.4.2)*	Vulnerability assessed on differences in estimated shrimp yield under historic and predicted air temperatures.	GCM-predicted air temperatures used directly.	Climate scenario sector.	Mean annual air temperatures for shrimp fishery area of concern.	Commercial shrimp yield for the shrimp fishery area of concern.
Sea Level	Estimate areal abundance of shrimp habitat from aerial photographs then predict annual shrimp yield from availability of shrimp habitat using regression analyses. *(Appendix J.4.3)*	Vulnerability assessed on differences in estimated annual shrimp yield for shrimp fishery areas of concern under historic and predicted sea levels.	Link GCM-climate changes to changes in sea level using Coastal Impact output. Link sea level changes to areal abundance of shrimp habitat using mapping techniques.	Coastal Zone Impact Sector to provide sea level change estimates.	Sea level.	Areal abundance of aquatic vegetation, depth requirements for vegetation and shrimp, and annual shrimp yield, for each shrimp fishery area of concern.
Sea Level	As above except in the absence of historic yield data, shrimp yield inferred qualitatively from availability of shrimp habitat. *(Appendix J.4.4)*	Vulnerability indirectly assessed on differences in quantitative estimates of habitat availability under current and predicted sea levels. Effects on yield inferred from differences in habitat availability.	Link GCM-climate changes to changes in sea level using Coastal Impact output. Link sea level changes to areal abundance of shrimp habitat using mapping techniques.	Climate Scenario and Coastal Zone Impact Sectors.	Sea level	Areal extent of existing intertidal and estuarine vegetation for each shrimp fishery area of concern.
Temperature and Sea Level	Develop species-specific habitat suitability models that incorporate water temperature sea level. *(Appendix J.4.5)*	Vulnerability assessed on differences in estimated habitat suitability under historic and predicted climate scenarios.	Link air temperature to water temperature using regression analysis. Link GCM-climate changes to changes in sea level using Coastal Zone Impact output. Link sea level changes to habitat availability using mapping techniques.	Same as above.	Physical descriptions (sea level, water temperature, salinity) of shrimp fishery areas of concern.	Shrimp species-specific habitat preferences and life–history requirements (such as temperatures for optimal reproduction or growth.

The assessment of potential climate change impacts on coastal marine fisheries resources (Table 5.8.3) focuses exclusively on the penaeid shrimp fishery. However, the assessment approach identified in Appendix J, which estimates natural mortality from annual water temperature, and the other approaches presented in Appendix J, which involve the development of habitat suitability models, can also be applied to fish species in coastal marine areas.

5.8.4 Economic Analysis

Commercial and recreational fisheries lend themselves to economic analysis. However, subsistence fisheries for the most part operate outside the money economy. Therefore traditional economic analysis is unlikely to be the most relevant approach for analyzing societal impacts of changes in these fisheries. One aspect that does lend itself to economics is analysis of the cost-effectiveness of potential adaptations in response to climate change, discussed in Section 5.8.5.

5.8.5 Adaptation Techniques

For freshwater (i.e., lacustrine and riverine) fisheries, adaptation in response to climate change might require increased stocking, perhaps with non-indigenous species. Habitat improvement could enhance the fish resource, while changes in equipment could enhance the efficiency with which the resource is utilized.

Adaptation techniques for species that depend on coastal wetlands, and would be adversely affected by sea level rise, include restoration of degraded wetlands and construction of new wetlands. In some places, it might be possible to prevent inundation of existing wetlands.

There are no adaptation techniques that would directly benefit open ocean fisheries. It might be possible, however, to supplement these resources through aquaculture. Using bigger boats or moving processing plants might enhance economic utilization of these resources.

5.8.6 Data Requirements

Data requirements depend on the type of fishery (i.e., lacustrine, riverine, coastal) and the climatic variable being evaluated (Tables 5.8.1, 5.8.2, and 5.8.3, and Appendix J). Some common physical data requirements include air and water temperatures; hydrological variables such as precipitation and runoff; and topographic and bathygraphic data. Fisheries and biological data, sometimes species-specific, will also usually be required.

5.8.7 Results Generated

As with the data requirements, the outputs from the vulnerability assessment will depend on the type of fishery (i.e., lacustrine, riverine, coastal) and the climatic variable being evaluated (Tables 5.8.1, 5.8.2, and 5.8.3, and Appendix J). Some common outputs include changes in growth, mortality, and yield, and changes in habitat extent and/or suitability.

5.8.8 Advantages and Disadvantages

The approaches identified for use in evaluating potential climate change impacts to freshwater lacustrine and riverine and coastal marine fisheries resources represent a dichotomy of methods. Methods such as the bioenergetics and habitat suitability models will permit the evaluation of potential climate change impacts to individual species. In contrast, other approaches are species-insensitive, evaluating potential impacts to overall yield or catch. These latter approaches cannot be used to identify potential effects of climate change on individual species of concern. None of the methods are capable of evaluating the potential for changes in biotic interactions due to climate change.

The species-specific approaches will be the most problematic to employ, largely due to the relative absence of species data on life history and physiology. In the absence of species-specific data, professional judgement may be used to estimate life history and bioenergetics variables. Difficulties associated with the species-insensitive approaches will be largely associated with the absence of hydrological and environmental data.

The overall approach for identifying potential impacts from climate change to fisheries resources follows a "weight-of-evidence" approach (US EPA 1992). This approach relies on multiple lines of evidence to evaluate the potential for adverse or beneficial impacts to fisheries from climate change. However, the methods identified in this report are not fully integrated and will likely give conflicting results. For example estimates of mortality using the relationships developed by Pauly (1980) may indicate adverse impacts under a particular climate-temperature scenario, while estimates of maximum sustainable yield using the relationships of Schlesinger and Regier (1982) may indicate a positive impact under the same temperature scenario. The evaluation of multiple results, some conflicting, may rely heavily on the professional judgement of the fishery biologists performing the vulnerability assessment. It should be remembered that the results of the fisheries vulnerability assessment is not intended to provide a quantitative, definitive identification of the nature and magnitude of impacts to fisheries resources that would occur for a particular change in climate. Rather, the assessment is meant to provide an indication of the potential of adverse impacts to fisheries resources, and provide a preliminary indication if, or which, fisheries resources may be at risk.

5.8.9 Supplementary Approaches

The methods described in this report represent "suggested" approaches, and these are summarized in Tables 5.8.1, 5.8.2, and 5.8.3, and described in Appendix J. Because the level of accuracy varies widely from method to method it is recommended that as many of the suggested approaches be implemented as possible to adequately evaluate impacts to fisheries. Short of extensive, long-term, experimental field and laboratory investigations, the approaches identified in this report provide the best predictions possible within the limitations of the methods. Although the confidence and accuracy of any one approach is limited, the confidence in the overall assessment will be greater if several separate methods give similar results.

The final choice of methods is at the discretion of the fisheries staff that will be performing the actual assessment. For each fishery type (lacustrine, riverine, and coastal marine), the in-country staff may select one or more of each of the suggested approaches, or may opt to use other methods not identified in this report.

5.8.10 References for Section 5.8

Christie, G.C. and H.A. Regier, 1988, *Measures of Optimal Thermal Habitat and Their Relationship to Yields of Four Commercial Fish Species*, Can. J. Fish. Aquat. Sci. 45: 301-314.

U.S. EPA, 1992, *Framework for Ecological Risk Assessment*, EPA/630/R-92/001, Risk Assessment Forum, U.S. Environmental Protection Agency, Washington, D.C., February.

Kennedy, V.S., 1990, *Anticipated Effects of Climate Change on Estuarine and Coastal Fisheries*, Fisheries 15(6): 16-24.

Lowe-McConnell, R.H., 1987, *Ecological Studies in Tropical Fish Communities*, Cambridge University Press, Cambridge, Great Britain.

Meisner, J.D. and B.J. Shuter, 1992, *Assessing Potential Effects of Global Climate Change on Tropical Freshwater Fishes*, GeoJournal 28: 21-27.

Meisner, J.D., J.L. Goodier, H.A. Regier, B.J. Shuter, and W.J. Christie, 1987, *An Assessment of the Effects of Climate Warming on Great Lakes Basin Fishes*, J. Great Lakes Res.: 13: 340-352.

Pauly, D., 1980, *On the Interrelationships Between Natural Mortality, Growth Parameters, and Mean Environmental Temperature in 175 Fish Stocks*, J. Cons. int. Explor. Mer. 39(2): 175-192.

Regier, H.A. and J.D. Meisner. 1990. *Anticipated Effects of Climate Change on Freshwater Fishes and Their Habitat.* Fisheries 15:10-15.

Schlesinger, D.A. and H.A. Regier, 1982, *Climatic and Morphoedaphic Indices of Fish Yields from Natural Lakes*, Trans. Am. Fish. Soc. 111: 141-150.

Shuter, B.J. and J.D. Meisner, 1992, *Tools for Assessing the Impact of Climate Change on Freshwater Fish Populations*, GeoJournal 28: 7-20.

Welcomme, R.L., 1976, *Some General and Theoretical Considerations on the Fish Yield of African Rivers*, J. Fish Biol. 8: 351-364.

Welcomme, R.L., 1985, *River Fisheries*, FAO Fisheries Technical Paper No. 262, Food and Agriculture Organization of the United Nations, Rome.

5.9 WILDLIFE AND BIODIVERSITY IMPACT AND ADAPTATION ASSESSMENT

Ecosystems, with their myriad plant and animal species, are complex systems whose structure is intimately determined by climate. Responses to climate change involve patterns and processes over a wide range of temporal and spatial scales, from the nearly instantaneous physiological responses of individual organisms, to broad-scale geographic shifts in biomes over decades and centuries, to changes in the genetic makeup of populations over millennia.

As befits a diverse and complex subject, many approaches have been used to assess the vulnerability of organisms and natural communities to climate change. In this section, we give an overview of the major types of analyses and some examples of their application. Additionally, we discuss management options that can be used to adapt to impending climate change. We proceed from the simplest and least data-intensive methods to those that require a detailed understanding of the relationship between climate and natural processes. Some of the techniques are applicable in many different kinds of ecosystems; however, our focus is on methodologies that are useful in terrestrial ecosystems. Investigators interested in riverine, lacustrine, and oceanic systems are directed to protocols in Sections 5.4 (Water Resources), 5.5 (Coasts), and 5.8 (Fisheries).

5.9.1 Primary Approach

5.9.1.1 Scope of the Analysis and Screening Techniques

A first task in any vulnerability study is to define the scope of the analysis to be undertaken. Resources of interest might include those of particular economic value, protected areas of national importance, selected faunal groups or ecosystems for which good data are available, or taxa that are ecologically important.

Once a set of resources is defined, screening techniques can be used to help identify those at greatest risk from climate change. This may be a first step in identifying species, ecosystems, or protected areas that are in need of further analysis, or when sufficient data are not available, a substitute for more quantitative procedures.

5.9.1.1.1 Species screening. Indications are that climate change in the next 100 years will occur at much higher rates than in the past (Crowley and North 1988, Crowley 1990, Hinckley and Tierney 1992), hence a major threat is the change itself (Peters and Darling 1985, Peters and Lovejoy 1992). One screening approach is to attempt to assess general vulnerability to change regardless of its exact nature. Dennis (1993) used this approach and scored species with respect to: 1) the capacity of the their resources and habitat to withstand climate change and 2) the flexibility of the their biology in the face of climate change. Species with narrow resource/habitat requirements and inflexible biology were assumed to be the most vulnerable.

A second species screening approach also avoids uncertainty as to the exact nature of future climate changes and focuses on species or systems already in danger. Many organisms and ecosystems are already under extreme pressure from human activities and their response to climate change will be in the context of these existing stresses (Peters 1990, Markham and Malcolm in press). Millsap et al. (1990) developed a method to assess the current vulnerability of species. Several criteria were used to rank vulnerability, including biological characteristics, existing knowledge about population status, and current management investments.

5.9.1.1.2 Ecosystems and Protected Areas. As summarized in Table 5.9.1, research has identified several biomes and ecosystem types that are thought to be at special risk from climate change (Houghton et al. 1990, Peters and Lovejoy 1992, Leemans and Halpin 1992, Markham et al. 1993). These biomes can be mapped at national or regional scales to identify priority areas for research, monitoring, and management. Similarly, a number of characteristics that predispose protected areas to climate change impacts have been suggested (Table 5.9.2) and can also be used to rank protected areas or other geographic units with respect to overall vulnerability.

Table 5.9.1 Examples of Ecosystem Types Sensitive to Climatic Change (after Markham and Malcolm in press)

Ecosystem/Biome	Key Climate Sensitivities	Authors
Coastal wetlands	Sea-level rise, storms	Reid and Trexler (1992)
Mangrove forest	Sea-level rise, storms	Ellison and Stoddart (1991)
Island ecosystems	Sea-level rise, temperature, storms	Rose and Hurst (1991)
Coral reefs	Sea surface temperature, storms	Smith and Buddemeieir (1992), Agardy (1994)
Arctic ecosystems	Temperature	Alexander (1990), Chapin et al. (1992)
Alpine/Montane ecosystems	Temperature, precipitation	Halpin (in press), Beniston (1994)
Boreal forest	Temperature, fire regime, soil moisture	Shugart et al. (1992)
Tropical forest	Drought, seasonality, fire regime, hurricanes	Hartshorn (1992), Bawa and Markham (in press)

Table 5.9.2 Some Characteristics of Protected Areas that Predispose Them to Climate Change Vulnerability

---

1) Presence of sensitive ecosystem type

2) Presence of species and/or ecosystems near the edges of their historic distributions

3) Presence of species and ecosystems that have geographically limited distributions

4) Geomorphological uniformity

5) Small size and high perimeter:area ratio

6) Isolation from other examples of component communities

7) Human-induced fragmentation of populations and ecosystems

8) Existing anthropogenic pressures within and close to borders

9) Presence of natural communities that depend on one or a few key processes or species

---

5.9.1.2 Eco-climatic Classification

A number of schemes have been developed that use climate variables to classify natural communities and to predict their geographic distributions (Holdridge 1947, 1967, Bailey 1979, Box 1981, Prentice et al. 1992). These methods are of particular value in climate impact and vulnerability studies because of their explicit reliance on climate variables and because organisms are typically restricted to one or a few biomes. In the final analysis, climate is only one of the important variables that determines plant and animal distributions and adaptations (soils being another obvious one), hence any classification method that relies solely on climate is doomed to failure. However, these techniques are often relatively accurate at large geographic scales and can usually predict the locations of major physiognomic shifts in vegetation. Perhaps most importantly, these classification schemes allow the investigator to focus on combinations of climate variables that are important to organisms. For example, water availability to a plant will depend not only on precipitation, but also on the rate of water loss via evaporation, which is partly a function of ambient temperature. Thus, a map that already combines these two variables in a realistic way will be of greater value to a plant ecologist than one that presents only one while excluding the other.

The Holdridge Life Zone system (Holdridge 1947, 1967) is a widely used index and serves as an example of the approach. In this scheme, global biomes are depicted graphically as a function of bio-temperature, mean annual temperature, and the ratio of potential evapotranspiration to precipitation (see Section 5.3 for details).

5.9.1.3 Expert Judgment

This is perhaps the most commonly used approach, and it is a relatively simple way to generate ideas and discussion about potential impacts. Typically, a scenario is assumed, the opinion of experts as to impacts on organisms or ecosystems is solicited, and a workshop or other forum is organized to present and discuss results. Sometimes, expert judgment is collated or organized in some predetermined fashion. For example, Herman and Scott (1993) assessed the vulnerability of selected vertebrate groups in eastern Canada by adopting a specific climate change scenario, enumerating resultant changes in the physical environment of possible importance to the organisms, and soliciting expert judgment to score potential impacts for each change. Another example is the Habitat Suitability Index (HSI) technique developed by the U.S. Fish and Wildlife Service (U.S. Fish and Wildlife Service 1981). In this system, a simple model is developed to describe how features in the habitat act together in determining overall suitability for a wildlife species. An expert or team of experts develops the model which is then used to judge the potential impacts of future climate change (Figure 5.9.1; see Section 5.8 for additional details).

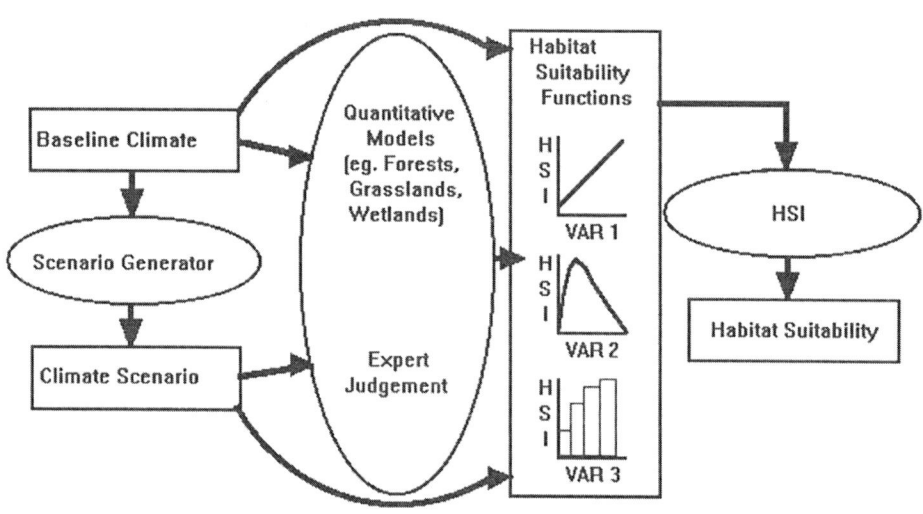

Figure 5.9.1. Flow Diagram Outlining Steps Involved in Implementing the Habitat Suitability Index (HSI) Technique.

5.9.1.4 Climate Envelopes and Profiles

This increasingly popular approach capitalizes on the fact that the geographic distributions of many species, communities, ecosystems, and biomes are strongly correlated with climate. In some cases, geographic limits may be caused by climatic thresholds; in others, climate may correlate with other factors that themselves determine the geographic limit (MacArthur 1972; Root 1988a, 1988b). The general idea is to model the relationship between baseline climate and the geographical distribution of interest and use the resultant model to predict the future distribution (Figure 5.9.2). For example, a species may currently occur where mean January temperatures are between 10 and 15 °C. Outputs from a climate model may indicate that the region of 10-15 °C has shifted poleward; accordingly, we can predict that the species will also shift poleward and, based on a map on temperature change, we can map its presumptive future distribution. Note that the predicted future distribution is after conditions have equilibrated—the species (or other geographical feature) is assumed to have somehow reached its new location. The analysis does not provide information on how far or fast it can move in reality, but is useful in assessing the amount and distribution of expected change.

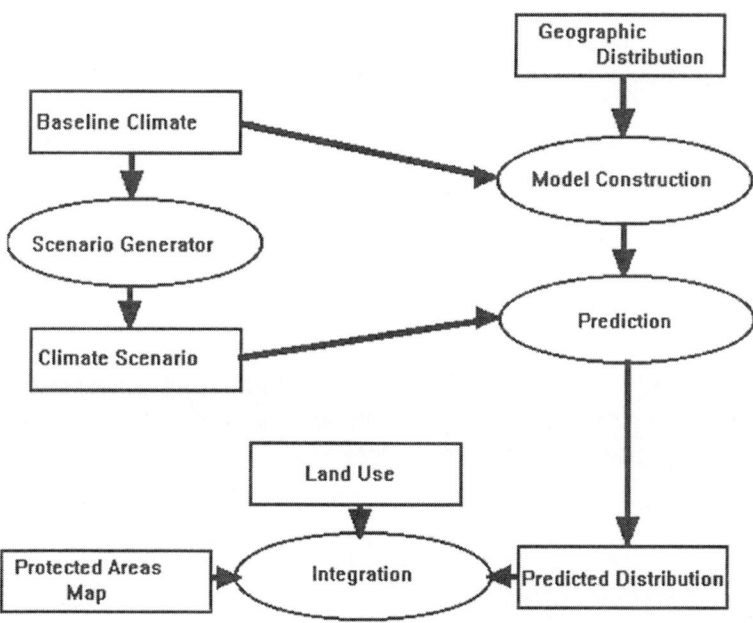

FIGURE 5.9.2 Flow Diagram Outlining Steps Involved in Implementing the Climate Envelope Techniques.

5.9.2 Description of the Methodology

5.9.2.1 Screening Techniques

5.9.2.1.1 Implementation. Both species screening techniques estimate vulnerability for each member of a group of species. Dennis (1993) designed his method with butterflies in mind; in an attempt to provide a more general formulation, we modified several categories so that they could be applied to other terrestrial organisms (Table 5.9.3). In both species screening techniques, each species is assigned a score for each variable and the scores are summed for groupings of the variables. Sums can subsequently be ranked, grouped, or mapped to prioritize adaptation and management investments (Millsap et al. 1990).

The scores are meaningful only in a relative sense, i.e. in comparison to scores from other taxa. Therefore, it is suggested that at least several taxa be scored. One or several of the following characteristics may be used in deciding whether or not a taxon is to be included in the analysis: taxonomic group, tropic level, feeding guild, economic value, geographic location, ecological importance, scientific importance.

Protocols to quantify vulnerability comparisons among protected areas or other geographic subregions have not been established. One possible procedure is to rank each protected area for each characteristic and sum the rankings to obtain an overall vulnerability index for each protected area. Alternatively, different characteristics can be differentially weighed to assign higher relative importance to some. Again, the final rankings obtained can be used to prioritize research, monitoring, or management.

5.9.2.1.2 Incorporation of Scenarios. Neither of the species screening methods incorporates climate scenarios. Instead, they attempt to assess overall vulnerability to future change regardless of its direction.

When biomes are classified as to their sensitivity (Table 5.9.1), implicit assumptions are made about the likelihood of particular climate change scenarios. Often, assumptions are also made about the way that climate interacts with organisms in the biome. Depending on current scientific understanding, certain climate change scenarios may be deemed more likely than others; hence, if desired, more attention can be devoted to certain biomes.

Some of the characteristics listed in Table 5.9.2 require assumptions about future climatic conditions, others are general measures of vulnerability to climate change irrespective of its direction, and some measure vulnerability irrespective of any climate change. Thus, the investigator can adjust the final set of characteristics to reflect knowledge about future conditions, or if desired, to make as few assumptions as possible.

5.9.2.2 Eco-climatic Indices.

5.9.2.2.1 Implementation. Typically, the indices are calculated under baseline conditions and under one or more scenarios. Useful analyses of the results include: (1) tabular and graphical presentation of biome types by geographic region under baseline conditions and under conditions of climate change, and (2) tabular and graphical presentation of areas in transition and areas of stability. In the latter analysis, it is useful in addition to tabulate the most extreme transitions. Eco-climatic analyses are especially useful when overlaid with land-use patterns and maps of protected areas (e.g. Leemans and Halpin 1992, Halpin and Secrett, 1995) and when used in concert with other vulnerability techniques (see Section 5.9.2.5).

Table 5.9.3 Species Vulnerability Characteristics Used to Screen Resource Capacity and Biological Flexibility in the Face of Climate Change (Modified from Dennis [1993])

Capacity Factors

A. Species range size, based on the percent latitudinal extent of the species on the continent.
B. Species distribution pattern, based on the percent of 10 km squares occupied within the range of the species.
C. Dietary specialization, based on the degree to which the taxon is restricted to certain food types and the way in which local populations respond to decreases in availability of preferred food type.
• Restricted to just one food type
• Restricted to just a few food types
• In response to shortage of preferred food, no substantial shift in diet, population number declines,
• In response to shortage of preferred food, substantial shift in diet, little change in population numbers,
D. Abundance of food resources.
• Food is substrate dependent
• Food is patchy within habitats
• Food is ubiquitous within habitats
• Food is ubiquitous across habitats
E. Vulnerability of major habitat seral stage occupied.
• Reaches maximum density at low-disturbance sites in climax habitat
• Reaches maximum density in naturally disturbed sites in climax habitat
• Reaches maximum density under moderate anthropogenic habitat modification
• Reaches maximum density under high anthropogenic habitat modification
F. Percent of the naturally occurring habitats in the region that the species occupies.

Flexibility Factors

G. Dispersal ability.
• Restricted population structure (e.g. colonies) with little evidence of movement outside
• Restricted population structure with evidence of movements outside
• Open population structures with evidence of frequent movements between habitats
• Migrants and species which are vagrant and known to engage in long-distance movements
H. Timing of reproduction.
• Highly seasonal and restricted to one attempt
• Highly seasonal with possibility of second attempt
• Two attempts common, broad season of reproduction
• Reproduction relatively aseasonal
I. Number of breeding attempts.
• Semelparous
• Two seasons of reproduction in life
• Multiparous, relatively high adult mortality between seasons
• Multiparous, relatively low adult mortality between seasons
J. Survival over yearly season of resource scarcity/climatic extremes.
• Population of harsh season comprised of immature individuals
• Population of harsh season comprised largely of immature individuals
• Population of harsh season comprised largely of adults, highly seasonal environment
• Population of harsh season comprised largely of adults, little seasonality in environment

5.9.2.2.2 Incorporation of climate change scenarios. Changes in biome types under incremental scenarios (e.g. + 3 °C, ± 10% precipitation) can be easily investigated. However, many meteorological variables are ignored in the classification schemes and can be investigated only indirectly. Several strategies have been developed to apply Global Circulation Models (GCMs) outputs to baseline climate in order to derive future scenarios suitable for eco-climatic analyses (e.g.. Smith et al. 1990).

5.9.2.3 Expert Judgment

5.9.2.3.1 Implementation. Prior to soliciting expert opinion, several decisions must be made. These include: the taxa or ecosystems of interest, the form of any assumed climate change scenario, and methods used to organize the resulting information. In the specific example of the HSI system, specific steps are required in order to create the model: (1) taxa and cover types of interest are chosen, (2) important life requisites and other limiting factors are identified, (3) single-variable relationships between habitat characteristics and habitat suitability are developed, and (4) relationships among the single-parameter models are established. Once a model has been constructed, various types of analyses are possible: habitat conditions under baseline and future climate conditions can be compared, sensitivity analyses can be used to identify key habitat features, impacts of management and adaptation strategies can be assessed, etc.

5.9.2.3.2 Incorporation of climate change scenarios. Typically, a climate change scenario is assumed. It may be phrased in general terms, such as "a general warming trend" or it may be more exact. In some HSI models, climate variables will number among the modeled single-parameter relationships, hence integration of HSI simulations and quantitative climate models will be straightforward. In many cases, however, climate will determine habitat suitability indirectly through its action on other habitat parameters. For example, island area is an important measure of the suitability of nesting and roosting habitat for the roseate spoonbill (Lewis 1983) and climate may indirectly influence island area through its effects on sea levels. Several approaches are available in these cases. One is to model the relationship between climate and the parameter of interest. For example, based on sea level and shore type, island area can be estimated for different climate scenarios (Section 5.5). Another is to use expert judgment to enumerate possible climate effects on the habitat parameters, and use HSI to perform sensitivity analyses to identify the most important ones.

5.9.2.4 Climate Envelopes and Profiles

5.9.2.4.1 Implementation. Several techniques incorporating various degrees of sophistication have been used to model the relationship between climate and geographic distributions. They differ in their choice of the original pool of climate variables, the method used to select the climate variables in the model, and the statistical form of the model itself (Table 5.9.4). All of these techniques are correlative and not causative. The usual caveat applies: as more variables are included in the model, the fit improves (that is, the model better describes the geographic distribution), but the model increasingly relies on spurious relationships. To take the extreme case, if we have as many climate variables as localities, then we can devise a model that perfectly distinguishes sites were a species does and does not occur, even if there is no correlation between the species distribution and climate in the first place. Under conditions of climate change, this model will predict extinction across the entire species range simply because of the new combination of climate variables at each locality. For these reasons, we recommend methods that choose some subset of the best climate variables. As a rule of thumb, we suggested that several models be constructed, including a simple one that uses only a few of the best predictors. The inclusion of extraneous non-predictive variables may lead to spurious results and is not recommended. An additional refinement is to use multivariate techniques to identify discrete "climate subspecies" (should they exist) and create separate climate profiles for each.

Table 5.9.4 Examples of Climate "Envelope" Approaches

Study	Taxa	Number of climate variables in pool	Variable selection technique	Models[a]
Morse et al. (1993)	North American vascular flora	1	N/A	1, 2, 3
Bennett et al. (1992)	Australian vertebrates	19	none	1, 4
Price (1995)	North American birds	18	logistic regression	5
Malcolm (in press)	Southern African ungulates	27	step-wise	6
Malcolm (in press)	Southern African ungulates	27	step-wise	7
Rogers and Williams (1993)	Tsetse fly	5[b]	step-wise	8

[a] 1 = observed range
2 = mean ± one standard error
3 = 16% trim from each tail of distribution
4 = 10% trim from tail of each distribution
5 = logistic regression
6 = discriminant analysis (homogeneous covariance matrices)
7 = multiple regression
8 = discriminant analysis (heterogeneous covariance matrices)

[b] Additional variables included elevation and the Normalized Difference Vegetation Index

5.9.2.4.2 Incorporation of climate change scenarios. Because of the direct reliance in this approach on climate variables, climate scenarios can easily be incorporated. Of course, the baseline climate variables used to model the distributions must be available in the scenarios.

5.9.2.5 Model Integration

The models can provide an integrated framework for assessing biological vulnerability to climate change. In some cases, only a subset of the methods can be used because data or resources are lacking or because the assumptions of the models are not met.

Several general strategies are available for integrating the models:

- Preliminary screening of reserves and protected areas can be coupled with detailed investigation of specific taxa and communities using screening techniques, expert judgment, and climate profiles.

- Maps of overall species vulnerability can be overlaid with the geographic boundaries of reserves and protected areas to assess the strength of the reserve system in protecting threatened organisms.

- Maps of distributional changes predicted using eco-climatic indices (e.g. Halpin and Secrett, 1995), the climate envelope approach, or spatial changes in habitat suitability (e.g. Lancia et al. 1986), can be overlaid with the geographic boundaries of reserves and protected areas to examine the extent to which taxa and communities will be represented in the protected areas in the future and to identify protected areas where climate change may be manifested.

- Key taxa from vulnerable eco-climatic regions can be investigated using screening techniques and process-based models to determine which might be impacted first or most (indicator species).

- Predictions from one model can be tested for general agreement with predictions from other models. For example, sites where a species is predicted to go extinct based on climate correlations should show a decrease in habitat suitability using HSI.

- Different levels of detail can be achieved depending on data availability. For species with poorly-known habitat requirements and geographic distributions, preliminary screening can be undertaken. For species or systems that are well understood and the subject of active research, detailed models can be constructed.

- Monitoring efforts can be preferentially devoted to distributional limits in areas of predicted change. Dynamic models and analog information can be used to study the time-frame of any predicted change.

- Research can be preferentially devoted to species and areas identified as being particularly vulnerable.

5.9.3 Economic Analysis

An economic analysis of the impacts of climate change on wildlife and biodiversity must consider impacts on the supply of the resource and secondary impacts on values and pricing that result from changes in supply. The techniques outlined above can provide information on potential changes in productivity and biomass. Examples of the integration of model outputs with economic analysis include:

- When species can be assigned direct economic values, for example based on their harvest value, the techniques can provide a measure of the economic costs associated with climate change.

- Given relationships between harvest rates, productivity, and habitat suitability, the influence of climate change on economic returns from harvests can be investigated using habitat suitability indices.

- Changes in species and habitat distributions can be overlaid with maps of regional resource utilization to identify regions of high impact and areas of likely shifts in resource utilization patterns.

Unfortunately, assigning realistic monetary values to wildlife resources is often a difficult task (McNeely 1988). In many cases, benefits fall to groups other than those who utilize or manage the resource and externalities are often subtle and difficult to calculate. Perhaps most seriously, long-term benefits are ignored in markets oriented to short-term gain. Economic analyses based on the market values of wildlife resources consider only a fraction of the national, regional, and global benefits accrued from natural communities.

5.9.4 Adaptation Techniques

Various strategies are available to aid in efforts to conserve wildlife and biodiversity, including the establishment and maintenance of protected areas (*in situ* preservation), the active management of wild populations outside of protected areas (*inter situ* management), and the maintenance of captive populations (*ex situ* methods) (Table 5.9.5). Of these, highest priority should be placed on *in situ* and *inter situ* conservation. Adaptation strategies should be developed within the context of global, regional, and national biodiversity conservation priorities. A protected areas review is a logical first step in defining conservation objectives and goals and in assigning adaptation priorities.

A large number of habitat management and intervention techniques can be used as part of an overall adaptation strategy. Many of these are already in use in protected areas and managed reserves throughout the world, and the techniques themselves can be adapted for use under a new set of climatic conditions (Table 5.9.6).

Often, strategies for adaptation to climate change must be developed in the absence of precise information about either the nature of the climate change itself or ecological responses. In the face of these uncertainties, several general principles apply (Markham and Malcolm, in press):

- Strategies should seek to maintain ecological structure and processes at all levels.

- Maintenance of ecological complexity is necessary in order to maximize evolutionary and ecological potential in species and ecosystems.

- Ecological resilience is the single most important factor influencing the ability of wildlife and natural habitats to withstand climatic change. Resiliency can be increased by:

  - Conserving biological diversity.

  - Reducing fragmentation and degradation of habitat

- Increasing functional connectivity amongst habitat blocks and fragments.

- Reducing anthropogenic environmental stresses. Species and ecosystems already under stress from environmental degradation and human pressure are likely to be the most vulnerable to the added stress of climatic change.

• Reducing vulnerability to climate change is likely to have strong additional benefits through the simultaneous reduction of vulnerability to other environmental and anthropogenic stresses.

Table 5.9.5. Typology of Biodiversity Conservation Strategies

Strategy	Types of Activity
In situ	Protected areas
Inter situ	Conservation outside protected areas, e.g. habitat conservation, development restrictions, buffer zones Extractive reserves Resource harvesting on a sustainable basis Ecological restoration Intensive management to restore degraded habitats and landscapes Zooparks Maintenance of artificial mixes of species under semi-natural conditions, e.g. game farms Agroecosystems and agroforestry High management production oriented systems, e.g. plantation forests, forest gardens
Living *ex situ*	Zoos, botanical gardens, aquaria
Suspended *ex situ*	Germplasm storage, e.g. seed banks

Table 5.9.6 Intervention Strategies (after Markham and Malcolm in press)

Strategy	Examples of current use	Potential use in a changing climate
Ecosystem restoration	Restoration of water meadows and riparian habitat along the Rhine in the Netherlands	Restoration of degraded land to provide connectivity between existing reserves.
Prescribed fire and fire exclusion	Use of fire to maintain age-suitability of jack pine habitat for Kirtlands warbler breeding in Michigan, USA	Prevent conversion of savannah to shrub-dominated communities.
Species relocation	Removal of elephants from Zimbabwe to South Africa to relieve population pressure in parks	Remove species from newly unsuitable habitat and relocate them in new areas. to which natural migration would be impossible
Removal of impediments to migration and colonization	Closure of logging roads in Western USA	Prepare land for colonization of desired species, e.g. by removal of scrub to allow forest recruitment. Remove dike or road systems preventing inland migration of coastal wetlands. Remove fences or provide bridges/tunnels.
Assisted migration or reintroduction	Reintroduction of wolves to Yellowstone National Park, USA	Capture and move animal species past obstacles to migration (e.g. agriculture, industrial developments) in direction of climatically forced population migration. Plant seeds, move soil.
Control of alien or invasive species	Eradication programs in native forest in Haleakala National Park, Hawaii	Monitor for new invasive species and prevent their spread. Minimize disturbance (e.g. canopy openings in tropical forest) to reduce susceptibility to invasion.
Control of disease	Tsetse fly control programs in southern Africa	Monitor for changes in disease distribution and expansion of range of disease vectors. Plan disease control strategies.
Irrigation or drainage	Creation of shallow, brackish coastal lakes at Minsmere Reserve (UK) to provide shorebird habitat	Use water management technologies to reduce impacts of drought or sea-level rise. Create new wetlands.
Food and water provision	Provison of water in game areas during droughts in Botswana	Plan or expand programs aimed at ameliorating impacts of drought or famine.

In addition to these general principles, several adaptation principles apply specifically to protected areas (Markham and Malcolm in press):

- Adaptation strategies should emphasize redundancy. Just as poorly enforced regulation or political instability can be arguments for increased number of reserves and therefore, greater redundancy, so too can the uncertainty of climate change impacts.

- Design considerations that promote the evolutionary potential of species within protected areas include maximization of reserve connectivity, size, and number.

- A protected areas network must balance preservation of ecological complexity with preservation of landscape diversity. Altitudinal range within reserves is important because species may be able to migrate upslope to avoid the consequences of warming (Peters and Darling 1985, McNeely 1990). Heterogeneity of topography, habitat, and microclimate in reserves allows for greater flexibility of organismal responses to climate change. Gap analysis systems based on representative ecosystem and enduring landscape features will have to become more dynamic (Halpin in press). Reserve planning in a changing climate must examine future needs in areas where present-day conservation needs are few, but where migration and dispersal of valued species may make protection necessary or desirable.

- Flexible zoning of reserve boundaries and the development of more effective buffer zone management will play an increasing role as climate change forces changes in species distributions (Peters and Darling 1985, Bennett et al. 1991, Parsons 1991).

- Fragmentation itself may be the single biggest barrier to ecosystem adaptation in a changing climate (Bawa and Markham, in press). Even where paleoecological studies suggest that species have been able to adapt to rapid climatic changes in the past, current habitat fragmentation patterns and human barriers may prevent range shifts. Edge effects that accompany fragmentation expose complex habitats to climatic extremes in adjacent simple habitats (Malcolm 1994). Thus, reduction of fragmentation rates is a critical climate mitigation strategy and increases in connectivity is a high priority adaptation response. Because range shifts are a likely response to climate change, corridors will need to function as habitat rather than as mere transit lanes (Simberloff et al. 1992).

5.9.5 Data and Computational Requirements

5.9.5.1 Screening Techniques

Because of the requirements for information on geographic distributions, the species screening techniques are implicitly regional in scope. Requirements include data on geographic distributions, population numbers, habitat and diet requirements, and reproductive seasons and parameters. In a few cases, data from manipulative or analog experiments is desired. Ultimately, the screening techniques are very flexible with regards to data requirements. Variables can be easily added or modified. As a final resort, variables for which data are lacking can be dropped from the analysis.

The screening of protected areas requires that biomes or ecosystems be mapped at a national or regional scale. Information from remote sensing and from maps of eco-climatic indices will aid in the creation of these maps. Other map layers that are useful include: (1) protected areas, (2) topography and soils, (3) geographic distributions of key taxa, (4) human population, and (5) land use classification. Geographical Information Systems (GIS) facilitate the storage, presentation, and analysis of these types of data.

5.9.5.2 Eco-climatic Indices

Global data sets of baseline monthly precipitation and temperature are currently available, however they are geographically imprecise because they have been interpolated to create grids. If available, locally collected data are preferred. The global data sets offer the advantage that various GCM scenarios have been created at the same resolution. Undertaking a regionally comprehensive analysis will require relatively sophisticated computation: calculating the eco-climatic indices, applying scenarios, and tabulating and graphing the results.

5.9.5.3 Expert Judgment

In general, data requirements will vary from expert to expert and from method to method. The HSI model is available as shareware written in BASIC, and will run on virtually any MS-DOS system.

5.9.5.4 Climate Envelopes and Profiles

Three data sets are required: baseline climate, the geographic distribution of the species or community of interest, and future climate scenarios. As a rule of thumb, climatic information from a minimum of 30 sites within a species or community range should be contrasted with data from at least as many points outside of the species range. Primary data from meteorological stations within the region of interest are preferred; however, global databases of mean monthly temperature and precipitation are available at a spatial resolution of $0.5° \times 0.5°$. Conveniently, several GCM scenarios are available at this same scale of resolution. The collection and compilation of numerous climate variables and statistics is encouraged because it increases the chances of finding a strong association between climate and the geographic distribution of interest.

Unfortunately, distributional information will usually be available for only a few groups of organisms. Because it is the boundaries of the distribution that are actually used to derive the model, the whole geographic range of a species or vegetation type is usually used in deriving "climatic profiles." Thus, although analysis of final results may be at a subregional level, calculation of the predictive model in most cases must be at a regional or even continental scale.

5.9.6 Results Generated

5.9.6.1 Screening Techniques

As an example, we show an analysis of the bird vulnerability in Florida as determined by Millsap et al. (1990) (Figure 5.9.3). Maps of vulnerability such as this one can be combined with protected areas maps to help set conservation priorities.

Halpin and Secrett (1995) recently conducted a vulnerability study of protected areas in Costa Rica. They used the Holdridge Life Zone scheme (see Section 5.3) to map vegetation types under current climate conditions and under two climate change scenarios. Areas of decreased life zone diversity and of changes in potential vegetation zones were plotted in several types of forest management areas.

5.9.6.2 Eco-climatic Indices

An illustrative example of the Holdridge method is shown in Figure 5.3.3.

5.9.6.3 Expert Judgment

Herman and Scott (1992) focused on eastern Canadian vertebrates thought to be highly susceptible to climate change: salamanders, shrews, and turtles. A regional climate change scenario based on consensus from global circulation models assumed decreased summer precipitation, increased winter precipitation (especially rainfall), and higher winter temperatures. Species were ranked with respect to overall climatic sensitivity and general vulnerability.

General vulnerability was assessed using Millsap et al. (1990). To score climatic sensitivity, a list of life history characteristics was developed and experts scored the sensitivity of each to the hypothesized climate changes. General vulnerability was high for the salamanders, turtles, and shrews, indicating that many populations were already in danger, irrespective of impending climate change. Also, for reptiles and amphibians, action scores were huge and revealed "monumental ignorance." The analysis suggested that shrews would be very sensitive to changes in winter conditions such as increased winter flooding and reduced snow and ice cover.

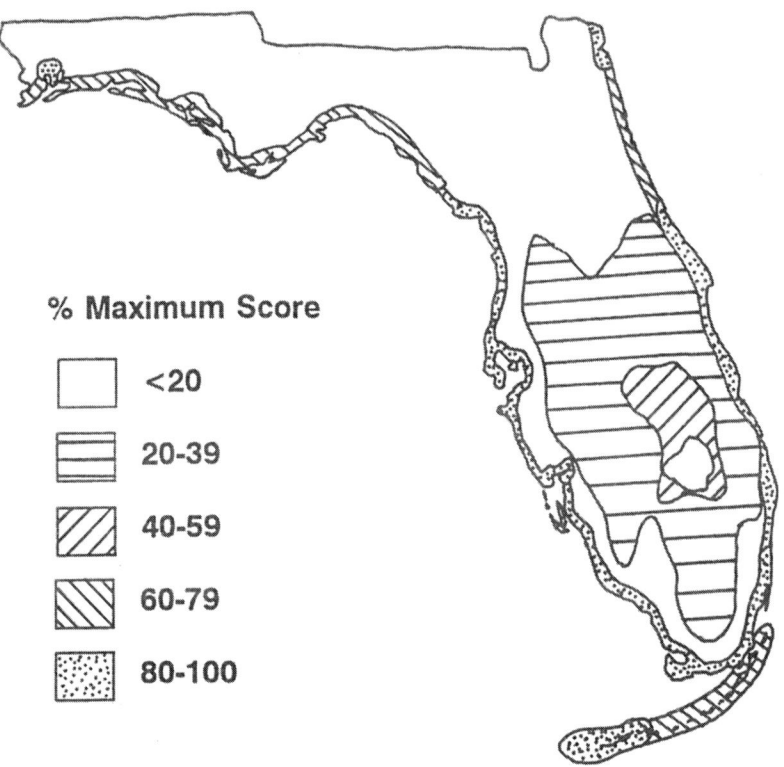

Figure 5.9.3. Geographic patterns in the vulnerability of bird species in Florida as determined by Millsap et al. (1990). Range maps of taxa with biological scores ≥24 were overlaid and scores were summed where ranges overlapped.

As an example of the HSI approach, we linked an HSI model with a forest gap model (see Brody et al. (1989) for another analysis that integrates forest succession models and HSI) and examined the implications of climate change in central Alaska for the marten (*Martes americana*), an important furbearer. The HSI model was developed by Allen (1982) for the marten's winter habitat in the western U.S. According to the model, marten required forested areas where canopy coverage was 50% or more and where succession was at least in the pole/sapling stage. Given these limits, optimal habitat was old or mature spruce/fir forests where fallen timber (>7.6 cm diameter) covered 20 to 50% of the ground surface. The model incorporated a variety of requirements for shelter, nesting habitat, and prey availability (Allen 1982, see also Laymon and Barrett 1986).

Results from a forest "gap" model under four GCM scenarios were expressed in terms of the biomass (t/ha) of three tree species (Figure 5.3.9). In the baseline simulations, the observed biomass relationships were maintained: the site was dominated by white spruce (*Picea glauca*) and birch (*Betula papyrifera*), and aspen (*Populus tremuloides*) was a minor component. When simulations were run under equilibrium $2 \times CO_2$ conditions from the GCMs, dramatic changes in community composition were observed primarily due to an increase in aridity. This result was in general agreement with a Holdridge analysis that predicted a shift from forest to steppe in the area.

We linked the marten model with the gap model by establishing quantitative relationships between species-specific tree biomass and the marten habitat variables. Habitat suitability was calculated at the start of the gap model simulations and at 25-year intervals thereafter (Figure 5.9.4). Under baseline climatic conditions and at 25 years into the gap models, the forest was nearly optimal for marten. However, by 50 years into the model habitat suitability had dropped to zero for the two scenarios with greatest drying trends. For the two more moderate GCM scenarios, habitat suitability was close to zero by 75 or 100 years into the simulations. Interestingly, even though stand dieback was prolonged for these latter two GCMs, the decline in marten habitat suitability was precipitous and occurred within 25 years. The dramatic decline can be attributed to the critical 50% canopy cover value posited in the HSI model.

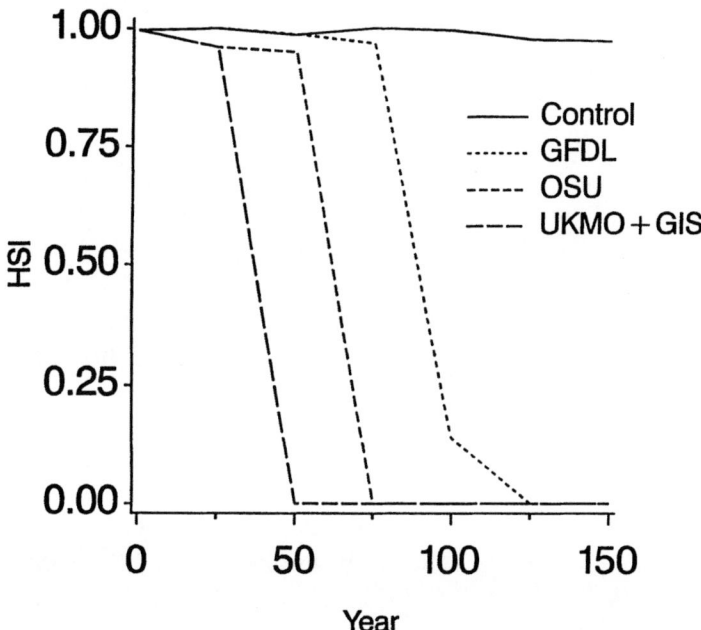

Figure 5.9.4. Simulated habitat suitability (HSI) for marten (*Martes americana*) based on outputs from a forest gap model at a site in central Alaska.

5.9.6.4 Climate Envelopes and Profiles

As an example, we modeled antelope diversity in the short-grass savannas of southern Africa (Figure 5.9.5). Based on the correlation between species diversity and climate, diversity in the area was predicted to decrease under three GCM scenarios.

5.9.7 Advantages and Disadvantages

5.9.7.1 Screening Techniques

The species screening techniques are easy to perform, they require relatively unsophisticated input information, and vulnerability can be assessed regardless of the exact nature of future climate change. However, generality is achieved at the expense of realism; for example, all variables are assumed to be equally important and interaction among variables is ignored. Also, the vulnerability scores are themselves relative measures and are useful only when compared with other scores. On the positive side, these techniques are of value irrespective of the eventual course of climate change, because they help to identify species that are already in trouble, or species that are likely to be in trouble in the face of additional negative human impacts. They can thus help in the establishment of national and regional priorities and action plans.

The vulnerability characteristics for protected areas listed in Table 5.9.2 are guidelines only and require refinement in order to form the basis for a quantitative, or even qualitative, ranking scheme. For example, to score protected areas according to "presence of taxa with geographically limited distributions," decisions must be made as to which floral and/or faunal groups are to be considered and how distributional sizes will be ranked. Moreover, for final analysis, decisions must be made as to the relative importance of the various characteristics. A detailed analysis of the vulnerability of protected areas requires the amassing of considerable amounts of data. Some is already available on a global basis albeit at low resolution. Much of the data will be of immediate use in other efforts to conserve natural resources; for example, in regional planning schemes or in plans to establish new protected areas.

5.9.7.2 Eco-climatic Indices

No eco-climatic classification system will accurately predict the distribution of biological communities, especially at fine spatial scales. Hence, the models are implicitly limited to analyses at the regional scale or above. However, the primary use of these models should not be to predict actual vegetation types, but to investigate directions of change. In essence, they simplify and make more understandable (at least from a biological viewpoint) complex meteorological change.

As techniques to predict actual vegetation or community changes, the eco-climatic classification techniques suffer from another defect: they assume that vegetation is in equilibrium with climatic conditions. That is, they provide no information on rates of change. They are thus static and non-interactive descriptions of a very dynamic process. However, the techniques can indicate directions of change, and hence can identify the biota in an area that might be at an advantage or a disadvantage under a certain scenario.

5.9.7.3 Expert Judgment

The compilation of expert judgment can be very useful if the scenario seems likely and if the biology of component organisms or ecosystems is well-known enough to judge likely impacts. The mere act of bringing together specialists from diverse disciplines can stimulate new approaches and ideas. Of course, if a particular scenario is assumed from the start and it later becomes necessary to consider a

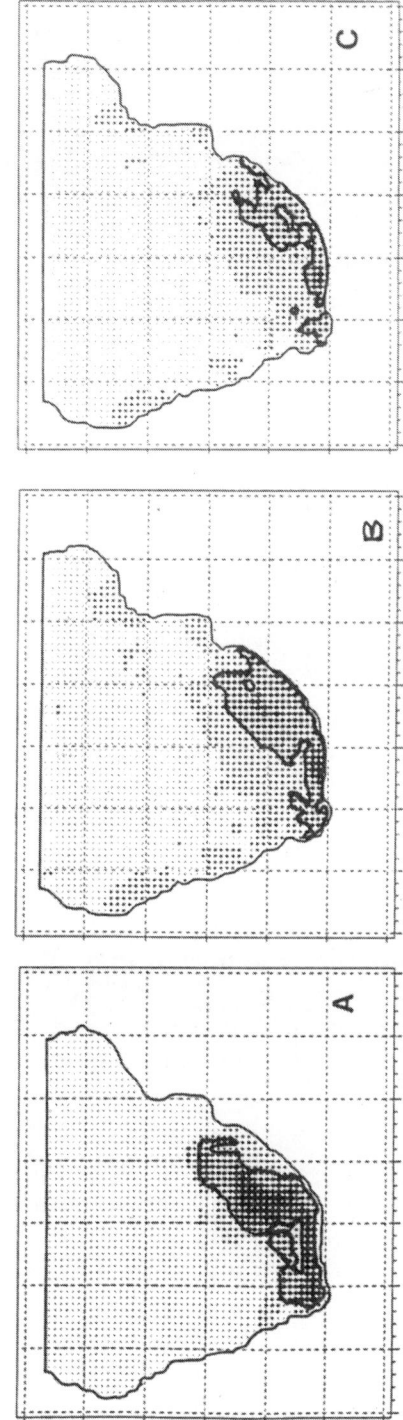

Figure 5.9.5. Diversity of "short-grass savanna" ungulates in southern Africa (after Malcolm in press). Outlined areas have two or more species.
A: Observed diversity
B: Modeled diversity under baseline conditions (step-wise multiple regression)
C: Predicted diversity under the CCC GCM scenario

different scenario, expert judgment must be solicited again. The expert judgment metho suffer from a lack of hereobjectivity—different experts will likely give different judgments as to climate impacts. Objectivity can be increased by including several experts and, if possible, by arriving at a consensus.

The HSI approach and others that are restricted to a particular class of models are seriously compromised if the class of models is unrealistic. The HSI approach suffers this defect—although model creation is very flexible, these are basically descriptive models of dynamic processes. Interactions among habitat variables cannot be incorporated nor can dynamic aspects of the responses to changes in habitat suitability (for example, feedback relationships between habitat variables and species abundance are ruled out). The focus is on pattern instead of process. The utility of the models in understanding species-habitat relationships is thus compromised and, perhaps as a result, models are rarely tested or refined (for example, of the 120 terrestrial models developed for the U.S., so far only 3 have been tested). From a climate change perspective, the models suffer an additional defect. For many organisms, climate does not directly determine habitat suitability, hence climate change scenarios can be incorporated only by making additional assumptions about the way in which climate interacts with the habitat characteristics of interest. On the positive side, the assumptions made are explicit and the simplicity of the approach allows users to easily create models that capture key elements of habitat suitability. As usual, simplicity is both an asset and a liability.

5.9.7.4 Climate Envelopes and Profiles

Climate correlation can be a useful tool in several types of analyses. Knowledge about climate associations can provide insight into a species biology and is potentially useful in habitat suitability analyses and in incremental climate change studies. Historic range shifts or manipulative experiments can be used in some cases to validate relationships. Also, predictions under baseline conditions can be mapped to test for correspondence with observed distributions. In some cases, this technique can locate previously unknown populations of a species (eg. Bennett et al. 1991). Expected distributions under climate change scenarios can be overlaid with landuse maps. Models can be created for several species in a region in order to investigate changes in biodiversity (e.g. Bennett et al. 1991).

A key assumption is that the geographic distribution of the element is in equilibrium with climate conditions. Thus, data from elements that are not in equilibrium with climate must be used with caution. Examples include introduced species that are currently expanding their geographic range, species whose distributions are importantly influenced by non-ecological factors (for example, those whose distributions are restricted by rivers, mountain ranges, or other geographic barriers), and species whose geographic distribution have been importantly modified by human activities. Whenever possible, *historical* distributions should be used, i.e. distributions that pre-date important human impacts. Ideally, geographic distributions should be from the same period as the climate baseline scenario, or from a period that had similar climate.

These techniques rely entirely on patterns of correlation, and as such they may not identify the key processes that drive wildlife responses to climate change. A host of potentially important limiting factors, for example species interactions (competition, predation, and parasitism) are ignored. Also, baseline or scenario data sets often do not include information on extreme climatic events, which can be fundamentally important in limiting distributions. No mechanism is provided for investigating the ability of species or ecosystems to track (migrate with) climate change. The models are static, equilibrium descriptions of a very dynamic process.

5.9.8 Supplementary Approaches

5.9.8.1 Research

Plant and animal populations serve as a barometer of ecosystem and global health. Extinction of species ultimately reduces the capacity of a system to respond to additional changes and its ability to provide useful services to human populations. Thus, losses of species are indicative of inappropriate and, in the final analysis, dangerous management practices.

Basic monitoring is required to establish population levels and distributional limits and to detect changes. In addition, research should be devoted to understanding the effects of temporal and spatial variability in climate on natural populations. These "analog" studies provide information on the kinds of climate changes and events that influence the resources of interest. In a very basic way, increased understanding of the role of climate in natural systems will improve our ability to detect and respond to global climate change. Many field studies have increased our understanding of the role of climate in natural systems (e.g., Foster 1982, Schreiber and Schreiber 1984, Hamburg et al. 1988, Laurie and Brown 1990, Overpeck et al. 1991, Stiles 1992, Pounds and Crump 1994).

5.9.8.2 Process-based models

Process-based models attempt to model key causative agents in a system and are very diverse in their approaches (see examples in Table 5.9.7). It is useful to distinguish two general sorts of models, what we may call tactical and strategic. The former are very detailed models of specific systems, and their objective is to mimic the behavior of the system as closely as possible. These models are probably the most common type in the climate change literature, and can provide very detailed predictions as to vulnerability and climate change impacts. Frequently, several models are linked to provide a more-or-less integrated whole. These sorts of models are especially useful in concrete problems of resource management. On the down side, they are so specific that it is difficult to extract general insight, and the calculations are so detailed that the model behavior is not transparent. Strategic models, on the other hand, sacrifice detail for generality, and attempt to incorporate the essences of a broader class of systems. Of course, there is a whole continuum of approaches lying between these two extremes that reflects attempts to balance thoroughness and realism on one hand and generality and workability on the other. Investment in modeling efforts helps to sharpen hypotheses by making predictions more precise and more easily tested. A second function is to generate new possibilities and ideas.

Table 5.9.7 Examples of Processed-Based Models in Climate Change Research

Model Type	Examples
Forest "Gap" Models	Botkin et al. 1972
	Botkin and Nisbet 1992
	Shugart and Smith 1992
	Shugart 1990
	Botkin et al. 1991
	Cambell and McAndrews 1993
Plant physiological models	Running and Coughlan 1988
	Running and Nemani 1991
Arctic plants	Reviewed in Reynolds and Leadley (1992)
Temperate Grasslands	Hunt et al. (1991)
Bird Populations	Rodenhouse (1992)
Prairie Wetlands	Poiani and Johnson (1991)
Coastal Ecosystems	Costanza et al. (1990)

5.9.9 References for Section 5.9

Agardy, M.T., 1994, "Advances in Marine Conservation: The Role of Marine Protected Areas," *Trends Ecol. Evol* 9:267-270.

Alexander V., 1990, "Impacts of Global Change on Arctic Marine Ecosystems," in McCullock, J. (ed.) *Global Change: Proceedings of the Symposium on the Arctic and Global Change*, The Climate Institute, Washington DC.

Allen, A.W., 1982, *Habitat Suitability Index Models: Marten*, U.S. Fish Wildl. Serv. Biol. Rep. FWS/OBS-82/10.11.

Bailey, H.P., 1979, "Semi-arid Climates: Their Definition and Distribution," in A.E. Hall, G.H. Cannell, and H.W. Lawton (eds.), *Agriculture in Semi-arid Environments*, Springer-Verlag, Berlin.

Bawa, K., and A. Markham, in press, "Tropical Forests and Climate Change," *Trends Ecol. Evol.*

Beniston, M., (ed.), 1994, *Mountain Environments in Changing Climates*, Routledge, London.

Bennett, S., R. Brereton, I. Mansergh, S. Berwick, K. Sandiford, and C. Wellington, 1991, *Enhanced Greenhouse Climate Change and its Potential Effect on Selected Victorian Fauna*, Arthur Rylah Institute Tech. Rep. No. 132, Department of Conservation & Environment, Victoria.

Botkin, D.B., and Nisbet, R.A., 1992, "Projecting Effects of Climate Change on Biological Diversity in Forests" in R.L. Peters, and T.E. Lovejoy (eds.), *Global Warming and Biological Diversity*, pp. 277-293, Yale University Press, New Haven.

Botkin, D.B., A.D. Woodby, and A.R. Nisbet, 1991, "Kirtlands Warbler Habitats: a Possible Early Indicator of Climatic Warming," *Biol. Conserv.* 56: 63-78

Box, E.O., 1981, *Macroclimate and Plant Forms: an Introduction to Predictive Modeling in Phytogeography*, Junk, Hague, Netherlands.

Brody, M., W. Conner, L. Pearlstine, and W. Kitchens, 1989, "Modeling Bottomland Forest and Wildlife Habitat Changes in Louisiana's Atchafalaya Basin," in R.R. Sharitz, and J.W. Gibbons (eds.), *Freshwater Wetlands and Wildlife*, CONF-8603101, DOE Symposium Series No. 61, DOE Office of Scientific and Technical Information, Oak Ridge, Tennessee.

Cambell, I.D., and J.H. McAndrews, 1993, "Forest Disequilibrium Caused by Rapid Little Ice Age Cooling," *Nature* 366: 336-338.

Chapin III, F.S., R.L. Jefferies, J.F. Reynolds, G.R. Shaver, and J. Svoboda, 1992, *Arctic Ecosystems in a Changing Climate: An Ecophysiological Perspective*, Academic Press, San Diego.

Costanza, R., F.H. Slar, and M.L. White, 1990, "Modeling Coastal Landscape Dynamics," *Bioscience* 40: 91-107.

Crowley, T.J., 1990, "Are There Any Satisfactory Geologic Analogs for a Future Greenhouse Warming," *J. Climate* 3: 1282-1292.

Crowley, T.J., and K. Kim, 1995, "Comparison of Longterm Greenhouse Projections With the Geological Record," *Geophysical Res. Let.* 22: 933-936.

Crowley, T.J., and G.R. North, 1988, "Abrupt Climate Change and Extinction Events in Earth History," *Science* 240: 996-1002.

Dennis, R.L.H., 1993, Butterflies and Climate Change, Manchester University Press, Manchester.

Ellison, J.C., and D.R. Stoddart, 1991, "Mangrove Ecosystem Collapse During Predicted Sea-Level Rise: Holocene Analogues and Implications," *J. Coastal Res.* 7: 151-165.

Emmanuel, W.R., H.H. Shugart, and M.P. Stevenson, 1985, "Climatic Change and the Broad-Scale Distribution of Terrestrial Ecosystem Complexes," *Clim. Change* 7: 29-43.

Halpin, P.N., in press, "Global Change and Natural Area Protection: Management Responses and Research Directions," *Ecol. App.*

Halpin, P.N., and C.M. Secrett, 1995, "Potential Impacts of Climate Change on Forest Protection in the Humid Tropics: a Case Study of Costa Rica," Impacts of Climate Change on Ecosystems and SpeciesII: Terrestrial Ecosystems. IUCN, Gland, Switzerland.

Hartshorn, G.S., 1992, "Possible Effects of Global warming on the Biological Diversity in Tropical Forests," in R.L. Peters, and T.E. Lovejoy (eds.), *Global Warming and Biological Diversity*, pp.137-46, Yale University Press, New Haven.

Herman, T.B., and F.W. Scott, 1993, "Protected Areas and Global Climate Change: Assessing the Regional or Local Vulnerability of Vertebrate Species," in J. Pernetta (ed.), *Impacts of Climate Change on Ecosystems and Species: Implications for Protected Areas*, IUCN, Switzerland.

Hinckley, D., and G. Tierney, 1992, "Ecological Effects of Rapid Climate Change," in K. Majumdar, L.S. Kalkstein, B. Yarnal, E.W. Miller, and L.M. Rosenfeld (eds.), *Global Climate Change: Implications, Challenges, and Mitigation Measures*, pp. 292-301, Pennsylvania Academy of Sciences, Pennsylvania.

Holdridge, L.R., 1947, "Determination of World Plant Formations from Simple Climate Data," *Science* 105:367-368.

Holdridge, L.R., 1967, *Life Zone Ecology*, Tropical Science Center, San Jose, Costa Rica.

Houghton, J.T., G.J. Jenkins, and J.J. Ephraums, 1990, *Climate Change: The IPCC Scientific Assessment*, Cambridge University Press, Cambridge, UK.

Hunt, H.W. et al., 1991, "Simulation Model of the Effects of Climate Change on Temperate Grassland Ecosystems," *Ecol. Model.* 53: 205-246.

Lancia, R., D.A. Adams, and E.M. Junk, 1986, "Temporal and Spatial Aspects of Species-habitat Models," in J. Verner, M.L. Morrison, and C.J. Ralph (eds.), *Wildlife 2000: Modeling Habitat Relationships of Terrestrial Vertebrates*, pp. 65-69, University of Wisconsin Press, Wisconsin.

Laurie, W.A., and D. Brown, 1990, "Population Biology of Marine Iguanas (*Amblyrhinchus cristatus*): III. Factors affecting survival," *J. Anim. Ecol.* 59: 545-568.

Laymon, S.A., and R.H. Barrett, 1986, "Developing and Testing Habitat-capability Models: Pitfalls and Recommendations," in J. Verner, M.L. Morrison, and C.J. Ralph (eds.), *Wildlife 2000: Modeling Habitat Relationships of Terrestrial Vertebrates*, pp. 87-91, University of Wisconsin Press, Wisconsin.

Leemans, R., and P.N. Halpin, 1992, "Biodiversity and Global Climate Change," in B. Groombridge (ed.), *Global Biodiversity: Status of the Earth's Living Resources*, Chapman & Hall, London.

Lewis, J.C., 1983, *Habitat Suitability Index Models: Roseate Spoonbill*, U.S. Dept. Int. Fish. Wild. Serv. FWS/OBS-82/10.50.

MacArthur R.H., 1972, *Geographical Ecology: Patterns in the Distribution of Species*, Princeton University Press, Princeton, New Jersey.

Malcolm J.R., in prep, refers to the upcoming UEA/WWF report on vulnerability to climate change in the southern African subregion.

Malcolm, J.R., 1994, "Edge Effects in Central Amazonian Forest Fragments," *Ecology* 75:2438-2445.

Markham, A., N. Dudley and S. Stolten, 1993, *Some Like it Hot: Climate Change, Biodiversity and the Survival of Species*, WWF-International, Gland, Switzerland.

McNeely, J.A., 1988, *Economics and Biological Diversity: Developing and Using Economic Incentives to Conserve Biological Resources*, IUCN, Gland, Switzerland.

McNeely, J.A., 1990, "Climate Change and Biological Diversity: Policy Implications," in M.M. Boer and R.S. de Groot (eds.), *Landscape Ecological Impacts of Climate Change*, pp 406-428, IOS Press, Amsterdam.

Millsap, B.A., J.A. Gore, D.E. Runde, S.I. Cerulean, 1990, "Setting Priorities for the Conservation of Fish and Wildlife Species in Florida," *Wildl. Monogr.* 111:1-57.

Morse, L.E., L.S. Kutner, G.D. Maddox, J.T. Kartesz, L.L. Honey, C.M. Thurman, and S.J. Chaplin. 1993. "The Potential Effects of Climate Change on the Native Vascular Flora of North America North of Mexico: A Preliminary Climate-Envelopes Analysis." Prepared by The Nature Conservancy, Arlington, VA, for the Electric Power Research Institute, Palo Alto, CA. September.

Overpeck, J.T., P.J. Bartlein, and T. Webb III, 1991, "Potential Magnitude of Future Vegetation Change in Eastern North America: Comparisons with the Past," *Science* 254: 692-695.

Peters, R.L., 1991, "Effects of Global Warming on Forests," *Forest Ecol. Manage.* 35: 13-33.

Peters, R.L., and J.D.S. Darling, 1985, "The Greenhouse Effect and Nature Reserves," *Bioscience* 35: 707-716

Peters, R.L. and T.E. Lovejoy (eds.), 1992, *Global Warming and Biological Diversity*, Yale University Press, New Haven.

Poiani, K.A., and W.C. Johnson, 1991, "Global Warming and Prairie Wetlands," *Bioscience* 41: 611-618.

Pounds, J.A., and M.L. Crump, 1994, "Amphibian Declines and Climate Disturbance: the Case of the Golden Toad and the Harlequin Frog," *Conserv. Biol.* 8: 72-85.

Prentice, I.C., W. Cramer, S.P. Harrison, R. Leemans, R.A. Monserud, and A.M. Solomon, 1992, "A Global Biome Model Based on Plant Physiology and Dominance, Soil Properties and Climate," *J. Biogeogr.* 19:117-134.

Prentice, I.C., M.T. Sykes, and W. Cramer, 1991, "The Possible Dynamic Response of Northern Forests to Global Warming," *Global Ecol. Biogeogr. Let.* 1: 129-135.

Price, J.T. Potential Impacts of Global Climate on the Summer Distribution of Some North American Grassland Birds. Ph.D. Dissertation. 1995. Wayne State University, Michigan.

Reid, W.V. and M.C. Drexler, 1991, "Drowning the National Heritage: Climate Change and U.S. Coastal Biodiversity," *World Resource Institute*, Washington, D.C.

Reynolds, J.F., and P.W. Leadley, 1992, "Modeling the Response of Arctic Plants to Changing Climate," in F.S. Chapin III, R.L. Jeffries, J.F. Reynolds, G.R. Shaver, and J. Svoboda (eds.), *Arctic Ecosystems in a Changing Climate* pp. 413-438, Academic Press, Inc., San Diego.

Rodenhouse, N.L., 1992, "Potential Effects of Climate Change on a Neotropical Migrant Bird," *Conserv. Biol.* 6: 263-272.

Rogers, D.J., and B.G. Williams, 1993, "Tsetse Distribution in Africa: Seeing the Wood and the Trees," in P.J. Edwards, R. May, and N.R. Webb (eds.), *Large-scale Ecology and Conservation Biology* pp. 247-271, Blackwell Scientific Publications, Oxford.

Root, T., 1988a, "Energy Constraints and Avian Distributions and Abundances," *Ecology* 69:330-339.

Root, T., 1988b, "Environmental Factors Associated with Avian Distributional Boundaries," *J. Biogeogr.* 15:489-505.

Rose, C., and P. Hurst, 1991, *Can Nature Survive Global Warming,?* WWF International, Gland, Switzerland.

Running, S.W., and J.C. Coughlan, 1988, "A General Model of Forest Ecosystem Processes for Regional Applications, I. Hydrologic Balance, Canopy Gas Exchange and Primary Production Processes," *Ecol. Model.* 42: 125-154.

Running, S.W., and R.R. Nemani, 1991, "Regional Hydrologic and Carbon Balance Responses of Forests Resulting from Potential Climate Change," *Climate Change* 19: 16: 31-51.

Schreiber, R.W., and E.A. Schreiber, 1984, "Central Pacific Seabirds and the El Nino Oscillation," *Science* 225:713-715.

Shugart, H.H., 1990, "Using Ecosystem Models to Assess the Potential Consequences of Global Climate Change," *Trends Ecol. Evol.* 5: 303-307.

Shugart, H.H., R. Leemans, and G.B. Bonan, 1992, *A Systems Analysis of the Global Boreal Forest*, Cambridge University Press, Cambridge, UK.

Shugart, H.H., and T.M. Smith, 1992, "Using Computer Models to Project Ecosystem Response, Habitat Change, and Wildlife Diversity" in R.L. Peters, and T.E. Lovejoy (eds.), *Global Warming and Biological Diversity*, pp. 147-157, Yale University Press, New Haven.

Simberloff, D., J.A. Farr, J. Cox, and D.W. Mehlman, 1992, "Movement Corridors: Conservation Bargains or Poor Investments?," *Conserv. Biol.* 6:493-504.

Smith, T.M., H.H. Shugart, and P.N. Halpin, 1990, "Global Forests" in *Progress Reports on International Studies of Climate Change Impacts.* USEPA-OPA, Washington, D.C.

Smith, S.V., and R.W. Buddemeier, 1992, "Global Change and Coral Reef Ecosystems," *Ann. Rev. Ecol. Syst.* 23:89-118.

Stiles, F.G., 1992, "Effects of a Severe Drought on the Population Biology of a Tropical Hummingbird," *Ecology* 73: 1375-1390.

Sutherst, R.W., and G.F. Maywald, 1985, "A Computerized System for Matching Climates in Ecology," *Agr. Ecosyst. Envir.* 13:281-299.

Sutherst, R.W., G.F. Maywald, and W. Bottomley, 1991, "From CLIMEX to PESKY, a Generic Expert System for Pest Risk Assessment," *EPPO Bull.* 21:595-608.

Thornley, J.H.M. 1976. "Mathematical Models in Plant Physiology." Academic Press. N.Y. 318pp.

U.S. Fish and Wildlife Service, 1981, *Standards for the Development of Habitat Suitability Index Models*, 103 ESM, U.S. Fish and Wildlife Service, Division of Ecological Services, Washington, DC.

6 INTEGRATION OF IMPACT ASSESSMENT RESULTS

Upon completion of the biophysical impact assessments in each of the sectors, as described in Section 5, it is recommended that countries combine the results of their work into an integrated assessment of these impacts. The integration process is one of the most difficult of the analytical steps, but it is one which best provides critical context for decision makers.

In integrating the sectoral results, countries must keep in mind the ultimate use of the analysis. That is, the analysis is designed to provide decision makers with information on the possible extent of the country's vulnerability to climate change and on the effectiveness of alternative adaptation techniques. In addition, integrated assessment results should provide a sense of the scope of the climate change issue and the priorities that should be assigned in the planning process. In general, the integrated results should be presented in a concise and understandable fashion so that they will be most useful to decision makers.

6.1 INTEGRATION APPROACHES

The results of the biophysical impact assessments obtained by means of the methods described in Section 5 can be integrated in a number of different ways. Each approach deals with different integration issues.

6.1.1 Integration across Sectors

This aspect of integration is designed to deal with the interactions across sectors. Table 6.1 shows some typical effects that a change in one sector (resulting from climate change impacts) can have on another sector. For example, altering water resources because of climate changes could impact agricultural productivity independently of the direct effects of climatic changes on the agricultural sector. As another example, in low-lying coastal agricultural regions, sea-level rise threatens to allow saltwater to find its way into fresh water supplies used to water crops. As agricultural production zones shift, natural ecosystems may be endangered. Interactions also exist between the water and grassland/forest sectors. Grasslands and forests function as watersheds that drain into aquifers or aboveground water storage for agricultural, industrial, municipal, recreational, and navigational uses. Changes in climate will affect not only grassland and forest production but will also impact water quantity and possibly water quality. Another key interaction occurs between the agricultural and grassland/livestock sectors, especially as they relate to the production of supplemental feed sources. Market and production areas may be forced to move because of changes in climate. This change could alter the supply and type of supplemental feed available to livestock producers.

With these types of cross-sector effects, the integration effort is designed primarily to ensure a consistent result for all sectors. Analysts in each sector should regularly communicate their results to other sectors so that a consistent, integrated picture of climate change impacts will emerge.

R. Benioff et al. (eds.), Vulnerability and Adaptation Assessments, 6-1–6-7.
© *1996 Kluwer Academic Publishers.*

TABLE 6.1 Typical Biophysical Impacts across Sectors: Cross-Sector Impact

Climate Change Impacts	Crops	Grasslands/ Livestock	Forests	Water Resources	Coastal Resources
Crops					
Change in agricultural yield	–	Land use conflicts	Land use conflicts	Water availability for agriculture	Loss of agricultural land
Grasslands and Livestock					
Change in grassland availability and livestock productivity	Land use conflicts; change in feed availability	–	Land use conflicts	Water availability for grass-lands/livestock	Loss of grazing/range lands
Forests					
Change in forest species and range	Land use conflicts	Land use conflicts	–	Water availability for forests	Loss of forest lands
Water Resources					
Change in water availability and water quality	Change in water availability for agriculture	Change in water availability for livestock	Changes in watersheds and in water quantity and quality	–	Change in salinity in coastal zones
Coastal Resources					
Loss of coastal land	Loss of agri-cultural land	Loss of rangeland	Loss of forest area	Change in water retention area	–

In addition, a biophysical impact that emerges from one sector as a product of an efficient adaptation to climate change could serve as a significant primary impact in another sector. For example, the land lost to sea-level rise could prove to be high-quality agricultural land, and this loss of prime cropland could significantly affect both the future of the agricultural sector and how this sector might plan to adapt to its own subset of climate change manifestations. Another example would be if the agricultural sector increased its use of irrigation as an adaptive measure. Such an increase in use would depend on the availability of adequate water supplies under climate change scenarios. Agricultural demand would have to compete with demands from urban and commercial users, thus requiring coordinated regional planning and management strategies. Similar interactions exist between the grassland/livestock and agricultural systems. For example, changes in climatic conditions may create economic incentives for displacing grazing systems with cropping systems, thereby shifting grazing to more marginal areas. Conversely, adapting to climate change may necessitate abandonment of farmland, with fields left to revert to natural vegetation and rangelands.

6.1.2 Integration into Common Metrics

This portion of the integration analysis combines the results from each sector into a set of common metrics so that tangible comparisons of impacts across sectors can be conducted. Table 6.2 shows some of the common metrics that can be considered.

6.1.2.1 Economic Costs

Using economic costs to evaluate a country's climate change vulnerability and adaptation options is a very effective integration analysis technique in cases where sufficient economic data exist. Costs used in this analysis should be developed by means of standard economic analysis and accounting procedures. For example, costs should be expressed in terms of net present value. Discount rates should be those used by the country in the analysis of development projects. Cost figures should include both expenses and revenues associated with the sectoral impact. Evaluation of economic impacts for each sector is described in Sections 5.1 through 5.6.

The cost figures from each of the sectors can be combined in several ways. The simplest way is to develop a total cost to the country of climate change, including damage costs (e.g., value of land lost to sea-level rise) and revenue increases (e.g., increased crop yields due to warmer temperatures). Analysis of adaptation options can be presented in terms of either a cost-effectiveness analysis (e.g., what is the cost of an adaptation technique per unit of decrease in damage) or a cost-benefit analysis (e.g., the comparison of total costs with total benefits).

By using a common metric of economic cost, it will be possible to compare the results from different sectors on a consistent basis. It will also be possible to determine the economic viability of alternative

TABLE 6.2 Typical Common Metrics Used to Integrate Results

Metric	Crops	Grasslands/ Livestock	Forests	Water Resources	Coastal Resources
Economic Cost					
Direct costs	Loss/gain in output	Loss/gain in output	Loss/gain of lumber, fuel	Increase/decrease in water supply costs	Values of land lost
Indirect costs	Impact on GNP	Impact on GNP	Impact on tourism and recreational value	Impact on energy resources, transportation	Increased price for remaining land
Population Affected	Farmer population	Rancher population	Forest and tourism industries	Urban water customers	Displaced population
Land Use	Agricultural land affected	Grassland affected	Forest area affected	Watershed area	Land lost

strategies for adapting to climate change impacts. However, the development of cost figures for the impacts in each sector may, in practice, be very difficult because some impacts are not easily expressed in terms of economic cost.

6.1.2.2 Populations Affected

Often the most important impact from a policy and/or societal perspective is that related to the extent of human populations affected by a given phenomenon. For example, sea-level rise, declines in agricultural productivity, reduction in available water, and deterioration of air quality due to climate change can all be integrated in terms of the number of people impacted by these changes.

6.1.2.3 Land Use

In many cases, impacts of climate change on various sectors and/or components of a given sector cannot be expressed in economic terms. In some of these cases, land use can be disrupted to integrate impacts across such sectors/components. For example, impacts to croplands, forests, coastal areas, and grasslands may best be evaluated in terms of the quantity of land disrupted, especially if it is difficult to estimate the monetary value for such lands. In addition, it may not be possible to place an economic value on the impacts of climate change on some individual components of a given sector such as destruction or deterioration of wildlife habitat because of changes in forest composition or drying up of wetlands or marshes. In such cases, quantity of land disrupted or altered can be used to integrate impacts.

In addition to the quantitative metrics, qualitative metrics can also be used. For example, impacts to each sector can be characterized as extensive, moderate, or minimal. The result is a qualitative identification of which sectors are likely to experience the most severe impacts from climate change. While not as satisfying as a quantitative measure, qualitative measures can provide useful decision-making information such as the prioritization of risks.

One of the problems in using qualitative measures is the difficulty in developing an overall evaluation. Numerical combination of results is not possible. An approach to dealing with this problem is to use multiattribute decision analysis. This technique combines items that are measured by different metrics, including both quantitative and qualitative, into an integrated picture. The decision analysis technique has been used in many other areas but not, to date, in climate change vulnerability and adaptation studies.

All of these approaches serve similar purposes—to support aggregation of impacts across sectors, to facilitate comparisons between alternative adaptive strategies, and to weigh the relative merits of different responses.

6.1.3 Integration with Mitigation Analyses

In some countries, the analysis of greenhouse gas mitigation options involves some of the impacted sectors included here. For example, planting more trees to sequester CO_2 is one mitigation option that can be considered. In many cases, the impact of climatic change on a given sector or resource would interfere with implementing a chosen mitigation strategy. For example, changes in forest growth and shifts in species composition because of changes in climate would affect the viability of reforestation strategies. Another example concerns fuel substitution. Increased dependence on hydroelectric power to decrease consumption of fossil fuels and reduce the production of CO_2 could be curtailed by decreased capacity of water reservoirs because of climate changes.

It is important for countries conducting mitigation analyses to be consistent with the biophysical impact results. Regular communication and sharing of results among analysts working in both areas is vital.

6.1.4 Integration with Other Government Programs

If possible, the results of the biophysical impact assessments should be factored into the general natural resource management plans and policies and into future research programs being planned by the government. Many of these other programs may not be driven by climate change considerations. Nevertheless, there may be important interactions and implications for these other programs and for climate change vulnerability and adaptation. Integrating consideration of these issues would improve the chances that climate-change-related measures would be implemented because they would be part of programs that had already been accepted and/or approved for other reasons.

6.2 DEALING WITH UNCERTAINTY

Vulnerability and adaptation analyses must deal with a wide range of uncertainty. Table 6.3 summarizes the principal sources of uncertainty and methods for dealing with them, as discussed in Sections 4 and 5.

TABLE 6.3 Sources of Uncertainty and Possible Methods for Dealing with Them

Source of Uncertainty	Possible Methods for Dealing with Them
Climate Change	
• Range of GHG emissions	• Use $2 \times CO_2$ and transient scenarios
• Climate response to GHG emissions	• Use outputs of several GCMs and non-GCM-based scenarios
Economic Conditions	
• Population growth	• Use high, medium, and low scenarios
• Economic growth	• Use high, medium, and low scenarios
Biophysical Impacts	
• Extent of impacted area	• Use best case, expected value, and worst case
• Severity of impacts	• Use sensitivity analysis
• Cross-sector impacts	• Use maximum, minimum interaction cases
Adaptation	
• Natural adaptive response	• Use upper limit, lower limit
• Effectiveness of adaption strategies	• Use upper limit, lower limit

The principal uncertainties deal with the range of possible changes in climate that may occur over the very long term. The basic method for handling these uncertainties suggests a few alternative scenarios of future climate change. The scenarios chosen as the basis for this current project are discussed in Section 4 in the context of the current state of knowledge about their inherent uncertainty. Care should be taken, however, to ensure that any sectoral analysis and integrating exercise can accommodate a wide range of different scenarios. Even those scenarios that are not now considered worthy of investigation (because they are very unlikely given the current state of knowledge) may be relevant at a future time as our understanding of the future evolves and improves over time. Decision makers will always want to know whether the newest findings change the outcome of decisions, and carefully constructed impact analyses will be able to accommodate future research designed to provide quick answers to policy questions.

Dealing with the uncertainty in population and economic growth are also discussed in Section 4. The use of scenarios was described as a means of dealing with this uncertainty.

In addition to its own population and economic growth, a country should study how its relationship with the rest of the world will evolve over time. This study should identify how a country might feel changes in international patterns of trade, migration, and social and political areas. These patterns must be consistent with the assumed climate change scenarios, and their explicit effects upon individual and correlated sectoral analyses must be recognized. Given that the future of human activity may be more uncertain than future climate changes, great care must be taken in selecting scenarios and accommodating sensitivity analyses in these areas.

Scientific understanding of the extent and severity of biophysical impacts that might be associated with any change in the climate is also unclear. Methods for dealing with these uncertainties, including sensitivity analysis for each sector, are described in Sections 5.1 through 5.6.

Adaptation is also a source of uncertainty. Sectoral assessments depend on adaptation, and adaptation depends on the physical ability of a sector to adapt, on value judgments, cultural factors, political structures, economic conditions, expectations, and how people learn about how the climate might be changing. In the initial stages of assessment, it is best to consider only the most obvious strategies and to design analytical methods within sectors, and correlated subsets of sectors, that can investigate their relative value, timing, and implementation, assuming the efficient processing of climate change information. These methods should be able, in the future, to accommodate subsequent analyses of adaptive decision making under uncertainty. However, such additional analyses should only be contemplated if downstream techniques can resolve the variability that such analyses might produce.

The range of possible outcomes to climate change can become very wide when all of these uncertainties are combined. However, in most cases, precise impact assessment will never be possible, and "order of magnitude" evaluation will likely suffice, especially with regards to ranking risks and for general decision-making purposes. In considering a range of adaptive responses, countries may want to focus primarily on robust, flexible, or contingency responses. Robust responses would be appropriate across a wide range of outcomes, while flexible responses can be adjusted to accommodate the future as it unfolds. Contingency responses display a special type of flexibility; they are evaluated, accepted, and planned in advance but are implemented only when and if required. In general, countries should focus on responses that are appropriate for a wide range of possible impacts and that help address current problems in addition to addressing likely future outcomes.

Scenarios, best-case/worst-case conditions, and sensitivity analyses can generate so much data with such widely varying implications as to be meaningless to the current decision-making process. The analyst must narrow the possible conditions to a meaningful subset for consideration by decision makers. Some of the steps that can be taken include the following:

- Use extreme conditions only to establish the bounds. Best-case/worst-case conditions are frequently beyond the range of reasonable expectation. Including these results will often provide a distorted picture that will cast doubts on the credibility of the entire analysis. Extreme assumptions (about climate change, about growth, about sectoral responses to climate change) should be used only to establish upper and lower bounds.

- Attempt to find robust results. Results that remain approximately consistent over a wide range of assumptions can be considered to be robust. These outputs will have the most certainty attached to them and can be identified as expected outcomes.

- Develop explicit measures of uncertainty. Where results vary widely over a range of scenarios, the extent of variation should be identified. This measure can be either quantitative or qualitative. By explicitly identifying the extent of the uncertainty, the results can be presented to decision makers with appropriate qualifiers.

- Develop "most likely" cases. Use expert judgment to select the conditions most likely to prevail and present the results for these conditions. This type of considered judgment is needed to give decision makers an informed opinion about the expected effects of climate change.

Given the large degree of uncertainty in predicting climate change and its effects over long times, analysts should try to avoid (1) being too definitive in providing exact answers to climate change questions, which may be flawed by poor assumptions, or (2) being too vague and uncertain in the results to provide decision makers with meaningful guidance.

7 ASSESSMENT OF ADAPTATION POLICY OPTIONS

7.1 INTRODUCTION

Nations participating in the Country Studies Program need to assess options for adapting to climate change because it is unlikely that climate change can be completely averted. Although measures are being taken to reduce greenhouse gas emissions, which will probably reduce the rate and magnitude of climate change, it is unlikely that greenhouse gas emissions can be reduced enough to stabilize climate. Therefore, adaptation will be necessary (IPCC 1990).

The vulnerability assessments, discussed in Sections 4 and 5, will address how natural and societal systems will adapt on their own to climate change. For example, they will address whether vegetation can migrate in pace with climate change or whether farmers can maintain production levels by changing their practices. Taking action in response to climate change is called "reactive adaptation." It is also sometimes referred to as "technical adaptation." For example, a farmer may start planting drought-tolerant crops after the climate becomes drier. In many cases, such reactive adaptations may completely or satisfactorily mitigate the effects of climate change. In other cases, relying on these reactive adaptations may result in:

- Unacceptable impacts, such as loss of human life or species diversity, and

- Missed opportunities to mitigate impacts by taking action before climate changes occur.

For example, increased monsoons could lead to increased loss of life, or higher temperatures could result in extinction of some species. Failure to anticipate these impacts by preparing flood evacuation plans or facilitating the transplantation of species could increase the damage that would occur if anticipatory measures had been taken.

This section addresses how to conduct an assessment of options for *anticipating* climate change. An "anticipatory adaptation" involves taking action before climate changes (Smith and Mueller-Vollmer 1993). For example, an oil drilling platform could be built 1 m higher than originally planned to allow for sea-level rise. This type of adaptation generally requires adopting a policy that recognizes the potential threat from a change in climate and acknowledges the need to commit current resources to alleviate future impacts.

This section focuses on governmental policies toward adaptation. Individuals, businesses, nongovernmental organizations, as well as plants and animals, will adapt or strive to adapt to climate change on their own. The emphasis here is on how, by implementing anticipatory adaptation options, government can make reactive adaptations as efficient and speedy as possible and mitigate damages that may occur.

The methods discussed in this section are not the only approaches for analyzing adaptation. Many formal approaches, such as benefit-cost analysis (e.g., Yohe 1991) or multi-attribute utility analyses (Smith et al. 1993), exist for evaluating policies. Researchers conducting adaptation assessments should select the method they feel most comfortable with. The goals stated here, however, are relevant regardless of the method used.

R. Benioff et al. (eds.), Vulnerability and Adaptation Assessments, 7-1–7-8.
© 1996 *Kluwer Academic Publishers.*

7.2 GOALS OF THE ADAPTATION POLICY ASSESSMENTS

The fundamental goal of the adaptation policy assessment is to thoroughly analyze *anticipatory* climate change adaptation options in a manner useful to policy makers.[8] In addition, the adaptation assessment should:

- Analyze the effectiveness of current policies in coping with climate change,

- Analyze the costs and benefits of alternative policies to anticipate the effects of climate change,

- Identify which policies are most in need of immediate implementation, and

- Involve policy makers in the assessment.

7.3 NECESSARY ELEMENTS OF ADAPTATION POLICY ASSESSMENTS

For the assessment to produce results that will be useful to policy makers, two elements are necessary:

- *Research on adaptation policy options should begin early in the project.* Assessments of adaptation policies should be analyzed in parallel with vulnerability assessments, not afterward. If adaptation analysis is not initiated until the end of a project, it will most likely face spent resources, time constraints, and exhausted researchers. Thus, researchers must be assigned to the policy assessment early in the country study.

- *The adaptation policy research should feed into an adaptation workshop (or series of workshops on different sectors) that will involve policy makers in specific discussions of adaptation policies.* The more that policy makers are involved in discussions and analysis of adaptation policy options, the better they will understand climate change effects and adaptation options.

7.4 PRIMARY APPROACH

The primary approach for evaluating adaptation policy options is shown in Figure 7.1. The analysis of adaptation policy options follows from the identification of vulnerabilities and technical adaptation discussed in Sections 4 and 5. The effectiveness of current policies in coping with climate change is then analyzed. If those policies are inadequate, alternative (adaptation) policies should be considered. If alternative adaptation policies are superior to current policies, they should be considered on the basis of their relative benefits and costs and the need to implement them in anticipation of climate change.

---

[8] A policy maker is defined as anyone with significant influence over the management of resources vulnerable to climate change, including national, regional, and local officials from the countries that are part of the assessment. Policy makers also include officials from donor nations or multilateral institutions that influence the management of these vulnerable resources.

7.4.1 Analyze Vulnerabilities and Technical Adaptation

The first step is to analyze the vulnerability of sectors of interest to climate change (e.g., agriculture, forest, or coastal resources). Methods for assessing vulnerability and technical adaptation are described in Sections 3 through 6. These assessments can be used to determine if there is a potential for significant negative effects of climate change; that is, if climate change could lead to loss of life, increase the incidence of diseases, or cause significant economic losses or harm to the environment, a sector should be considered vulnerable. The vulnerable sectors would then be subjects of the adaptation policy assessments.

7.4.2 Define Scope of Adaptation Policy Assessment

The second step is to define the scope of the adaptation policy assessments. Within the sectors with potential for significant negative effects of climate change, regions that appear to be most vulnerable should be the subject of the adaptation policy assessments. For example, one part of a country's coastline could be low-lying and have valuable resources, such as properties or wetlands at risk from sea-level rise, while another part of the coast could have cliffs and few valuable resources at

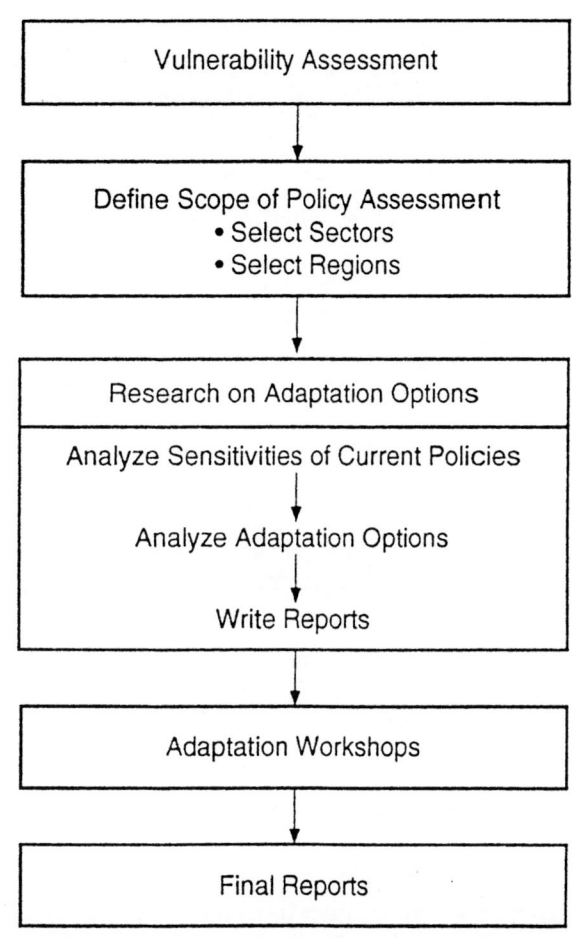

FIGURE 7.1 Adaptation Assessment Steps

sea level. The adaptation policy assessment need only focus on the low-lying part of the coast, not the area with cliffs.

7.4.3 Analyze Sensitivities of Current Policies

Following the analysis of vulnerability and definition of the scope, the adaptation policy assessment then identifies the sensitivity of the current management approach for the resources in question. Current policies on potentially vulnerable resources should be inventoried. For example, the water resources component could include policies on water allocation, water quality, reservoir operation, dam construction, and riparian land development.

The objectives of policies involving vulnerable resources are then identified. For example, researchers identify whether reservoirs are operated to supply drinking water, water for irrigation, flood control, hydroelectric power, navigation, minimum flow for fish and wildlife, and recreational benefits. Stakhiv (1993) notes that it is important that researchers quantify these objectives so that the relative success of different adaptation policies can be measured. Quantitative measures of success or failure should be developed for all policy objectives. These measures can also be assigned weights that reflect their relative importance.

A matrix can be developed that includes policy objectives, weights, and the degree to which objectives are met under current climate change scenarios. Preferably, a similar metric will be used for each objective (e.g., monetary values, number of days meeting a standard), which will allow for quantitative comparison of results. The use of such an approach is demonstrated with a hypothetical example in Appendix I. It is also possible to use matrixes to compare the relative effect of different climate change scenarios on meeting objectives (Smith et al. 1993).

The preferred method for identifying policy objectives and weights is to conduct a half-day to day-long workshop with policy makers. The list of objectives should be based on a consensus reached among policy makers at the workshop. If it is not feasible to hold a workshop, researchers should determine the objectives on the basis of interviews with policy makers about objectives and analysis of laws and regulations.

The effectiveness of current policies should be evaluated by means of quantitative models from the vulnerability assessments. A measure that could be used to evaluate the effectiveness of a policy is the number of days on which a certain standard is not met (failure to fulfill objectives). This method helps ensure consistency between this assessment and the vulnerability assessment. If models are not available or cannot be used for policy analysis, the adaptation researchers must rely on their own judgment and that of experts (such as those conducting the vulnerability assessments). The adaptation policy assessment should also use the same climate change scenarios used in the vulnerability assessments. This approach will enable results from the vulnerability assessment to be used in the adaptation policy assessment.

7.4.4 Analyze Adaptation Policy Options

Assuming that current policies become less effective in fulfilling their objectives under climate change scenarios than they are under current climate, the next task is to analyze alternative policies that eliminate or minimize the failure to fulfill objectives.

Researchers should identify alternative policies that anticipate climate change. A list of suggested adaptation policies to consider is provided in Table 7.1, with additional details in Appendix I. An alternative adaptation policy may meet the objectives better than current policies. A test for determining which policy best meets the objectives involves evaluating whether the alternative policy comes closer than current policies in fulfilling objectives under various climate change situations.

Researchers should analyze whether the alternative policies that anticipate climate change improve upon current policies on the basis of the following:

- *Flexibility.* Do policies minimize the failure to meet objectives (or even result in increased attainment of objectives) under a variety of climate change situations?

- *Efficiency.* Does the gain in meeting objectives (i.e., benefits) exceed increases in costs (Titus 1990)?

TABLE 7.1 Typical Adaptation Policies

Category	Adaptation Policy Option
General	• Assess current practices of crisis management • Inventory existing practices and decisions used to adapt to different climates • Promote awareness of climatic variability and change
Agriculture	• Develop new crop types and enhance seed banks • Liberalize agricultural trade • Avoid tying subsidies or taxes to type of crop and acreage • Promote agricultural drought management • Increase efficiency of irrigation • Disperse information on conservation management practices
Forests	• Encourage diverse management practices • Reduce habitat fragmentation and promote development of migration corridors • Enhance forest seed banks • Establish flexible criteria for intervention
Water Resources	• Use river basin planning and coordination • Adopt contingency planning for drought • Make marginal changes in construction of storage and distribution facilities • Maintain options to develop new dam sites • Conserve water • Allocate water supplies by using market-based systems • Use interbasin transfers • Control pollution
Sea-Level Rise	• Adopt coastal zone management • Use presumed mobility • Plan urban growth • Discourage permanent shoreline stabilization • Incorporate marginal increases in the height of coastal infrastructure • Preserve vulnerable wetlands • Decrease subsidies to sensitive lands • Tie disaster relief to hazard-reduction programs • Promote public education
Ecosystems	• Integrate ecosystem planning and management • Protect and enhance migration corridors or buffer zones • Enhance methods to protect biodiversity off-site

Special consideration should be given to analyzing adaptation policy options that have a high priority for implementation in anticipation of climate change. Those policies would increase flexibility and have a net increase in meeting objectives over cost, as well as mitigate the following types of climate change effects:

- *Irreversible or catastrophic impacts.* Because loss of life or species extinction is irreversible, anticipatory actions must be taken to avoid these outcomes. Extensive loss of property or destruction to valued resources such as forests may be a sufficient reason to justify anticipatory actions.

- *Long-term projects.* Some projects today, such as building dams, can have long lifetimes and will be affected by climate change. Climate change considerations should be taken into account in designing these long-term projects (National Academy of Sciences 1992).

- *Unfavorable trends.* Continuation of some trends may make adaptation more difficult. For example, continued development of vulnerable shorelines may increase the damage from rises in sea level (Smith and Mueller-Vollmer 1993).

Researchers should compile a list of adaptation options and enter these options into the matrix format, using a different matrix for each sector (i.e., forest, water resources, etc.). The options should then be ranked based on flexibility and the degree to which they maximize meeting objectives compared with costs. High-priority adaptation policy options should be singled out, either by listing them in a separate list or by clearly identifying them with, for example, asterisks or italics.

7.4.5 Document Results

Researchers should issue reports that describe the sensitivity of current policies, identify adaptation options, identify high-priority adaptation options, and quantitatively analyze the options. A report should be prepared for each sector, such as agriculture, coastal resources, or water resources. These reports should be distributed to workshop participants in advance of the meeting.

7.4.6 Conduct Adaptation Policy Workshop

The adaptation assessment should culminate in a one- to two-day workshop with natural resource policy makers to discuss sensitivity to climate change and adaptation policy options. It is important that policy makers or their high-ranking advisers attend the workshop. As noted, representatives from donor agencies should also participate. The major goal of the adaptation workshop should be to engage policy makers in a comprehensive discussion of adaptation options.

It is imperative that the workshop be very structured. This plan requires a chairperson who uses a "strong hand" in running the meeting. Participants should not be allowed to digress from the discussion on topics such as the likelihood of climate change or its impacts. Such diversions result in valuable time lost from discussion of adaptation. However, participants should have a basic understanding of climate change theory and impacts analysis.

In addition, participants may argue that managers of natural resources cannot cope with known variability and should not be concerned about an unknown variable such as climate change. In that case, the chairperson should state that preparing for known variability is a paradigm for preparing for climate change. Enhancing the ability to cope with known extreme events usually involves increasing flexibility in a manner in which benefits exceed costs—precisely the criterion for adaptation policies.

If a country assesses a number of sectors, it may be difficult to have a full plenary discussion of adaptation policy options in all sectors in a one- to two-day workshop. Either separate workshops could be held for different sectors, or participants could attend breakout sessions to address specific sectors.

7.4.6.1 Workshop Presentations

The following presentations should be part of the workshop:

- A brief introduction to climate change science. It is important to have this presentation so that all participants have at least a minimal understanding of climate change.

- Results from the vulnerability assessments.

- Results from the adaptation policy assessments described in Section 7.4.

7.4.6.2 Workshop Discussions

The chairperson should lead participants in a discussion of adaptation policy options. To involve the participants, the discussion should be specific and follow similar steps to those taken in the research. For each of these steps, the chairperson could either add or subtract from the matrix developed in the research, or fill in a blank matrix. In this discussion, participants should:

- Review the objectives of current policies;

- Assign weights to objectives to reflect their relative merit;

- Review the sensitivities of current policies;

- Identify adaptation policies and their effectiveness;

- Identify high-priority adaptation policies;

- Discuss how policies can be integrated into current policies, plans, programs, and activities; and

- Review issues for future research and implementation.

A review of the sensitivities of current policies, identification of adaptation policies, and identification of high-priority policies could be done in breakout sessions to concentrate on specific sectors.

7.4.6.3 Reports

The researchers could act as rapporteurs, record the discussion, and document the results in a report summarizing the workshop. The report from the workshop could be incorporated into the reports called for in Section 7.4.5 and be part of the country's final report.

7.5 REFERENCES FOR SECTION 7

IPCC, 1990, *Policy Makers Summary of the Formulation of Response Strategies*, Intergovernmental Panel on Climate Change, World Meteorological Organization and United Nations Environment Programme, Geneva, Switzerland.

National Academy of Sciences, 1992, *Policy Implications of Greenhouse Warming*, National Academy Press, Washington, D.C.

Smith, A.E., et al., 1993, *Adaptation Strategy Evaluator (2.0): Quick Reference Pamphlet*, Report to the Office of Policy Planning and Evaluation, U.S. Environmental Protection Agency, Decision Focus, Inc., Washington, D.C.

Smith, J.B., and J. Mueller-Vollmer, 1993, *Setting Priorities for Adapting to Climate Change*, Report prepared for the Office of Technology Assessment, Washington, D.C., by RCG/Hagler, Bailly, Inc., Boulder, Colo.

Stakhiv, E.Z., 1993, *Evaluation of Adaptation Strategies*, Institute for Water Resources, U.S. Army Corps of Engineers, Fort Belvoir, Va. (draft).

Titus, J.G., 1990, "Strategies for Adapting to the Greenhouse Effect," *APA Journal* 311, Summer.

Yohe, G.W., 1991, "Uncertainty, Climate Change, and the Economic Value of Information: An Economic Methodology for Evaluating the Timing and Relative Efficacy of Alternative Response to Climate Change with Application to Protecting Developed Property from Greenhouse Induced Sea Level Rise," *Policy Sciences* 24:245-269.

8 DOCUMENTATION AND PRESENTATION OF RESULTS

Many considerations will go into the documentation and presentation of the results of the vulnerability and adaptation assessments. It is not possible to prescribe a single process that will fit the needs of all countries participating in the program. Only general guidelines can be given.

Table 8.1 gives a general outline for a report on the vulnerability and adaptation assessments that can be submitted as part of the overall documentation of a country study. **This outline should be used only as a general guide and should be adapted to the specific needs of a country.**

R. Benioff et al. (eds.), Vulnerability and Adaptation Assessments, 8-1–8-3.
© 1996 *Kluwer Academic Publishers.*

TABLE 8.1 General Outline for a Report on Vulnerability and Adaptation Assessments

---

1 BACKGROUND

- Provide general background on the vulnerability and adaptation assessments.
- Identify the organizations involved.
- Give any limitations to the scope of the assessments.

2 COUNTRY SECTORS VULNERABLE TO CLIMATE CHANGE

- Identify sectors considered to be vulnerable to climate change (e.g., agriculture, forests, water resources).
- Give the current status and condition of each sector.
- Describe the geographical extent of the sector in the country.
- Describe the time frame of expected effects.

3 SCENARIOS AND ASSUMPTIONS

- Describe the scenarios and assumptions used in the assessments.

    3.1 CLIMATE CHANGE SCENARIOS

    3.2 NON-CLIMATE-RELATED SCENARIOS

    3.3 OTHER ASSUMPTIONS

4 SECTORAL IMPACTS AND ADAPTATION OPTIONS

- For each sector studied, provide the following structure. (Agriculture impacts are used as an example. Other sectors are structured similarly.) Use only relevant sectors.

    4.1 AGRICULTURAL IMPACTS

        4.1.1 Impacts of Climate Scenarios
- Describe the effects of each of the climate scenarios studied.
- Give the effect on crop productivity.
- Estimate the cost of the effect.
- Describe any secondary effects.

        4.1.2 Adaptation Options
- Identify options for adaptation to climate change in the sector.
- Present the results of evaluation of these options.

        4.1.3 Conclusions and Observations for the Sector
- Present conclusions for the sector.
- Identify uncertainties.
- Identify additional work needed in the sector analysis.

TABLE 8.1 (Cont.)

4.2 GRASSLAND AND LIVESTOCK IMPACTS

4.3 FOREST IMPACTS

4.4 WATER RESOURCE IMPACTS

4.5 COASTAL RESOURCE IMPACTS

4.6 OTHER IMPACTS

5 INTEGRATION OF SECTORAL RESULTS

- Describe the interactions between sectors considered.
- Describe the results of integrating each of the sectoral impacts into an overall evaluation.
- Provide a summary of integrated implications for the country.

6 POLICY RECOMMENDATIONS

- Describe policies that can be considered for each sector and across sectors..
- Describe the results of the evaluation of the effectiveness of the policies for adapting to climate change.
- Identify barriers to implementing the policies.
- Identify policy priorities.

7 CONCLUSIONS AND RECOMMENDATIONS

- Give overall conclusions of the assessments.
- Give policy recommendations.
- Identify additional work needed.

APPENDIX A:

BASELINE SCENARIOS

R. Benioff et al. (eds.), Vulnerability and Adaptation Assessments, A-1–A-26.
© 1996 *Kluwer Academic Publishers.*

APPENDIX A NOTATION

The following is a list of the acronyms, initialisms, and abbreviations (including units of measure) used in this appendix. Some acronyms used only in tables are defined in those tables.

ACRONYMS, INITIALISMS, AND ABBREVIATIONS

ASF   Atmospheric Stabilization Framework
FAO   Food and Agriculture Organization
GDP   gross domestic product
GNP   gross national product
IPCC   Intergovernmental Panel on Climate Change
VOC   volatile organic compound
WRI   World Resources Institute

CHEMICALS

CH_4 methane
CO   carbon monoxide
CO_2 carbon dioxide
N_2O nitrous oxide
NO_x nitrogen oxides
SO_x sulfur oxides

UNITS OF MEASURE

ha   hectare(s)
yr   year(s)

CONTENTS

TABLES

FIGURES

APPENDIX A:

BASELINE SCENARIOS

A.1 INTRODUCTION

This appendix describes the method and sources of data that countries use to create baseline scenarios. These scenarios depict the way socioeconomic and environmental conditions will change by 2075 in the absence of changes in climate. This document supplements Section 4.2.

Baseline scenarios estimate non-climate-related changes in socioeconomic and environmental conditions for comparison with scenarios that include climate change. Comparison with future baseline conditions yields an estimate of the sensitivity of socioeconomic and environmental conditions to climate change alone. Baseline scenarios are highly uncertain because they rely primarily on projections of socioeconomic factors such as population growth, gross national product (GNP), and technological improvement. Any forecast of these factors over more than two decades is highly uncertain. Despite the uncertainties, an estimate of baseline conditions will be useful for comparison with climate change scenarios.

A.1.1 Recommended Source of Data

The recommended baseline scenarios rely heavily on a study prepared by Pepper et al. (1992) for Working Group I of the Intergovernmental Panel on Climate Change (IPCC) (Houghton et al. 1992). This study estimated emissions of greenhouse gases through 2100 and was used by the IPCC to estimate changes in climate. Thus, the economic forecasts by Pepper et al. (1992) are consistent with the scenarios of climate change used in this assessment. Pepper et al. (1992) also estimated energy use, gross domestic product (GDP), population, and industrial growth to project changes in air pollutant emissions, deforestation, and growth in agricultural production through 2100; however, the estimates are for broad regional areas, such as continents. Therefore, the projections may be used as a basis for determining baseline conditions in specific countries, but expert judgment or additional country-specific data would be required to apply the projected trends.

Pepper et al. (1992) provide an internally consistent forecast of conditions in 2100. If more appropriate sources are identified, they should be used. The projections by Pepper et al. (1992) may be optimistic; other sources, such as Meadows et al. (1992), present a more pessimistic view of GNP and population growth (i.e., lower GNP, higher population growth, and limited availability of resources) by 2075. Therefore, the estimates by Pepper et al. (1992) may need to be adjusted. Again, expert judgment should be used.

A.1.2 Summary of Method

The method of Pepper et al. (1992) relies on the U.S. Environmental Protection Agency's Atmospheric Stabilization Framework (ASF), which is an integrated set of computer models that use common assumptions of population growth, economic growth, and structural change to develop scenarios of greenhouse gas emissions. The ASF calculates scenarios of future energy use, energy supply, and greenhouse gas emissions. The framework includes the following:

- *The energy module.* This module is a partial equilibrium model within the ASF. It combines detailed country models of total energy use, energy supply and demand, and energy use by sector, fuel, and technology to estimate future emissions. Outputs include carbon dioxide (CO_2), methane (CH_4), nitrous oxide (N_2O), carbon monoxide (CO), nitrogen oxides (NO_x), sulfur oxides (SO_x), and volatile organic compounds (VOCs) by country group through 2100. The results are highly sensitive to the assumptions on population and income.

- *The agricultural module.* This module estimates future regional consumption and production. It also estimates land used for rice production and associated CH_4 emissions, production of meat and dairy products and associated CH_4 emissions, fertilizer use and N_2O emissions from that use, and N_2O emissions from the production of nitric acid. Pepper et al. (1992) based future agricultural activity on scenarios developed by the Center for Agriculture and Rural Development for future production and consumption; the scenarios embody various assumptions on population and income. The scenario used by Pepper et al. (1992) matched the assumptions on population from IPCC Scenario IS92a. Forecasts of agricultural production are not shown in the report but are reproduced here (Section A.3).

- *The tropical deforestation module.* This module combines estimates of current rates of deforestation with assumptions about the way that forest-clearing rates may change in the future, about uses of deforested land, and about rates of carbon uptake and emissions. Outputs from this module include (1) forest clearing (ha/yr) by forest type (closed broad-leaved, open broad-leaved, and coniferous) through 2100 for Brazil, other Latin American countries, Asia, and Africa and (2) gaseous emissions from deforestation. The IPCC Scenarios IS92a-f relate changes in the amount of forest cleared after 1990 to changes in population. In IPCC Scenario IS92d, greater per capita income is assumed to lead to voluntary actions to halt deforestation, and forest clearing in all countries is assumed to decline at more than 2.5%/yr after 1990.

- *The industry module.* This module combines population, economic growth assumptions, and growth in energy consumption to estimate emissions from industrial activities.

- *The atmospheric composition model.* This model is integrated with a model of CO_2 and heat uptake by the oceans.

A.1.3 Method and Organization

This appendix presents methods for estimating population, GNP, agricultural production and crop land area, desertification, air pollutant emissions, deforestation, and extinction of species in 2075. Côte d'Ivoire is used as an example to demonstrate these methods. For each area where sufficient data are available, the change in baseline conditions for Côte d'Ivoire is estimated. The estimates developed for Côte d'Ivoire should not be interpreted as a prediction of the future for Côte d'Ivoire; rather, they are examples of the way in which the method for developing baseline scenarios can be applied to a particular country.

A wide disparity exists among potential baseline scenarios because of the uncertainty about future socioeconomic and environmental conditions. No scenario is necessarily right or wrong, but a scenario must be internally consistent; for example, estimates of population must be consistent with GNP estimates, agricultural production, and deforestation. A change in assumptions about one variable, such as population, can affect many other variables.

Section A.2 discusses the development of population and GNP forecasts. Section A.3 presents sources of data and an approach for estimating growth in agricultural production and crop land area. Section A.4 discusses desertification of dry land areas. Section A.5 discusses tropospheric air pollution, and Section A.6 presents methods for estimating future deforestation. Finally, Section A.7 discusses the extinction of species, which is primarily caused by deforestation.

A.2 POPULATION AND GROSS NATIONAL PRODUCT

Pepper et al. (1992) have developed regional estimates of population, GNP, and GNP per capita through 2100. Although a number of socioeconomic growth scenarios are presented in their study, IPCC Scenario IS92a is used because it is based on the most likely[1] estimates of population and income growth. This scenario is based on the World Bank's forecast of a global population of 11.3 billion people by 2100 (Pepper et al. 1992). Economic growth through 2025 is based on a reference scenario from the Energy and Industry Subgroup of the Response Strategy Working Group, which was adjusted downward for Eastern Europe, the former Soviet Union, and the Persian Gulf because of recent political events.

---

[1] Given the long time frame for the baseline scenarios, little confidence exists in even the most likely scenario.

A.2.1 Applying Data on Population and National Income

The following estimates on population and national income should be applied:

- Estimates of national population from the World Bank should be used directly.

- National income (GNP or GDP) should be calculated by multiplying regional growth rates from Pepper et al. (1992) by current income. Expert judgment should be used to adjust the growth rates on the basis of particular national circumstances.

- Estimates of the standard of living (GNP per capita) should be calculated by dividing GNP by population. (A forecast of growth in productivity or technology has not been identified.) Long-term forecasts of participation by the workforce do not appear to be available. The GNP per worker is a measure of productivity that may be able to be estimated. Estimates of changes in technology, standard of living, and worker productivity may need to be made by using expert judgment.

A.2.2 Example: Growth in Population, GNP, and GNP per Capita in Côte d'Ivoire

Figure A.1 presents population growth in Côte d'Ivoire on the basis of projections by the World Bank. (From a population of 12 million in 1990, the World Bank [1992, 1993] projects 56 million in 2075.) The growth in GNP in Côte d'Ivoire (from $8,337 \times 10^6$ in 1990 to a projected $152,806 \times 10^6$ in 2075) is presented in Figure A.2 on the basis of the following estimates of annual percent growth in GNP from the IS92a scenario for other countries (Pepper et al. 1992): 1990-2000, 3.6%; 2000-2025, 4.2%; 2025-2050, 3.4%; and 2050-2075, 2.8%. Figure A.3 presents growth in the standard of living in Côte d'Ivoire, as measured by the growth in GNP per capita (from $695 [$8,337/12] in 1990 to a projected $2,729 [$152,806/56] in 2075).

A.3 AGRICULTURE

With large increases in population, a significant increase in agricultural production will be required to meet the demand for food. Increases in agricultural production can be achieved by increases in efficiency (i.e., technological advances), in cropping intensity, or in the amount of land applied to agriculture. The growth in agricultural land area is limited by the amount of arable land available. Changes in agriculture are likely to significantly affect land use, which will directly impact deforestation, desertification, and water use and indirectly impact biodiversity.

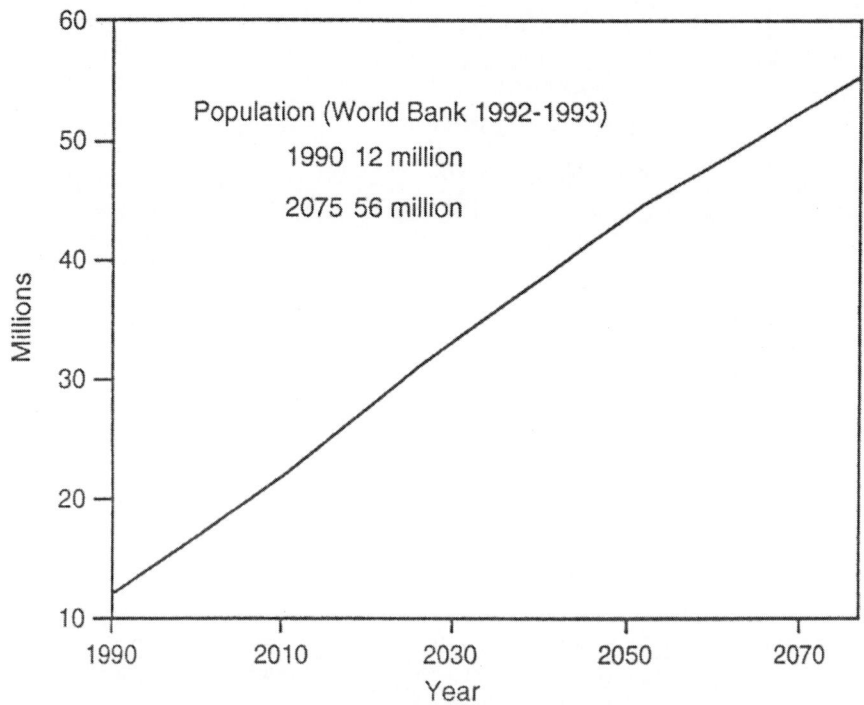

FIGURE A.1 Growth in Population in Côte d'Ivoire

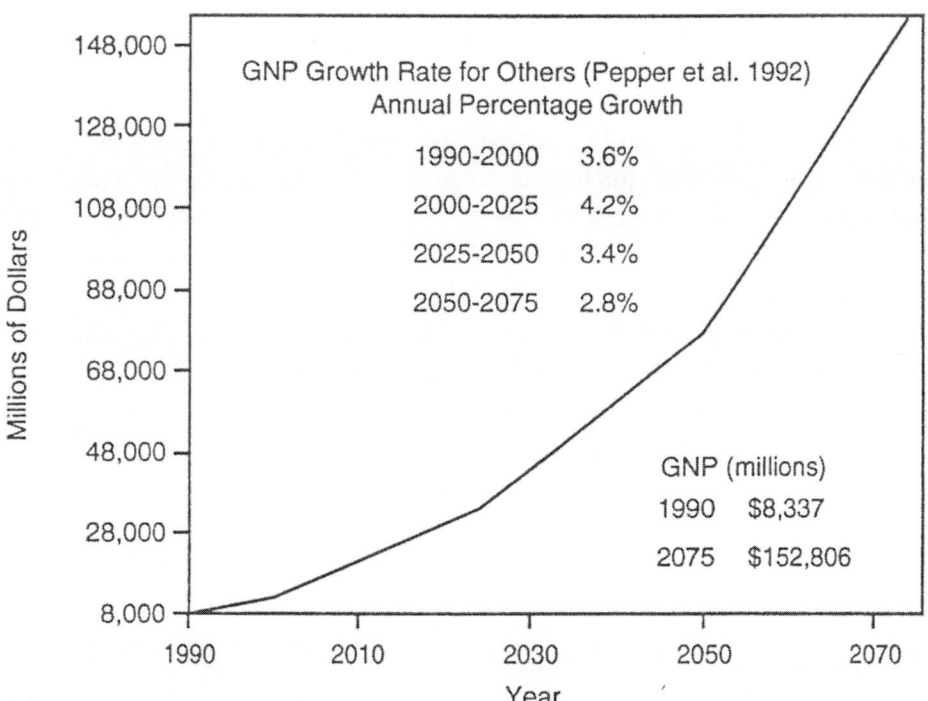

FIGURE A.2 Growth in GNP in Côte d'Ivoire

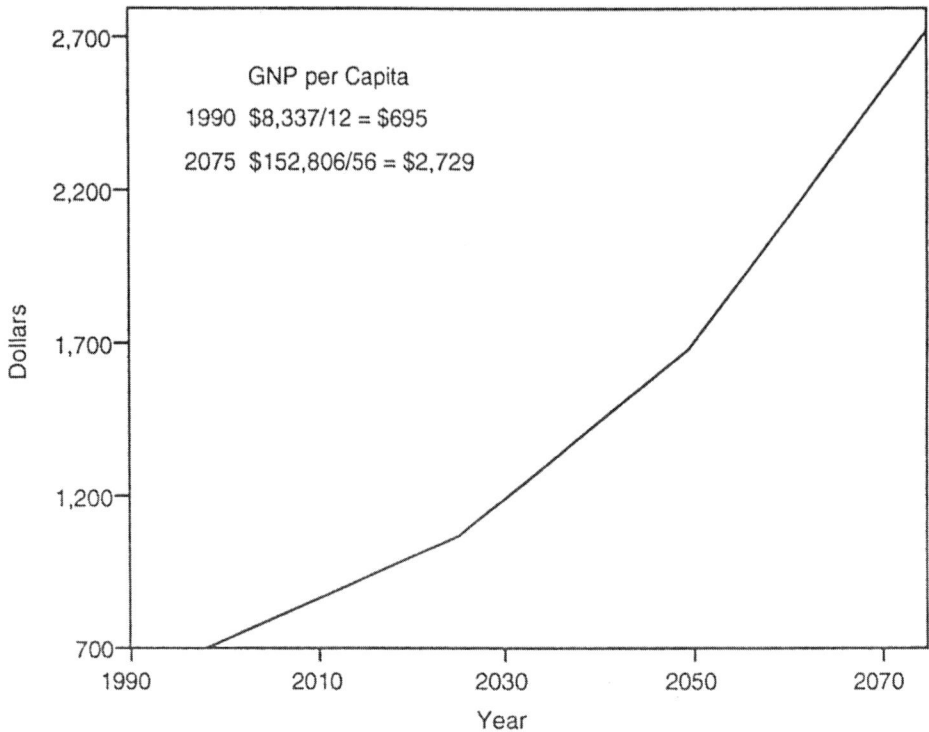

FIGURE A.3 Growth in the Standard of Living in Côte d'Ivoire

A.3.1 Predicting Growth in Agricultural Production

Regional forecasts of growth in agricultural production may be used to project growth through the next century for individual countries. Data on actual country-specific production should be obtained, if possible. Estimates of annual growth in agricultural production are provided for regional areas in Table A.1 (Pepper 1994). According to the Food and Agriculture Organization (FAO) of the United Nations, two-thirds of the growth in all agricultural production is expected to come from increases in yield; the rest is expected to come from the expansion of harvested area (FAO 1993). Regional forecasts of agricultural production have been developed for coarse grains, dairy, meats, other animals, other foods, protein feeds, rice, and wheat.

To determine how the data from Pepper (1994) compare with data on agricultural production provided by the FAO, projections of wheat production in developing countries were compared. The FAO estimates that wheat production is expected to grow 2.1%/yr through 2010 for developing countries (FAO 1993).[2] The FAO projected growth only through 2010.

---

[2] Developing countries in sub-Saharan Africa, Near East/North Africa, East Asia, South Asia, Latin America, and the Caribbean are included. China was not included in the FAO estimate because data on arable land and cropping patterns were unavailable. To facilitate comparison, centrally planned Asia was not included in calculations of the data of Pepper et al. (1992).

**TABLE A.1 Annual Growth Rate in Agricultural Production
(Scenario IS92a)**

Agricultural Production and Area[a]	Annual Growth Rate (%)				
	1990- 2000	2000- 2010	2010- 2025	2025- 2050	2050- 2100
Coarse Grains					
United States	1.13	0.39	0.39	0.14	-0.13
OECD Europe/Canada	2.85	1.46	1.15	-0.29	-0.03
OECD Pacific	5.43	2.54	3.09	1.86	-0.17
USSR/Eastern Europe	1.45	1.23	1.02	0.61	0.01
China/centrally planned Asia	1.06	0.36	0.48	0.27	0.11
Middle East	3.05	1.88	1.51	0.85	0.76
Africa	2.92	1.62	1.72	0.79	0.84
Latin America	2.92	2.75	2.67	2.63	0.25
South and East Asia	1.53	0.81	1.01	0.44	0.39
Dairy					
United States	0.79	1.57	1.49	0.98	-0.20
OECD Europe/Canada	0.64	0.41	0.37	0.13	-0.08
OECD Pacific	1.23	0.84	0.47	-0.06	-0.24
USSR/Eastern Europe	0.89	0.17	0.23	-0.22	-0.05
China/centrally planned Asia	2.46	1.68	1.18	0.78	0.04
Middle East	3.06	2.36	1.64	1.07	0.75
Africa	3.06	2.33	2.01	1.35	0.74
Latin America	2.96	2.35	1.85	1.21	0.18
South and East Asia	2.96	2.83	2.65	2.53	0.32
Meats					
United States	1.23	-0.15	0.05	-2.18	-0.22
OECD Europe/Canada	0.59	0.43	0.67	0.63	-0.09
OECD Pacific	1.10	0.98	1.14	0.90	-0.26
USSR/Eastern Europe	0.90	0.28	0.38	-0.14	-0.05
China/centrally planned Asia	1.21	0.92	0.71	0.52	0.02
Middle East	3.57	2.72	2.02	1.45	0.73
Africa	3.46	2.55	2.33	1.62	0.76
Latin America	2.33	1.81	1.49	0.92	0.16
South and East Asia	2.96	2.46	2.13	1.42	0.30
Other Animals					
United States	0.39	0.34	0.02	-0.38	-0.15
OECD Europe/Canada	1.16	0.80	0.39	-0.38	-0.02
OECD Pacific	2.18	2.02	2.34	2.62	-0.18
USSR/Eastern Europe	0.90	0.54	0.11	-0.42	0.02
China/centrally planned Asia	2.93	2.82	2.71	2.67	0.08

TABLE A.1 (Cont.)

Agricultural Production and Area[a]	Annual Growth Rate (%)				
	1990-2000	2000-2010	2010-2025	2025-2050	2050-2100
Middle East					
Africa	5.27	3.49	1.90	0.84	0.82
Latin America	3.48	2.63	1.74	0.91	0.81
South and East Asia	3.86	3.18	2.48	1.86	0.23
	3.39	2.62	2.06	1.59	0.35
Other Foods					
United States	0.60	0.76	0.34	-0.11	-0.39
OECD Europe/Canada	1.00	1.30	1.61	1.76	-0.33
OECD Pacific	1.41	1.07	0.40	-0.15	-0.43
USSR/Eastern Europe	0.82	0.96	0.49	0.60	-0.25
China/centrally planned Asia	1.48	1.60	1.11	0.94	-0.15
Middle East	3.52	2.68	1.88	1.21	0.58
Africa	3.63	2.69	2.27	1.72	0.62
Latin America	2.24	1.97	1.48	0.94	-0.01
South and East Asia	3.03	2.90	2.40	2.44	0.12
Protein Feeds					
United States	1.03	1.00	0.78	0.47	-0.34
OECD Europe/Canada	0.90	0.94	0.62	0.13	-0.24
OECD Pacific	2.55	2.48	2.25	2.28	-0.38
USSR/Eastern Europe	1.00	0.68	0.27	-0.20	-0.19
China/centrally planned Asia	1.78	1.62	1.08	0.87	-0.11
Middle East	2.50	2.03	1.65	1.38	0.59
Africa	3.66	2.51	1.98	1.50	0.66
Latin America	2.14	1.66	1.58	0.91	0.04
South and East Asia	1.95	2.02	1.98	1.37	0.17
Rice					
United States	2.55	2.18	1.76	1.44	-0.25
OECD Europe/Canada	1.52	1.54	1.22	0.62	-0.05
OECD Pacific	-0.12	-0.19	-0.65	-0.39	-0.29
USSR/Eastern Europe	0.90	0.53	0.45	0.25	-0.10
China/centrally planned Asia	0.75	0.40	0.20	0.03	-0.01
Middle East	2.20	1.71	1.09	0.38	0.71
Africa	4.11	2.90	2.25	1.47	0.76
Latin America	2.10	1.60	1.11	0.41	0.13
South and East Asia	2.74	2.03	1.96	1.12	0.2

TABLE A.1 (Cont.)

Agricultural Production and Area[a]	Annual Growth Rate (%)				
	1990- 2000	2000- 2010	2010- 2025	2025- 2050	2050- 2100
Wheat					
United States	1.71	1.13	1.07	0.48	-0.25
OECD Europe/Canada	0.87	0.86	1.65	0.18	-0.24
OECD Pacific	2.67	0.91	2.05	0.64	-0.29
USSR/Eastern Europe	0.90	0.59	0.58	0.43	-0.13
China/centrally planned Asia	0.93	0.90	0.40	0.43	-0.03
Middle East	0.88	0.51	0.93	0.55	0.67
Africa	2.02	1.32	1.70	1.03	0.68
Latin America	1.87	1.34	2.04	2.75	0.13
South and East Asia	4.04	2.37	1.78	1.17	0.27

[a] OECD = Organization for Economic Cooperation and Development.

Source: Pepper (1994).

On the basis of the Pepper (1994) data, wheat production is expected to grow 1.8%/yr over the same period for developing countries in the Middle East, Africa, Latin America, and South and East Asia (Table A.1). Therefore, the growth rates appear to be similar.

A.3.2 Predicting Growth in Crop land

Growth in agricultural lands can be based on estimates provided by the FAO (1993). According to the FAO, the area of harvested land in developing countries (except China) is expected to increase by 1.0%/yr through 2010 (Table A.2). Some of this growth is attributed to increased use of arable land (12%), while the rest is attributed to an increase in cropping intensity[3] (79-85%) (FAO 1993). Increases in the actual area of crop land used will likely come from increases in arable land, because changes in cropping intensity will occur on lands currently in use. Therefore, estimates of change in the area of crop land are based on growth rates for arable land. Additional data on growth in arable land can be obtained from FAO (1994).

After 2010, growth in crop land is likely to decline as growth in agricultural production slows, and efficiency and technology are improved. Regional FAO estimates of growth in arable land through 2010 may be used to project future trends, but these trends

---

[3] Increases in cropping intensity can be achieved by reducing fallow periods or by practicing multiple cropping.

TABLE A.2 Growth in Arable Land in Developing Countries, except China

Area	Arable Land (10^6 ha)		Annual Growth in Arable Land, 1990-2010 (%)
	1988/1990	2010	
Developing countries	757	850	0.6
Africa (sub-Saharan)	212	255	0.9
Near East/North Africa	77	80	0.2
East Asia	77	88	0.7
South Asia	201	210	0.2
Latin America/Caribbean	189	217	0.7

Source: FAO (1993).

should be adjusted to account for historical trends and likely future trends. Expert judgment should be used whenever possible. Changes in the rate of growth in agricultural production after 2010 could also be used to adjust crop land growth rates.

A.3.3 Example: Future Crop Land in Côte d'Ivoire

Table A.3 presents potential growth in the area of crop land in Côte d'Ivoire. Through 2010, projections are based on the annual rate of growth of arable land in sub-Saharan Africa (Table A.2). After 2010, the annual growth rate is scaled to account for declines in the growth of agricultural production, according to data provided by Pepper (1994) (Table A.1). On the basis of trends in the production of coarse grains, wheat, and rice in Africa, growth is estimated to decline by about 50% after 2010. Therefore, the predicted growth is adjusted to 0.45% accordingly. To date, Côte d'Ivoire supports 4 million ha of crop land. By 2075, the area of crop land is expected to increase to 6.4 million ha.

A.4 DESERTIFICATION

Desertification is defined as land degradation in arid, semiarid, and dry subhumid areas. Moist and tropical areas are not considered here. The land use factors that cause desertification include irrigation, through salinization and waterlogging; rainfed cropping, through increased water and wind erosion; livestock, through overgrazing and removing trees; mining; and (locally) tourism.

TABLE A.3 Estimated Area of Crop Land in Côte d'Ivoire through 2075

Variable	1990	2010	2025	2050	2075
Annual growth in arable land area (%)	0.9	0.9	0.45	0.45	0.45
Crop land area (10^6 ha)	4.0	4.8	5.1	5.7	6.4

Source: FAO (1993).

A.4.1 Estimating Current Levels of Desertification

Estimates of desertification in irrigated areas, **rainfed areas,** range land areas, and total dry land areas are presented in Dregne and Chou (1992). These data are provided for only one year, and no information on past or future trends in desertification is presented. According to Dregne (1994), much of the existing desertification has been caused by overgrazing.

A.4.2 Predicting Future Desertification

Future desertification is likely to be caused primarily by wind and water erosion because of the expansion of cultivated lands. Although overgrazing has contributed significantly to current desertification, the trend is not likely to continue in the future because most of the arid, semiarid, and subhumid lands available for grazing are already being used. Therefore, the trend through 2075 is likely to be driven largely by the expansion of cultivated lands (Dregne 1994).

If data on past trends in desertification can be found, those data should be used as a basis for extrapolating to future trends. If no data are available, the current percentage of desertified agricultural lands (provided by Dregne [1994]) could be applied to the future. The FAO (1993) study gives an estimate of future (through 2010) growth in the area of arable land. However, this figure represents growth in arable dry land areas and all other arable lands. Because desertification will occur only in dry land areas (by definition), the rate of growth should be adjusted to account only for applicable dry areas. Expert judgment will likely be required to determine potential growth in cultivated dry land areas (Dregne 1994).

In addition, the FAO projections of growth in land area will need to be adjusted because projections are through 2010, not 2075. As discussed in Section A.3, growth in arable land area will likely diminish over time as a result of a decline in the growth of agricultural production, increased efficiency, or improved technology. Historical trends, likely future trends, expert judgment, or estimates of agricultural production through 2100 provided by Pepper (1994) could be used to scale the rate of growth in agricultural lands.

A.4.3 Example: Desertification in Côte d'Ivoire

Currently, desertification occurs primarily in range land. Table A.4 presents current levels of desertification in Côte d'Ivoire. To date, 17,000 ha (or 16%) of the agricultural land (not including range lands) exhibit desertification. Because most future desertification is expected to result from the expansion of agricultural dry lands, the current percentage of desertification of agricultural lands is applied to projections of future agricultural dry lands.

In Côte d'Ivoire, the expansion of crop lands is likely to occur largely in forested areas. On the basis of expert judgment, the growth in dry land areas converted to crop lands will be one-third the rate of growth in total crop lands (Dregne 1994). The FAO estimates that growth in the harvested land area in sub-Saharan Africa will be 0.9% annually (Table A.2) through 2010 (FAO 1993). One-third (0.3%) of the FAO estimate of annual growth in arable land area for Africa is applied to determine the cultivated dry lands in Côte d'Ivoire in 2010. After 2010, growth in crop lands will likely decline in Africa because growth in agricultural production is expected to decrease (Pepper 1994). Pepper's estimates (1994) of agricultural production (Table A.1) form the basis for the assumption that crop land growth will decline by approximately 50%. Therefore, an annual growth rate of 0.15% is applied after 2010.

Table A.5 presents the estimated future desertification in Côte d'Ivoire. The estimates presented here could underestimate desertification, because the less productive lands that will be converted to agricultural lands in the future may be more vulnerable to desertification.

TABLE A.4 Estimated Current Desertification in Côte d'Ivoire

Severity of Desertification	Area (10^3 ha)		
	Irrigated Land	Rainfed Crop Land	Range Land
Slight	64	24	535
Moderate	--	15	50
Severe	--	2	300
Very severe	--	--	10
Total of moderate+	0	17	360
Total area	64	41	895

Source: Dregne and Chou (1992).

TABLE A.5 Estimated Future Desertification in Côte d'Ivoire[a]

Variable	1990[b]	2010	2025	2050	2075
Total agricultural dry lands (10^3 ha)	105	108	110	114	118
Total desertification (10^3 ha)	17	17	18	18	19

[a] Initial growth is assumed to be one-third of the annual growth rate in arable land (0.3%) for sub-Saharan Africa estimated by FAO (1993) (Section A.3). After 2010, growth is assumed to be 0.15% on the basis of the approximate 50% decline in agricultural production estimated by Pepper (1994) after 2010. A constant rate of desertification of agricultural land of 16% is assumed.

[b] Source: Dregne and Chou (1992).

A.5 AIR POLLUTION

This section covers tropospheric air pollutants such as VOCs, NO_x, SO_x, and CO. Tropospheric air pollution is caused primarily by the production and consumption of energy, biomass burning, industrial processes, and motor vehicles.

A.5.1 Estimating Current Levels of Air Pollution

Several sources provide historical and current emissions or concentrations of air pollutants for selected countries. Pepper et al. (1992) report 1990 emissions of VOCs for five regions and emissions of other pollutants for the world. The World Resources Institute (WRI 1992) gives data on 1970-1989 emissions of NO_x, SO_x, CO, hydrocarbons, and particulates for European and selected Asian countries. The World Bank (1992) provides annual growth rates and concentrations of sulfur dioxide (SO_2) and particulate matter from 1979 through 1990 for selected cities. Data on emissions are likely to be available through in-country sources.

A.5.2 Predicting Future Levels of Air Pollution

Pepper et al. (1992) also estimated air pollution levels in 2050 and 2100 for the world. The annual rates of growth in emissions levels are provided in Table A.6. These estimates could be used to predict trends in emissions at the national level. Because these estimates are global, expert judgment should be used to adjust these growth rates for national circumstances. No example of air pollution in Côte d'Ivoire is presented here because data on baseline levels of air pollution were not identified.

TABLE A.6 Annual Growth in Selected Global Emissions for 1990-2075

Type of Emission	Growth per Year (%)			
	1990-2000	2000-2025	2025-2050	2050-2075
NO$_x$	0.53	0.98	0.56	0.40
VOCs	0.41	1.08	0.75	0.57
Sulfur	0.30	1.34	0.87	-0.07
CO	0.45	0.38	0.53	0.43

Source: Pepper et al. (1992).

A.6 DEFORESTATION

Deforestation is primarily caused by the use of fuel wood and other timber products, the conversion of forests to crop land and range land, and urban development (WRI 1992). Globally, 17 million ha of tropical forest is lost each year. Latin America, Asia, and Africa account for a significant portion of these losses (WRI 1992).

A.6.1 Estimating Current Levels of Deforestation

Several sources provide current and past data on the total deforested area. Pepper et al. (1992) provide data on the deforested area and annual rates of deforestation for selected Latin American, African, and Asian countries for 1980-1990. The WRI study (WRI 1992) presents current forest area and annual rates of deforestation by country. Both Pepper et al. (1992) and WRI (1992) present data based on reports by the FAO (1988, 1991). Relying on data from other sources, Dixon et al. (1994) provide alternate estimates of current forest area and changes in forest area over the last 20 years for selected regions.

A.6.2 Predicting Future Deforestation

Pepper et al. (1992) have estimated tropical deforestation in 2100 based on broad geographic regions. These estimates can provide a basis for country-specific estimates. Expert judgment should be used to adjust them. Pepper et al. (1992) also provide data on tropical deforestation during the 1980s for specific countries. Although the forecasts of deforestation are presented for regions, they can be scaled to the country level on the basis of the country's 1980 portion of the regional forest area.

A.6.3 Example: Deforestation in Côte d'Ivoire

Côte d'Ivoire has one of the most rapid rates of deforestation in the world because of shifting cultivation, logging, and converting land to agricultural uses. Deforestation is enhanced by government policies and taxes related to clearing land and to plantations. Programs for reforestation have not been successful because of fires and poor implementation and maintenance of projects (WRI 1993).

Estimates of future deforestation in Côte d'Ivoire are based on projections of forest clearing by Pepper et al. (1992) for Africa (Scenarios IS92a, b, and e). Average annual rates of deforestation during 1981-1985 in Côte d'Ivoire (5.1%) are much higher than those for Africa as a whole (0.56%); therefore, the estimates of deforestation in Côte d'Ivoire are likely to be very conservative (Pepper et al. 1992). The regional rates for Côte d'Ivoire were not adjusted because in-country experts were not available; however, adjustments to account for these types of differences should be made, when possible.

To determine tropical deforestation in Côte d'Ivoire, it is assumed that losses are proportional to the current forested area. Currently, Côte d'Ivoire contains 2% of African closed broad-leaved forests and 1% of African open broad-leaved forests. Therefore, on the basis of the estimate by Pepper et al. (1992) that 2.1 million ha of closed broad-leaved forest and 3.9 million ha of open broad-leaved forest will be lost in Africa in 1990, it is estimated that 42,000 ha (2%) of the closed broad-leaved forest and 39,000 ha (1%) of the open broad-leaved forest will be lost in Côte d'Ivoire in 1990.

Future estimates of deforestation are presented in Table A.7. The total forest area lost by 2075 can be determined by interpolating between the forecasted rates.

A.7 EXTINCTION OF SPECIES

Species become extinct primarily because of loss of habitat (e.g., deforestation), excess harvesting of species, pollution (water and, to a lesser extent, air pollution), and land development. The loss of species can be defined as either local or global loss of organisms. Endangered or threatened species may be at greater risk than species that are currently plentiful; however, large-scale deforestation can lead to the loss of many species that are currently not in danger of extinction.

A.7.1 Estimating Future Loss of Species

Extinction of species can be estimated by using projections of expected tropical deforestation and rates of species loss associated with deforestation. Table A.8 summarizes estimates of global loss of species. Table A.9 presents estimates of loss of species by 2040 in Africa, Asia, and Latin America; however, significant uncertainty exists in these projections, and they should be applied with caution.

TABLE A.7 Estimated Annual Deforestation in Côte d'Ivoire[a] on the Basis of Scenarios IS92a, b, and e

Forest Type and Area	Deforestation (10^3 ha/yr)					
	1990	2000	2025	2050	2100	Total Area Lost by 2075
Closed broad-leaved						
Lost	42	44	62	46	0	3,977 (89%)
Remaining	4,458	4,028	2,708	1,358	208	-
Open broad-leaved						
Lost	39	46	75	51	0	4,475 (83%)
Remaining	5,376	4,521	3,009	1,434	159	-
Total						
Lost	81	90	137	97	0	8,452 (86%)
Remaining	9,834	8,549	5,717	2,792	367	-

[a] Based on current percentage of African forests in Côte d'Ivoire applied to predicted deforestation in Africa.

Source: Pepper et al. (1992).

Information on the number of known species per country is provided by the WRI (1992, Table 20.4); however, these data appear to be highly uncertain, and additional information about the number of tropical species and the applicability of estimates of species may be required to calculate any loss in species. Relying on the judgment of in-country or global experts to calculate the loss of species is recommended.

A.7.2 Example: Loss of Species in Côte d'Ivoire

According to estimates (Scenarios IS92a, b, and e) by Pepper et al. (1992), tropical deforestation in Africa will range from 6 to 11 million ha/yr. Reid (1992) estimates that the loss of species in Africa will be 3-6% if tropical deforestation occurs at a rate of 5 million ha/yr or 6-13% if 10 million ha is deforested each year. Therefore, on the basis of estimates by Pepper et al. (1992) of deforestation in Africa, the loss of species may be on the order of 3-13% in Africa by the year 2040. The assumption is made that the same rate of extinction will apply directly to Côte d'Ivoire, which is likely to be a conservative estimate because rates of deforestation in Côte d'Ivoire are higher than those for the rest of Africa. Again, expert judgment should be used, if possible, to adjust regional rates, although such judgment is not applied here.

TABLE A.8 Global Rates for Extinction of Species

Estimate of Loss of Species	Species Loss per Decade (%)	Method	Source[a]
1 million species, 1975-2000	4	Extrapolation of past trends	Myers (1979)
15-20% of species, 1980-2000	8-11	Species area curve, Global 2000 forest loss estimates	Lovejoy (1980)
12% of plant species in neotropics; 15% of bird species in Amazon basin	-	Species area curve	Simberloff (1986)
2,000 plant species per year in tropics and subtropics	8	50% species loss in areas deforested by 2015	Raven (1987)
25% of species, 1985-2015	9	50% species loss in areas deforested by 2015	Raven (1988a,b)
7% of plant species	7	50% species loss in 10 hotspots (deforested areas)	Myers (1988)
0.2-0.3% of species per year	2-3	50% of rain-forest species lost with deforestation	Wilson (1988, 1989)
2-13% of species, 1990-2015	1-5	Species area curve, current rate of deforestation, and 50% increase	Reid (1992)

[a] As cited in Reid (1992).

According to the WRI (1992), Côte d'Ivoire provides habitat for more than 1,100 species of mammals, birds, and freshwater fish and for more than 3,600 species of plants. On the basis of these estimates of known species, Côte d'Ivoire could possibly lose 140 to 610 species by the year 2040. The following assumptions were made regarding Côte d'Ivoire: (1) the rate of deforestation in Africa ranges from 6 to 11 million ha/yr (Pepper et al. 1992); (2) the rate of loss of species is 3-13% (Reid 1992); and (3) the number of known species is 4,700 (WRI 1992). The loss of species to the year 2075 is not estimated because of the high level of uncertainty in extrapolation.

TABLE A.9 Predicted Loss of Species by Region by 2040

Region	Percent Loss		
	Low Scenario[a]	Central Scenario[b]	High Scenario[c]
Africa	3-6	6-13	10-21
Asia	5-11	12-26	28-53
Latin America	3-8	8-18	15-32
All tropics	4-8	9-19	17-35

[a] Assumes tropical deforestation of 5 million ha/yr.

[b] Assumes tropical deforestation of 10 million ha/yr.

[c] Assumes tropical deforestation of 15 million ha/yr.

Source: Reid (1992).

A.8 APPENDIX A REFERENCES

Dixon, R.K., et al., 1994, "Carbon Pools and Flux of Global Forest Ecosystems," *Science* 263:185-190.

Dregne, H.E., 1994, personal communication from Dregne (Texas Technical University, Lubbock, Tex.; telephone: 806-742-2218; fax: 806-742-5063).

Dregne, H.E., and N. Chou, 1992, "Global Desertification Dimensions and Costs," in H.E. Drege (ed.), *Degradation and Restoration of Arid Lands*, Texas Technical University, Lubbock, Tex., pp. 249-282.

FAO, 1994, Economics and Social Policy Division (N. Alexandratos or J. Bruinsme), Food and Agriculture Organization of the United Nations, Rome, Italy (fax: 39-6-5225-3152; e-mail: esdgs@irmfa001.bitnet).

FAO, 1993, *Agriculture: Towards 2010*, presented at the 27th Session, Nov. 6-25, Food and Agriculture Organization of the United Nations, Rome, Italy.

FAO, 1991, *Forest Resources Assessment 1990 Project*, Forestry No. 7, Food and Agriculture Organization of the United Nations, Rome, Italy.

FAO, 1988, *An Interim Report on the State of the Forest Resources in the Developing Countries*, Food and Agriculture Organization of the United Nations, Rome, Italy.

Houghton, J.T., et al., 1992, *Climate Change 1992—The Supplementary Report to the IPCC Scientific Assessment*, Intergovernmental Panel on Climate Change, Cambridge University Press, Cambridge, England.

Leatherman, S.P., 1984, "Coastal Geomorphic Responses to Sea-Level Rise: Galveston Bay, Texas," in *Greenhouse Effect and Sea-Level Rise: A Challenge for This Generation,* Van Nostrand Reinhold, New York, N.Y.

Lovejoy, T.E., 1980, "A Projection of Species Extinctions," in G.O. Barney (ed.), *Global 2000 Study, The Global 2000 Report to the President*—Entering the 21st Century: A Report, Vol. 2, pp. 328-331, U.S. Department of State, Council on Environmental Quality, Washington, D.C.

Meadows, D.H., et al., 1992, *Beyond the Limits: Confronting Global Collapse, Envisioning a Sustainable Future*, Chelsea Green Publishing, Post Mills, Vt.

Myers, N., 1988, "Threatened Biotas: 'Hotspots' in Tropical Forests," *Environmentalist* 8(3):1-20.

Myers, N., 1979, *The Sinking Ark: A New Look at the Problem of Disappearing Species*, Pergamon Press, Oxford, England.

Pepper, W., 1994, personal communication from Pepper (ICF, Inc.).

Pepper, W., et al., 1992, *Emission Scenarios for the IPCC: An Update: Assumptions, Methodology, and Results*, prepared for the Intergovernmental Panel on Climate Change, Working Group 1, May.

Raven, P.H., 1988a, "Biological Resources and Global Stability," in S. Kawano et al. (eds.), *Evolution and Coadaptation in Biotic Communities*, University of Tokyo Press, Tokyo, Japan, pp. 3-27.

Raven, P.H., 1988b, "Our Diminishing Tropical Forests," in E.O. Wilson and F.M. Peter (eds.), *Biodiversity*, National Academy Press, Washington, D.C., pp. 119-122.

Raven, P.H., 1987, "The Scope of the Plant Conservation Problem World-wide," in D. Bramwell et al. (eds.), *Botanic Gardens and the World Conservation Strategy*, Academic Press, Inc., London, England, pp. 19-29.

Reid, W.V., 1992, "How Many Species Will There Be?" in T.C. Whitmore and J.A. Sayer (eds.), *Tropical Deforestation and Species Extinction*, Chapman and Hall, New York, N.Y., pp. 55-73.

Simberloff, D., 1986, "Are We on the Verge of a Mass Extinction in Tropical Rain Forests?" in D.K. Elliott (ed.), *Dynamics of Extinction*, John Wiley, New York, N.Y., pp. 165-180.

Wilson, E.O., 1989, "Threats to Biodiversity," *Scientific American* 108-116 (Sept. 1990).

Wilson, E.O., 1988, "The Current State of Biological Diversity," in E.O. Wilson and F.M. Peter (eds.), *Biodiversity*, National Academy Press, Washington, D.C., pp. 3-18.

World Bank, 1993, *Global Economic Prospects and the Developing Countries*, prepared for the World Bank, Washington, D.C.

World Bank, 1992, *World Development Report 1992: Development and the Environment*, Oxford University Press, New York, N.Y.

WRI, 1993, *Environmental Almanac*, World Resources Institute, Houghton Mifflin Company, New York, N.Y.

WRI, 1992, *World Resources 1992-93*, World Resources Institute, United Nations Environment Programme and United Nations Development Programme, Oxford University Press, New York, N.Y.

A.9 APPENDIX A BIBLIOGRAPHY

A.9.1 Population and Gross National Product

United Nations, 1990, *Overall Socio-Economic Perspective of the World Economy to the Year 2000*, ST/ESA/215, prepared by the Department of International Economic and Social Affairs, United Nations, New York, N.Y.

United Nations, 1993, *World Population Prospects: The 1992 Revisions.*

World Bank, 1993, *Global Economic Prospects and the Developing Countries.*

Zachariah, K.C., and M.T. Vu, 1988, *World Population Projections, 1987-1988 Edition*, World Bank, Johns Hopkins University Press, Baltimore, Md.

A.9.2 Agriculture

Rosenzweig, C., et al., date unknown, *Climate Change and World Food Supply.*

A.9.3 Desertification

United Nations Environment Programme, 1992, *World Atlas of Desertification*, Edward Arnold Publishers, New York, N.Y.

A.9.4 Deforestation and Extinction of Species

Grainger, A., 1993, "Rates of Deforestation in the Humid Tropics: Estimates and Measurements," *The Geographic Journal* 159(1):33-44.

Groombridge, B., 1992, *Global Biodiversity: Status of the Earth's Living Resources*, Chapman and Hall, New York, N.Y.

Sayer, J.A., and T.C. Whitmore, 1991, "Tropical Moist Forests: Destruction and Species Extinction," *Biological Conservation* 55:199-213.

A.9.5 Other References

Cubasch, U., et al., 1992, "Time Dependent Greenhouse Warming Computation with a Coupled Ocean Atmosphere Model," *Climate Dynamics*.

Hansen, J., et al., 1988, "Global Climate Changes as Forecast by the GISS 3-D Model," *Journal of Geophysical Research* 93:9341-9364.

Jäger, L., 1976, "Monatskarten des Niederschlags für die ganze Erde," *Berichte des Deutschen Wetterdienstes* 139.

Jenne, R.L., 1992, "Climate Model Description and Impact on Terrestrial Climate," in Majumdar et al. (eds.), *Global Climate Change,* pp. 145-164, Pennsylvania Academy of Sciences.

Karl, T.R., et al., 1990, "A Method of Relating General Circulation Model Simulated Climate to the Observed Local Climate. Part I: Seasonal Statistics, *Journal of Climate* 3:1053-1079.

McFarlane, N.A., G.J. Boer, J.P. Blancet, and M. Lazare, 1991, "The CCC Second Generation GCM and Its Equilibrium Climate," *Journal of Climate* (submitted for publication).

Mitchell, J.F.B., C.A. Senior, and W.J. Ingram, 1989, "CO_2 and Climate: A Missing Feedback?," *Nature* 341:132-134.

Ramanathan, V., 1988, "The Greenhouse Theory of Climate Change: A Test by an Inadvertent Global Experiment," *Science* 240:293-299.

Schlesinger, M.E., 1988, "Model Projections of the Climatic Changes Induced by Increased Atmospheric CO_2," presented at the Symposium on Climate and Geosciences, May 22-27, Reidel Publishing Company, Dordrecht, the Netherlands (in press).

Schlesinger, M.E., and Z.-C. Zhao, 1988, *Seasonal Climate Changes Induced by Doubled CO_2 as Simulated by the OSU Atmospheric GCM Mixed-Layer Ocean Model,* CRI Report and submitted to Climate for publication.

Stouffer, R.J., et al., 1989, "Interhemispheric Asymmetry in Climate Response to a Gradual Increase of Atmospheric CO_2," *Nature* 342:660-662.

Strzepek, K.M., and J.B. Smith (eds.), (forthcoming), *As Climate Changes: The Potential International Impacts of Climate Change,* Cambridge University Press, New York, N.Y.

Wetherald, R.T., and S. Manabe, 1992, *A Reevaluation of CO$_2$-Induced Hydrological Change as Obtained from Low and High Resolution Versions of the GFDL General Circulation Model,* July, 1990-91 (in preparation); part of the text was published in the 1991 IPCC report.

APPENDIX B:

CLIMATE CHANGE SCENARIOS FOR VULNERABILITY ASSESSMENTS

R. Benioff et al. (eds.), Vulnerability and Adaptation Assessments, B-1–B-20.
© 1996 *Kluwer Academic Publishers.*

APPENDIX B NOTATION

The following is a list of the acronyms, initialisms, and abbreviations (including units of measure) used in this appendix.

ACRONYMS, INITIALISMS, AND ABBREVIATIONS

GCM	general circulation model
GFDL	Geophysical Fluid Dynamics Laboratory
GISS	Goddard Institute for Space Studies
NCAR	National Center for Atmospheric Research
$1XCO_2$	single levels of carbon dioxide
$2XCO_2$	doubling of carbon dioxide levels

CHEMICAL

CO_2	carbon dioxide

UNITS OF MEASURE

°C	degree(s) Celsius
d	day(s)
km	kilometer(s)
mbar	millibar(s)
mm	millimeter(s)
mo	month(s)
yr	year(s)

CONTENTS

FIGURES

APPENDIX B:

CLIMATE CHANGE SCENARIOS FOR
VULNERABILITY ASSESSMENTS

This appendix provides details on the approaches that the Country Studies Management Team recommends for developing climate change scenarios. The reasons for this recommendation are discussed in Sections 4.2 and 4.3.

The scenarios recommended for use in the Country Studies Program are based on general circulation model (GCM) output and incremental changes in climate variables (e.g., +2°C, +4°C). The GCM scenarios are based on the difference between model estimates doubling of carbon dioxide (CO_2) (referred to as $2XCO_2$) and single levels of CO_2 (referred to as $1XCO_2$) and on the difference between GCM transient output and $1XCO_2$ estimates. Specifically,

- Data for 1951-1980 are used as baseline climate;

- Average monthly differences or ratios between $2XCO_2$ and $1XCO_2$ estimates by GCM are combined with baseline data;

- Differences or ratios from the GCM transient are combined with baseline data; and

- Incremental scenarios of +2°C, +4°C, and +6°C are combined with precipitation changes of 0, ±10%, and ±20%.

B.1 BASELINE CLIMATE

Climatic data for the years 1951-1980 should be gathered from observation stations in the geographic area of analysis. If the assessment of vulnerability is limited to a specific site, as in an agricultural study, weather observations should be gathered from the nearest weather station. If the analysis encompasses a broader area, such as a river basin, data on weather should be collected from stations in the region. Data should be collected for all climatic variables needed for the assessment; however, in many cases, only data on temperature and precipitation will be available.

Where data are missing from the 1951-1980 record or for certain variables (such as solar radiation), researchers can use a weather generator to generate missing data for the baseline climate. The process for using a weather generator is described in Section 5 and Appendix D.

B.2 CLIMATE CHANGE SCENARIOS FROM GCMs

The National Center for Atmospheric Research (NCAR) will provide researchers participating in the vulnerability assessment with monthly data from several GCMs. These outputs will be combined with the 1951-1980 data to produce climate change scenarios. These scenarios will serve as inputs to the vulnerability assessment.

B.2.1 Selecting GCMs

Researchers should select at least three GCMs that best represent current climate. Current output from the GCMs ($1XCO_2$) should be compared with a *spatially averaged* database of observed climate, not with site-specific climatic data.[1] The NCAR will make available an observed global temperature and precipitation database (which combines data from Crutcher and Meserve [1970] and from Jäger [1976]).

Researchers should compare seasonal patterns and approximate magnitudes of temperature and precipitation. No GCM will exactly replicate current climate. Comparisons of GCM estimates of current climate with observed climate over a wide spatial area, perhaps as broad as a continent, are vitally important. A comparison of GCM output for only one or a few grid boxes with observed data may give an inaccurate sense of how well that GCM replicates current climate. This inaccuracy occurs because regional climates are strongly influenced by broad patterns of circulation. Looking at an area of at least 40° latitude by 40° longitude is better for determining which three GCMs have the greatest accuracy in estimating current climate.

If all the GCMs contain major errors in estimating current climate, using only annual average output or regional averages from GCM grid boxes may seem reasonable. In general, countries should not use these averages until they have consulted with technical advisers from the Country Studies Program.

The data and the methods for creating the $2XCO_2$ and transient scenarios are described subsequently. A scheme for interpolation is then used to develop site-specific information on climate change from the GCMs.

B.2.2 Creating $2XCO_2$ Scenarios

The following data are needed:

- GCM $2XCO_2$ monthly means,

- GCM $1XCO_2$ monthly means, and

- Daily data from 1951 to 1980.

---

[1] Expecting that a GCM grid box as large as 8° × 10° can represent climate at any particular point would be unreasonable.

B.2.2.1 Compute the Difference between 2XCO$_2$ and Current or 1XCO$_2$

First, average monthly statistics from 2XCO$_2$ runs are used. The GCMs are run for 10-15 simulated years under 2XCO$_2$ conditions. Average monthly statistics are saved. January temperatures are calculated by averaging the temperatures in a particular grid box predicted for all Januarys in the run.

Second, 1XCO$_2$ data are interpolated to a specific point by using GCM data from the four nearest grid points, on the basis of linear averaging by the inverse of distances between the specific point and the GCM grid points. The NCAR will provide a disk with a subroutine to interpolate the GCM data. The basic formula for temperature is as follows:

$$\frac{\sum\limits_{1}^{i} \frac{(T_{i\,1x})}{D_i}}{\sum\limits_{1}^{i} \frac{(1)}{D_i}} , \qquad (B.1)$$

where

D_i = distance from the site to grid point i, and

$T_{i\,1x}$ = 1XCO$_2$ value for the temperature at grid point i.

The same procedure is used for precipitation or solar radiation by using the absolute amount of precipitation or solar radiation.[2] Figure B.1 gives an example of the interpolation technique. If the interpolated point is located on or very close to a GCM grid point, it is simpler to use the GCM grid point than to interpolate among the four closest grid points.

Figure B.1 is an example of the use of the interpolation technique to calculate precipitation for a site that is not on a grid point. Site X is between grid points A, B, C, and D. It is assumed that the grid points are 500 km apart. Site X is 100 km north and 150 km west of the nearest grid point, C. Next, 1XCO$_2$ precipitation is calculated at site X, assuming that average daily precipitation is 8 mm/d at point A, 10 mm/d at point B, 5 mm/d at point C, and 4 mm/d at point D. Site X is 532 km from point A, 427 km from point B, 180 km from point C, and 364 km from point D. The average daily precipitation at a particular grid point is multiplied by the ratio of the inverse of the distance from the grid point to site X divided by the total of the inverse of the distances between site X and all the grid points. This calculation yields an interpolated 1XCO$_2$ precipitation of 5.4 mm/d at site X.

Third, the 2XCO$_2$ data are interpolated by the same formula as used previously, but $T_{i\,2x}$ (i.e., 2XCO$_2$ output from the GCMs) is used instead of T_{i1x}.

---

[2] As described subsequently, differences between 2XCO$_2$ and 1XCO$_2$ values are used for temperatures, but 2XCO$_2$/1XCO$_2$ ratios are used for all other variables. The differences in temperature could be interpolated, but interpolating ratios would be incorrect.

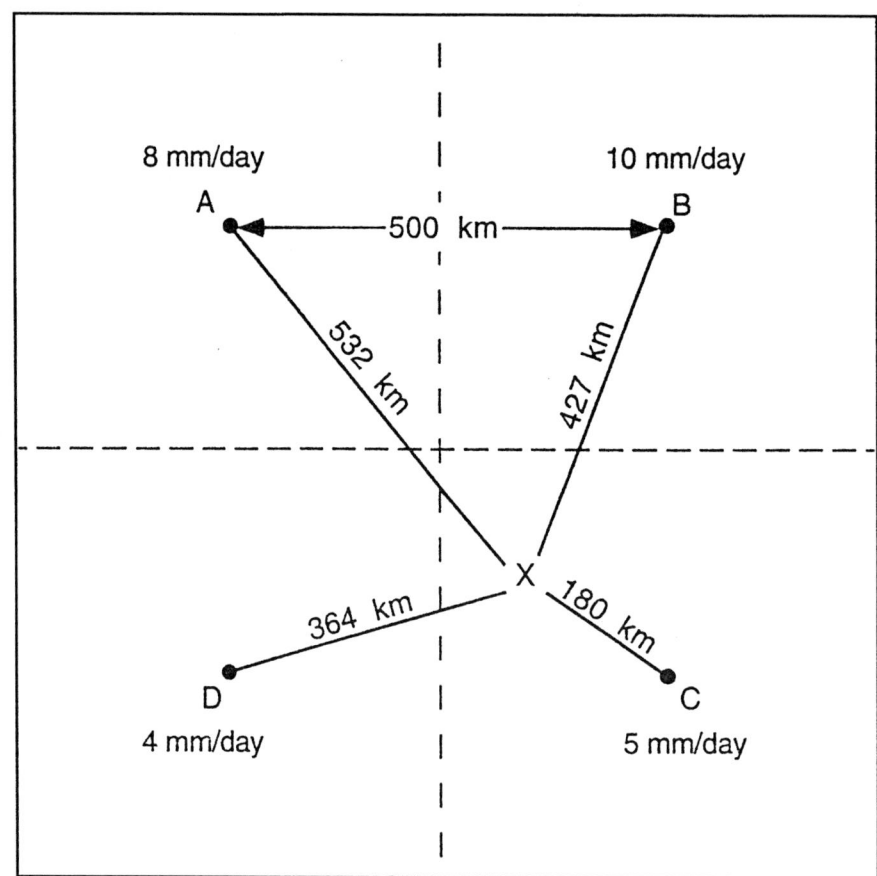

FIGURE B.1 January 1XCO$_2$ Precipitation

Fourth, the difference between 2XCO$_2$ and 1XCO$_2$ (the "adjustment statistics") is calculated by subtracting the weighted average 1XCO$_2$ temperature from the weighted average 2XCO$_2$ temperatures. For all other variables, the ratio between 2XCO$_2$ and 1XCO$_2$ is used. For example, in calculating the change in precipitation, P_{2x}/P_{1x}, where P is precipitation, is used.

Example 1. Assume that the weighted average 1XCO$_2$ temperature for January for Warsaw estimated by a GCM is -10°C. If the weighted average 2XCO$_2$ January temperature is -5°C, the adjustment statistic for January is the difference between the 2XCO$_2$ temperature and the 1XCO$_2$ temperature (-5°C − [-10°C] = +5°C).

Example 2. Assume that the weighted average 1XCO$_2$ precipitation for January for Warsaw estimated by a GCM is 100 mm/mo. If the weighted average 2XCO$_2$ precipitation for January is 80 mm/mo, the adjustment statistic for January is the ratio (80 mm/mo)/(100 mm/mo) = 0.8.

B.2.2.2 Apply the Adjustment Statistic to the Base Period

Multiply the historical daily (or weekly or monthly, depending on the input requirements for the model) weather statistic at a particular weather station by the monthly adjustment statistic for the appropriate grid box. For temperature, add the historical data to the adjustment statistic.

1. Multiply the January adjustment statistic for precipitation for the site by the observed precipitation at the site for January 1, 1951. This product gives a scenario precipitation at the site for year 1, day 1. If rain did not occur in the observed record, assume no rain in the climate change scenario.

2. Multiply the same statistic by the historical precipitation for the site for January 2, 1951, and repeat for all January days (Figure B.2).

Figure B.2a displays the effect of a 2°C temperature increase on a historical record of temperature. It is assumed that in the scenario, each day is 2°C warmer than in the historical record. The day-to-day pattern of temperature change is maintained, but the average monthly temperature becomes 2°C higher.

Figure B.2b displays the effect of a 25% increase in precipitation on a historical record of daily rainfall. Every day it rained in the historical record, the scenario has a 25% increase in precipitation. The historical pattern is maintained because no additional days of precipitation occurred, only increases in intensity. In addition, the relative amounts of precipitation remain the same.

3. Use the February adjustment statistic for the historical data for February and work on through the first year.

4. Start again with the January adjustment statistic for January 1952 observed data and work through the entire 30-yr period. In other words, the adjustment statistic for January is applied to all observed January data, the adjustment statistic for February is applied to all observed February data, and so on. This calculation yields a 30-yr climate scenario for the site under $2XCO_2$ conditions. It is assumed this 30-yr scenario is representative of the climate in the year 2075.

B.2.3 Creating GCM Transient Scenarios

The following data are needed:

- GCM transient monthly statistics,

- GCM $1XCO_2$ monthly statistics, and

- Daily data from 1951 to 1980.

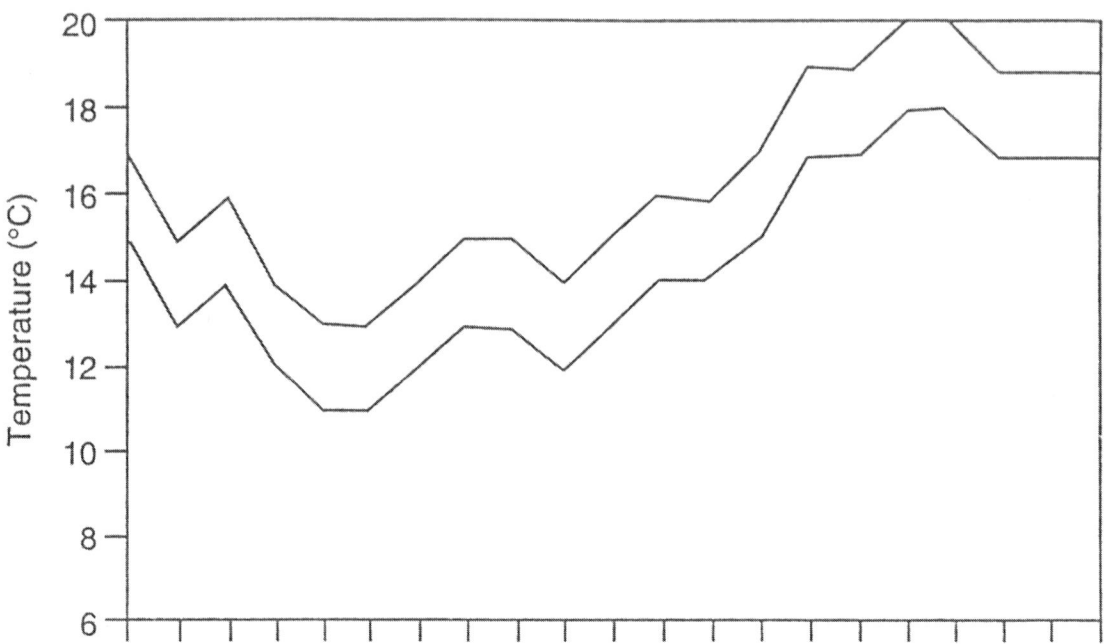

FIGURE B.2a Effect of a 2°C Temperature Increase on a Historical Temperature Record

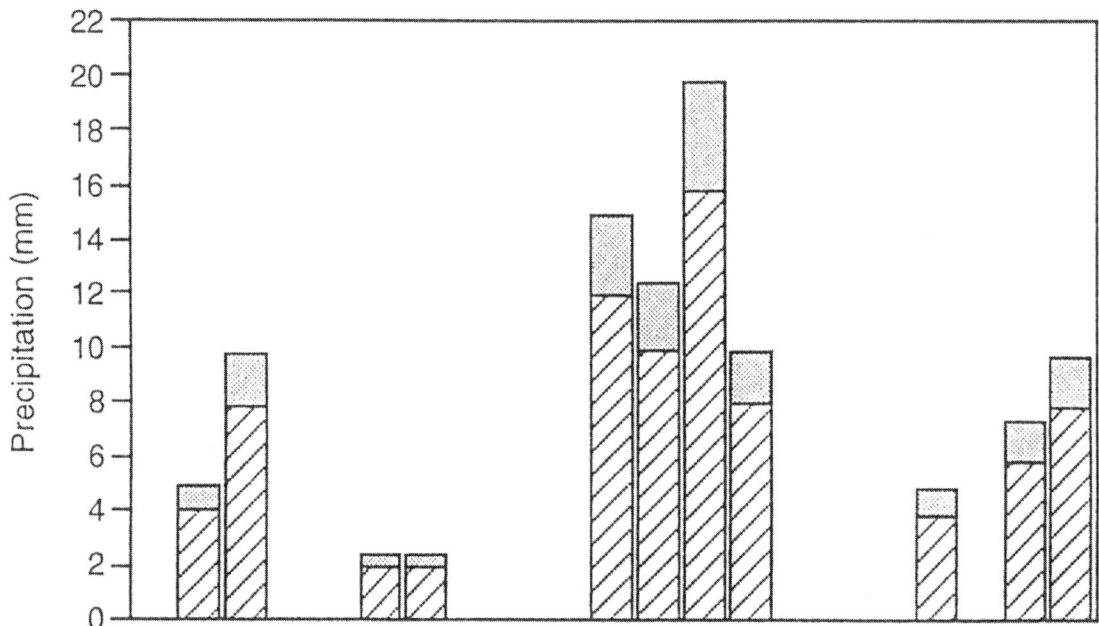

FIGURE B.2b Effect of a 25% Increase in Precipitation on a Historical Precipitation Record

The basic procedure is to take the underlying warming trends from the GCM transient scenarios and combine them with a baseline (observed) climate. Researchers should construct a transient run from 1991 through 2070. This run stops 5 yr before the $2XCO_2$ of 2075.

B.2.3.1 Continuous Transient

The NCAR will provide average monthly data for the first, fourth, seventh, and tenth decades from the Geophysical Fluid Dynamics Laboratory (GFDL) transient (Stouffer et al. 1989). Transient runs are not necessarily set at a particular time, but the beginning of most transient runs is nearly consistent with the 1951-1980 climatic normal data. It is assumed that the transient run begins in the middle of the 1951-1980 period. Thus, rounding off to the nearest decade, it is assumed that the transient begins in 1971; so the first decade of the transient run corresponds to the 1970s, the fourth corresponds to the 2000s, the seventh to the 2030s, and the tenth to the 2060s. It is assumed that the data supplied by NCAR are for the midpoint of the decade (e.g., 2026) (decades run from a year ending in 1 to a year ending in 0). Yearly data on climate change for a particular month can be calculated by linearly interpolating between these points in time; for example, to obtain data for January, for the 2010s, interpolate between the January data for the 2000s and the January data for the 2030s.

The adjustment statistics are calculated in the same manner as for the $2XCO_2$ scenarios. The same technique of linear interpolation should be used to create adjustment statistics for specific sites. The $1XCO_2$ temperatures for the respective month are subtracted from transient temperatures; transient data for other variables are divided by $1XCO_2$ data.

Baseline data are determined by taking the period 1951-1980 and repeating it two and two-thirds times to represent the baseline scenario of the years 1991-2070 without climate change. This procedure may repeat slight patterns of warming or cooling in baseline climate, but these oscillations are likely to be very small in comparison with the GCM transient changes in climate.

The adjustment statistics for temperature are added to baseline data on temperature and, for other variables, are multiplied by the baseline data.

A transient scenario is created as follows:

1. To get a transient scenario starting on day 1 of year 1 (1991) and going through the last year of the transient, first multiply the interpolated January adjustment statistic for the first year of the transient by the observed January 1951 daily data. Multiply the interpolated January adjustment statistic by January 1, 1951, data; then multiply by January 2, 1951, data; and then multiply by the rest of the January days. This process gives a transient scenario for a particular weather station for the first month of the first year of the transient.

2. Multiply the interpolated February adjustment statistic for the first year of the transient by the February 1951 daily historical record.

3. Work through the 1951 daily data.

4. Apply the interpolated January adjustment statistic for the second year of the transient to the January 1952 data.

5. Work through the baseline data in this manner.

6. Repeat using the historical record by applying the interpolated January adjustment statistic for the 31st year of the transient (2021) to January 1951 data.

7. Work through all of the transient monthly adjustment statistics. This process gives the full transient scenario.

Figure B.3 displays the transient scenario for global annual average temperatures obtained by using the GFDL transient (Stouffer et al. 1989). Figure B.3a displays the baseline climate for 1991-2070 (Jones et al. 1990). This figure contains 1951-1980 global average temperature anomalies repeated for 80 years. The first 30 years are the 1951-1980 record, the next 30 years are the same record, and the final 20 years are the 1951-1970 record. This information is the baseline climate for 1991-2070.

Figure B.3b displays the interpolated change in global average annual temperature from the GFDL transient for 1991-2070. The figure was created by plotting the four data points supplied by NCAR. They are for the first decade (1970s), the fourth decade (2000s), the seventh decade (2030s), and the tenth decade (2060s). It is assumed that the data point is for the year ending in a 6 (e.g., the first data point is for 1976). It is also assumed that the first point equals 0, but that point is not displayed because the graph starts in 1991. To develop the year-by-year temperature changes, straight lines were drawn between each data point. The resulting line shows the underlying warming trend from the transient.

Figure B.3c displays the transient scenario. It was created by adding the data from Figure B.3a to the data from Figure B.3b for each respective year. For example, the anomaly for 1991 from Figure B.3a was added to the change in temperature for 1991 from Figure B.3b and entered in 1991 in Figure B.3c. Although warming and cooling periods are present, an underlying warming trend quickly becomes apparent.

B.2.3.2 Equilibrium Points in Time

A second option for using the transient data is to use the decadal averages provided by NCAR as equilibrium points. This option is a much easier method for examining potential climate change at a particular time. It will give 30 years of data representative of average conditions for a future time. It does not make sense to use this approach to examine the way in which a system would be affected by a gradual change in temperature over time.

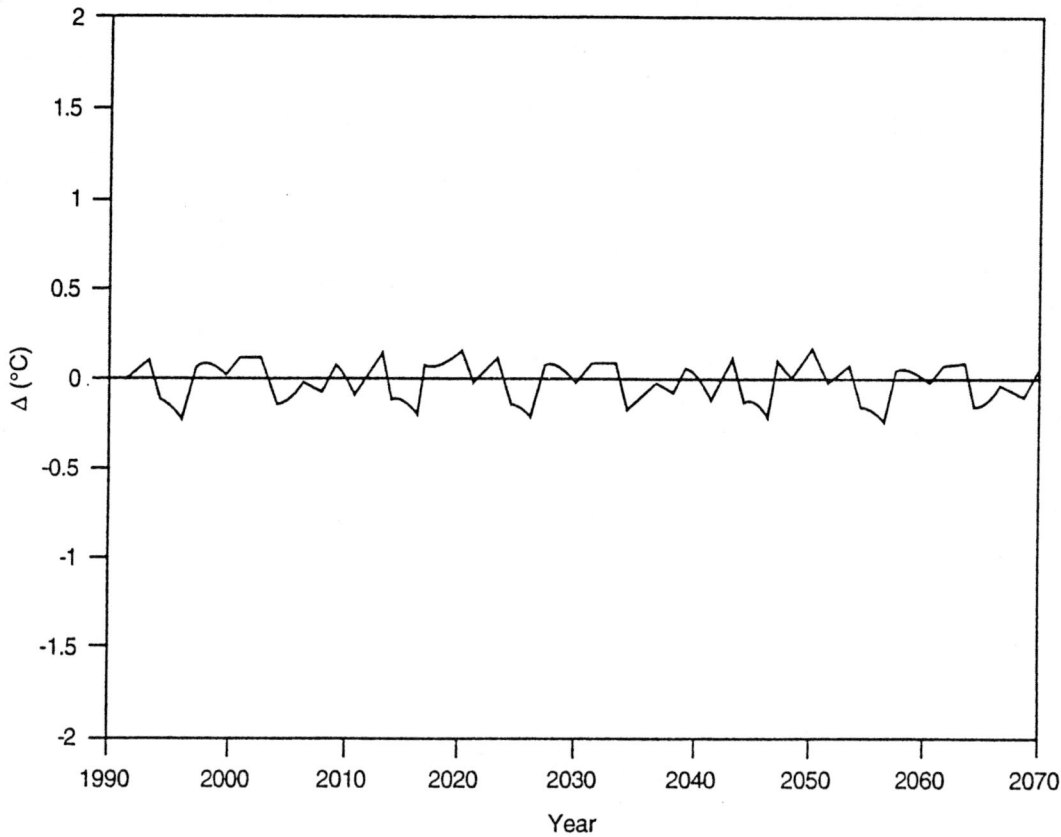

FIGURE B.3a Baseline Climate Deviations from the 1950-1979 Reference Period for 1991-2070

The difference or the ratio between the transient data is taken from the decadal midpoint and the $1XCO_2$ data in the same manner as was done for creating the $2XCO_2$ adjustment statistics. Add or multiply the adjustment statistics by the 30 years of baseline data; for example, use the fourth decade from the GFDL transient, which has a global warming of 0.8°C. By assuming an average warming rate of 0.3°C per decade, this warming would be representative of the 2020s. A 30-year scenario could be constructed for this equilibrium analysis because the adjustment statistics could be applied to the entire baseline of 1951-1980. Interpolation between the transient output could be done to calculate scenarios in other years. To avoid confusion about using 30 years of data to examine potential impacts of climate change in a particular decade, vulnerability researchers should report only the results for means and variances.

B.3 INCREMENTAL SCENARIOS

The incremental scenarios provide a wide range of potential regional climate changes and help to identify sensitivities to changes in temperature and precipitation. Increases in temperature (+2°C, +4°C, and +6°C) should be combined with no change and with ±10% and

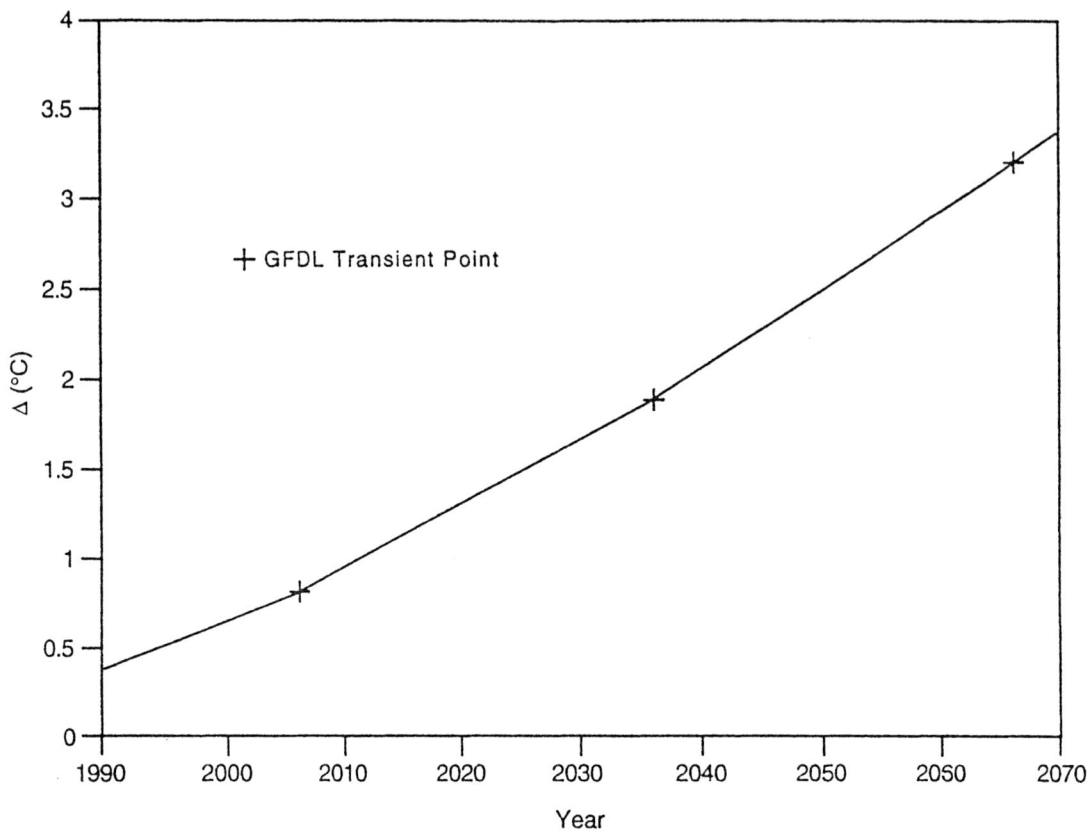

FIGURE B.3b Smoothed GFDL Transient

±20% changes in precipitation. For example, a 2°C increase in temperature could be combined with no change and with -20%, -10%, +10%, and +20% changes in precipitation. The same procedure should be done for the 4°C and 6°C temperature changes. For countries in latitudes between 30°N and 30°S, the 6°C temperature rise need not be used because this rate of warming from a $2XCO_2$ appears unlikely in the tropics. For countries in latitudes above 60°N and below 60°S, an 8°C warming, in addition to the other three temperature changes, may also be examined.

Incremental scenarios are created by adding the same temperature change to all the observed temperatures in the 1951-1980 period. Precipitation is calculated by converting the percent change to ratios (e.g., a 10% increase is 1.1; a 10% decrease is 0.9) and multiplying by the rainfall amounts in the baseline period.

B.4 ALTERNATIVE SCENARIOS

The approaches described previously for developing scenarios based on GCM output have the advantage of being relatively easy to apply. Therefore, they will consume few resources and be less likely to be applied incorrectly; but these approaches also have the disadvantage of assuming no change in daily or interannual variability and no change in

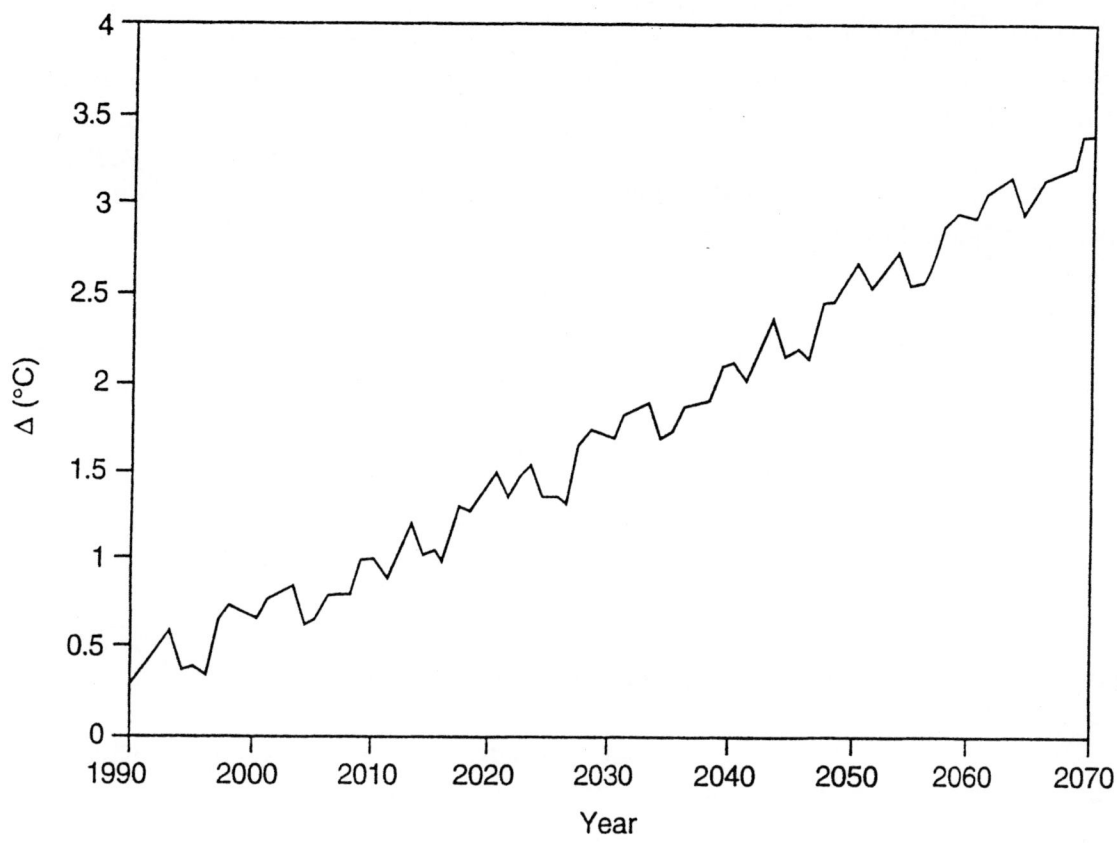

FIGURE B.3c Transient Scenario

spatial variability within GCM grid boxes. A number of articles have been published that suggest alternative ways of using GCM output to create climate change scenarios; some articles question whether GCMs should be used at all. These articles are summarized in the following sections.

B.4.1 Scenarios of Spatial Variability

Wigley et al. (1990) developed a statistical technique for examining spatial variance for temperature and precipitation within GCM grid boxes. These investigators used area-averaged temperature and precipitation, mean sea-level pressure, and 700-mbar heights as their predictor variables. Regression techniques estimated the relationship between the predictor variables and spatial variations in temperature and precipitation.

The approach was tested by using observed climatic data from Oregon, USA. The approach did very well in predicting observed January temperatures, predicting more than 80% of spatial variance, and reasonably well for July temperatures, predicting more than 58% of variance. The approach worked less well on precipitation, predicting only 40-50% of the spatial variation.

Wigley et al. (1990) then used this approach to examine whether significant differences in changes in temperature and precipitation might occur across four sites in Oregon under climate change scenarios. Using a scenario of a 1.0-2.6°C warming and a 20% increase in precipitation and the results of the regression described previously, Wigley et al. (1990) found that noticeable spatial variation could occur in changes in precipitation; for example, although a 20% increase in precipitation for all months was assumed, some sites could have as much as a 20% decrease in precipitation in some months. The spatial variance for temperature change was relatively small and could be ignored in constructing temperature change scenarios (Wigley 1994). This approach is promising for developing sub-grid-scale scenarios; however, it will require a fair amount of data analysis and time to develop. In general, this approach is not recommended for creating scenarios of spatial variability.

B.4.2 Scenarios of Temporal Variability

Mearns et al. (1992) examined the effect of changes in monthly variability on crop yields in Kansas, USA. These investigators modified interannual variability of a historical record (1951-1980) of monthly temperature and precipitation and developed monthly time series with double, quadruple, half, and one-quarter of observed interannual variance; for example, Mearns et al. (1992) changed the variance of monthly January temperatures and used the changed interannual monthly statistics to develop a daily climate change scenario. Daily temperatures were adjusted by the modified change in mean monthly temperature (similar to the approach recommended in this document). Adjusting monthly precipitation was more complicated. If increased variability resulted in a negative value for precipitation in a month, Mearns et al. (1992) assumed no precipitation. To ensure that the change in annual precipitation was unaffected, they subtracted the absolute value of any excess monthly negative value from the other months. They then multiplied the ratios of new precipitation values to old values by daily data. This approach is similar to the recommended approach, except that interannual variability is changed.

Wilks (1992) developed a technique for introducing daily and interannual variability for maximum and minimum temperatures, daily precipitation, and solar radiation, on the basis of GCM output. The method adapts stochastic weather-generating models to generate a daily time series consistent with average monthly GCM estimates. Wilks (1992) focused on estimating the likelihood of precipitation on day t as a function of whether precipitation occurred on day $t - 1$. He developed an example for Ithaca, New York, USA, on the basis of the Goddard Institute for Space Studies (GISS)-A transient scenario. In this example, Wilks (1992) assumed that monthly precipitation and interannual variation of monthly precipitation would increase.

The Mearns et al. (1992) approach is relatively easy to apply to single sites. However, to apply this approach across many sites could be time-consuming and expensive. The Wilks (1992) approach would be more difficult to apply and, therefore, may not be appropriate for the Country Studies Program.

B.4.3 Relating Large-Scale Circulation Patterns to Local-Scale Climate

Karl et al. (1990) examined the possibility of taking grid-point-free atmospheric statistics from GCMs to estimate local daily maximum and minimum temperatures, daily precipitation, and cloud ceilings. These investigators used a statistical procedure called "climatological projection by model statistics" to relate observed free atmospheric variables to observed local climate. They examined a number of sites across the United States and used observed climatic data from 1966 to 1986. They found that, for most of the cases studied, the local climatic variables could be explained well by the larger-scale free atmospheric variables. When Karl et al. (1990) tried to estimate local climate from GCM-derived free atmospheric data (using the Oregon State University GCM; Schlesinger and Gates 1980), the results were not as good. This discrepancy was due primarily to biases in the model control estimates of free atmospheric means and variances and to internal inconsistencies between these variables. It can be concluded that this method could be used to develop local climatic information from grid-scale GCM output, provided model biases are minimized; however, this approach would be very complex and time-consuming to use in the Country Studies Program.

Mika (1993) examined the relationship between large-scale circulation patterns and local climatic anomalies. If a statistically significant relationship could be established, estimates of large-scale circulation patterns from GCMs could be used to estimate local-scale changes in climate. Mika (1993) used 1951-1980 observed climatic data from Europe. He examined the relative predictive capability of large-scale circulation patterns, physical patterns (i.e., noncirculation patterns), and the mixture of large-scale circulation and physical patterns to estimate variations in local climate. The large-scale circulation patterns were based on a $5° \times 10°$ grid pattern, similar to GCM grid boxes. He then examined anomalies of temperature and precipitation for one site in Hungary and found that the physical forces explained many more of the temperature and precipitation anomalies than did large-scale circulation patterns. The one exception was extreme decreases in precipitation, where circulation explained a substantial portion of the variance. This finding indicates that using large-scale circulation patterns from GCMs to estimate spatial variance of local climate could be difficult.

In contrast, Wigley et al. (1990) found that large-scale circulation could explain a significant amount of local climatic variation, particularly for temperature; and Karl et al. (1990) found that free atmospheric statistics could explain local climatic variations. It is not clear whether the differences between these studies are due to the application of different methods or to the selection of different sites to test the methods. More research is needed to resolve this difference.

B.4.4 Composite Standardized GCMs

Santer et al. (1990) compared regional differences in estimates of change in temperature and precipitation from $1XCO_2$ to $2XCO_2$ among five GCMs. These investigators analyzed whether GCMs in some regions consistently have more or less warming than other regions or whether some regions are more or less likely to have less precipitation than others. The GCM output was converted to a standard 4° × 5° grid. To ensure that the results were not dominated by models with relatively large amounts of temperature increase, Santer et al. (1990) normalized GCM output by dividing regional change in temperature by global average change. This process yields the relative change in regional temperature or precipitation per degree of global warming. The results are then averaged across the GCMs to produce a composite change in regional temperatures. For any change in global temperatures, the relative change in regional temperature can be multiplied by the change in global temperature to calculate the absolute change in temperature. Santer et al. (1990) also examined the probability of regional decreases in precipitation. This approach could also be used to calculate a percent change in precipitation as a function of a 1°C increase in global temperature.

This approach will allow researchers to examine physically plausible regional climates on the basis of arbitrary increases in global temperatures. Unfortunately, the composite standardization has not yet been developed for the countries that are part of the Country Studies Program. It may be possible to use this approach in the future.

B.4.5 Using Expert Judgment to Evaluate Regional Application of GCMs

Ackerman and Cropper (1988) advocated comparing $1XCO_2$ estimates of current climate with observed climatic data and using expert judgment to determine whether GCMs simulate current climate reasonably. These authors were vague about the size of the geographic area over which this comparison should be conducted. If GCMs compare well with current climate, they advocate using a transfer function to combine GCM output with baseline climate. Such a transfer function might be the application of a temperature change from a GCM uniformly across the respective grid box (as recommended in this document). If GCMs do not compare well with current climate, they recommend conducting a sensitivity analysis or using expert judgment to assess impacts.

Robock et al. (1993) applied this approach and determined that GCMs did not replicate current climate well in several areas around the world. These investigators then used sensitivity analyses based on the GCMs.

The GCM estimates of current climate should be compared with observed normal climatic values over a very large area, such as a continent. Unlike the Ackerman and Cropper (1988) approach, this approach does not advocate choosing between using GCMs or sensitivity analysis. Using both approaches is preferable because each has its strengths and weaknesses (Section 4).

B.4.6 Conclusions

A number of interesting alternative approaches for creating GCM-based scenarios are described in the literature; however, the use of these approaches in other than isolated cases may be premature. Few of these approaches have been applied for climate change vulnerability analysis. Furthermore, because so much uncertainty exists about regional and temporal climatic change, using the simplest applications of GCM data — the approach described in the first part of this appendix — is best for now.

B.5 APPENDIX B REFERENCES

Ackerman, T.P., and W.P. Cropper, Jr., 1988, "Scaling Global Climate Projections to Local Biological Assessments," *Environment* 30(5):31-34.

Crutcher, H.L., and J.M. Meserve, 1970, *Selected Level Height Temperatures for the Northern and Southern Hemispheres*, Naval Weather Service Command, Washington, D.C.

Jäger, L., 1976, "Monatskarten des Niederschlags für die ganze Erde," *Berichte des Deutschen Wetterdienstes* 139(18).

Jones, P.D., et al., 1990, "Global and Hemispheric Temperature Anomalies," in T.A. Boden, P. Kanciruk, and M.P. Farrell (eds.), *Trends '90*, ORNL/CDIAC-36, Carbon Dioxide Information Analysis Center, Oak Ridge National Laboratory, Oak Ridge, Tenn.

Karl, T.R., et al., 1990, "A Method of Relating General Circulation Model Simulated Climate to the Observed Local Climate: Part I. Seasonal Statistics," *Journal of Climate* 3:1053-1079.

Mearns, L.O., et al., 1992, "Effect of Changes in Interannual Climatic Variability on CERES-Wheat Yields: Sensitivity and $2XCO_2$ General Circulation Model Studies," *Agricultural and Forest Meteorology* 62:159-189.

Mika, J., 1993, "Effects of the Large-Scale Circulation on Local Climate Anomalies in Relation to GCM Outputs," *IDÓJÁRÁS* 97:21-34.

Robock, A., et al., 1993, "Use of General Circulation Model Output in the Creation of Climate Change Scenarios for Impact Analysis," *Climatic Change* 23:293-335.

Santer, B.D., et al., 1990, *Developing Climate Scenarios from Equilibrium GCM Results*, report 47, Max-Planck-Institute für Meteorologie, Hamburg, Germany.

Schlesinger, M.E., and W.L. Gates, 1980, "The January and July Performance of the OSU Two-Level Atmospheric General Circulation Model," *Journal of Atmospheric Sciences* 37:1914-1943.

Stouffer, R.J., et al., 1989, "Interhemispheric Asymmetry in Climate Response to a Gradual Increase of Atmospheric CO_2," *Nature* 342:660-662.

Wigley, T.M.L., 1994, personal communication from Wigley (National Center for Atmospheric Research).

Wigley, T.M.L., et al., 1990, "Obtaining Sub-Grid-Scale Information from Coarse-Resolution General Circulation Model Output," *Journal of Geophysical Research* 95:1943-1953.

Wilks, D.S., 1992, "Adapting Stochastic Weather Generation Algorithms for Climate Change Studies," *Climatic Change* 22:67-84.

APPENDIX C:

GENERAL CIRCULATION MODELS AND DATA

R. Benioff et al. (eds.), Vulnerability and Adaptation Assessments, C-1–C-28.
© 1996 Kluwer Academic Publishers.

APPENDIX C NOTATION

The following is a list of the acronyms, initialisms, and abbreviations (including units of measure) used in this appendix.

ACRONYMS, INITIALISMS, AND ABBREVIATIONS

ASCII	American Standard Code for Information Interchange
CCC	Canadian Climate Centre
CD-ROM	compact disk with read-only memory
GCM	general circulation model
GFDL	Geophysical Fluid Dynamics Laboratory
GISS	Goddard Institute for Space Studies
GrADS	Grid Analysis and Display System
NCAR	National Center for Atmospheric Research
PC	personal computer
U.K.	United Kingdom
UKMO	United Kingdom Meteorological Office
$1XCO_2$	current climate for carbon dioxide
$2XCO_2$	doubled atmospheric concentration of carbon dioxide

UNITS OF MEASURE

bpi	bits per inch
cm	centimeter(s)
cm^2	square centimeter(s)
g	gram(s)
Gbyte	gigabyte(s)
h	hour(s)
in.	inch(es)
kbyte	kilobyte
km	kilometer(s)
m	meter(s)
m^2	square meter(s)
mbar	millibar(s)
Mbyte	megabyte(s)
W	watt(s)
yr	year(s)

CONTENTS

CONTENTS (Cont.)

TABLES

FIGURES

APPENDIX C:

GENERAL CIRCULATION MODELS AND DATA

C.1 INTRODUCTION

This appendix presents some of the technical details about climate models, compares some models, and uses observed climate data and general circulation model (GCM) output data and software to extract and display GCM data. It also discusses the variability of natural and GCM climate data.

The analyses to be conducted as part of the Country Studies Program need monthly average data to simulate single levels of carbon dioxide (referred to as $1XCO_2$) and doubling of carbon dioxide (referred to as $2XCO_2$), plus monthly average data from several decades for a few transient model runs. Table C.1 shows the main climate model runs at the National Center for Atmospheric Research (NCAR) that will be available for the Country Studies Program. Four decades of data from the 100-year German transient runs plus three $1XCO_2$ and $2XCO_2$ model runs are available on diskettes (see the last column in Table C.1).

The seven diskettes have a volume of about 7.7 Mbyte and contain three variables: average monthly temperature, precipitation, and downward solar radiation at the surface. In addition, they have climatological data for temperature and precipitation. The tapes at NCAR have much more data than these. Additional model simulations will arrive at NCAR from June through December 1994, and NCAR will gradually make data available from other model runs.

The NCAR usually obtains the model data in various formats, with binary formats being the most common. Because the volume of the data subset on the diskettes is relatively small, NCAR put it all into the American Standard Code for Information Interchange (ASCII). The format description is included on the diskettes.

The physical processes simulated by the climate models are published by the climate modeling groups in detail. Jenne (1992) gives an overview of the physical processes, without all of the technical details.

C.1.1 Overview of Capabilities of Data Access Programs

Section C.6 gives detailed information about using the data access software, indicates what data are on the diskettes, and discusses how to access and display the data. This section provides a brief overview of these issues. One access program delivers monthly data for $1XCO_2$ and $2XCO_2$ runs for one location; Section C.6 shows the example output. Another program gives back all gridded data within a latitude/longitude box specified by the user. This program can also be used to print out temperature differences between $1XCO_2$ and

TABLE C.1 Model Data for Use in the Country Studies Program

Climate Model	Date Model Run Was Finished	Date Model Run Was Released	Model Resolution (latitude × longitude)	Vert. Model Levels	ΔT for $2XCO_2$	Country Studies Program (CSP) Data Availability
$1XCO_2$, $2XCO_2$						
GISS[a]	1982	Now	7.83° × 10.0°	9	4.2°	On disk Feb. 1994
GISS	Nov. 1994	March 1995	3.90° × 5.00°	12-18	TBD[b]	Release too late for CSP
GFDL R-30[c]	May 1989	Now	2.22° × 3.75°	9	4.0°	On disk Feb. 1994
GFDL R-30 V2	March 1994	Nov. 1994	2.22° × 3.75°	14	3.2°	Available when released
U.K.[d]	Nov. 1989	Aug. 1994	2.50° × 3.75°	11	3.5°	Available when released
NCAR T42	Sept. 1994	Jan. 1995	2.81° × 2.81°	18	TBD	Release too late for CSP
CCC T32[e]	Nov. 1989	Now	3.75° × 3.75°	10	3.5°	On disk Feb. 1994
German	1990	Nov. 1994	5.63° × 5.63°	19	2.6°	Available when released
Transients						
U.K.	1990	April 1994	TBD	TBD	TBD	Available on request
German	1990	March 1994	5.60° × 5.60°	19	(2.6°)[f]	Available on request
GFDL	1991	Jan. 1994	4.44° × 7.50°	9	(e 4.0°)	On disk Feb. 1994[g]

[a] GISS = Goddard Institute for Space Studies.

[b] TBD = to be determined.

[c] GFDL = Geophysical Fluid Dynamics Laboratory.

[d] U.K. = United Kingdom.

[e] CCC = Canadian Climate Centre.

[f] Transient runs do not really have a sensitivity. The numbers show the temperature sensitivity when the model or its near equivalent is used in $1XCO_2$ and $2XCO_2$ runs.

[g] The diskettes only contained four decades of this 100-yr run (averages for control run plus the decades 2010, 2030, and 2050).

$2XCO_2$ runs (or between two decades in a transient run) or to print the ratio of $2XCO_2/1XCO_2$ precipitation for the array of model grid points. In general, the main tools in the access software allow the following capabilities:

- Obtain all grid points of a specified type (e.g., temperature from a model) within a latitude/longitude box that the user specifies.

- Obtain a gridded field of model data for either the whole world or a latitude/longitude region, which allows users to work with the data in their own programs.

- Obtain data at a given location, for all 12 months, from a $1XCO_2$, $2XCO_2$ run specified by the user. This capability provides the data necessary to obtain a complete annual cycle of data from a point. The program uses linear interpolation to calculate data between four grid points. One option is to provide data from the closest grid point only, but this option is not recommended for general use.

- Enable the user to calculate the difference between the $1XCO_2$ and $2XCO_2$ temperature grids from one model, and view this new array over a selected region; enable the user to calculate the ratio of the $2XCO_2$ precipitation to the $1XCO_2$ run.

- Compare a model field to climatology. This capability becomes more difficult because the grid points in the two fields will not be at the same location; therefore, some interpolation is needed (by the software).

- Display data using a display system called the Grid Analysis and Display System (GrADS), prepared by the University of Maryland.

- The user can activate a built-in demonstration of what can be displayed. For example, this system can display a contoured map of temperature difference, in color, with the grid point values printed.

If the software changes or additions are made, updates will be made available to participants in the Country Studies Program.

C.1.2 A General View of the Climate Model Archives at NCAR

For the present stage of the Country Studies Program, analysts need only be concerned about the data obtained on the NCAR diskettes. However, for some purposes, it may be useful to know what data are in the rest of the NCAR archives. Section C.5 summarizes the data contained on the diskettes and outlines the contents of the other archives at NCAR. One tape at NCAR (132 Mbyte) contains more data than can be sent on a few diskettes. Therefore, only the text about this tape (Joseph 1991) is included on the NCAR diskettes. The text includes formats and tables to show the models and some of the variables for each model. The text is not up to date; some data on the floppy disks are not yet noted in the text.

C.1.3 Handling High-Volume Data and Plans for a CD-ROM

The volume of the data is a problem, especially from newer model runs at a higher resolution. If diskettes are used for more model runs, and if more variables are included, the number of diskettes soon becomes unreasonable. NCAR plans to continue to provide a selection of variables for various models on diskettes. Countries that want a larger set of data will have to obtain it on 0.5-in. 6,250 bpi tapes, Exabyte tapes, or IBM 3480 cartridges. The data may be provided in this way at a later stage of the program so that NCAR has more of the data from newer climate model experiments. Near the end of 1994 or in 1995, it may also be possible to give countries the option of obtaining the model data on a compact disk with read-only memory (CD-ROM), which will hold up to 660 Mbyte.

C.1.4 On-Line Information at NCAR

The NCAR has an on-line data information system that can be used free of charge by anyone who can use the Internet communications system. The system will contain information about the model data for the Country Studies Program. NCAR will put new information about the models and about the status of the data into this system as it becomes available. The method for accessing this information is provided in Section C.6.3.

C.1.5 Restrictions on Use of Data

The model data on the February 1994 diskettes are unrestricted. NCAR will soon receive the U.K. run of November 1989. The U.K. data are restricted for commercial use but not for research use.

C.1.6 Volume of Data

It is useful for users to know the resolution of each model, the data points in one global grid, and the volume of a global grid. Table C.2 gives this information for various models. The 1XCO$_2$ and 2XCO$_2$ runs have two time periods and three variables; therefore, the volume is six times the numbers given in Table C.2. The volume of one grid is 360 bytes of header information and 6 bytes for each point. The personal computer (PC) diskettes also have a line feed and a carriage return for each 120 bytes of data. These data are in an ASCII format for simplicity.

C.2 GENERATING CLIMATE CHANGE SCENARIOS

This section briefly discusses how climate change scenarios are generated. More detailed information can be found in Appendix B.

TABLE C.2 Dimensions of Grids and Data Volume

Climate Model	Resolution (latitude × longitude)	Points in Global Grid	Data Volume for 12 Months, One Variable, Global Grid (kbyte)[a]
GISS	7.83° × 10.0°	36 × 24	67
GISS	3.90° × 5.00°	72 × 48	256
GFDL R-15	4.44° × 7.50°	48 × 40	144
GFDL R-30	2.22° × 3.75°	96 × 80	562
U.K.	5.00° × 7.50°	48 × 36	130
U.K.	2.50° × 3.75°	96 × 72	506
CCC, T32	3.75° × 3.75°	96 × 48	340
German	5.63° × 5.63°	64 × 32	173

[a] This volume assumes 6 bytes per point and no further packing or compression.

To study how crops, forests, rivers, and other natural resources might change if the climate changes, daily or monthly observations (for a 30-yr period) are used along with climate model data. Several realities must be accounted for:

- The climate model does not perfectly simulate the present climate. The model changes from the present to the future climate are more reliable than either the present or the future simulation alone.

- Models are relatively low resolution (they were 500 to 1,000 km; some models from about 1989 on have had a resolution of 350 to 400 km). The terrain that a model sees is very smooth compared with the hills and mountain ranges of the real earth.

- The temperature in the model may be at a different elevation above the model ground than the observations, which are at 1.5 m.

The assessments of crops and natural resources for the present climate are conducted by using daily observations of temperature, precipitation, and other climatic variables for the site in question. This site might be in a valley or on a hill, depending on where the crop or forest is growing. To simulate the crop for the new climate, the present-day daily observations are modified according to the changes between the present and future climate model runs. The crop model is then run on this new data series (Figure C.1).

One option for testing the sensitivities of crops and natural resources to climate change is to calculate their responses to incremental changes in climatic variables such as temperature and precipitation and insert the results into a matrix (Appendix B). This method is very useful for obtaining a general view of the effects of climate change. However, the models may provide changes in the seasonal cycle that are not given in this simple picture (where the changes are the same for every month).

For a particular location in a $1XCO_2$, $2XCO_2$ run, Figure C.1a shows the temperature difference between the control ($1XCO_2$) and the $2XCO_2$ run. Figure C.1b gives similar temperature change (ΔT) information from control and transient for each decade in a transient run.

Figure C.1c shows observations near a model grid point in a country. Assessment studies use actual observations (in a valley or on a hill) to simulate present-day crops or trees. To simulate crops for a future climate, the observations of temperature are then modified by the ΔT from the model, and observations of precipitation are modified by the percentage increase (or decrease) in precipitation given by the model.

C.3 OBSERVED DATA FOR USE IN ASSESSMENT STUDIES

Most assessment studies require daily observations for a recent 30-yr period (a few studies require only monthly average data). Most studies require observations of daily high and low temperatures and daily precipitation. Because plant growth and the evaporation of

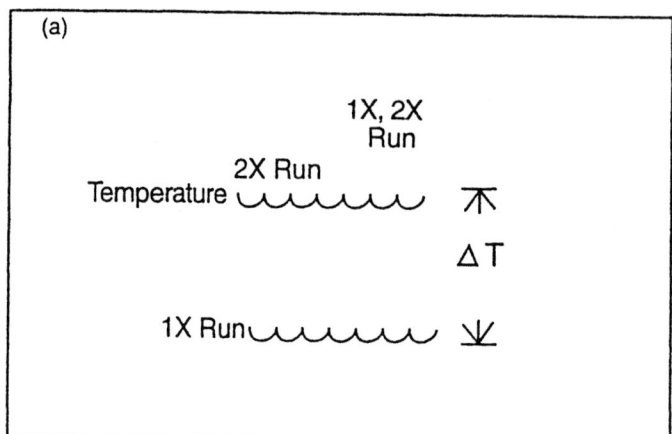

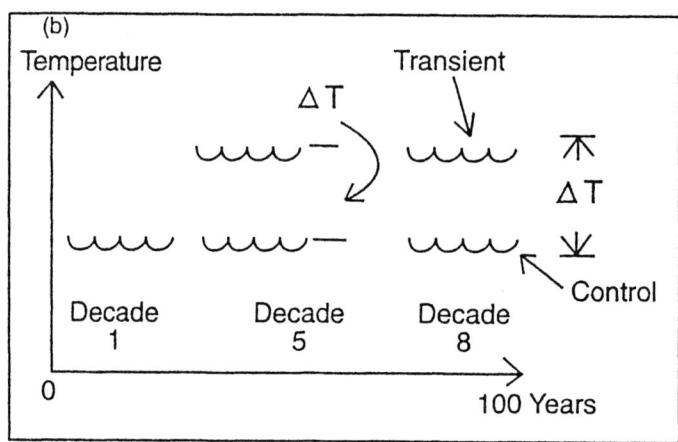

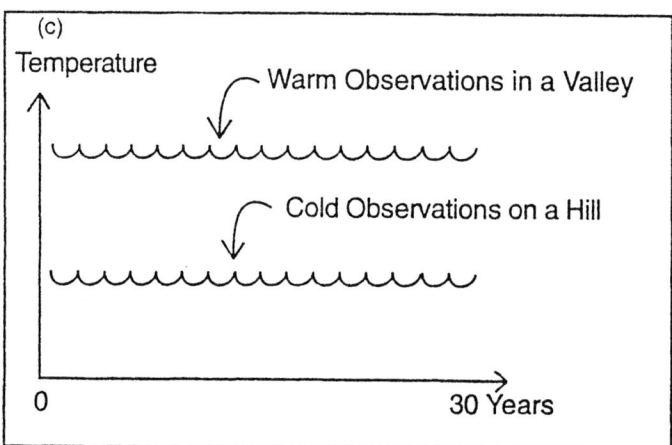

FIGURE C.1 (a) Temperature Difference Between Control ($1XCO_2$) and $2XCO_2$ Runs for a Particular Location; (b) Information from Control and Transient for Each Decade in a Transient Run; and (c) Observations near Model Grid Point in a Country

water depend on the input of solar energy, daily solar energy input (or sunshine data) is also desirable from at least a few stations. It is also desirable to have wind data each six hours and air dew point or relative humidity to help estimate evaporation.

Data for enough stations need to be available to sample the major climatic regions in a country. Countries with land areas similar to Poland or France require preparation of the following observed data:

- Daily precipitation (about 5 to 20 stations),

- Daily high and low temperature (5 to 20 stations),

- Daily solar radiation or sunshine data (3 to 10 stations), and

- Wind measurements and air moisture measurements for each 6-h period (3 to 5 stations).

Daily data for precipitation and temperature are essential for almost all studies (three to five stations). The other observations are desirable, but weather generation models can produce these daily variables in a degraded form, so that the assessment models have the data they need, even though all observations are not available. For example, crop models require 30 years of daily observations. It is generally expected that the local daily observations needed (especially maximum/minimum temperature and precipitation) will be obtained from local or national archives.

Each country usually has some stations that measure solar radiation at the surface (i.e., "global horizontal radiation"). Data from these stations can be compared with model data for the present climate. Measurements of sunshine can also be used as a proxy to derive radiation. Using satellite data to derive radiation at the surface is possible. NCAR hopes to obtain some of these global data sets derived from satellites by December 1995. Some text at NCAR already describes data sets for clouds and radiation.

C.4 GCM OUTPUT DATA FOR USE IN ASSESSMENT STUDIES

C.4.1 Surface Air Temperature

Most measurements of surface air temperature around the world are taken at about 1.5 m above ground. The air temperature from a model often is not at this height; for example, in the nine-level GFDL models, the lowest level is about 10 mbar (80 m) above ground. Because the boundary-layer physics in the model is primitive, calculating a 1.5-m temperature does not make sense. An observed surface temperature differs from a model temperature because of this difference in height, the difference between station elevation and model elevation, and other reasons. The way in which the model temperature data are used (e.g., the differences between the $1XCO_2$ and $2XCO_2$ runs) is more important than the actual temperature. Therefore, these differences in height are not important considerations.

The GISS model carries the surface skin temperature and the temperature for the lowest model sigma level. From these variables, a temperature is calculated that is valid at about 30 m over land and at 10 m over the ocean and sea ice.

C.4.2 Precipitation

The typical climate model outputs are both total precipitation and a convective component of precipitation. The data on total precipitation are on the NCAR diskettes. Small changes in the convective algorithm in a model may substantially change the convective component, but the total precipitation will not change very much. Therefore, it is hard to compare convective precipitation between different models, which is why it has not been included on the NCAR diskettes.

C.4.3 Downward Solar Radiation at the Surface

Downward solar radiation at the surface is an interesting variable from models because that variable is the energy input that plants use for growth. Downward solar radiation also strongly affects the entire surface energy budget and the ability to evaporate water. The climate models tell whether a change in future solar energy at the surface should be expected.

This section addresses the solar constant because that constant also affects the radiation at the ground. The solar constant gives the intensity of sunlight at the top of the earth's atmosphere when the earth is about an average distance from the sun. The best measured value (by using satellites) is about 1,365 to 1,367 W/m^2. This value changes 1 or 2 W over an 11-year solar cycle. The intensity of 1,367 W/m^2 is equivalent to an average of 342 W on each square meter of the earth because the surface area of a sphere is four times the area of the inscribed circle.

The earth is presently closest to the sun on about January 4. The solar intensity at that time is about 7% greater than it is in July. All of the models change the intensity through the year in a proper way.

C.4.3.1 Solar Constant in the GISS Model

The GISS model (1982) uses 1,367 W/m^2 for the annual average solar constant, which is like the observed data. The GISS model would naturally tend to warm up too much, probably because not enough radiation is reflected by clouds or from the surface. To compensate, a global annual average of 5 W/m^2 is removed from the energy balance at the surface. This value compares with an annual average short wave absorption at the surface of 173 W/m^2. The average net radiation at the surface is 123 W/m^2. The special removal of energy is only over ocean areas; the rate of removal is a function of the intensity of solar incident energy at the surface. This removal is part of the "fine-tuning" of the model to give a better comparison with the present climate.

C.4.3.2 Solar Constant in the GFDL Model

The GFDL model (1985 run) uses 2 calories/cm^2 per minute (or 1,443.7 W/m^2) for the solar constant. This figure which is much higher than what is observed and causes too much solar energy to reach the ground. The solar constant is used to fine-tune the model to give a good simulation of the present climate. The run completed in March 1994 is the first run that uses a proper (observed) solar constant. The data on downward solar radiation at the surface are affected by this high value, but the ratio between the 1XCO$_2$ and the 2XCO$_2$ runs is probably unaffected.

The GISS model uses a proper solar constant so that model downward radiation at the surface can be compared with observed data. The high solar constant used by GFDL may cause too much solar radiation to reach the ground. In either case, the ratio between the 1XCO$_2$ and 2XCO$_2$ runs indicates whether solar radiation is expected to increase in the future.

C.4.3.3 Incident Solar Radiation at the Surface

The energy flux of the direct beam of solar radiation on a clear day near sea level is about 1,000 W/m^2. The intensity on a horizontal surface is less; also, the sun only shines for part of the day, and clouds reflect part of the radiation. Thus, the average intensity on a horizontal surface in the middle latitudes is about 60 W/m^2 in January and 260 W/m^2 in the summer. This downward solar radiation is given by a climate model.

C.4.4 Soil Moisture and Runoff from Climate Models

The climate models include a general description of soil type, vegetation, and topography for large grid squares. The soil moisture and runoff calculated for a grid box in a climate model are usually not accurate enough for local studies. This inaccuracy occurs because the box is not representative of a local area, and there are difficulties in the calculation. Soil moisture and runoff are sensitive to surface evaporation, energy budgets, and vegetation. The climate model calculation of these variables usually will not be as good as for temperature and precipitation. The climate model description of soil types, vegetation, and topography also will usually not be valid — even for mid-sized river basins and farm areas. In the GFDL Q-flux model (run finished February 1988), runoff occurs only when the soil water bucket is full. Other models allow for some runoff, even if the soil is not saturated.

Assessment studies for areas such as crop yields and water resources should first incorporate basic variables like temperature, precipitation, downward solar radiation, and wind, and then use these data to remodel terms such as evaporation, soil moisture, and runoff for their local region of interest.

C.4.4.1 Soil Moisture in 1985 GFDL Run

Soil moisture is given in millimeters of water in the soil. Consider also that 1 g/cm^2 water is the same for practical purposes as 1 cm of water depth. The soil moisture applies to a variable depth of soil, depending on the type of soil. The soil "bucket" of water becomes full when it has 15 cm of water. At that time, any additional water goes into runoff.

C.4.4.2 Soil Moisture in 1982 GISS Run

The GISS model has two soil layers. The water storage capacity of the top layer is 20 cm for rain forests, 3 cm for other forests and crops, about 20 cm for grass and crops, and 1 cm for deserts. The second layer can hold about 30 to 45 cm of water for all forests, about 20 cm for grass and crops, and 1 cm for deserts. When the ground is frozen, water cannot be taken out of the soil.

C.4.4.3 Soil Moisture in the Canadian Model

The soil moisture is determined in a one-layer soil model with varying water capacity and a soil and vegetation type that modifies the evaporation potential and albedo.

C.4.5 Calculation of Model Data for a Given Location

The upper-air gridded data from models are rather smooth and continuous fields of numbers. When data are needed at a given location, it makes sense to use the four surrounding grid points (and bilinear interpolation) to derive a value at the given location (Appendix B).

For the surface variables (e.g., temperature and precipitation) needed for assessment studies, the model output is more strongly influenced by the condition of the surface (e.g., trees, grass, ice, or water) that the model used. For this reason, some modelers advise use of data at the nearest grid point without interpolation; however, this advice leads to inconsistencies for studies of small regions that overlap two grid boxes. Some model outputs such as precipitation from the GFDL model also need light smoothing of the output; therefore, interpolation makes the best sense. The use of simple interpolation between grid points is recommended for obtaining data for a given location at the surface. This kind of interpolation is built into the data access programs.

C.5 GCM OUTPUT DATA AVAILABLE AT NCAR

C.5.1 Data Available on Diskettes

A set of model data from relatively recent model runs is available from NCAR (Table C.1). Eight diskettes are available, including seven diskettes containing model data (1989 GFDL R-30, 1989 CCC T32, 1982 GISS, and 1991 GFDL transient) and one diskette with graphics software to help view the data. The data diskettes only have three variables: average monthly temperature, precipitation, and downward solar radiation at the surface. This radiation is given because it is what plants feel, and it is a primary term in the surface energy budget.

In 1990, NCAR prepared a set of three diskettes to provide input data for several international assessment projects (25 countries for agriculture, 10 for forests, and 10 for rivers). These diskettes have average monthly data for only three variables (temperature, precipitation, and downward solar radiation) from three models (1982 GISS, 1988 GFDL, and 1986 U.K.). Access software is available that will select a latitude/longitude window from any of the global grids. Another program selects data for all 12 months from both $1XCO_2$ and $2XCO_2$ for one grid point.

Data for decade averages for three main transient model runs will be made available by NCAR. These models all have dynamic oceans, which is a much better procedure than the earlier transient runs using slab oceans (no dynamics). The volume is too large to put every decade on diskette. Data for every decade will be included on a CD-ROM about December 1994. A subset (four decades) from the GFDL run is included in the NCAR set of diskettes.

C.5.2 Data Available on Primary Tape

This tape has long-term monthly means for a number of $1XCO_2$, $2XCO_2$ runs and the decade means for two GISS transients and some year-month data. Data are given for 7 to 24 surface variables such as air temperature, precipitation, and runoff. The tape content was chosen to fill the needs of most assessment studies (total volume, 94.4 Mbyte). This set is the data set to use if more variables are needed than are practical to put onto floppy disks. The tape contents is as follows:

- *$1XCO_2$ and $2XCO_2$ runs, slab ocean.* Each run has 7 to 24 variables, such as surface air temperature and humidity, precipitation, runoff, evaporation, surface wind, and downward surface radiation. Monthly averages over about 10 years are given for runs such as the 1982 GISS[1],

---

[1]   Used for U.S. Environmental Protection Agency studies conducted October 1987 through May 1988, published in 1990.

1985 GFDL R-15, 1988 GFDL R-15, 1985 Oregon State University (OSU), 1986 U.K., and 1989 CCC.

- *Transient runs.* Two long runs from GISS (one is for 1958 through 2062). Only the average data for each decade is in the standard format. The basic data includes every year but is stored on other tapes. A program interpolates annual values between the decade means for the user.

- *Short text.* A short text lists variables and data volume on this tape. It also describes the formats.

C.5.3 Daily Model Data Available at NCAR

The volume of daily data tends to become high, which is why NCAR does not have very much of it in these special climate model archives. Many model runs have been made without even saving the daily data. Let us consider the size of a daily archive of six variables for 10 years each of $1XCO_2$ and $2XCO_2$. Assume that the resolution is fairly high ($3° \times 3°$), and that only 16 bits (2 bytes) are used to pack each number. The volume is 631 Mbyte (6×20 years $\times 120 \times 60$ points $\times 365$ days $\times 2$ bytes). Even with packing and with few variables, the volume is quite high, but it is manageable if some of the newer technologies are used (CD-ROMs or Exabyte tape).

The GFDL sent NCAR a long sample of three years of daily data from the $1XCO_2$ run and three years from $2XCO_2$, for the Q-flux run (February 1988; total of 24 tapes, 171 variables). The basic data are in a difficult format. Extracted data for the six years (only 21 variables) are in an easy format stored on six tapes. Also, twice-daily data from the CCC model (November 1989 run) are at NCAR, but only for North America. These data are for 10 years for each run — $1XCO_2$ and $2XCO_2$.

The NCAR also has daily climate model data from the CCC model. This data set for North America (20°N to the North Pole, 40°-150°W) has 12-hour model output for a large set of surface variables, and it includes height data for 500 mbar (the only upper-air grid). The grid is 30×19 points ($3.75° \times 3.75°$ resolution). There are 10 years of simulations of the present climate and 10 years for the $2XCO_2$ climate. The volume is 20 tapes, each with 134 Mbyte (the total volume is 2.68 Gbyte). The data set has surface temperature, humidity, and wind, plus daily maximum/minimum temperature, surface pressure, albedo, precipitation, evaporation, surface heat radiation fluxes, and snow information. This is from the same model run as the 3.75° global data set.

C.5.4 Access to Large Datasets on Other Tapes

The primary tape has most of the variables that assessment studies need, and the data are in one easy format. It is more difficult to provide access to the other data, but it is possible. NCAR cannot provide free access to the larger data sets, but prices can be given. Models for which NCAR has rather full sets of output from the statistics (in native format), not just a few selected variables, are given below:

- 1982 GISS $1XCO_2$, $2XCO_2$ run; 54 variables, has height levels to 30 mbar, and two wind levels.

- 1985 GISS transients; 56 variables.

- 1985 GFDL $1XCO_2$, $2XCO_2$ run; 160 variables.

- 1988 GFDL $1XCO_2$, $2XCO_2$ run; 171 variables; contains the 10-yr monthly means.

C.6 SOFTWARE TO ACCESS DATA STORED ON DISKETTES

Each model run on the NCAR diskettes contains data for three variables — surface temperature, precipitation, and incoming solar radiation at the surface. In addition, current climate data are provided for temperature and precipitation. All model and climate data are global coverage, but grid resolutions vary from model to model.

To use the information, all files from the eight diskettes need to be copied to one directory on a PC hard disk. About 10 Mbyte of space for these files will be needed. Two diskettes (5 and 6) contain subdirectories that can be ignored unless problems occur in getting the display package (GrADS) to work on the computer. Once the files are copied to one directory, all programs may be executed while in that directory. Following is a list of the programs likely to be used most often. More detail is given in a later section.

- GRDPTI.EXE. This program generates data values at a longitude-latitude point specified by the user. Values are given for 12 months of the year; $1XCO_2$ and $2XCO_2$ and a difference or ratio; and for the three variables. This output is often used as input to assessment models.

- GRDFILE.EXE. This program creates a simple tabular output for data points within a longitude-latitude window specified by the user. Windows from all records in the file are output and if the file contains $1XCO_2$ and $2XCO_2$ equilibrium run outputs, a difference or ratio is computed and output. The output is put in a file which can be printed, browsed by an editor, or used as input to another program.

- GRD2GRAD.EXE. This code reads model or climate files and creates files that can be read by the GrADS.

- GRDINT.EXE. This code allows interpolation from one global grid to another. The output can then be used as input to any other program that reads the basic model format. For easiest compatibility, an output file name should be chosen to look like file names used in the original model files (i.e., cccxtmp.dat).

C.6.1 GCM Data Stored on NCAR Diskettes

Each file contains data for one variable from one climate model run. All data files are ASCII and can be easily read by user programs, but the files are stored in a compact form and are not intended to be browsed directly by editors or other general-purpose PC applications. The programs provided will create files that are easily read.

All data files include 12 monthly grids. The CCC and the GFDL R-30 data include three files (one for each variable) and each file contains data for $1XCO_2$ and $2XCO_2$ values for each of the 12 months. The GFDL transient data include three files, and each file contains monthly decadal means from decades 1, 4, 7, and 10. The climate data include two files (temperature and precipitation), and each file contains one set of 12 monthly values. The first four characters of each file name indicate the model run, and the next three characters indicate the variable. Table C.3 gives the full list of data files. In addition, the file ntyplsti.dat contains header information used by some of the basic programs.

Models:

- cccm - CCC model
- gfd3 - GFDL R-30 model
- gf01 - GFDL 1% per year transient
- giss - GISS 1982 model run
- clim - climate data

Variables:

- tmp - surface temperature
- pcp - precipitation
- sol - incoming solar radiation at the surface

TABLE C.3 List of Data Files on February 1994 Diskettes

Climate Model	Precipitation File	Solar File	Temperature File	File Size (byte)	Total Model Volume (byte)
CCC	cccmpcp.dat	cccmsol.dat	cccmtmp.dat	685,152	2,055,456
GFDL R-30	gfd3pcp.dat	gfd3sol.dat	gfd3tmp.dat	1,133,136	3,399,408
GISS	gisspcp.dat	gisssol.dat	gisstmp.dat	292,434	877,302
GFDL transient	gf01pcp.dat	gf01sol.dat	gf01tmp.dat	579,744	1,739,232
Observed data	climpcp.dat	--	climtmp.dat	246,840	493,680

C.6.2 Programs for Manipulating Model Data

Source code and executable versions are provided for the codes that read the basic model data. The source code is FORTRAN, and the executables were created by compiling with Microsoft FORTRAN 5.0.

C.6.2.1 GRDPTI

This code creates a file containing 12 monthly values of $1XCO_2$, $2XCO_2$, and differences or ratios for the three variables at a longitude-latitude point specified by the user. A typical execution of this program generates the following output screen (> indicates response keyed by user). By default, this code will interpolate a value to longitude-latitude coordinates specified.

```
WOULD YOU RATHER JUST USE THE NEAREST POINT?(N,Y)

            > <cr>
Enter longitude of the point (west is negative)
            > 30
Enter latitude of the point (south is negative)
            > -10
Enter model as gfd3, cccm , gf01, clim, etc.
            > cccm
Enter output file name
            > cccmpt.out

            Reading cccmtmp.dat
            Reading cccmpcp.dat
            Reading cccmsol.dat
            PROCESSING COMPLETED, OUTPUT IN
            cccmpt.out
```

The output in the file named cccmpt.out would appear as follows:

```
        VALUES at 30.0  -10.0 from cccm

   MONTH  TEMPERATURE (C)        PRECIPITATION (MM/DAY) SOLAR (W/M**2)
          1xCO2  2xCO2  Diff     1xCO2  2xCO2  Ratio    1xCO2  2xCO2  Ratio
      1    19.6   22.2  2.58      6.5    6.3    0.98     222.   211.   0.95
      2    20.1   22.5  2.42      8.0    7.8    0.98     217.   207.   0.95
      3    19.7   22.1  2.38      6.0    7.1    1.18     224.   201.   0.90
      4    17.7   20.4  2.64      2.1    2.7    1.30     240.   223.   0.93
      5    15.4   17.8  2.43      1.1    1.3    1.12     244.   237.   0.97
      6    14.4   16.9  2.50      1.1    1.3    1.10     231.   223.   0.97
      7    14.3   17.2  2.84      1.0    1.0    0.94     236.   234.   0.99
      8    16.1   19.2  3.11      0.6    0.6    0.88     271.   268.   0.99
      9    19.7   23.0  3.34      0.3    0.4    1.41     306.   298.   0.98
     10    22.3   25.5  3.14      2.4    3.5    1.46     285.   270.   0.95
     11    21.3   23.5  2.19      9.4    8.8    0.94     224.   212.   0.94
     12    20.1   22.3  2.19      9.1    6.8    0.75     213.   206.   0.97
```

C.6.2.2 GRDFILE

This program creates an easily readable or printable file for a longitude-latitude window as specified by the user. A typical execution of this program generates the following output on your screen (> indicates response keyed by user). GRIDS WRITTEN includes the generated difference grids and is therefore 12 larger than GRIDS READ.

```
Select a subarea grid and put it in a file for printing or browsing

Uses the file NTYPLSTI.DAT

Define a window smaller than the whole grid? (y,n)
        >   y
Enter minimum longitude, (west is negative)
        >   10
Enter maximum longitude
        >   40
Enter minimum latitude, (south is negative)
         >   -20
Enter maximum latitude
        >   20
Long/Lat limits =        10.00     40.00    -20.00     20.00
Enter input file name
          >   cccmtmp.dat
Enter output file name
          >   cccmtmp.out

        Processing month   1
        Processing month   1
        Processing month   2
        Processing month   2
        Processing month   3
            "        "
        Processing month  11
        Processing month  12
        Processing month  12
        GRIDS READ =   24   GRIDS WRITTEN =   36
```

The resulting output file will contain 36 output grids, which look like the following:

```
NR,NYR,NMO,NTYP,NRUNCD,NLEVT,XLV1       1 9999    1   16    1    3    0.0
GRID      2001   ST         1          97          48        0        1                    14SEP93

         JAN       ATMOSPHERIC TEMPERATURE       C              0.000 METERS FROM SFC
1XCO2

         LONG    7.50   11.25   15.00   18.75   22.50   26.25   30.00   33.75   37.50   41.25
    LAT
    20.41     4.25    3.92    4.37    5.87    5.87    4.61    8.19   15.08   24.63   11.72
    16.70     9.29   10.21   10.84   10.75    9.33    7.98   10.69   16.18   18.88   24.61
    12.99    14.63   15.51   16.08   14.79   13.31   13.21   15.63   16.80   15.66   14.40
     9.28    17.86   17.43   18.89   18.46   16.38   18.19   21.11   17.46   13.97   12.57
     5.57    22.09   18.73   20.23   21.94   19.57   19.33   20.54   17.97   15.72   15.56
     1.86    27.22   20.80   20.29   23.83   21.41   18.72   18.07   17.36   17.32   21.02
    -1.86    26.67   22.77   22.58   24.36   23.12   20.05   17.46   16.39   18.60   26.29
    -5.57    26.40   26.26   24.55   23.47   21.89   19.58   18.10   18.01   20.49   27.00
    -9.28    25.22   25.61   24.23   22.13   20.48   20.02   19.50   20.20   22.93   27.55
   -12.99    22.85   23.56   22.70   21.97   21.94   21.27   20.07   20.66   23.51   27.42
   -16.70    20.92   21.19   22.75   23.46   24.77   23.78   20.54   20.96   23.96   27.38
   -20.41    20.23   19.71   23.52   24.69   26.47   24.65   21.47   21.99   26.60   27.41
```

C.6.2.3 GRD2GRAD

This program creates files for input into the GrADS package. The output files are called xxxx.ctl and xxxx.grd, where xxxx is the model output the user specifies. See Section C.6.2.6 and the GrADS documentation on the diskettes for use of these output files. A typical execution of this program generates the following output on the screen. Most of the output is not of interest to users and simply indicates the program is continuing execution (> indicates response keyed by user).

```
Prepare Climate Model Data for GrADS Display
Enter Model Name, (GFD3, GF01, CLIM or CCCM)
       > cccm
NR,NYR,NMO,NTYP,NRUNCD,NLEVT,XLV1      1 9999      1    16    1    3    0.0
NR,NYR,NMO,NTYP,NRUNCD,NLEVT,XLV1      2 9999      1    16    2    3    0.0
NR,NYR,NMO,NTYP,NRUNCD,NLEVT,XLV1      3 9999      2    16    1    3    0.0
NR,NYR,NMO,NTYP,NRUNCD,NLEVT,XLV1      4 9999      2    16    2    3    0.0
NR,NYR,NMO,NTYP,NRUNCD,NLEVT,XLV1      5 9999      3    16    1    3    0.0
    "        "        "
NR,NYR,NMO,NTYP,NRUNCD,NLEVT,XLV1     66 9999      9   175    2    3    0.0
NR,NYR,NMO,NTYP,NRUNCD,NLEVT,XLV1     67 9999     10   175    1    3    0.0
NR,NYR,NMO,NTYP,NRUNCD,NLEVT,XLV1     68 9999     10   175    2    3    0.0
NR,NYR,NMO,NTYP,NRUNCD,NLEVT,XLV1     69 9999     11   175    1    3    0.0
NR,NYR,NMO,NTYP,NRUNCD,NLEVT,XLV1     70 9999     11   175    2    3    0.0
NR,NYR,NMO,NTYP,NRUNCD,NLEVT,XLV1     71 9999     12   175    1    3    0.0
NR,NYR,NMO,NTYP,NRUNCD,NLEVT,XLV1     72 9999     12   175    2    3    0.0
END-NR,NST       72   1
```

C.6.2.4 GRDINT

This program interpolates data values from one model grid to another. The output grid desired is specified by giving the name of a data file that contains data on the desired output grid. The following example would interpolate the CCC temperature data to the grid of the climatology data. The file name for output is chosen specifically to have the same form as other data file names so as to have greater compatibility with the other programs (> indicates response keyed by user).

```
Enter source file name
       >    cccmtmp.dat
Enter file name which defines destination grid
       >    climtmp.dat
Enter file name for output
       >    cccxtmp.dat
       Processing month    1
       Processing month    1
       Processing month    2
        "        "
       Processing month   11
       Processing month   12
       Processing month   12
       RECS INTERPOLATED =       24
```

C.6.2.5 GRDRD, GRDPT

The grdrd program is a basic read program intended for use as a starting point for developing user codes. The grdpt program is a previous version of the grdpti program.

C.6.2.6 GrADS

The GrADS is a product of the Center for Land-Ocean-Atmosphere Interactions at the University of Maryland. It is a powerful, yet compact, tool for displaying and analyzing grid fields. It can also work with station data, but this capability is beyond the scope of Country Studies Program usage. The GrADS can produce PostScript files so that hard copies can be created on PostScript printers.

Included on the NCAR diskettes is the basic PC GrADS 1.4 distribution. Source code is not available, but compiled versions for other systems can be made available. For a basic PC without extended memory, executables from version 1.3 of the GrADS are included on diskettes 5 and 6. If the user encounters trouble with 1.4, version 1.3 should be tried. Both of these versions require at least VGA and a math co-processor. Full documentation of the GrADS is included on diskette 7. For a quick demonstration, go to the climate model directory and type demo. This demonstration runs the grd2grad program, enters grads, and does some sample displays. This demonstration is not intended to show any particular data; rather, it is intended to give a quick display of the types of graphics available through GrADS. For an interactive session that allows the user to choose grids for display:

```
1. Run grd2grad on the model of choice (say cccm)
2. When this is finished:
        type grads
        type <cr> (at prompt for portrait mode).
        type run display.run
        type (respond to prompts to choose grids for display)
        type <cr> (to move back up to higher order selection)
        type <cr> (on model prompt to exit to grads)
        type quit (at ga> prompt to exit grads)
```

In this process, a GrADS script file is running that prompts for user specification of area, model, date, CO_2 level, etc. The user can only ask for models for which grd2grad has been run first.

C.6.3 Internet Access to NCAR Data Support Section Information

The NCAR Data Support Section maintains an anonymous ftp area that contains information on the holdings in its archive, some small data set files, and special project support areas. Users having access to the Internet can ftp to ncardata.ucar.edu (128.117.8.111) and log in as "anonymous." At the connect level, the file README gives an introduction to the area. If trouble is encountered or questions arise, send e-mail to joseph@ncar.ucar.edu or datahelp@ncar.ucar.edu. Data that directly apply to the Country Studies Program will reside in a directory called "pub/country_studies." This directory currently contains the same information included on the NCAR diskettes.

C.7 THE VARIABILITY OF MODEL OUTPUT DATA

The real atmosphere has a natural variability that may be very high for short time periods. In middle latitudes, it is recognized that a very cold period of a few days may be followed very quickly by a warm spell. Climate models also have a natural variability that occurs whether or not greenhouse gases are changed. One question that must be addressed is whether the decade means for a small region are fairly stable or whether they still include a significant natural variability, because a 10-yr period is rather short.

Ten decades of monthly temperature and precipitation are plotted in Figure C.2 for one grid point in the United States. The first 12 points are the monthly 10-yr averages (January, February, etc.) for the first decade, followed by monthly data for the next decade. The 10-yr averages of temperature are long enough that the curves are well behaved. With precipitation, a good deal of variation in the monthly averages still occurs because of the short period. The data can be prefiltered to effectively obtain longer-period averages to dampen the noise, but this task cannot be performed quickly. Because of this noise, an assessment modeler should feel free to do some reasonable smoothing on the annual cycle of precipitation at a grid point. For example, a 1-2-1 smoother could be used on the annual cycle of monthly values of precipitation to dampen month-to-month variations. In this case, the data for each month are replaced with one-half of their present value, plus one-fourth of the values for the preceding and following months. To fix the problem will take more research, perhaps longer runs, and better ways to prepare the statistics.

Climate model temperature varies a few tenths of a degree (for annual means) on time scales of 5 to 20 years. Consider the natural variability of observed data inferred from tree growth. Summer temperature has been derived for a 3,620-yr period (1634 BC to 1987 AD), from about 5 to 20 tree series in South America (35° to 44° South). It appears from figures in Lara and Villalba (1993) that the variations of seasonal temperature means over 5- to 20-yr periods are about -0.7° to +0.7°C.

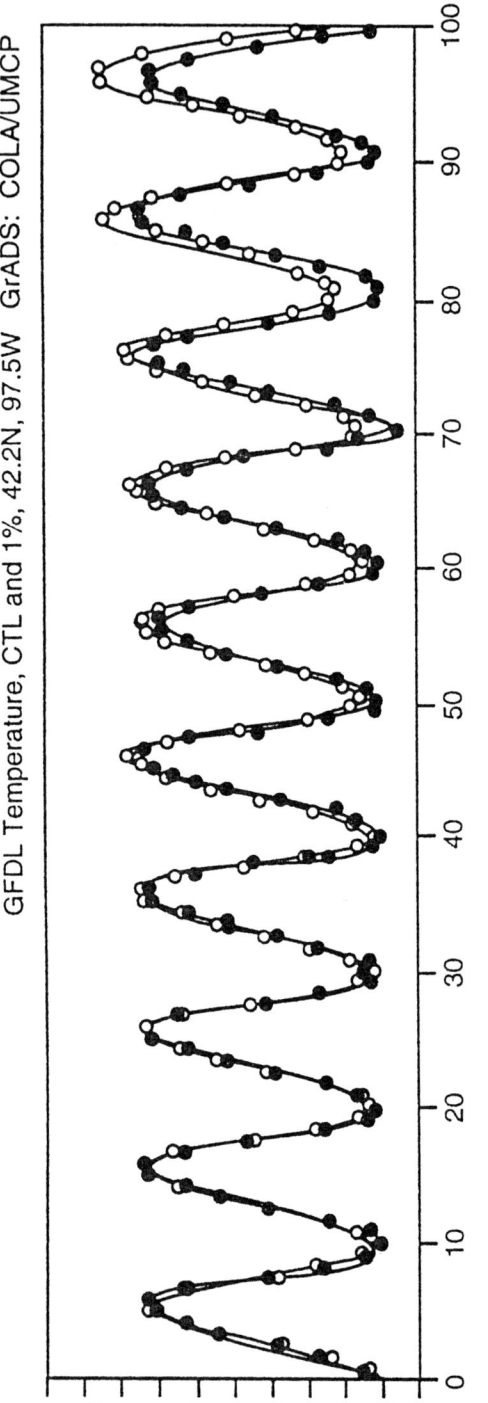

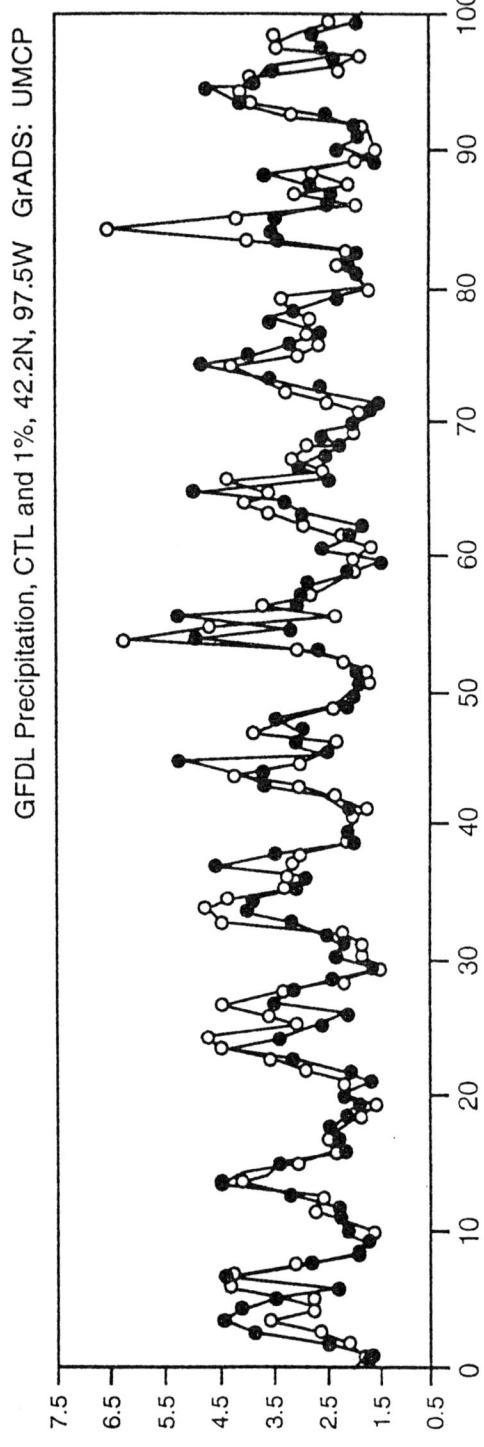

FIGURE C.2 Averages of Temperature and Precipitation for the 100-Year Control and Transient by GFDL (1991 run) (solid circles = monthly averages for each of 10 decades for the GFDL) (1991 run, open circles = transient run)

C.8 COMPARISONS BETWEEN MODELS AND OBSERVED DATA

Sections C.8.1 and C.8.2 indicate the results of simulations from climate models. Space is not available to show many variables and regions. More information (and the digital data) is available from NCAR. The 1990 Intergovernmental Panel on Climate Change documents also give more detailed comparisons for variables, such as temperature, precipitation, soil moisture, and sea-level pressure. A selection of figures in Jenne (1992) compares the output of temperature and precipitation from several models for regions of North America, Africa, and East Asia.

C.8.1 Comparison of Temperature — Model vs. Observed

The main conclusion from the comparison of simulated near-surface temperatures with observed temperature is that, while each model has systematic errors, the models have few errors in common. One characteristic error is that temperatures over East Asia are too cold in winter. The zonal mean temperature is the average around the world at one latitude. The spread in the zonal mean between models at various latitudes is usually about 4 to 6°C in winter and 4°C in summer; but at 50° North, it increases to 10°C. However, by omitting one model that will not be used in the Country Studies Program, the spread is only about 4°C at 50° North. It appears that temperature errors in simulating the temperature for present-day climate should not be a concern, both because they are relatively small and because the assessment models will use the actual observed temperatures instead of the model temperatures.

C.8.2 Comparison of Temperature — Model vs. Model

Two regions are briefly discussed: the United States and East Asia. For the United States, the models show the following warming for $2XCO_2$: GISS (3.8° to 5.3°C), United Kingdom Meteorological Office (UKMO) (5.1° to 9.5°C), GFDL (3.6° to 5.5°C), and GFDL high resolution (3.6° to 5.7°C). The CCC model gives 3.2° to 6.9°C.

The warming of the winter climate over East Asia in the $2XCO_2$ model runs was inspected. The high-resolution GFDL model warms the region by about 4.0°C compared with the control run. This result is similar to the older GFDL run, but somewhat less. The CCC model warms this region by 4.0° to 9.0°C, which is more than the overall sensitivity of the model (only 3.5°C). The GISS model warms the region by about 3.0° to 7.0°C, and UKMO by 4.0° to 8.0°C. These figures represent large changes.

C.9 APPENDIX C REFERENCES

IPCC, 1990, *Climate Change: The IPCC Scientific Assessment*, Houghton et al. (eds.), Cambridge University Press, New York, N.Y.

Jenne, R.L., 1992, "Climate Model Description and Impact on Terrestrial Climate," in Majumdar et al. (eds.), *Global Climate Change: Implications, Challenges, and Mitigation Measures*, Pennsylvania Academy of Sciences, Chap. 10, pp. 145-164.

Joseph, 1991 *Climate Model Output Data*, Internal Publication, National Center for Atmospheric Research; this report describes data and formats on the main model.

Lara, A., and R. Villalba, 1993, "A 3,620-Year Temperature Record from *Fitzroya cupressoides* Tree Rings in Southern South America," *Science* 260:1104-1106.

Schlesinger, M.E., 1988, "Model Projections of the Climatic Changes Induced by Increased Atmospheric CO_2," Symposium on Climate and Geo-Sciences, May 22-27, 1988, Reidel Publishing Company, Dordrecht, Holland (in press).

C.10 APPENDIX C BIBLIOGRAPHY

The key references for each model are given first. The more general references are then presented.

C.10.1 GFDL 1985 Model Run

Manabe, S., and R.T. Wetherald, 1987, "Large-Scale Changes in Soil Wetness Induced by an Increase in Carbon Dioxide," *Journal of Atmospheric Science* 44:1211-1235.

Wetherald, R.T., and S. Manabe, 1988, "Cloud Feedback Processes in a GCM," *Journal of Atmospheric Science* 1397-1415, April.

C.10.2 GFDL Q-Flux Model Runs

Wetherald, R.T., and S. Manabe, 1990, *A Reevaluation of CO_2-Induced Hydrologic Change as Obtained from Low- and High-Resolution Versions of the GFDL General Circulation Model* (in preparation), July.

C.10.3 GFDL Transient Runs

Stouffer, R.J., S. Manabe, and K. Bryan, 1989, "Interhemispheric Asymmetry in Climate Response to a Gradual Increase of Atmospheric Carbon Dioxide," *Nature* 342:660-662.

C.10.4 CCC Model Runs

Boer, G.J., N. McFarlane, and M. Lazare, 1991, "Greenhouse Gas-Induced Climatic Change Simulated with the CCC Second-Generation GCM," *Climate* (in preparation).

C.10.5 U.K. Model Runs

Wilson, C.A., and J.F.B. Mitchell, 1987, "A Doubled CO_2 Climate Sensitivity Experiment with a Global Climate Model Including a Simple Ocean," *JGR* 92(D11):13315-13343.

Mitchell, J.F.B., C.A. Senior, and W.J. Ingram, 1989, "CO_2 and Climate: A Missing Feedback?" *Nature* 341:132-134.

C.10.6 GISS Model (1XCO$_2$, 2XCO$_2$, Transient)

Hansen, J., G. Russell, D. Rind, P. Stone, A. Lacis, S. Lebedeff, R. Ruedy, and L. Travis, 1983, "Efficient Three-Dimensional Global Models for Climate Studies: Models I and II," *Monthly Weather Review* III(4):609-662.

The control runs for the GISS model sensitivity tests (including the diurnal cycle) were described in a *Monthly Weather Review* article in April 1983 as above. The 2XCO$_2$ run was made in 1983 and described in "Climate Processes and Climate Sensitivity," *Geophysical Monograph*, Vol. 29, 1984.

Rind, D., 1988, "Dependence of Warm and Cold Climate Depictions on Climate Model Resolution," *Journal of Climate* 1:965-997.

Hansen, J., I. Fung, A. Lacis, D. Rind, S. Lebedeff, R. Ruedy, G. Russel, and P. Stone, 1988, "Global Climate Changes as Forecast by Goddard Institute for Space Studies Three-Dimensional Model," *JGR* 93:9341-9364.

C.10.7 OSU Model Runs

Schlesinger, M.E., and Z.C. Zhao, 1988, "Seasonal Climate Changes Induced by Doubled CO_2 as Simulated by the OSU Atmospheric GCM/Mixed-Layer Ocean Model," *Journal of Climate* 2:459-495.

Ghan, S.J., et al., 1982, "A Documentation of the OSU Two-Level Atmospheric GCM," CRI Report 35, Oregon State University.

C.10.8 Comparison of Models

The output from GFDL, NCAR, and GISS (1982) models are compared in the following paper. This paper does not include the OSU and U.K. (Mitchell) models.

Schlesinger, M.E., and J.F.B. Mitchell, 1987, "Model Projections of the Equilibrium Climate Response to Increased Carbon Dioxide," *Review of Geoph.* 25:760-798.

A paper in the *Bulletin of the American Meteorological Society* compares July soil moisture from all five models. Schlesinger (1988) also compares all five climate models. Further information can be obtained from M. Schlesinger at the University of Illinois.

C.10.9 Other References

Grotch, S.L., 1987, *Regionally Intercomparing GCM Predictions with Historical Climate Data*, Lawrence Livermore National Laboratory, Livermore, Calif.

Ramanathan, V., 1988, "The Greenhouse Theory of Climate Change: A Test by an Inadvertent Global Experiment," *Science* 240:293-299.

Schlesinger, M.E., 1988, "Model Projections of the Climatic Changes Induced by Increased Atmospheric CO_2," Symposium on Climate and Geo-Sciences, May 22-27, 1988, Reidel Publishing Company, Dordrecht, the Netherlands (in press).

APPENDIX D:

CROP IMPACTS

R. Benioff et al. (eds.), Vulnerability and Adaptation Assessments, D-1–D-38.
© 1996 *Kluwer Academic Publishers.*

APPENDIX D NOTATION

The following is a list of the acronyms, initialisms, and abbreviations (including units of measure) used in this appendix.

ACRONYMS, INITIALISMS, AND ABBREVIATIONS

CCCM	Canadian Climate Centre model
DSSAT	Decision Support System for Agrotechnology Transfer
GCM	general circulation model
GFDL	Geophysical Fluid Dynamics Laboratory
GIS	geographic information system
GISS	Goddard Institute for Space Studies
IBSNAT	International Benchmark Sites Network for Agrotechnology Transfer
ICASA	International Consortium for Application of Systems Approaches to Agriculture
NCAR	National Center for Atmospheric Research
v2.9	version 2.9

CHEMICAL

CO_2	carbon dioxide

UNITS OF MEASURE

ha	hectare(s)
kByte	kilobyte(s)
kg	kilogram(s)
m	meter(s)
Mbyte	megabyte(s)

CONTENTS

CONTENTS (Cont.)

TABLES

FIGURE

APPENDIX D:

CROP IMPACTS

D.1 INTRODUCTION

Simulating the effects of climate change on agricultural production in different countries requires a coordinated effort in which data, software, and expertise from various disciplines, institutions, and countries are integrated and used. The International Benchmark Sites Network for Agrotechnology Transfer (IBSNAT) and the International Consortium for Application of Systems Approaches to Agriculture (ICASA) have identified data requirements, developed computer programs, modified other programs, and assembled data so that each Country Studies participant can use specific information on weather, soil, and crops to simulate the effects of climate change in their country.

The objectives of the analysis system for crop impacts are to:

- Summarize scientific and technical information needed for simulating impacts on crops,

- Provide references to more detailed information in user manuals for the Decision Support System for Agrotechnology Transfer (DSSAT) comprehensive software system and in the scientific literature, and

- Present exercises to teach fundamentals of modeling the impacts of climate change on crop production.

The DSSAT system integrates crop growth models with crop, weather, and soil data and with various application programs. The system is the main tool used for studying climate change and integrating databases and crop models. The IBSNAT/ICASA Project developed the software to assist scientists who are studying crop growth, development, and yield responses to various soil, weather, and management conditions.

This appendix explains the general tenets of the recommended approach, the data requirements for DSSAT input, basic techniques for running DSSAT, and the crop models within it. Guidance on the interpretation of results and the limitations of the approach are also addressed. Many features available in DSSAT version 2.9 (v2.9) are not fully explained because the focus is on the use of DSSAT for climate change studies. These features are explained in detail in the DSSAT v2.9 technical documentation that accompanies the software. Users are encouraged to "explore" the many functions of the software and to communicate their questions concerning all possible applications of the software to the technical assistance team.

The test version distributed at the U.S. Country Studies Workshop (DSSAT v2.9) is intended for participating countries only. Participants will automatically receive the DSSAT version 3.0 software, documentation, and all updates as IBSNAT/ICASA releases and publishes them. The IBSNAT/ICASA network invites all participants in the Country Studies Program to contribute their suggestions and comments for improving performance and ease of use.

D.1.1 DSSAT

The DSSAT v3.0 microcomputer software program combines soil, crop, and weather databases with crop models and application programs to simulate multiyear outcomes of climate change scenarios and crop management strategies. Crop models for maize, wheat, sorghum, millet, barley, soybeans, peanuts, and dry beans are included. As a software package integrating the effects of soil, crop cultivars, weather, and management options, DSSAT allows you to ask "what if" questions in regard to climate change and seek answers by conducting simulated experiments. The DSSAT software also validates crop models for agronomic and climate current conditions. This element is a key part of the climate change impact studies, because confidence in the models' simulation of current conditions is needed to engender confidence in predicting crop responses to changed conditions. Crop models are validated by comparing simulated outcomes with observed results digitally stored in the DSSAT format.

D.1.2 Data Requirements

Weather, soil, and crop parameters are needed to simulate crop growth with DSSAT v3.0. Additional data are required for crop model validation.

D.1.2.1 Climate Data

Each site simulation requires daily maximum and minimum air temperature, daily precipitation, and daily solar radiation. The time series of data should ideally cover from 1950 to 1980 (or as many years of daily climate records as available) to adequately characterize the baseline climate, including interannual variability, and generate the climate change scenarios. National weather services and agricultural research centers usually collect and maintain such climate data. If solar radiation data are unavailable, they may be generated. Once collected or generated, the weather data must be arranged in the DSSAT weather file format and checked for quality. Programs may be found in DSSAT (Section D.3) to generate weather data from monthly means, generate solar radiation (if unavailable), arrange climate data into the DSSAT format, and check for missing data and values out of normal ranges.

D.1.2.2 Soil Data

Site soil characteristics specified for each horizon (up to 10 layers) include horizon thickness; initial soil water content; and water content at saturation, drained upper limit, and lower limit of plant extraction. Additional characteristics are soil albedo, stage 1 soil evaporation, profile drainage coefficient, and run-off curve number. Root traits needed are root length per unit root weight, rate of root depth increase, and root growth preference factor to allocate roots by horizon. Initial amounts of nitrogen in the forms of soil nitrate and ammonium, pH, organic matter, and fertilizer application date(s), amount applied, depth of application and type of fertilizer are also specified as inputs, if the simulation includes nitrogen balance.

National soil surveys and/or records at agricultural research centers may have these data, or they may be characterized by local soil physicists. The Country Studies technical advisers can also provide data for generic soil types, characterized by texture and depth (e.g., deep sandy loam), if detailed soil profile data are not available, for use in the climate change simulations. The DSSAT formats for soil data are discussed below and in the DSSAT documentation.

D.1.2.3 Crop Management Data

Planting date, plant population, cultivar, and irrigation dates and amounts (if applicable) are needed in the simulations to characterize crop management. These data should represent regional practices and can be obtained from regional agricultural specialists. The DSSAT formats for crop management data are discussed below and in the DSSAT documentation.

D.1.2.4 Calibration and Validation Data

To calibrate and validate the crop models, periodic measurements (every two to four weeks) during a crop growing season are needed. Data to be gathered during the growing season include leaf area index, leaf weight, stem weight, seed weight, and timing of vegetative and reproductive stages. It is essential to know the final seed yield and seed weight. These data are collected by research agronomists at field experiment stations and may be obtained from them. The IBSNAT/ICASA network has gathered many calibration and validation data sets; these data are also available through the Country Studies technical advisory team. The DSSAT methods and formats for calibration and validation are discussed below and in the DSSAT documentation.

D.2 INSTALLATION OF DSSAT SOFTWARE

D.2.1 Computer Requirements

The DSSAT software requires installation on personal computers that use DOS version 3.3 or higher. The DSSAT v2.9 requires a 286, 386, or 486 microprocessor chip and a math coprocessor to run its various programs. In addition, the software requires at least 590 kbyte of random-access memory and 14.0 Mbyte of hard-disk memory. The graphics display must be a video graphics array color system.

D.2.2 Computer Disk List

The analysis system for crop impacts contains 18 disks:

- DSSAT v2.9 (eight disks);

- Climate change example, Memphis, Tennessee (two disks); and

- Output and programs of the National Center for Atmospheric Research (NCAR) global climate models (eight disks).

D.2.3 Installation Procedure

The software is installed into the computer by following the instructions below:

1. Turn on the computer. Check to see that the current drive is C:\.

2. Place the "DSSAT v2.9 DISK 1" into the "A" disk drive and type **a:**.

3. Type **install** and follow the directions given by the program. To install the program into the default directories provided by the DSSAT (this is the recommended technique), press <**ENTER**> when prompted and change the disks to install the proper sections.

4. When the installation program has finished, make sure the PATH statement in the AUTOEXEC.BAT file includes C:\DSSAT3; for example, PATH=C:\;C:\DOS;C:\WP should be changed to **PATH=C:\;C:\DOS;*C:\DSSAT3*;C:\WP**.

5. Add the following statement to the AUTOEXEC.BAT file:

   SET DSSAT3=C:\DSSAT3.

6. Add the following statement to the CONFIG.SYS file in the C:\ directory if it is not already there:

   DEVICE=C:\DOS\ANSI.SYS.

D.2.4 Country Studies: Example Files

Use the following procedures to add the files that will be used in the example of climate change:

1. Put the disk titled CLIMATE CHANGE EXAMPLE — DISK 1 into the A drive. Use the **A:** command to switch to the **A:** prompt.

2. Use the **COPY** command to transfer files into the proper DSSAT directories:

   COPY A:\MAIZE*.* C:\DSSAT3\MAIZE
   COPY A:\SEASONAL*.* C:\DSSAT3\SEASONAL

3. Type **C:** to switch back to the C:\ prompt. Then create the subdirectories needed for this study by using the MD command:

   MD NCAR.

   Type **CD NCAR** to change to the NCAR directory. Next, make the directory WEATHER in NCAR by typing **MD WEATHER**.

4. Finally, place the disk titled CLIMATE CHANGE EXAMPLE — DISK 2 into the A drive, and copy disk 2 to hard disk C:

   Copy A:*.* C:\NCAR\WEATHER*.*

Once these steps have been completed, all of the necessary programs and weather data are installed onto the hard disk. You are now ready to start the procedures for creating climate change scenarios and running simulations.

D.2.5 Starting the Decision Support System for Agrotechnology Transfer

Starting the DSSAT after installing the software requires the following commands:

1. Always start the DSSAT from the directory C:\DSSAT3. To go to this directory from the C:\ prompt, type **CD DSSAT3**.

2. To start the DSSAT, type **DSSAT3**.

D.3 DECISION SUPPORT SYSTEM FOR AGROTECHNOLOGY TRANSFER

The DSSAT is a user-friendly system in which a "shell" program resides in computer memory and provides a wide range of functions through "pop-up" menus on the screen. The system has three main components: crop models for many major crops, weather and soil databases, and crop management and biophysical data. The crop data are referenced by site. Users can enter site-specific experimental data for crop model validation and sensitivity analyses. Utility programs assist users in data entry, graphic display, and linkage to crop models contained in the system.

D.3.1 Basic Concepts in DSSAT: Experiments and Treatments

Because DSSAT was created as a tool for agricultural scientists, the crop models can simulate either observed conditions or hypothetical situations. Simulations are termed "experiments." An experiment can have several different treatments that have slightly varied conditions to explore the different aspects of an experiment. When simulating crop growth and development, the concept of treatments is very important. *Within the DSSAT, a treatment is a set of environmental and management variables used to simulate crop growth and development.*

D.3.2 Preliminary Exercise

The primary goal of the preliminary exercise is to accustom new users to the modeling approach used in DSSAT. The following simulation is of an experiment with maize conducted in 1982 in Gainesville, Florida. The experiment combines different levels of irrigation and fertilizer to test the performance of maize in north central Florida. Because the experiment was actually conducted in the field, the model's graphic output includes field results. You will conduct similar model runs when you are validating the model in your own countries.

1.   From the DSSAT shell, choose the **MODELS** menu item.

2.   Choose **CEREALS**.

3.   Choose **MAIZE**.

4.   Press **ENTER** to start the model.

5.   Choose the Gainesville, Florida, experiment:

     N X IRRIGATION, GAINESVILLE.

6.   Choose a treatment to simulate **1. RAINFED LOW NITROGEN**.

7.   When asked to run the simulation or to change simulation inputs, press **0** to run the simulation.

8.   When asked to name the simulation run, press **ENTER** to use the default name **RAINFED LOW NITROGEN**.

9.   When the model has finished simulating, choose **Y** (Yes) when asked if you wish to simulate again.

10. Choose the same experiment as before:

     N X IRRIGATION, GAINESVILLE

11. Now choose a different treatment from the one chosen before:

    2. RAINFED HIGH NITROGEN.

12. When asked to run the simulation or change simulation inputs, press **0** to run the simulation.

13. When asked to name the simulation run, press **ENTER** to use the default name **RAINFED HIGH NITROGEN**.

14. When the simulation has ended, type **N** (No) to return to the DSSAT shell.

When you have finished the model simulations, you can graph the model results by following these instructions:

1. Choose **GRAPH** to compare the two simulations graphically.

2. Choose **GROWTH** to view the possible growth data to graph.

3. From the Variables list, choose

   GRAIN DRY WEIGHT (kg/ha).

4. From the Run Numbers list, choose

   RAINFED LOW NITROGEN and
   RAINFED HIGH NITROGEN.

5. At the bottom of the screen, choose **GRAPH** to create the graph showing grain dry weight (in kilograms [kg] per hectare [ha]) versus time for both model runs.

6. Press any key to return to the previous screen.

7. Press the **ESC** key to clear the screen.

8. Choose **EXIT** to return to the DSSAT shell.

D.4 CLIMATE CHANGE EXERCISES

Climate change exercises are intended to help participants to simulate crop responses to climate change scenarios by using the different parts of DSSAT. The problem statement is as follows:

> *Given maize crop management, weather, and soils data for a site in Memphis, Tennessee, USA, analyze the potential crop impacts of a climate change scenario.*

D.4.1 Single-Year Simulation for Climate Change

This exercise simulates observed climate for 1977 and a climate change scenario for a maize crop. All management practices between the two treatments are the same; the only variables that change are the doubling of the atmospheric concentration of carbon dioxide ($2XCO_2$), temperature, precipitation, and solar radiation (climate change projections from the Canadian Climate Centre model [CCCM] for Memphis, Tennessee).

1. From the DSSAT shell, choose the **MODELS** menu item.

2. Choose **CEREALS**.

3. Choose **MAIZE**.

4. Press **ENTER** to start the model.

5. Choose the Memphis, Tennessee, experiment:

   MEMPHIS, TENNESSEE CLIMATE CHANGE EXAMPLE.

6. Choose a treatment to simulate

   1. NORMAL CLIMATE, RAINFED.

7. When asked to run the simulation or change simulation inputs, press **0** to run the simulation.

8. When asked to name the simulation run, press **ENTER** to use the default name **NORMAL CLIMATE, RAINFED**.

9. When the model has finished simulating, choose **Y** (Yes) when asked if you wish to simulate again.

10. Choose the same experiment as before:

    MEMPHIS, TENNESSEE CLIMATE CHANGE EXAMPLE.

11. Now choose a different treatment from the one chosen before:

    3. CCCM CLIMATE 2XCO2 RAINFED.

12. When asked to run the simulation or change simulation inputs, press **0** to run the simulation.

13. When asked to name the simulation run, press **ENTER** to use the default name **CCCM CLIMATE 2XCO2 RAINFED**.

14. When the simulation has ended, type **N** (No) to return to the DSSAT shell.

When you have finished the model simulations, you can graph the model results by following these instructions:

1. Choose **GRAPH** to graphically compare the two simulations.

2. Choose **GROWTH** to view the possible growth data to graph.

3. From the Variables list, choose

    LEAF DRY WEIGHT (kg/ha) and
    STEM DRY WEIGHT (kg/ha) and
    GRAIN DRY WEIGHT (kg/ha).

4. From the Run Numbers list, choose

    NORMAL CLIMATE, RAINFED and
    CCCM CLIMATE 2XCO2, RAINFED.

5. At the bottom of the screen, choose **GRAPH** to create the graph showing

    Leaf dry weight (kg/ha) vs. Time
    Stem dry weight (kg/ha) vs. Time
    Grain dry weight (kg/ha) vs. Time

    for both model runs.

6. Press any key to return to the previous screen.

7. Press the **ESC** key to clear the screen.

8. Choose **EXIT** to return to the DSSAT shell.

D.4.2 Multiyear Simulation for Climate Change

Most climate change simulations are run for a period of years to allow study of both mean and variance changes in output variables. The DSSAT can be used for multiyear simulations in its "Analyses" section. The multiyear simulation runs all treatments within an experiment file for a number of years. Soil moisture conditions are reinitialized before each season's crop simulation, as prescribed by the user in the input file.

The use of the multiyear simulation makes available a wider variety of options for analysis. Once a seasonal simulation has been completed, you can display statistics on the biophysical crop outputs. Results of the simulation can be plotted as box plots, cumulative function plots, or mean-variance plots. Results can be printed or saved as computer files.

In addition, you can conduct an economic analysis to calculate statistics on economic returns. Results of the simulation can be plotted as box plots, cumulative function plots, or mean-variance plots. Prices can be changed, or the price-cost variability can be calculated. Results can be printed or saved as computer files.

In this exercise, the treatments listed in the previous exercise are simulated for 30 years. Crop growth is simulated for the baseline-climate data set from Memphis, Tennessee, and for several climate change scenarios from general circulation models (GCMs). Throughout this exercise, whenever a blinking sentence appears on the screen, you can press any key to advance the program. These screens appear often but are not listed in the instructions.

1. From the DSSAT shell, choose the **ANALYSES** menu item.

2. Choose **SEASONAL**.

3. Choose the Memphis, Tennessee, file for maize simulation:

   MEMPHIS, TENNESSEE CLIMATE CHANGE EXAMPLE.

4. The DSSAT may ask if you wish to overwrite the file because the simulation has been run previously. If you have this screen, choose **Y** (Yes).

5. The model will run all of the treatments in the experiment.

6. When the model runs are finished, messages will inform you of the output file name. Press any key to continue.

7. When asked if you wish to simulate another file, choose **N** (No) to return to the DSSAT shell.

The next procedure allows you to analyze the summary data from the numerous outputs simulated by the multiyear runs:

1. Choose **ANALYZE**.

2. Choose a cropping season file for analysis:

   MEMPHIS, TENNESSEE CLIMATE CHANGE EXAMPLE.

3. Choose **Y** (Yes) to accept this output name.

4. The screen will show summary data statistics. Press any key to continue.

5. Choose **BIOLOGICAL** analyses.

6. Choose **HARVEST DAY** by pressing the space bar. A check will appear in the [ ]. Press **ENTER** to continue.

7. The screen will show summary data statistics. Press any key to continue.

8. Choose **BOX PLOT** to graph the data.

9. Choose any combination of treatments to graph. The default setting is all treatments. Press **ENTER** to continue.

10. The graph shows harvest day (days after planting) versus time for all six treatments. Press any key to continue.

11. Choose **QUIT** to exit.

12. By pressing **ESCAPE** and answering **Y** (Yes), you have the option of printing a report summary of the data. Answer **N** (No) for now.

D.4.3 Adaptations to Climate Change

Several methods simulate the response to adaptive measures. You can alter treatments in the experiment file to reflect growers' responses to changing environmental conditions. In addition, you can alter inputs within the crop models when prompted by the program.

Standard adaptive changes include changing the planting date, adding irrigation, and switching crop cultivars or species. You are encouraged to try many different strategies that might be practiced within your country in response to changing climatic conditions.

To conduct a simulation of a planting-date adaptation:

1. From the DSSAT shell, choose the **MODELS** menu item.

2. Choose **CEREALS**.

3. Choose **MAIZE**.

4. Press **ENTER** to start the model.

5. Choose the Memphis, Tennessee, experiment:

MEMPHIS, TENNESSEE CLIMATE CHANGE EXAMPLE.

6. Choose a treatment to simulate:

3. CCCM CLIMATE 2XCO2 RAINFED.

7. When asked to run the simulation or change simulation inputs, press **0** to run the simulation.

8. When asked to name the simulation run, press **ENTER** to use the default name **CCCM CLIMATE 2XCO2 RAINFED**.

9. When the model has finished simulating, choose **Y** (Yes) when asked if you wish to simulate again.

10. Choose the same experiment as before:

MEMPHIS, TENNESSEE CLIMATE CHANGE EXAMPLE.

11. Choose the same treatment to simulate:

3. CCCM CLIMATE 2XCO2 RAINFED.

12. When asked to run the simulation or change simulation inputs, press **1** to select **SENSITIVITY ANALYSIS OPTIONS**.

13. When shown a list of the model inputs, choose **7** for **PLANTING**.

14. Choose **1** for **PLANTING DATE**.

15. Change planting date to **MAR 14**.

16. Press **0** to return to **PREVIOUS MENU**.

17. Press **0** to return to **MAIN MENU**.

18. When asked to name the simulation run, type in a different name, and press **ENTER**.

19. When the simulation has ended, type **N** (No) to return to the DSSAT shell.

20. Choose **GRAPH** to compare the two simulations.

D.5 CREATING CLIMATE CHANGE SCENARIOS

This section focuses on how to generate climate change scenarios from observed weather data and GCM output. Participants use a sample weather file from Memphis, Tennessee, and output from four GCMs to make crop model input files for baseline and climate change scenarios. The weather file is named METN.DAT and has weather data for 30 years (1950-1980).

D.5.1 Using Output from General Circulation Models

Each GCM produces data files of climate changes by month for temperature, precipitation, and solar radiation for each grid point around the globe. Two programs in the C:\NCAR\WEATHER directory (GRDPTI.EXE and CLIMV29.EXE) produce monthly GCM data. They are then combined with user-defined observed daily data to create a climate change scenario. The final products of these programs are DSSAT weather data files that can be used to run climate change simulations.

The climate change files are stored on three floppy disks, referred to as NCAR/DSS disks 1-7. These data must be copied, one disk at a time, to the hard disk in a subdirectory named C:\NCAR\WEATHER. Each participant must create this subdirectory before copying. Also contained on this disk is a program called GRDPTI.EXE, which must be run for creating site-specific weather changes on a monthly basis. After copying the three NCAR/DSS disks, run GRDPTI.EXE. This run will result in four output files, one for each GCM (i.e., CCCM???.DAT, GFD3???.DAT, GF01???.DAT, and GISS???.DAT, where the ??? refers to a user-named site code). This three-character site code (e.g., MPT for Memphis, Tennessee) is important for subsequent steps.

D.5.2 Creating Annual Weather Data Files for GCM Scenarios

A program, also on disk 2 in subdirectory A:\NCAR\WEATHER, was developed to create GCM scenario weather data files. This program (CLIMV29.EXE) reads the monthly site-specific weather change values from the files created by the GRDPTI.EXE program described earlier and the participant's daily weather data for a site after the data have been reformatted. The program creates annual weather data files for standard weather and for each GCM scenario (CCCM; Geophysical Fluid Dynamics Laboratory [GFDL] R-30 [GFD3]; GFDL 1%/yr transient [GF01]; and Goddard Institute for Space Studies [GISS]).

Each file has the form ???_XX01.WTH, where the ??? is the three-character site code (MPT for Memphis, Tennessee), the _ is the scenario number (0, 1, 2, 3, or 4 for BASELINE, CCCM, GFD3, GF01, and GISS scenarios, respectively), and the XX represents the year; for example, the weather file MPT25101.WTH is the file containing the GFDL scenario weather data for Memphis, Tennessee, for the year 1951, months 01 through 12. As described previously, 30 years of weather data for a site will produce 120 files (30 for each of three scenarios plus the 30 for the BASELINE weather).

In summary, participant's weather data for each site will be formatted and read by a second program (CLIMV29.EXE) to create yearly data files for the site for the following:

- BASELINE weather, ???0XX01.WTH;

- CCCM, ???1XX01.WTH;

- GFD3, ???2XX01.WTH;

- GF01, ???3XX01.WTH; and

- GISS, ???4XX01.WTH.

Until this point in the procedure, all of the participant's weather data and programs for the manipulation of the data have been located in the subdirectory C:\NCAR\WEATHER. After confirming that the files have been created correctly, all of the yearly files must be copied into C:\DSSAT3\WEATHER.

Each participant will have several (and possibly many) locations for which simulations are to be performed. Filling up the hard disk C: is a danger (if hard disk space is limited), unless some precautions are taken. Participants should consider running simulations for one site, backing up the climate change scenarios and baseline weather data for that location on disk, and then erasing the weather data from subdirectory C:\NCAR\WEATHER. Another site's weather can then be copied and used to create the next set of climate change scenarios.

D.5.3 Climate Change Scenario Exercise

This exercise uses GRDPTI.EXE and CLIMV29.EXE to create GCM weather files for the Memphis, Tennessee, site.

1. Exit DSSAT, and set the directory to **C:\NCAR\WEATHER**.

2. Type **GRDPTI** and press **ENTER**.

3. This program provides climate change data in two ways. One method uses data from the nearest GCM grid point; the other interpolates between the two nearest GCM grid points. The program asks, "Would you rather just use the nearest point?" The interpolated GCM data are recommended. Therefore, answer the question with **N** (No).

4. Enter the longitude, as prompted, in decimal form.

5. Enter the latitude, as prompted, in decimal form.

6. Enter the following output names for the specific model used:

   a. For the CCCM model, name the output file: **CCCMMPT.DAT**.
   b. For the GFD3 model, name the output file: **GFD3MPT.DAT**.
   c. For the GF01 model, name the output file: **GF01MPT.DAT**.
   d. For the GISS model, name the output file: **GISSMPT.DAT**.

7. Once the GRDPTI program has created the monthly data, type **CLIMV29**.

8. Type **MPT** for the site code for Memphis, Tennessee, and then type **METN.DAT** as the input data file.

9. The program immediately converts the one file METN.DAT into 30 yearly files with the DSSAT weather file names.

10. The program will then ask if you wish the model data for each climate change scenario. Type **Y** (Yes) for the files to be made.

D.6 BASELINE WEATHER DATA

For climate change studies, participants must supply their own weather data for representative agricultural sites, with up to 30 years of daily data for precipitation, maximum and minimum temperature, and solar radiation. These data are usually obtained from government ministries, agricultural experimental stations, or university research projects. If data on solar radiation are not available, sunshine hours or percent sunshine can be used. If no information is available on solar radiation or sunshine, published values of percent sunshine can be used. WeatherMan, within DSSAT, will compute solar radiation from the best available source of information and produce an output file with data arranged in the proper format for the IBSNAT/ICASA crop models. You must copy your own weather data into the subdirectory C:\NCAR\WEATHER. Later, this file could be deleted after IBSNAT/ICASA yearly weather data files are created for each scenario.

D.6.1 WeatherMan Utility Program

The DSSAT "Weather" section has several options for importing, verifying, and exporting weather files. In the "List/Edit" section, you can locate, edit, search, or get institutional information concerning specific weather files by choosing "Daily Data," "Monthly Mean," "Generated Data," or "Climate Data." Within the "Utilities" section, a program named WeatherMan is activated. WeatherMan is a useful program for managing and generating daily weather data. While all of the features of WeatherMan are explained in the DSSAT v2.9 Users Guide, the main features of the program allow you to

- Import or export weather data to any format,

- Check for file format or data range errors,

- View weather data graphically or in statistical reports, and

- Generate missing weather data or create new sets of weather data with two different weather generators.

D.6.2 WeatherMan Exercise

In WeatherMan, each weather station is identified and characterized. First, enter general information about the site (e.g., name, latitude, longitude) and then import the available weather data for the site. The data are then ready to be manipulated.

1. In the DSSAT shell, choose **DATA**.

2. Choose **EXPERIMENTS**.

3. Choose **WEATHER**.

4. Choose **UTILITIES**; this starts the program WeatherMan.

5. Choose **STATION**, and type in the new station name using the letters **METN** for Memphis, Tennessee. Choose **OK** when finished.

6. WeatherMan will automatically prompt you for additional data for the weather station. For each category, enter the following data to reflect the Memphis, Tennessee, weather data:

   a. Enter **MEMPHIS, TENNESSEE** in the title area. Press **TAB**, or use the mouse to highlight the next item.

   b. Enter a latitude of **35.0** (positive for north and negative for south) and a longitude of **-89.6** (positive for east and negative for west).

   c. Enter an elevation of **300.0** m.

   d. Do not change angstrom coefficient A or B.

   e. Set the reference height at **1.50** m and a wind reference height at **2.0** m.

   f. Do not change the other values. These values will be changed when weather data are imported into the METN.CLI climate file.

7. Choose **OK** to save data.

The next important procedure in using WeatherMan is to import weather data to be checked, cleaned, and exported to files compatible with DSSAT v2.9.

1. Start WeatherMan as explained previously, and choose **METN** as a weather station.

2. Choose **IMPORT/EXPORT**.

3. Choose **EDIT RANGE CHECKS** to verify the desired upper, lower, and rate change limits. Choose **OK** when finished.

4. Choose **IMPORT SINGLE FILE**.

5. When asked for the file name that you want to import, type **C:\NCAR\WEATHER\METN.DAT**, and then choose **OK**.

6. Specify a file format for the incoming file. Because the weather data are already in a format used by DSSAT software, choose **IBSNAT3**, and then choose **OK**. WeatherMan will begin to read the data and check the data for the ranges that you specified in item 3.

7. When WeatherMan has finished its range checking, it shows a summary of the range check. To continue, choose **OK**.

8. To fill in missing data at a station, choose **STATION** and then **CALCULATE MONTHLY MEANS**. This command uses the imported weather data to calculate monthly means for use with the weather generators in WeatherMan. (See manual for background information on how to check and fill in missing weather data.)

9. To export data for use in DSSAT v2.9, choose **IMPORT/EXPORT** again, and then choose **EXPORT YEARLY FILES**.

10. Choose specific starting dates, or use the dates already listed.

11. Choose a file format for the export files. Choose **IBSNAT3** as the export format.

12. WeatherMan will write the files into the C:\DSSAT3\WEATHER directory unless you specify another directory.

13. The created weather files can now be accessed by the crop models.

D.7 CROP MODELS

D.7.1 Input Files

The DSSAT crop models require input files for weather, soils, and management information. The basic file structure is displayed in Figure D.1. In addition, the list (.LST) files organize the different files to be read by the DSSAT program; for example, all soil files available for use in DSSAT are listed in the SOIL.LST file.

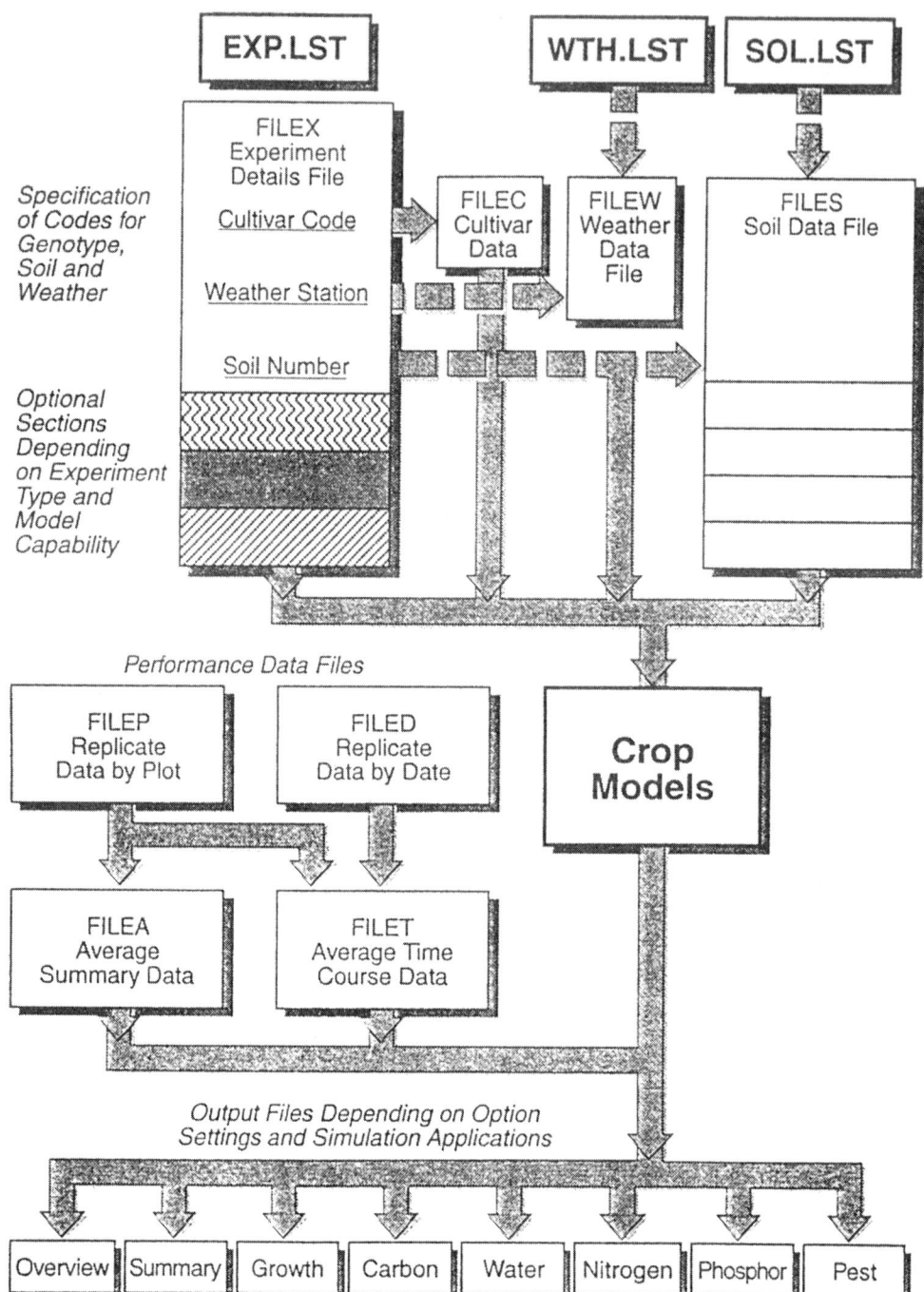

FIGURE D.1 Overview of Input and Output Files Used by Crop Models (Source: DSSAT v3 User's Guide)

To make identifying different types of files easier, naming conventions were adopted for the file names. The prefix (the first eight letters before the period in the file name) is constructed from the institute code (two characters), the site code (two characters), the year of the experiment (two characters), and the experiment number (two characters).

For example, an experiment conducted by the University of Florida (UF) at Gainesville (GA) in 1988 (88) would make a prefix of UFGA8801. In the climate change exercise, crop experiments conducted in Memphis, Tennessee, will have the first four characters of METN.

The suffix (the last three characters after the period in the file name) has a variety of possible codes, depending on the information the file contains. In files describing a specific crop, the first two letters refer to the crop code. The crop codes are listed in Table D.1. The last letter in the suffix refers to the type of crop information stored in the file. The different letters are displayed in Table D.2, where cc represents the crop code.

In continuing with the file example, a file that contains experimental details for a maize trial grown at the University of Florida, Gainesville, in 1988 would have a file name of "UFGA8801.MZX."

The minimum files needed to run the crop models are an X (.ccX) file (Section D.9), a weather (.WTH) file (Sections D.6 and D.7), a soil file (.SOL), and a genetic information (.CUL) file. The other files (designated the .ccP, .ccA, and .ccT files) compare simulated results with crop experiment trials in model validation runs (Section D.8).

D.7.2 Crop Codes

Table D.1 lists the crop codes used in the DSSAT file-naming convention.

D.7.3 Experiment Files

The files listed in Table D.2 contain different types of data for each crop. The characters "cc" represent the crop codes listed in Table D.1.

TABLE D.1 Crop Codes for DSSAT

Crop	Code
Alfalfa	AL
Aroid	AR
Barley	BA
Dry beans	BN
Broad-leaved weeds	BW
Cotton	CO
Cassava	CS
Fallow	FA
Grass weeds	GW
Pearl millet	ML
Maize	MZ
Peanuts	PN
Potatoes	PT
Rice	RI
Soybeans	SB
Sugarcane	SC
Sorghum	SG
Wheat	WH

D.7.4 XCREATE Exercise

The next exercise uses the program XCREATE to make a management input file that corresponds to the desired simulation or field experiment. The program XCREATE can be run from several locations within the DSSAT shell. (Recall that a treatment is a set of environmental and management choices used to conduct a crop experiment or simulate crop growth and development within the DSSAT modeling system). As in crop experiments, a variety of treatments can be conducted within one simulation. Treatment information is stored in the .ccX files, where the "cc" represents the crop code. For example, the file METN7801.MZX contains treatment data for maize crops at the Memphis, Tennessee, site. More than one treatment can be stored in each .ccX file.

TABLE D.2 Codes for DSSAT Experimental Data

Code[a]	Description
.ccX	Experimental details file
.ccP	Observation data
.ccA	Average values of observation data
.ccT	Time course data

[a] cc = crop code.

Creating an X file consists of (1) selecting an existing experiment as a "template," or base file; (2) removing any unwanted treatments; (3) editing sections as required until complete; and (4) saving the new X file under a new name.

To edit a .ccX file, follow these steps:

1. From the DSSAT shell, choose the **DATA** menu item.

2. Choose **EXPERIMENT**.

3. Choose **CREATE**.

4. Choose an experiment to serve as a template for the file you will save later under a different file name.

5. Define the type of file to create. Choose **EXPERIMENT**, and then choose **OK**.

6. Set up the correct number of treatments by adding or deleting. First choose **MANAGEMENT** and then **TREATMENTS**.

7. Use this same menu to choose a treatment to alter. The chosen treatment appears in a small box in the upper right of the computer screen.

8. Use the menus, windows, and dialogue boxes to alter each chosen treatment.

9. Save the file under a new name by pressing the **F7** key and entering in the following data to define the Memphis experiment.

10. XCREATE will write the new file into EXP.LST for use in the "Models" section.

D.7.5 Soils Data

Choosing the "Soils" option allows you to view, edit, or import soil information for use in the crop models. The "Soils Data" section is divided into three options: "List/Edit," "Create," and "Utilities." The "List/Edit" section allows you to search for specific soil files, edit soil files, or find institutional information. The "Create" section allows you to enter soil information from experimental data forms or append other soil information into the DSSAT. The program writes the soil name into the overall soil list SOIL.LST. Within the "Utilities" section, a graphics program allows you to compare soil properties for several different soils.

D.7.6 Crop Cultivar Data

The model uses genetic coefficients to describe specific cultivar traits for each crop. The handling of the genetic coefficients varies, depending on which crop model is used. The cereal models (maize, wheat, etc.) have one cultivar (.CUL) file for each crop; for example, the maize component has the file MZCER920.CUL to store its genetic coefficients.

The legume model (soybean) has three files to store the genetic information for the crop. This structure groups genetic traits into sets so that you can alter only the parameters that directly apply to your crop of interest. For legumes, the .CUL file holds genetic parameters for each specific cultivar. The .ECO file contains coefficients that change among groups of cultivars, but not for every cultivar. The .SPE file contains parameters that change only for different species within the model; for example, the soybean crop would have a different .SPE file than peanut (groundnut). Each legume crop species has its own .CUL, .ECO, and .SPE files. For exact file formats, consult the DSSAT v2.9 documentation.

The "Genotype" section of DSSAT allows you to view and edit information about crop cultivars. The species, ecotype, and cultivar files can be viewed and edited in the same method as data on soils and weather.

Many possible cultivars exist for each crop in the DSSAT; for example, cultivars vary in the length of crop development stages (in degree-days) and in the number of grains per stalk.

D.7.7 Crop Cultivar Exercise

This exercise shows how to check for a cultivar name against the current cultivar lists for each crop in DSSAT. In setting up the Memphis, Tennessee, climate change experiments, genetic coefficients must be reviewed to be sure of the cultivar and its growing characteristics. In this exercise, the user is checking for the maize cultivar Pioneer 3382 (PIO 3382), which is one of the main cultivars grown in Memphis, Tennessee.

1. From the DSSAT shell, choose the **DATA** menu item.

2. Choose **GENOTYPE**.

3. Choose **LIST/EDIT**.

4. Move the highlighted bar to the file **MZCER930.CUL**, which is listed as **GENETICS PARAMETER INPUT FILE — MAIZE GENOTYPE COEFFICIENTS**.

5. Press the **F8** key to edit the file.

6. Verify that the cultivar **PIO 3382** is listed in the file.

7. Depending on the editor, either

    a. Press the **ESC** key to return to the file list, or
    b. Press **ALT** and then **F**, and then choose **EXIT**.

D.8 CALIBRATION AND VALIDATION

Calibrating and validating the IBSNAT/ICASA crop models are essential steps in studying the impacts of climate change. To have confidence in the models' simulation of potential responses to future climate and CO_2 conditions, simulation of crop responses to current conditions must be as accurate as possible. The DSSAT documentation contains detailed instructions for coordinating field experiments with model simulation. A brief summary is presented here; but for complete documentation, refer to *Technical Report 2, Field & Laboratory Methods for the Collection of the IBSNAT Minimum Data Set for the Decision Support System for Agrotechnology Transfer (DSSAT v2.1)*.

D.8.1 Use of Field Experiments

The description and results of crop experiments can be recorded on the data forms included with the DSSAT v2.9 software. The forms that record crop experiments are also used to input the data into computer files that can then be accessed from the DSSAT for model calibration and validation. Because all of the input files are in ASCII text format, a convenient method of creating a new experiment file is to load an old file into any text editor, enter the data, and then save as a new file name.

In the DSSAT shell's "DATA" pull-down menu, the "Background" section contains database information for different researchers, institutes, and experimental sites. Participants can search for the names and addresses of scientists or research sites. In addition, the "Codes" section contains definitions for all input codes used in the data files. These sections are good resources for help and information if the data entry process for model validation is confusing.

Within the "Experiment" section, several functions allow manipulation of experimental (.ccX) files and crop performance (.ccA and .ccT) files. The "List/Edit" option allows listing, searching, and editing of .ccX, .ccA, and .ccT files through a database program. The "Create" option uses the program XCREATE to edit .ccX files. Within the "Utilities" section, several methods analyze crop trial performance data. Also included is the CONVERT program, which converts older DSSAT information (in the version 2.1 format) to the new DSSAT v2.9 format.

D.8.2 Experimental Data Exercise

Verifying the experimental data files with experimental data forms that have been filled out by hand is important. You can view crop experiment and performance files from several locations within the DSSAT. This exercise presents one method of accessing these files.

1. From the DSSAT shell, choose the **DATA** menu item.

2. Choose **EXPERIMENT**.

3. Choose **LIST/EDIT**.

4. Use the arrow keys to move the highlighted bar to **UFGA8201.MZX**.

5. Press **F8** to edit the file.

6. Move the highlighted bar to choose the **TIME COURSE** file and then press **ENTER**.

7. **F1** gives help information for the editor.

8. Depending on the editor, either

   a. Hit the **ESC** key to leave the editor without changing the file, or
   b. Press **ALT** and then **F**, and then choose **EXIT**.

By using the L key and rebuilding the database (by using the F9 and F10 keys), you can include specific experiments to be simulated by the crop models.

D.9 AGGREGATION OF CROP MODEL RESULTS

It is useful in climate change studies to aggregate the site-specific results obtained from crop models to regional and national estimates of impacts. One method for aggregating model results is presented as an example. First, identify the major agricultural regions within the country that produce each crop to be modeled and the percentage of total crop production produced in each region. Within these identified regions, designate the soil types and weather stations to be used as crop model inputs. Crop management practices (e.g., cultivar, planting date, irrigation regimes) within these regions must also be specified for use as model inputs. Baseline climate and climate change scenario simulations are then run, and estimates of changes in crop growth, development, and yield are obtained by comparing the results from the climate change and baseline simulations. When the model results have been generated, the changes in crop yields and other output by site can be aggregated by weighing according to the regional percentage of the national crop production.

Table D.3 gives an example of the aggregation method described previously. In the example country, almost all maize is grown in two regions. One region in the southeast grows 35% of the total maize production, while the other region in the northwest grows the remaining 65%. In the southeast region, modelers have identified one weather station (station 101), two major soil groups (A and B), and two management practices (high-input commercial and subsistence) as representative of the region. In the northwest region, modelers have identified two weather stations (stations 333 and 334), three major soil groups (C, D, and E), and three management practices (high-input commercial, medium-input commercial, and subsistence) as representative of the region.

After simulating each group of weather, soil, and management inputs under baseline and changed climate conditions, a percent change in crop yields (or other variable of interest) is calculated by subtracting the climate change scenario model results from the baseline climate model results and dividing by the baseline model results. The percent change in yield

TABLE D.3 Aggregation Example

Region	Weather Station	Soil Group	Management Practice	Percent of Total Area	Change in Crop Yield (%)
Southeast	101	A	High-input commercial	17	+5
Southeast	101	B	Subsistence	18	-30
Northwest	333	C	High-input commercial	5	-10
Northwest	333	D	Medium-input commercial	10	-25
Northwest	333	E	Subsistence	40	-35
Northwest	334	D	High-input commercial	6	-10
Northwest	334	E	Medium-input commercial	4	-15

for each different condition is then production-weighted on the basis of the production percentages given in Table D.3, according to the following: change in national production = the sum of the percent total production of growing condition 1 times the percent change in yield of growing condition 1 + the percent total production of growing condition 2 times the percent change in yield of growing condition 2 + ...

This procedure results in regional and national estimates of potential impacts of climate change. Regional and national results derived from crop model results can be mapped by a geographic information system (GIS).

D.10 ECONOMIC ANALYSIS

As a starting point for such countries, researchers should construct a simple checklist of available economic and agricultural information for the country. The availability and quality of such information can help determine the existence and possible magnitude of economic vulnerability in the agricultural sector. This information falls into three general categories: production information, consumption information, and policy information.

Production information can suggest the vulnerability of agriculture to climate changes or other environmental stress. This type of information includes the following:

- Number of alternative crops,

- Number of production techniques (e.g., irrigation),

- Nature and extent of resource use in agricultural production, and

- Cost of production information.

The first three categories of production information suggest the substitution possibilities in the agricultural sector. Specifically, the greater the number of crop alternatives and production techniques in a country, the greater the likelihood of mitigation/adaption possibilities. Conversely, for countries with few crops and few production techniques, the greater the potential vulnerability. The cost of production information can help define the costs of such adaptation as well as the conditions under which supplies of each commodity are likely to increase (decrease) with increases (decreases) in prices of commodities.

Information dealing with the consumption (demand) aspects of agriculture is also useful in understanding possible economic consequences and vulnerabilities. Three important types of economic information are included:

- Role of each crop in the country's overall food consumption,

- Percent of crops consumed domestically vs. exported crops, and

- Price movements of commodities.

The extent to which a crop contributes to domestic consumption has two potentially important effects. For example, a crop produced primarily for domestic consumption (i.e., a staple) is likely to be important to the well-being of both producers (in terms of their own consumption) and consumers. Reduction in the supply of such crops implies some increased vulnerability in terms of diet. A crop that is largely exported is typically a major source of foreign exchange earnings for the country. Reduction in exports can reduce economic well-being of both producers and others in the economy who depend on those export earnings (e.g., beneficiaries of government programs funded by the export earnings). As with production, countries with more crop alternatives in terms of both domestic and export use are less likely to be vulnerable to climate change.

These consumption and production effects are reflected in markets through the movements in prices for each commodity. Information on domestic price movements can reflect the degree to which crops are staples (highly inelastic demands). Prices in export markets reflect international conditions (unless the country in question is the major supplier of a commodity, in which case the changes in prices reflect both international consumption patterns as well as domestic production). The nature of price and consumption information can thus signal the degree of vulnerability in a particular country.

Information on government policies toward the agricultural sector can indicate the extent to which economic adaptations are encouraged. Specifically, government intervention in agriculture (or other economic sectors) typically distorts economic processes, resulting in less efficient resource allocation decisions than would be achieved in the "free market" situation. Government intervention is common to both developed and developing economies, although the goals of the intervention may differ.

Most developed countries are attempting to reduce government intervention *and* distortions in agriculture, in part because of the direct cost of such intervention to national treasuries. Government policies tend to build rigidities into the agricultural sector by protecting producers from events that would be expected to guide producers toward alternative uses of their resources. Removal of some government policies is thus expected to facilitate more rapid adjustment to environmental change. Conversely, increased government intervention implies less flexibility and perhaps greater vulnerability to climatic stress. Hence, some understanding of the present and future government involvement in the agricultural sector can be useful in forecasting potential vulnerabilities.

D.11 INTERPRETATION OF RESULTS

Within a country study, analysis of agricultural effects is composed of a diverse suite of climate change scenarios, their biophysical effects on important crops (both subsistence and commercial), potential farm-level adaptations to the projected changes, and economic adjustments. The study planning decisions for which processes to include at each stage of analysis should be arrived at by an interdisciplinary team of climatologists, agrometeorologists, agronomists, economists, and policy makers. Results should also be integrated with results from parallel studies of water resources, coastal resources, and natural ecosystems, where appropriate.

The results from the crop modeling studies should be reviewed in the context of the physical, economic, and institutional characteristics of the agricultural sector in a given country. Some countries will need to include production and consumer adjustments, and associated market effects. In the absence of off-the-shelf economic models, "rules of thumb" gleaned from the extant literature can be applied to "bound" or otherwise adjust the agricultural effects suggested by the biophysical modeling approach. For other countries, economic adjustments may play a lesser role in accurately measuring the vulnerability of the agricultural sector.

These procedures lead to an assessment of the vulnerability of the agricultural system under study. The approach used in the Country Studies vulnerability assessments efficiently uses techniques and understanding gained from past experience, crop modeling studies, and economic analysis. Agricultural producers, domestic and international consumers, and policy makers play important roles in defining the ultimate consequence of an environmental change. Consequences to all groups should be considered.

D.12 SOURCES OF UNCERTAINTY

The primary uncertainties in the recommended approach depend on assumptions embedded in the IBSNAT/ICASA crop models, the methods by which the IBSNAT/ICASA models are used in the study of climate change impacts, and the difficulty of estimating future technological improvements in agriculture.

The IBSNAT/ICASA models contain many simple, empirically derived relationships that do not completely mimic actual plant processes. These relationships may or may not hold under differing climatic conditions, particularly the higher temperatures predicted for global warming. For example, most of the data used to derive the relationships in the crop models were obtained with temperatures below 35°C, whereas the project temperatures for $2XCO_2$ are often 35 or even 40°C during the growing period. Other simplifications of the crop models are that weeds, diseases, and insect pests are controlled; there are no problem soil conditions such as high salinity or acidity; and there are no catastrophic weather events such as heavy storms.

Several key points must be made regarding the improved use of the IBSNAT/ICASA models for climate change impact studies. Climate data may be of differing quality in various study sites, and length of record may vary; quality control of climate data is highly desirable. Model calibration and validation should be pursued rigorously. Furthermore, changes in climate variability, levels of fertilization, and choices in cultivar should be tested wherever possible. Finally, the crop models simulate current available agricultural technologies. They do not include potential improvements in such technologies, although they may be used to test some potential improvements, such as improved varieties and irrigation schedules.

Some of these limitations may be overcome by using a set of crop models and crop modeling techniques developed from a range of sources (Section D.13).

D.13 GENERAL METHOD AND ADDITIONAL CROP MODELS

While the IBSNAT/ICASA DSSAT crop modeling system is described in this appendix, many other methods exist for various types of analysis. Crop models span a wide range of differing basic assumptions and levels of complexity. Simple agroclimatic indexes, such as growing degree-days and the length of the growing season, are useful for determining how agroecological zones may shift and should be included in the early stages of a country study. Mapping of these and other results is an important analytic tool. Dynamic process models of crop growth are useful for analyzing more detailed agronomic responses. A selected list of such models is given in Table D.4; references associated with this table are given in Section D.17.

Questions help participants to assess whether a model or method is potentially useful for their climate change research. The first question is, *"What are the major assumptions of the model or method?"* Most crop models have three or four major processes that interact to simulate crop growth and development. The processes usually include a water balance model, a growth and development routine, and a yield prediction function. The basis for these assumptions is important to the success of modeling climate change impacts. The next question is, *"Will these assumptions limit my capacity to model climate effects on the crop?"* For example, a model that calculates yield by accumulating daily evapotranspiration amounts until the end of the growing season may not be adequate for simulating the effects of increasing atmospheric CO_2 on plant growth, transpiration, and yield.

Another assumption that can limit the effectiveness of models is whether the model can be applied to geographic regions with differing climates. Many crop models were developed in the United States or Europe and have assumptions based either on the climate or on the cropping practices of a particular country. The use of such a model that is not applicable geographically may bias the simulated results. Contacting the model's developers or the crop-impacts technical support staff can be helpful in deciding whether a certain crop model will provide useful simulations of climate change impacts. In addition, scientific journals may also provide information on the methods used by other scientists to conduct similar studies.

A final question is, *"Can I provide all the necessary model inputs?"* Information on weather, soils, and crop management is required by almost all crop models; however, the level of complexity varies according to the model. Sometimes simplifying assumptions can be made with regard to certain inputs without biasing the results. Once again, conducting a search of the scientific literature and consulting the model's developers or the technical support staff are recommended.

TABLE D.4 Selected Crop Models

Crop	Model Name	Reference
Alfalfa	ALSIM (Level 2) ALFALFA	Fick (1981) Dennison and Loomis (1989)
Barley	CERES-Barley	Ritchie et al. (1989a)
Cotton	GOSSYM COTCROP COTTAM	Baker et al. (1983) Brown et al. (1985) Jackson et al. (1988)
Dry beans	BEANGRO	Hoogenboom et al. (1989)
Maize	CERES-Maize	Jones and Kiniry (1986); Ritchie et al. (1989b)
	(unnamed) CORNF SIMAIZ CORNGRO CORNMOD (unnamed) VT-Maize GAPS CUPID	Stockle and Campbell (1985) Stapper and Arkin (1980) Duncan (1975) Childs et al. (1977) Baker and Horrocks (1976) Morgan et al. (1980) Newkirk et al. (1989) Buttler (1989) Norman and Campbell (1983)
Peanuts	PNUTGRO (unnamed)	Boote et al. (1989) Young et al. (1979)
Pearl millet	CERES-Millet RESCAP	Ritchie and Alagarswamy (1989) Monteith et al. (1989)
Potatoes	(unnamed) SUBSTOR	Ng and Loomis (1984) Griffin et al. (1993)
Rice	CERES-Rice RICEMOD (unnamed)	Godwin et al. (1990) — Horie (1988)
Sorghum	SORGF CERES-Sorghum SORKAM RESCAP	Arkin et al. (1976) Ritchie and Alagarswamy (1989) Rosenthal et al. (1989) Monteith et al. (1989)
Soybeans	SOYGRO GLYCIM REALSOY SOYMOD	Wilkerson et al. (1983); Jones et al. (1989) Acock et al. (1983) — Curry et al. (1975)
Sugarcane	CANEMOD	Inman-Bamber (1991)

TABLE D.4 (Cont.)

Crop	Model Name	Reference
Wheat	CERES-Wheat	Ritchie (1985); Godwin and Vlek (1985)
	(unnamed)	Stockle and Campbell (1989)
	TAMW	Maas and Arkin (1980)
	(unnamed)	Aggarwal and Penning de Vries (1989)
	(unnamed)	van Keulen and Seligman (1987)
	SIMTAG	Stapper (1984)
General model	EPIC	Williams et al. (1984)

Source: Adapted from Jones and Ritchie (1990).

D.14 SELECTED CROP MODELS

Table D.4 and its references are from Jones and Ritchie (1990), with the addition of the sugarcane model. This list of dynamic process crop models is not complete. It is provided to allow participants more choice in selecting models that correspond to the research needs of their country study.

D.15 RESOURCE REQUIREMENTS

A team approach is highly recommended to accomplish a national climate change crop impacts study. Two senior professionals are needed to lead the project, one from the field of agronomy and one from the field of resource economics. An agrometeorologist, a water resource scientist, and an agricultural policy analyst should be members of the team, on a consultation basis. Computer specialists from the fields of agronomy and economics are useful for running the crop models and compiling the economics databases. A minimum of one year is needed to successfully complete a crop impacts study.

Cost considerations should include salaries for the above personnel, data collection costs, the purchase of the DSSAT package (about $400), and a personal computer powerful enough to run the many simulations in a reasonable amount of time. Computer requirements are given in Section D.2.1.

D.16 ADDITIONAL TRAINING

Further training on the use of the crop modeling system is available through workshops organized by IBSNAT/ICASA. These workshops are usually two weeks long and are conducted in various locations in the United States and abroad. The extended length of these workshops allows detailed and personalized instruction and training for workshop participants.

D.17 APPENDIX D REFERENCES

Acock, B., et al., 1983, *The Soybean Crop Simulator GLYCIM: Model Documentation 1982*, report 2, U.S. Department of Energy, Carbon Dioxide Research Division, Office of Energy Research, Washington, D.C.

Aggarwal, P.K., and F.W.T. Penning de Vries, 1989, "Potential and Water-Limited Yields in Rice-Based Cropping Systems in Southeast Asia," *Agricultural Systems* 30:49-69.

Arkin, G.F., et al., 1976, "A Dynamic Grain Sorghum Growth Model," *Transactions of the ASAE* 19(4):622-626, 630.

Baker, C.H., and R.D. Horrocks, 1976, "CORNMOD, A Dynamic Simulator of Corn Production," *Agricultural Systems* 4:57-77.

Baker, D.N., et al., 1983, *GOSSYM: A Simulation of Cotton Crop Growth and Yield*, tech bulletin 1089, South Carolina Agricultural Experimental Station, Clemson, S.C.

Boote, K.J., et al., 1985, "Modeling Growth and Yield of Groundnut," in *Proceedings of the International Symposium on Agrometeorology of Groundnut*, ICRISAT, Sahelian Center, Niamey, Niger.

Brown, L.G., et al., 1985, *COTCROP: Computer Simulation of Growth and Yield*, information bulletin 69, Mississippi Agricultural and Forestry Experimental Station, Mississippi State University, Mississippi State, Miss.

Buttler, I.W., 1989, *Predicting Water Constraints to Productivity of Corn Using Plant-Environmental Simulation Models*, Ph.D. dissertation, Cornell University, Ithaca, N.Y.

Childs, S.W., et al., 1977, "A Simplified Model of Corn Growth under Moisture Stress," *Transactions of the ASAE* 20(5):858-865.

Curry, R.B., et al., 1975, "SOYMOD I: A Dynamic Simulator of Soybean Growth and Development," *Transactions of the ASAE* 18(5):963-968.

Dennison, R.F., and R.S. Loomis, 1989, *An Integrative Physiological Model of Alfalfa Growth and Development* (complete citation unavailable).

Duncan, W.G., 1975, "SIMAIZ: A Model Simulating Growth and Yield in Corn," in D.N. Baker et al. (eds.), *An Application System Method to Crop Production*, Mississippi Agricultural and Forestry Experimental Station, Mississippi State University, Mississippi State, Miss.

Fick, G.W., 1981, *ALSIM I (Level 2) User's Manual*, agronomy mimeo 81-35, Department of Agronomy, Cornell University, Ithaca, N.Y.

Godwin, D.C., et al., 1990, *A User's Guide to CERES-Rice-V2.10*, International Fertilizer Development Center, Muscle Shoals, Ala.

Godwin, D.C., and P.L.G. Vlek, 1985, "Simulation of Nitrogen Dynamics in Wheat Cropping Systems," in W. Day and R.K. Arkin (eds.), *Wheat Growth and Modeling*, Plenum Press, New York, N.Y.

Griffin, T.S., B.S. Johnson, and J.T. Ritchie, 1993, *A Simulation Model for Potato Growth and Development: SUBSTOR-Potato Version 2.0*, Research Report Series 02, IBSNAT, Department of Agronomy and Soil Science, College of Tropical Agriculture and Human Resources, University of Hawaii, Honolulu, Hawaii.

Hoogenboom, G., et al., 1989, "A Computer Model for the Simulation of Bean Growth and Development," in *Advances in Bean* (Phaseolus vulgaris L.) *Research and Production*, publication CIAT No. 23, Cali, Colombia.

Horie, T., 1988, "Simulated Rice Yield under Changing Climatic Conditions in Hokkado Island," in M.L. Parry et al. (eds.), *Assessment of Climate Effects on Agriculture*, Vol. 1, Kluwer Academic Publishers, Dordrecht, the Netherlands.

Inman-Bamber, N.G., 1991, "A Growth Model for Sugar-Cane Based on a Simple Carbon Balance and the CERES-Maize Water Balance," *South African Journal of Plant Science* 8(2):93-99.

Jackson, B.S., et al., 1988, "The Cotton Simulation Model 'COTTAM': Fruiting Model Calibration and Testing," *Transactions of the ASAE* 31(3):846-854.

Jones, C.A., and Kiniry (eds.), 1986, *Ceres-Maize: A Simulation Model of Maize Growth and Development*, Texas A&M University Press, College Station, Tex.

Jones, J.W., and J.R. Ritchie, 1990, "Crop Growth Models," in G.J. Hoffman et al. (eds.), *Management of Farm Irrigation Systems*, Chap. 4, American Society of Agricultural Engineers Monograph, St. Joseph, Mo.

Jones, J.W., et al., 1989, *SOYGRO v5.42 Soybean Crop Growth Simulation Model: User's Guide*, Florida Agricultural Experimental Station Journal 8304, University of Florida, Gainesville, Fla.

Maas, S.J., and G.F. Arkin, 1980, *TAMW: A Wheat Growth and Development Simulation Model*, Research Center Program and Model Development No. 80-3, Texas Agricultural Experimental Station, Blackland Research Center, Temple, Tex.

Monteith, J.L., et al., 1989, "RESCAP: A Resource Capture Model for Sorghum and Pearl Millet," in S.M. Virmani et al. (eds.), *Modeling the Growth and Development of Sorghum and Pearl Millet*, research bulletin 12, ICRISAT, Andhra Pradesh, India.

Morgan, T.H., et al., 1980, "A Dynamic Model of Corn Yield Response to Water," *Water Resources Research* 16(10):59-64.

Newkirk, K.M., et al., 1989, *User Guide to VT-Maize Version 1.0 (R)*, Virginia Water Resources Resource Center, Virginia Polytechnic Institute and State University, Blacksburg, Va.

Ng, E., and R.S. Loomis, 1984, *Simulation of Growth and Yield of the Potato Crop*, PUDOC, Wageningen, The Netherlands.

Norman, J.M., and G. Campbell, 1983, "Application of a Plant-Environment Model to Problems on Irrigation," *Adv. Irrig.* 2:155-188.

Ritchie, J.T., 1985, "A User-Oriented Model of the Soil Water Balance in Wheat," in W. Day and R.K. Arkin (eds.), *Wheat Growth and Modeling*, Plenum Press, New York, N.Y.

Ritchie, J.T., and G. Alagarswamy, 1989, "Simulation of Sorghum and Pearl Millet Phenology," in *Modeling the Growth and Development of Sorghum and Pearl Millet*, research bulletin 12, S.M. Virmani et al. (eds.), ICRISAT, Andhra Pradesh, India.

Ritchie, J.T., et al., 1989a, *Development of a Barley Yield Simulation Model*, USDA 86-CRSR-2-2867, Michigan State University, East Lansing, Mich.

Ritchie, J.T., et al., 1989b, *A User's Guide to CERES-Maize-V2.10*, International Fertilizer Development Center, Muscle Shoals, Ala.

Rosenthal, W.D., et al., 1989, *SORKAM: A Grain Sorghum Crop Growth Model*, MP-1669, Texas A. Experimental Station, College Station, Tex.

Stapper, M., 1984, *SIMTAG: A Simulation Model of Wheat Genotypes*, University of New England, Department of Agronomy and Soil Sciences, Armidale, New South Wales, Australia.

Stapper, M., and G.F. Arkin, 1980, *CORNF: A Dynamic Growth and Development Model for Maize* (Zea mays L.): *Program and Documentation*, No. 80-2, Texas Agricultural Experimental Station, College Station, Tex.

Stockle, C.O., and G.S. Campbell, 1989, "Simulation of Crop Response to Water and Nitrogen: An Example Using Wheat," *Transactions of the ASAE* 32(1):66-74.

van Keulen, H., and N.G. Seligman, 1987, *Simulation of Water Use, Nitrogen Nutrition, and Growth of Spring Wheat Crop*, Centre for Ag. Publ. and Doc., Wageningen, The Netherlands.

Wilkerson, G.G., et al., 1983, "Modeling Soybean Growth for Crop Management," *Transactions of the ASAE* 26(1):63-73.

Williams, J.R., et al., 1984, "A Modeling Approach to Determining the Relationship between Erosion and Soil Productivity," *Transactions of the ASAE* 27(1):129-144.

Young, J.H., et al., 1979, "A Peanut and Development Model," *Peanut Science* 6:27-36.

APPENDIX E:

GRASSLAND/LIVESTOCK IMPACTS

R. Benioff et al. (eds.), Vulnerability and Adaptation Assessments, E-1–E-58.
© 1996 *Kluwer Academic Publishers.*

APPENDIX E NOTATION

The following is a list of the acronyms, initialisms, and abbreviations (including units of measure) used in this appendix.

ACRONYMS, INITIALISMS, AND ABBREVIATIONS

ACTMO   Agricultural Chemical Transport Model
ARS   Agricultural Research Service
CBCPM   Colorado Beef Cattle Production Model
CREAMS   Chemicals, Runoff, and Erosion from Agricultural Management Systems
DM   dry matter
ELM   Ecosystem Level Model
EPIC   Erosion/Productivity Impact Calculator
GCM   general circulation model
GFDL   Geophysical Fluid Dynamic Laboratory
GIS   geographic information system
GISS   Goddard Institute for Space Studies
LAI   leaf area index
SPUR   Simulation of Production and Utilization of Rangelands
TAMU   Texas A&M University
USDA   U.S. Department of Agriculture
USLE   universal soil loss equation
VGA   Video Graphics Array
WEPP   Water Erosion Prediction Project

CHEMICALS

C   carbon
CO_2 carbon dioxide
K   potassium
N   nitrogen

UNITS OF MEASURE

°C   degree(s) Celsius
cm   centimeter(s)
d   day(s)
ft   foot (feet)
g   gram(s)
h   hour(s)
ha   hectare(s)
in.   inch(es)
kg   kilogram(s)
km   kilometer(s)

km^2	square kilometer(s)
m^2	square meter(s)
Mbyte	megabyte(s)
mg	milligram(s)
mm	millimeter(s)
ppm	part(s) per million

CONTENTS

TABLES

TABLES (Cont.)

FIGURES

APPENDIX E:

GRASSLAND/LIVESTOCK IMPACTS

E.1 DESCRIPTION OF SPUR SIMULATION MODELS

Much of the text in this section has been taken (all or in part) from the *SPUR2 Documentation and User's Guide* (Hanson et al. 1992).

E.1.1 Introduction and Evolution of SPUR2

Rangelands are agricultural systems with relatively low productivity that make up about 40% of the earth's land surface. The host of plant species populating native range provides food and habitat for domestic livestock and many species of wildlife. Complex interactions between system components make understanding the structure and function of rangelands difficult. In recent years, national legislation has induced the U.S. Department of Agriculture (USDA) Agricultural Research Service (ARS) to develop several systems-level computer models from which SPUR2 has evolved (Figure E.1).

Clean-water legislation led to the development of the Agricultural Chemical Transport Model (ACTMO) in 1975 and, subsequently, the Chemicals, Runoff, and Erosion from Agricultural Management Systems (CREAMS) in 1979 (Knisel 1980). In 1984, the Erosion/Productivity Impact Calculator (EPIC) was completed (Williams et al. 1983) in response to the Soil and Water Conservation Act of 1977. The ALMANAC model, an enhancement of EPIC, was developed to analyze regional phenomena. In 1985, the ARS began work on the Water Erosion Prediction Project (WEPP) in an effort to respond to several conservation laws. In 1978, a range-modeling workshop was held to begin planning the development of a general grassland model; the model was officially started in 1981 and given the name SPUR (Simulation of Production and Utilization of Rangelands). Some of the theories used in developing SPUR were modifications and improvements from the Ecosystem Level Model (ELM), which was developed by Innis (1978) under the International Biological Program, and from the CREAMS model.

E.1.2 Description of the SPUR Model

The goals for the SPUR model were to (1) evaluate rangeland systems and provide a basis for decisions on management, (2) optimize systems of rangeland management for desired multiuse products, (3) plan and evaluate practices of land improvement, (4) provide a computational basis for investigating the impacts of environmental modifications on alternative management strategies, and (5) forecast the effects of climate changes on range ecosystems. The model (as released in 1987) was a general grassland simulation model composed of five basic components: hydrology, plant growth, animals (domestic and wildlife),

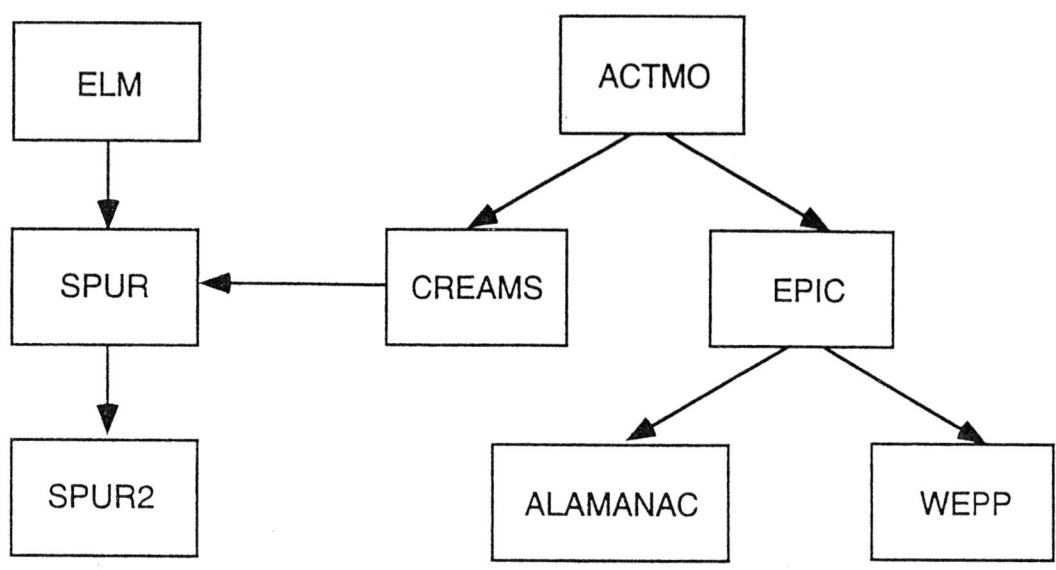

FIGURE E.1 Evolution of the USDA-ARS SPUR2 Rangeland Ecosystem Model

and economics. The model was driven by daily inputs of data on precipitation, maximum and minimum temperatures, solar radiation, and wind run. These variables were derived either from existing weather records or from the use of a stochastic weather generator.

The hydrology component calculated upland surface runoff volumes, peak flow, snowmelt, upland sediment yield, and channel stream flow and sediment yield. Soil-water tensions, used to control various aspects of plant growth, were generated by using a soil-water balance equation. Surface runoff is estimated by the Soil Conservation Service curve number procedure, and soil loss was computed with the modified universal soil loss equation (USLE) (Williams 1975). The snowmelt routine used an empirical relationship between air temperature and energy flux of a dynamic snowpack.

In the plant component, carbon and nitrogen were cycled through several compartments, including standing green, standing dead, live roots, dead roots, seeds, litter, and soil organic matter. Soil inorganic nitrogen was also simulated. The model simulated competition between plant species and the impact of grazing on vegetation. Required initial conditions included the initial biomass content for each compartment.

Domestic livestock physiology and forage harvesting by wildlife were calculated in the animal component. Detailed information on growth was provided for cattle on a steer-equivalent basis by using an adaptation of the Texas A&M University (TAMU) Beef Simulation Model (Sanders and Cartwright 1979). Consumption of forage was calculated for all classes of animals. A unique feature of the model was the development of preference vectors on the basis of forage palatability, abundance, and location to control plant use by animals. Species of wildlife, including insects, were considered as fixed consumers and were given first access to the available forage. Animal production and net gain were used by the economic component to estimate the benefits and costs of alternative grazing practices, range improvements, and animal management.

Initially, two versions of SPUR were developed. Version 1 — a grazing unit or field-scale version — allowed grazing animals to graze different parts of the unit according to various location-preference vectors. Plant dynamics and the impact of grazing on rangeland were simulated in the field-scale version. The field-scale version was capable of simultaneously simulating up to seven plant species growing on up to nine sites, with no restrictions on the size of the sites. Version 2 of SPUR — the basin-scale version — required considerable averaging of animal and plant attributes. The basin-scale version used the watershed as a management unit. Version 2 was intended to give high resolution for runoff, peak flow, sediment yield, and channel hydrology. This version predicted quantities of runoff and sediment yield for basins of up to 26 km^2, containing up to 27 hydrologic units. SPUR Phase I was a prototype of the desired rangeland simulation model and was designed to be used as a research-and-development tool. The projected users were research personnel primarily associated with the USDA, the ARS, and various universities.

E.1.3 Applications of SPUR

SPUR has been subjected to numerous validation tests and to investigative research. As an example, results from studies describing the abiotic and biotic components are discussed subsequently.

E.1.3.1 Abiotic Components

Richardson et al. (1987) successfully validated the weather generator. Renard et al. (1983) tested the hydrology module by conducting a 17-yr (1965-1981) simulation for two Arizona watersheds and obtained high correlation between observed and predicted runoff ($r = 0.88$ and 0.94). Springer et al. (1984) used the hydrology component of SPUR to predict the hydrologic response of three Idaho watersheds over a 5-yr period; the explained variance ranged from 63 to 85%. Springer et al. (1984) concluded that the model did a good job of simulating the timing and amount of monthly runoff but was a poor predictor of erosion and sedimentation. Cooley et al. (1983) tested the snow accumulation and snowmelt components in a watershed in southwestern Idaho. By using parameters derived from the 1980 snow season, these investigators obtained a correlation of 0.91 between observed data (from 1970 to 1972 and from 1977) and model predictions.

E.1.3.2 Biotic Components

Skiles et al. (1983) simulated the growth of the two dominant grasses in the shortgrass steppe of Colorado and concluded that the SPUR plant growth module adequately reproduced the biomass production of the grasses and matched the dynamics of the growing season. In a test of the plant-animal interface, Hanson et al. (1988) showed that the SPUR model correctly predicted weight gains for domestic animals as a function of stocking rate for a Colorado grassland, which is an indirect validation of the plant module. SPUR has successfully predicted animal gains on pastures in West Virginia (Stout et al. 1990). Urie (1990) used the SPUR model to evaluate early- and late-season grazing of crested wheatgrass in central Nevada. The SPUR model has also been used to provide simulated forage for a modified heifer module (Field 1987) and to estimate biomass for a grazing behavior model (Baker et al. 1992).

E.1.4 Description of the SPUR2 Model

Every attempt was made to make SPUR2 look and operate like the SPUR model; however, many changes were made to the code to increase its utility. SPUR2 consists of 20,962 lines of FORTRAN computer code (of which 35% are comment statements). Not all of the routines were modified, and much of the way in which a user interacted with SPUR has been left intact in SPUR2. Following is a discussion of the modules present in the model.

E.1.4.1 Hydrology Submodel

The hydrology component was not changed. Evaporation and transpiration are calculated by using the Ritchie (1972) equations. Potential evaporation (E_o) is computed as a function of daily solar radiation, mean air temperature, and albedo. Potential evaporation at the soil surface (E_{so}) is computed as a function of the leaf area index (LAI) of the field and the amount of litter that covers the soil surface. Actual soil evaporation (E_s) is computed in two stages on the basis of the soil moisture status of the upper layer. Stage 1 evaporation is limited by energy; that is, the soil profile is saturated, and the process depends on the amount of solar radiation reacting with the soil surface. Stage 2 evaporation depends on the movement of water to the soil surface and the amount of water in the soil profile; stage 2 evaporation is probably the more common of the two mechanisms in operation in the arid and semiarid regions of the western United States. Potential plant transpiration (E_{po}) is calculated as a function of E_s and LAI (if LAI < 3.0) or as a function of E_o and E_s (if LAI > 3.0). If soil water is limited, plant transpiration (E_p) is computed as a function of E_{po}, soil water in the root zone, and the field capacity. Solar radiation and the soil-surface area covered by aboveground biomass govern water loss from the soil. The water balance equation used in SPUR follows (Renard et al. 1987). Water balance for a single time step at a single point on a watershed is

$$SW = SW_o + P - (Q + ET + PL + QR) \ , \tag{E.1}$$

where

> SW =   current soil water content,
>
> SW_o =   previous day's soil water content,

P = cumulative precipitation,

Q = cumulative amount of surface runoff,

ET = cumulative amount of evapotranspiration, and

PL = cumulative amount of percolation loss to deep groundwater.

For evaporation from the soil to occur, water must infiltrate through the soil surface and into the soil profile. Any soil water that is not lost to percolation, root uptake, or evaporation remains in the interstitial area between soil particles and is termed "storage." The process of infiltration reduces the amount of water available for the overland flow component of runoff (Q) and increases the amount of water for evapotranspiration (ET) and deep percolation (PL). Once water has percolated through the root zone of the soil, the water is essentially lost to this hypothetical single point on the watershed. Such percolated water may return to the surface down the slope as return flow to another point on the watershed or as flow into a channel. Otherwise, this water remains in an aquifer but is not available for transpiration or evaporation. If precipitation occurs while air temperatures are below freezing, that moisture will not be available for any of the hydrologic processes mentioned previously because it will be tied up in the snowpack. The consequence is a reduction in the effective precipitation for a given time step. This water may show up later once the snow melts; the water will then be available for infiltration and for runoff.

E.1.4.2 Plant Submodel

The SPUR plant model remains largely intact (Hanson et al. 1988). The major enhancements include the development of a one-day time step model and the ability to simulate plant response to elevated concentrations of carbon dioxide (CO_2).

Photosynthesis is the basis for C budgeting in many plant growth models, especially those operating at the ecosystem level (Innis 1978; Detling et al. 1979; Hanson et al. 1988). In these cases, diurnal patterns of the net CO_2 assimilation rate, often based on the rectangular hyperbola (Hanson 1991), are numerically integrated. The model has the form

$$P_c = \frac{a \cdot I_g \cdot P_{max}}{a \cdot I_g + P_{max}} \; . \tag{E.2}$$

Simpson's rule was used in SPUR to integrate Equation E.2. Hanson (1991) demonstrated how the commonly used Thornley light equation (Thornley 1976) could be analytically solved, thus eliminating the use of inefficient numerical integrators. In this model, solar radiation was assumed to be symmetrically distributed around noon. Thus, daily crop photosynthesis was estimated by

$$P_c = P_{max} \int_u^d \frac{S(t)}{P_{max} + a \cdot S(t)} \, dt \quad , \tag{E.3}$$

where

P_{max} = theoretical maximum net assimilation rate;

u and d = times representing the start and end, respectively, of the photosynthetic period; and

a = light-use efficiency coefficient.

The value of $S(t)$ is defined as

$$S(t) = R_{max} \sin\left[\frac{\pi \cdot (t - u)}{PP}\right] \quad , \tag{E.4}$$

where

$S(t)$ = light-flux density of the canopy at time t,

PP = photoperiod, and

R_{max} = average maximum light-flux density for the day within the canopy.

Integrating Equation E.1 results in a three-part solution that depends on the relationship between P_{max} and $a \cdot R_{max}$ (Hanson 1991). Let $r = (P_{max})/a \cdot R_{max}$. Then,

if $\rho > 1$,

$$\beta = \frac{1 - \rho}{\sqrt{\rho^2 - 1}} \cdot \left[\frac{1 - \tan^{-1} \cdot \left(\frac{1}{\sqrt{(\rho^2 - 1)}}\right)}{5.0 \cdot \pi}\right] \quad ; \tag{E.5}$$

if $\rho = 1$,

$$\beta = 1 - \frac{2}{\pi} \quad ; \tag{E.6}$$

or if $\rho < 1$,

$$\beta = 1 + \rho \cdot \frac{\ln\left[\frac{1 - \sqrt{(1 - \rho^2)}}{1 - \sqrt{(1 - \rho^2)}}\right]}{\pi \cdot \sqrt{(1 - \rho^2)}} \quad . \tag{E.7}$$

The total daily crop photosynthesis (P_c), as a function of light-flux density and scaled for the appropriate day length, is calculated as

$$P_c = \beta \cdot R_{max} \cdot PP \ \ .$$
(E.8)

Because the estimate of P_c is calculated on a unit leaf-area basis, the unit ground-area estimate of P_c, as required by large-scale simulation models, can be estimated by multiplying P_c by the total leaf area. To solve the problem, R_{max} can be estimated by using either the Monsi-Saeki equation (Monsi and Saeki 1953) to ignore the transmission coefficient of the leaf (m) or the Saeki equation (1960) to include m. First, if m is ignored, the amount of light at any point within the canopy (I_y) is

$$I_y = I_o e^{-\kappa L} \ \ ,$$
(E.9)

where

I_o = total daily solar radiation incident on the surface of the canopy,

κ = extinction coefficient, and

L = canopy LAI (Monsi and Saeki 1953).

By applying the mean value theorem, the average light throughout the canopy (I_y) is defined as

$$I_y = I_o \frac{\int_0^1 e^{-\kappa L} \, dt}{L} \ \ .$$
(E.10)

Equation E.9 can be integrated to give

$$I_y = I_o \frac{1 - e^{-\kappa L}}{\kappa L} \ \ .$$
(E.11)

Then, from Hanson (1991),

$$R_{max} = \frac{\pi \cdot I_y}{2 \cdot PP} \ \ .$$
(E.12)

Elevated CO_2 was assumed to affect only the net photosynthetic rate of plants. The following assumptions were made for the SPUR2 model:

- Net photosynthesis will increase by 35% as the CO_2 concentration doubles.

- The base ambient CO_2 concentration is 330 parts per million (ppm).

- The intercellular CO_2 concentration is 70 ppm.

- The response is linear.

This routine is rather simple, and a copy is provided.

E.1.4.3 Grazing Animals

Many changes and additions have been made within the animal component for SPUR2. These changes include the addition of a new plant-animal interface, correction of an error in the steer model, inclusion of the Colorado Beef Cattle Production Model (CBCPM), development of a new wildlife model and an appropriate interface, and addition of a two-species grasshopper model.

E.1.4.4 FORAGE Interface Model

FORAGE is a deterministic simulation model that has been developed as an interface between the plant and animal components of a general rangeland ecosystem model (SPUR2). The model predicts intake of forage and the selection of diet by grazing beef cattle by simulating the mechanistic components of grazing behavior. Bite size, rate of biting, rate of intake, and grazing time are modified by changes in the characteristics of the sward. Sensitivity analyses revealed that the model is sensitive to underestimation of critical parameters. The model suggests that, when the availability of forage is limited, intake is limited by rate of intake and grazing time (Baker et al. 1992).

The model was constructed on the basis of the following primary assumptions:

- Forage consisted of a homogeneous mixture of the functional plant groups (i.e., warm- and cool-season grasses, warm- and cool-season forbs, and shrubs) simulated by the plant component of the SPUR2 model. Horizontal distribution of plants within the sward was expressed as the amount of standing biomass.

- Competition among animals did not exist. The amount of standing crop was equal for all animal classes during a time step.

- Distance between feeding stations was not considered in the model framework. Grazing time was assumed to be the amount of time that the animal spent in the head-down grazing position. Intake was not influenced by the distance that the animal would have to travel to reach the next feeding station.

- The time needed to drink, ruminate, and rest occurred after grazing.

- Dietary proportions of functional plant groups were independent of an animal's demand for forage.

The percentage of nitrogen for a forage class was used to calculate a preference ranking for each functional plant group. The equations used in the model to calculate preference are modifications of those derived by Senft (1984) from data on cattle grazing on a shortgrass steppe. The nitrogen content of aboveground standing biomass (in kilograms per hectare) for each functional group is converted to **crude protein** by multiplying the percentage of nitrogen of the available forage by 6.25 (Lloyd et al. 1978).

To allow the user control over the model, an **adjusted preference equation** was constructed to control the selection of forage by the animal. The equation is expressed as

$$ARP_i + RP_i + \delta(MRP - RP_i) \; , \qquad \text{(E.13)}$$

where

ARP_i = adjusted preference,

RP_i = relative preference for the i'th forage class, and

MRP = mean of the relative preferences.

The parameter δ controls selectivity during grazing. With the parameter set to zero, the animal selects forage strictly on the basis of the preference equation described by Senft (1984). If the parameter is set to one, preferences for all forage groups will be the same, and the animal will select its diet on the basis of available forage.

Availability of a forage can be separated into two components: forage that is apparently available, and forage that is actually available to the animal. Standing crop (SC_i) is expressed as the amount of biomass (in kilograms per hectare) of the i'th functional group that is apparently available to the grazing animal. The value of the effective standing crop (ESC_i) (in kilograms per hectare) represents the effective availability of the forage and is calculated by adjusting standing crop for each functional group by the animal's ability to differentiate the i'th functional group from the j'th functional group when selecting a bite.

Therefore, the ESC_i is a mixture of functional groups and is calculated as

$$ESC_i = \sum_{j=1}^{n} (MIXMAT_{ij} \cdot SC_j) \ , \tag{E.14}$$

where $MIXMAT_{ij}$ is the probability that a bite of the j'th functional group will also contain the i'th functional group when densities (in kilograms per hectare) are equal. Values in the $MIXMAT$ matrix range from zero to one and represent bite selection among functional groups.

Relative preference for the i'th mixture ($MIXPREF_i$) is then calculated as a weighted average of adjusted preferences expressed by

$$MIXPREF_i = \frac{\sum\limits_{j=1}^{n} (MIXMAT_{ij} \cdot SC_j \cdot ARP_j)}{ESC_i} \ . \tag{E.15}$$

Bite size is a function of the animal and the physical characteristics of the sward. Under conditions of unrestricted availability, an animal's bite size is limited only by the size of its mouth (Stobbs 1973b; Black and Kenney 1984). Consequently, every animal has a maximum bite size that is a function of the animal's structural size. Maximum bite size ($BSMAX$) is expressed as

$$BSMAX_k = \theta \cdot GCW_k^{0.75} \ , \tag{E.16}$$

where GCW_k ($k = 1, 2, \ldots m$) is the growth curve weight of the k'th animal for that day of the simulation, and θ is the maximum metabolic bite size. Little evidence is found in the literature to suggest what maximum bite size may be. Therefore, maximum metabolic bite size in the model is expressed as a parameter with a value of 11.14 mg of dry matter (DM) per bite per unit of metabolic weight.

When intake is restricted by availability, the animal no longer is able to obtain a maximum bite, and the size of the bite decreases with decreasing availability (Stobbs 1973b; Chacon and Stobbs 1976; Forbes and Hodgson 1985). A rectangular hyperbola calculates the bite size for each functional group mixture:

$$BS_{ik} = \frac{BSMAX_k \cdot ESC_i}{\lambda + ESC_i} \ , \tag{E.17}$$

where BS_{ik} is the bite size for the k'th animal for the i'th functional group, and λ is a shape parameter.

The rate of biting is the number of bites per minute that an animal takes during head-down grazing. The biting rate for the current day of simulation is calculated as

$$RB_{ik} = \psi - \tau \cdot BS_{ik} , \qquad (E.18)$$

where

RB_{ik} =   rate of biting for the k'th animal for the i'th forage mixture,

ψ =   maximum rate of biting, and

τ =   shape parameter.

The rate of intake is subsequently the product of bite size and rate of biting:

$$RI_{ik} = BS_{ik} \cdot RB_{ik} , \qquad (E.19)$$

where RI_{ik} is the rate of intake for the k'th animal for the i'th functional group mixture. The shape of the function for RI_{ik} is the same as that for bite size.

The proportion ($PROP$) of the i'th mixture in the diet is a function of the animal's preference for a mixture ($MIXPREF$), the availability of the mixture (ESC), and the rate of intake (RI) for the mixture. The equation for diet proportion ($PROP_{ik}$) is expressed as

$$PROP_{ik} = \frac{MIXPREF_i \cdot ESC_i \cdot RI_{ik}}{\sum\limits_{i=1}^{n} MIXPREF_i \cdot ESC_i \cdot RI_{ik}} . \qquad (E.20)$$

The mean rate of intake for the k'th animal ($FINRI_k$) for all functional group mixtures selected is calculated as

$$FINRI_k = \sum_{i=1}^{n} (PROP_i \cdot RI_{ik}) . \qquad (E.21)$$

Both final bite size and bite rate are calculated by substituting RI_k for BS_k and RB_k in Equations E.20 and E.21.

Grazing time is calculated as a function of three limits. The first limit is determined by the amount of forage required ($FGDMND_k$) by the animal. Potential grazing time for the k'th animal ($POTGT_k$) is calculated on the basis of the limit imposed by forage demand:

$$POTGT_k = \frac{FGDMND_k}{FINRI_k} . \qquad (E.22)$$

The next limit set by the model is the time that an animal will devote to grazing. During any 24-hour period, a grazing animal has limited time to graze, rest, ruminate, and socialize. Few observations have been made of cattle grazing longer than 13 hours (Stobbs 1970, 1974; Cowan 1975; Arnold and Dudzinski 1978). Consequently, maximum grazing time (ϕ) was set equal to 13 hours. The time limit to grazing ($TLIMGT_k$) is subsequently the minimum (MIN) of the potential grazing time and maximum grazing time:

$$TLIMGT_k = MIN\ (\phi,\ POTGT_k)\ .\qquad\text{(E.23)}$$

The last limit to grazing is bite limit. The number of bites (NOB) that the animal takes is a function of both the time limit to grazing and the final rate of biting. The maximum number of bites recorded in a 24-hour period for cattle is approximately 36,000 bites (Stobbs 1973a; Chacon and Stobbs 1976). $ACTNOB_k$ is the actual number of bites taken by the k'th animal in a day. The bite limit to grazing time for the k'th animal ($BLIMGT_k$) is expressed as

$$BLIMGT_k = \frac{ACTNOB_k}{FINRB_k}\ .\qquad\text{(E.24)}$$

Final grazing time ($FINGT$) is subsequently calculated as

$$FINGT_k = MIN\ (TLIMGT_k,\ BLIMGT_k)\qquad\text{(E.25)}$$

and is multiplied by the final rate of intake to calculate intake for the k'th animal ($INTAKE_k$):

$$INTAKE_k = FINGT_k,\ \cdot\ FINRI_k\ .\qquad\text{(E.26)}$$

Intake is passed back to the animal model, where actual amounts of crude protein and dry matter intake are used for simulating production of body tissues. The forage harvested summed over all animals is subtracted, and results are passed back to the plant model.

E.1.4.5 Steer (Stocker) Model

The basic structure of the steer model has been adapted from the TAMU Beef Simulation Model (Sanders and Cartwright 1979). Further modifications used herein were incorporated into the TAMU Beef Simulation Model by Notter (1977). The grazing season is defined by Julian dates of turnout to pasture and of removal from the pasture. The initial physical and physiological status of the steers is inferred from their age and weight at turnout. Supplemental feed can be offered between input Julian dates. Steers consume all supplemental feed before eating any of the available herbage. Except for the calculation of forage intake, the SPUR2 steer model is identical to the model in SPUR. Forage intake and diet selection in SPUR2 are calculated by the FORAGE interface.

E.1.4.6 Cow-Calf (CBCPM) Model

Recently, the CBCPM, a second-generation beef-cattle production model, which is a modification of the TAMU Beef Simulation Model (Sanders and Cartwright 1979; Bourdon 1983; Bourdon and Brinks 1987), was linked with the SPUR2 model. The CBCPM is a herdwide, life-cycle simulation model and operates at the level of the individual animal. The biological routines of the CBCPM simulate animal growth, fertility, pregnancy, calving, death, and demand for nutrients. Currently, 18 genetic traits related to growth, milk, fertility, body composition, and survival can be studied. The user can determine the size and age distribution of the herd. The breeding season, calving season, date of castration, and day of weaning are also defined by the user within SPUR2. The phenotype of the cattle is a function of user-input breeding values for birth weight, yearling weight, mature weight, milk production, average age of puberty, and length of gestation. The model can be run either stochastically or deterministically by making the appropriate selection in the user interface. Users experienced with the CBCPM can manipulate data files and FORTRAN code for advanced applications. These manipulations include cross-breeding systems, selection studies, management policies, and others.

E.1.4.7 Wildlife Model

SPUR2 allows the user to specify up to 10 different species of wildlife. Dry matter intake (*DMI*) is calculated as

$$DMI = \frac{WT^{0.9} \cdot k}{1,000} \quad ,$$

where *WT* is the mature body weight of the simulated wildlife species, and *k* is the daily intake of DM per unit of metabolic weight (in milligrams per kilogram). The animal is assumed to eat 25 mg/kg of body weight (Minson 1990). Preferences for site and for forage are defined by the user. Wildlife is not physically limited from grazing specific sites. Data for mature body weights in the user interface were obtained from Chapman and Feldhamer (1982).

E.1.4.8 Grasshopper Model

The grasshopper population dynamics submodel is called on a daily basis from SPUR2. The model simulates daily activities of two types of spring-emerging grasshoppers: "grass feeders," which feed exclusively on grass species present, and "mixed feeders," which feed on either grasses or forbs in proportion to their relative frequency. Grasshoppers are more competitive grazers than cattle are and can drastically reduce carrying capacity. The demand of grasshoppers for forage is likely to be highest when production of forage is lowest. SPUR2 simulates a dynamic grasshopper population. The user must specify the date of emergence and the density for grasshopper larvae. Grasshopper density within the year and subsequent grasshopper damage exponentially decrease. Therefore, if control is necessary, action should be taken early in the season. SPUR2 allows control by using either carbaryl or malathion.

E.1.5 Resource Requirements

The resource requirements necessary to conduct each step of the climate change analysis are presented in Table E.1. The SPUR2 model is truly an interdisciplinary model; therefore, a modeling team should be established with members representing the scientific disciplines of animal sciences, range science, soil science, and agricultural economics. This team should be responsible for providing the scientific guidance and technical expertise during the modeling effort, conducting the biophysical and economic analyses, and testing adaptive management strategies.

E.2 METHOD FOR USE OF SPUR2 MODEL

E.2.1 Installation of SPUR2

The SPUR2 distribution package consists of a *Documentation and User's Guide* and one 3.5-in. high-density disk. Make sure that \SPUR2 is in your path. The installation package did not update your AUTOEXEC.BAT file, so you must do that. (See the subsequent note concerning compiling and linking the SPUR2 program.)

TABLE E.1 Resource Requirements for Each Step of the Climate Change Analysis

Step	Personnel Skill Level[a]		Time Requirement (person-hours)
	Technical	Scientific	
Site selection	(+)	+	40
Input data collection	(+)	+	80
Model parameterization and validation	(+)	+	120
Simulation runs	+	-	20
Biophysical analysis	(+)	*	80
Economic analysis	(+)	*	80
Testing adaptive strategies	(+)	*	40

[a] - = not necessary, + = necessary, (+) = desirable but not essential, and * = absolutely necessary.

The installation software copies SPUR2 programs and data to your hard drive and creates all necessary subdirectories automatically. The package requires you to specify the device for the floppy and the hard drives for installation:

1. Insert the disk into a high-density floppy disk drive.

2. From DOS, select the 3.5-in. drive as the default device. With the assumptions that your 3.5-in. device is B: and that the hard drive onto which you want to install SPUR2 is device C, type

B:INSTALL B C <ENTER>

Wait for the installation to be complete.

All SPUR2 programs and supporting data will be loaded onto the hard disk into the SPUR2 directory. Subdirectories will be created for the system software, default data files, and other support programs (\BIN); for the source code (\CODE); for user projects (\PROJECTS); for climate files (\CLIMATE); and for other important documentation (\DOCUMENT). Figure E.2 shows the subdirectories for SPUR2.

If you are running DOS 6.0 or higher, you can delete the files \spur2\choice.com and \spur2\move.exe. If you have FORTRAN Powerstation on your machine, you can delete the files \spur2\dosxnt.386 and \spur2\dosxmsf.exe.

To execute SPUR2 from DOS, type:

SPDOS *projname projfile,*

where *projname* is the name of the project you wish to use, and *projfile* is the SPUR2 input data file. Each project will be stored in a subdirectory under the name *projname*. Default data will be copied into the project subdirectory. These data can be edited by using the data file generator program.

E.2.1.1 Naming Conventions

The SPUR2 input data file is named *<projfile>*.PDF. Other input data to SPUR2, primarily for the cow-calf (CBCPM) model, are in various files of the form TAPE?.DAT. Output files for SPUR2 have an .SDF extension.

E.2.1.2 Compiling and Linking SPUR2

SPUR2 was written with Microsoft FORTRAN Powerstation. Powerstation is a full-featured, 32-bit FORTRAN programming language. Powerstation supports all of the language features of 16-bit FORTRAN but runs in Microsoft Windows and produces executables for a 32-bit MS-DOS-extended environment. Data file handling routines for Powerstation FORTRAN have been included in the code.

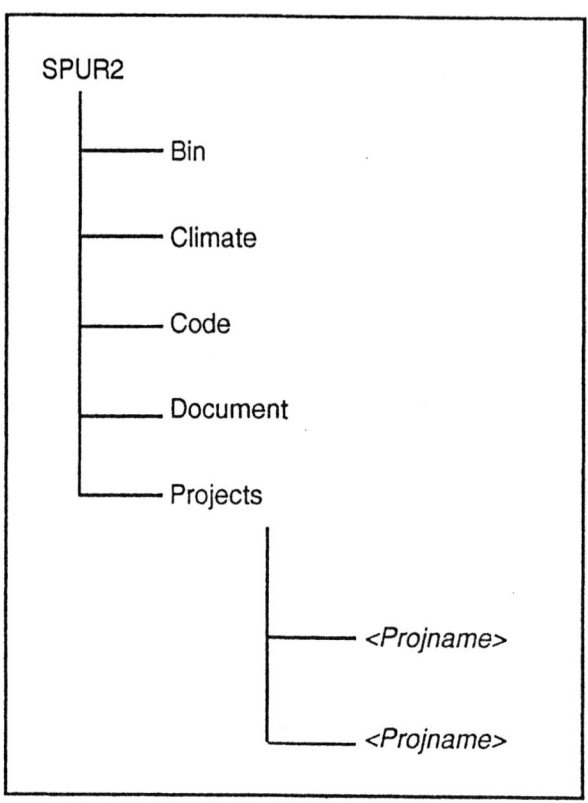

FIGURE E.2 Subdirectory Structure for SPUR2

Therefore, to modify the code, recompile, and link, the user must use Powerstation FORTRAN. To run Powerstation FORTRAN from DOS, the following statements should be added to the AUTOEXEC.BAT file:

SET LIB=<*device*>:\F32\LIB
SET INCLUDE=<*device*>:\F32\INCLUDE ,

where <*device*> is the drive unit where Powerstation FORTRAN was installed. (Powerstation FORTRAN supplies a BAT file called F32ENV.BAT in \F32\BIN that sets these environment variables. It is recommended that the statements be added directly to the AUTOEXEC.BAT file.) Once code has been modified, execute MAKE to recompile and link the new code. MAKE will execute the Microsoft program NMAKE and use the make file SPUR2.FMK. Only modified files will be recompiled. If Powerstation FORTRAN is being run from the Windows environment and memory errors occur, it is recommended that Powerstation from DOS be used. In case a different FORTRAN compiler is being used, the include files from Powerstation FORTRAN have been placed in \SPUR2\BIN.

E.2.1.3 Windows

SPUR2 can be run from a DOS prompt in Windows. DOSXNT.386 is a Windows device driver that is necessary if you want to run SPUR2 in an MS-DOS session under

Windows. To install DOSXNT.386, you need the following entry in the SYSTEM.INI file (usually in c:\windows) under the section [386Enh]:

$$\text{device=}<device>:\backslash\text{spur2}\backslash\text{dosxnt.386}\ ,$$

where <*device*> is the drive onto which SPUR2 was installed (e.g., c or d).

E.2.1.4 Hardware Requirements

The hardware requirements follow:

- 386/33 with 4 Mbytes of RAM,

  (**Note:** If Microsoft Windows is to be used, a minimum of 8 Mbytes of RAM is required. Also, a 486/66 CPU computer with 8 Mbytes of RAM would enable faster processing of the model.)

- 200-Mbyte hard drive,

- 3-1/2-in. floppy drive,

- 5-1/4-in. floppy drive, and

- Super Video Graphics Array (VGA) monitor and Trident or Trident-compatible video card.

E.2.1.5 Software Requirements

The software requirements follow:

- DOS 6.0 or higher,

- Microsoft Windows 3.1 (optional but recommended), and

- Microsoft FORTRAN Powerstation for Windows.

  (**Note:** Other FORTRAN compilers may be used, but they must be able to create a virtual link. If other compilers are used, then *major code revisions* will be required in MAIN.FOR source code.)

E.2.2 Input and Output Data

This section describes the data required and output files generated by the SPUR2 suite of models.[1] A list of the input and output files either required or generated by the model is found in Table E.2.

---

[1] Much of the text in this section has been taken (all or in part) from the original SPUR documentation and user guide (Wight and Skiles 1987).

TABLE E.2 Input and Output Files Generated or Required by SPUR2

File No.	File Name	File No.	File Name
1	TAPE1.DAT	46	T46.SDF
2	TAPE2.DAT	48	T48.SDF
3	T^APE3.DAT	50	T50.SDF
4	TAPE4.DAT	52	T52.SDF
5	TAPE5.DAT	54	T54.SDF
6	TAPE6.DAT	56	T56.SDF
8	TAPE8.DAT	58	T58.SDF
10	DATEFILE	60	T60.SDF
11	HOPPER.SDF	62	T62.SDF
12	CARBON.SDF	70	SCRATCH.SDF
13	PJFILE	71	T71.SDF
14	NITRO.SDF	72	T72.SDF
16	SUMARY.SDF	74	YEARSUM.SDF
18	T18.SDF	76	T76.SDF
20	OUTPUT.SDF	78	T78.SDF
21	IDENT.SDF	80	T80.SDF
22	DISP.SDF	82	T82.SDF
23	SINGLE.SDF	84	T84.SDF
24	MULT.SDF	86	T86.SDF
25	NUTRCOW.SDF	88	T88.SDF
26	NUTRCLF.SDF	90	T90.SDF
27	NUTRFP.SDF	91	WEIGHT.SDF
28	REPORT.SDF	92	SOILMST.SDF
29	CATINOUT.SDF	93	COMMAND
30	NUTRIN.SDF	94	PLNTOUT.SDF
42	T42.SDF	95	PEAKPLT.SDF
44	T44.SDF	99	COMMAND

E.2.2.1 Definitions

A *simulation study area* is a region or area where a simulation study is to be conducted.

A *range site* is an area of rangeland where climate, soil, and relief are sufficiently uniform to produce a distinct plant community. A range site is the product of all of the environmental factors responsible for its development. The site is typified by an association of species that differ from those on other range sites in kind or proportion of species or total production.

A *functional group* is a group of plant species that respond in a physiologically uniform manner to abiotic and biotic factors in the environment. Groupings are made by photosynthetic pathway and physical structure.

E.2.2.2 Overview of Data Input

SPUR2 allows the simulation of 15 plant species or functional groups (instead of 7) on 36 range sites (instead of 9). The default plant parameter set has been modified. The number of data files has also been condensed for SPUR2. In SPUR, data files were required for hydrology, plants, animals, and climate. In SPUR2, model parameters are included in a single, documented data file created by a DOS-Windows user interface. Thus, a climate file and a parameter are all that are necessary for the model.

E.2.2.3 User Interface

The interface has several menus that allow for user-specific input. Data entry is broken into five basic categories:

- Control parameters,

- Non-site-specific parameters,

- Site-specific parameters,

- Plant growth parameters, and

- Animal growth parameters.

E.2.2.3.1 Control Parameters. This set of menus is used to set up the simulation and to control which output files are printed.

E.2.2.3.2 Non-Site-Specific Parameters. Data entered in this set of menus affect the entire simulation site. Categories of parameters in these fields include positional information, snowmelt parameters, decomposition parameters, and climatic adjustment factors.

E.2.2.3.3 Site-Specific Parameters. As the name implies, these parameters are specific to each range site that is being simulated. In this menu, range sites can be added, modified, or deleted. All soil parameters and initial conditions for the range site are entered in these menus. If a new soil is defined, it can be saved to the "SOIL.DAT" database for future use.

E.2.2.3.4 Plant Growth Parameters. Plant species or functional groups are added, modified, or deleted in this menu. If a new plant or functional group is added or an existing one is modified, the user will have the option of accepting the default parameter set or entering new plant parameters. After the modifications, the new data set can be stored in the "PLANT.DAT" database for future use. As a cautionary note, parameters should only be changed if the current value is invalid for the plant being simulated. Further information is provided in Section E.2.4.1 on the parameterization process of the plant growth model.

E.2.2.3.5 Animal Growth Parameters. All parameters for the steer (stocker), cow-calf, wildlife, and insect models are entered in these menus. If the cow-calf model is used, additional parameters can be edited with the DOS editor. It is strongly recommended that the files for the cow-calf model (TAPE1.DAT through TAPE8.DAT) be manipulated only through the interface or by users experienced with the CBCPM. (**Note:** The data in TAPE?.DAT are read with *formatted* read statements in the model; therefore, if the data are changed, they need to be right justified in the column.)

E.2.2.4 Input Data Necessary for SPUR2

Data for climate, soils, plants, and animals are needed as input data for the SPUR2 model. The following subsections describe the type of data needed for each category. The quality of the data should be assessed before the simulations are conducted. The user should ensure that all data are in the correct units and that values are reasonable for the parameter being used. However, SPUR2 has an error-check subroutine that will provide the user with a message should an unreasonable value be entered.

E.2.2.4.1 Climate Data. The SPUR2 model requires five daily input weather variables for the length of the simulation. These variables are maximum and minimum temperatures (in degrees Celsius), precipitation (in centimeters), solar radiation (in langleys), and wind run (in kilometers per day). All climate files must reside in the <*device*>:\SPUR2\CLIMATE directory.

E.2.2.4.2 Soil Characteristic Data. As stated previously, SPUR2 can simulate up to 36 range sites. The following data are needed for each range site being simulated:

- Soil texture class and representative composition of clay, silt, and clay (percentages of each);

- Estimate of canopy type and height, percent cover, and percentage of cover that contacts the ground;

- Estimate of soil organic matter (in grams per square meter);

- Number of soil layers;

- Slope, slope length, and aspect;

- Elevation; and

- Latitude and longitude.

For each soil layer on each range site, the following data are needed:

- Depth of each layer,

- Soil texture class (percent composition), and

- Estimate of rooting depth.

The following data can be estimated from soil texture if empirical data are not available. The procedure for estimating these parameters is discussed in Section E.2.4 in more detail.

- Estimate of saturated-soil hydraulic conductivity (in centimeters per hour),

- Estimate of bare-soil evaporation (in millimeters per day$^{1/2}$), and

- Estimates of total porosity and of water-holding capacity at -1/3 and -15 bar.

E.2.2.4.3 Plant Data. SPUR2 can simulate 15 species or functional groups; however, the default set of parameters is set to five functional groups that include C_3 and C_4 grasses, C_3 and C_4 forbs, and shrubs. Following are the minimum site-specific data required for using the default set of parameters; we highly recommend that novice users of SPUR2 use the default number of functional groups:

- Julian dates for beginning of seed production and for beginning and ending of senescence for each functional group;

- Estimates of initial root, propagule, standing live, and standing dead biomass (in grams per square meter) by functional group;[2] and

- Estimates of initial soil inorganic nitrogen, dead roots, litter, and soil organic matter (in grams per square meter).[2]

E.2.2.4.4 Animal Data. As stated previously, the beef-cattle livestock models in SPUR2 are the CBCPM and a steer version of the TAMU model. The CBCPM simulates a cow-calf production system at the level of individual animals. The stocker beef-cattle or steer

---

[2]   These parameters can be estimated from model runs if data do not exist (Sections E.2.4.1.2 and E.2.4.1.3).

model will simulate the growth of animals that are approximately 6-18 months of age. Parameter data needed for these models include the following:

- Birth weight, yearling weight, and mature weight;

- Milk production, average age at puberty, and length of gestation;

- Management dates for weaning, calving, and marketing; and

- Estimate of supplemental forage fed during the year.

E.2.2.5 Output Data Files Created by SPUR2

Output data files are controlled from the user interface under the menu title "Control Parameters."[3]

E.2.2.5.1 Printer Flag Control, Page 1. Output for plant and hydrology data are generated from the following menus.

1. Print Switch for Daily Report (=1)

    I. Unit = 16 File = Sumary.Sdf

    II. Output: Day, Year, Precipitation, Max and Min Temperature, Solar Radiation, Snow Water Equivalent, a Report on a Per-Site Basis for Potential Evaporation, Leaf-Area-Index, Potential Transpiration, Potential Soil Evaporation, Root Depth, Available Water, and Aboveground Standing Biomass Live and Dead

2. Print Switch for Monthly and Annual Report

    I. Unit = 16 File = Sumary.Sdf

    II. Output: Monthly and Yearly Totals on a Field and for Each Site for Precipitation, Infiltration, Potential Evaporation, Transpiration, Potential Soil Evaporation, Deep Percolation, and Plant-Available Water; Month-End Aboveground Live Biomass, Month-End Livestock and Wildlife Intake, Livestock Weight, and Economics Report

3. Print Switch for Water Flow Output

=1 Evaporation and Transpiration

---

[3] Much of the text in this subsection has been taken (all or in part) from the original SPUR documentation and user guide (Wight and Skiles 1987).

I.   Unit = 18   File = T18.Sdf

II.   Daily Output for Year, Site, Day, Eo, Eso, Es, Epo, Ep, and Total Site Leaf Area

=2 Moisture Tension and Temperature

I.   Unit = 71   File = T71.Sdf

II.   Daily Output for Year, Site, Day, Moisture Tension in Upper 15 cm, Moisture Tension in Wettest Layer, 10-Day Running Average Air Temperature, and 5-Day Running Average Soil Moisture Tension in Upper 15 cm

=3 Soil Water Content

I.   Unit = 72   File = T72.Sdf

II.   Output is written on a per-layer, per-site, per-day basis in centimeters. The upper two 3-in. layers are summed and written in the first column.

4.   Print Switch for Plant Carbon

=1 Yearly Summary of Plant Production

I.   Unit = 74   File = Yearsum.Sdf

II.   Output is written on a species-per-site basis.

=2 In addition to 1, intermediate variables are printed.

=3 Daily Output of Plant Parameters

I.   Units = 76, 78, ... 90   Files = T76.Sdf, T78.Sdf, ... T90.Sdf

II.   Daily Output Written to Respective Files: Translocation from Roots to Shoots, Translocation from Shoots to Roots, Net Photosynthesis, Effect of Moisture on Photosynthesis, Effect of Temperature on Photosynthesis, Effect of Soil Nitrogen on Photosynthesis, Effect of Temperature on Decomposition, and Effect of Moisture on Decomposition

=4 Daily Output of Carbon States

I.   Unit = 12   File = Carbon.Sdf

II. Daily Output Written on a Per-Species Basis for Site, Standing Live, Live Roots, Propagules, and Standing Dead Carbon; and on a Per-Site Basis: Dead Roots, Litter, Organic Matter, and Inorganic Nitrogen

>4 Interval for Output of Carbon State Variables

I. Same as previous, but output is on simulated days that are multiples of the switch value.

E.2.2.5.2 Printer Flag Control, Page 2. Output files for plant N, livestock, wildlife, and insect reports are generated from these menus.

1. Print Switch for Plant Nitrogen

=1 Yearly Summary

I. Unit = 42 File = T42.Sdf

II. Output Written on Per-Site Basis

=2 Daily Output of Selected Variables

I. Unit = 44, 46, ... 52 File = T44.Sdf, T46.Sdf, ... T52.Sdf

II. Daily Output Written to Respective Files: Effect of Moisture on Nitrogen Uptake, Effect of Temperature on Nitrogen Uptake, Nitrogen Fixed for the Atmosphere, Effect of Moisture on Denitrification, and Mineralized Nitrogen

=3 Daily Output of State Variables

I. Unit = 14 File = Nitro.Sdf

II. Output Written Daily per Site per Species: Site, Species, Nitrogen in Standing Live, N:C Ratio of Standing Live, Nitrogen in Live Roots, and N:C in Live Roots; and on a Per-Site Basis: Nitrogen in Dead Roots, Nitrogen in Litter, Nitrogen in Organic Matter, and Inorganic Nitrogen

>3 Interval for Output of State Variables

I. Same as previous, but output is on simulated days that are multiples of the switch value

2. Print Switch for Livestock Component

=1 Yearly Summary

    I.   Unit = 54 File = T54.Sdf

    II. Output Written on a Yearly Basis: Total Weight Gain for Season, Summed Digestible-Dry-Matter Intake, and Dry Matter Harvested

=2 Report of Harvested Forage by Forage Class

    I.   Unit = 56 File = T56.Sdf

    II. Daily Output for the Amount of Forage Harvested from Each Species

=3 Daily Output of Diet (for Steer Model Only)

    I.   Unit = 58 File = T58.Sdf

    II. Daily Output for Year, Day, Percent Crude Protein, Digestibility of Dry Matter Harvested, Amount of Forage Harvested, and Average Steer Weight

=4 Daily Output of Live and Dead Forage Harvested

    I.   Unit = 60 File = T60.Sdf

    II. Year, Site, Day, and Harvest of Forage per Forage Species per Site

3. Print Switch for Wildlife Component

=1 Daily Report of Forage Harvested by Site and Species

    I.   Unit = 62 File = T62.Sdf

4. Print Switch for Insect Component

=1 Yearly Report of Densities, Forage Harvested, and Survival

    I.   Unit = 11 File = Hopper.Sdf

E.2.2.5.3 Printer Flag Control, Page 3. Selected reports for the cow-calf model are generated from these menus.

1. Identify Report

=1 Daily Output Written for Individual Animals: Id Number, Dam and Sire Id Numbers, Sex, Birth Day and Year, and Birth Weight and Gestation Length

     I.   Unit = 21   File = Ident.Sdf

2. Dispose Report

=1 Output for the Disposition of Individual Animals

     I.   Unit = 22   File = Disp.Sdf

3. Nutrition Report

=1 Output Cow Nutrition

     I.   Unit = 25   File = Nutrcow.Sdf

     II.   Output on a Per-Day Per-Animal Basis for Id, Year, Month, Day, Percent Chemical Fat, Intake Grazed Forage, Weight, GCW, Delta Weight, Intake of Hay, Intake Protein Supplement, Intake Energy Supplement, Milk Production, and Forage Digestibility

E.2.3 Parameterization Process

Much of the text in this subsection has been taken (all or in part) from the original SPUR documentation and user guide (Wight and Skiles 1987).

E.2.3.1 Hydrology Submodel

Except for the snowmelt parameters, all of the hydrology model can be parameterized from the "Site Specific Parameters" screen. The following discussion lists the steps to parameterize a site.

E.2.3.1.1 Soil Descriptors and Hydrologic Parameters. The following steps are used to determine soil description and hydrologic parameters.

1. Select area, and enter the value in acres.

2. Identify the number and soil texture class for each layer. **Note:** Soil texture class can be estimated from Table E.3 if percentages of clay, silt, and sand are known.

3. Either add or modify a layer (maximum of eight layers).[4]

4. Values for porosity and for volumetric water content at -1/3 bar and -15 bar are taken from Table E.4.

5. The value for saturated hydraulic conductivity (K_{sat}) is taken from Table E.3. **Note:** The values for K_{sat} in Table E.3 are presented in centimeters per hour, and the value needed in the user interface is required as inches per hour.

6. Record the depth of the layer in inches.

7. Repeat steps 4-6 for each layer.

TABLE E.3 Selected Soil Properties on the Basis of Soil Texture Class

Soil Texture	Representative Composition (%)			Saturated-Soil Hydraulic Conductivity (cm/h)			Bare-Soil Evaporation (mm/d$^{1/2}$)		
	Clay	Silt	Sand	Avg[a]	Low	High	Avg	Low	High
Sand	3	7	90	23.0	11.7	43.21	3.3	3.05	3.32
Loamy sand	5	15	80	6.1	3.6	11.76	3.3	3.05	3.32
Sandy loam	10	20	70	2.2	1.7	3.6	3.5	3.10	4.06
Loam	20	40	40	1.3	0.91	1.7	4.5	3.20	4.57
Silt loam	15	65	20	0.69	0.46	0.91	4.5	3.20	4.57
Silt	5	87	8	0.51	0.30	0.61	4.0	3.15	4.40
Sandy clay loam	30	10	60	0.30	0.25	0.46	3.8	3.15	4.32
Clay loam	35	35	30	0.20	0.19	0.25	3.8	3.15	4.32
Silty clay loam	35	55	10	0.18	0.15	0.19	3.8	3.15	4.32
Sandy clay	45	5	50	0.13	0.11	0.15	3.4	3.10	3.56
Silty clay	45	50	5	0.10	0.09	0.11	3.5	3.10	3.81
Clay	65	20	15	0.08	0.06	0.09	3.4	3.10	3.56

[a] Avg = average.

---

[4] To remove the impact of moisture inputs of relatively small magnitude, the soil between 0 and about 15 cm (0.0 and 6.0 in.) should be divided into two layers. The preferred thickness of each layer is about 7.5 cm (3.0 in.). This division provides a smoother response of the soil-water tension and a more appropriate response of the plant model (Springer and Lane 1987).

TABLE E.4 Hydraulic Properties

Soil Texture	Total Porosity			-1/3-bar Water-Holding Capacity			-15-bar Water-Holding Capacity		
	Avg[a]	Low	High	Avg	Low	High	Avg	Low	High
Sand	41	39	43	9	7	15	3	2	6
Loamy sand	43	39	45	12	10	20	6	4	8
Sandy loam	45	39	52	20	14	29	9	5	12
Loam	47	45	52	26	20	36	12	9	18
Silt loam	50	49	55	31	20	36	13	7	20
Silt	51	49	55	28	26	30	9	6	12
Sandy clay loam	42	38	45	27	17	34	17	11	21
Clay loam	47	40	51	34	29	38	20	16	34
Silty clay loam	47	46	51	36	33	40	21	18	24
Sandy clay	42	40	44	31	27	40	21	18	30
Silty clay	48	46	49	40	35	46	27	23	32
Clay	49	44	52	42	34	49	29	23	38

[a] Avg = average.

E.2.3.1.2 Initial Nitrogen and Biomass Conditions. The parameters for initial soil inorganic N, dead roots, litter, and soil organic matter for a particular site either (1) come from literature values or field measurements from the site or (2) can be estimated by running the model with a long climate (≥30 years) for the site. To perform this estimate:

1. Make sure that Section E.2.4.1.4 has been completed before continuing this step.

2. Once the soil characteristics for the site have been parameterized, run the model for 30 or more simulated years.

3. Edit the NITRO.SDF file from the simulation.

4. Use the ending values for soil inorganic N, dead roots, litter, and soil organic matter as initial conditions.

E.2.3.1.3 Initial Carbon State Variables. The parameters for initial root, propagule, standing live, and standing dead biomass for a particular site either (1) come from literature values or field measurements from the site or (2) can be estimated by running the model with a long climate (≥30 years) for the site. To perform this estimate:

1. Make sure that Section E.2.4.1.4 has been completed before continuing this step.

2. Once the soil characteristics for the site have been parameterized, run the model for 30 or more simulated years.

3. Edit the CARBON.SDF file from the simulation.

4. Use the ending values for root, propagule, standing live, and standing dead biomass as initial conditions.

E.2.3.1.4 Universal Soil Loss Equation Parameters. The USLE was developed by the USDA to predict average annual soil losses from sheet and rill erosion by rainfall and its associated runoff on specific field slopes (USDA 1976). The process for estimating the USLE parameters needed for SPUR2 is outlined as follows:

1. The condition 1 curve number is taken from Table E.5 for the specific range site.

2. The *K* factor is taken from Table E.6.

3. The *P* factor is always set to equal 1.0.

4. The *C* factor is taken from Table E.7.

5. The slope length factor is obtained from Table E.8.

6. Rooting depth is the effective depth over which plants can extract water by transpiration. The values must be less than or equal to the top of the bottom soil layer.

7. The soil evaporation factor is obtained from Table E.3. **Note:** The value for bare-soil evaporation is presented in Table E.3 as millimeters per $day^{1/2}$ and must be converted to inches per $day^{1/2}$ for entry into the user interface.

8. The aspect of the simulation site is obtained from field observations.

9. The slope of the site is obtained from field observations.

10. The fraction of field capacity in initial storage, the value for crop factor, and crack factors remain as the default unless empirical evidence exists to change these parameters.

TABLE E.5 Runoff Curve Number

Range Site	Runoff Curve Number for Range Condition		
	Poor	Fair	Good
Wetland	95	95	95
Very shallow	95	90	85
Saline subirrigated	90	90	85
Subirrigated	90	90	85
Shale	90	85	80
Dense clay	90	85	80
Alkali clay	90	85	80
Saline upland	90	95	90
Igneous	90	90	75
Shallow clayey	85	80	75
Shallow sandy	80	75	70
Shallow loamy	80	75	70
Shallow igneous	80	75	70
Steep clayey	80	75	70
Clayey	80	75	65
Gravelly loamy	80	75	65
Steep loamy	80	75	65
Overflow	80	70	60
Loamy overflow	80	70	60
Clayey overflow	80	70	60
Coarse upland	80	70	60
Limy upland	80	70	60
Shallow breaks	80	70	60
Stony	80	70	60
Steep stony	80	70	60
Lowland	80	70	60
Saline lowland	80	70	60
Loamy lowland	80	65	55
Loamy	80	65	55
Sandy lowland	75	60	50
Sandy	75	60	50
Gravelly	70	55	45
Sands	70	55	40
Choppy sands	80	55	40

TABLE E.6 Gross Approximations of *K* Factor

Texture Class	*K* Factor for Organic Matter Content		
	<0.5% K	2% K	4% K
Sand	0.05	0.03	0.02
Fine sand	0.16	0.14	0.10
Very fine sand	0.42	0.36	0.28
Loamy sand	0.12	0.10	0.08
Loamy fine sand	0.24	0.20	0.16
Loamy very fine sand	0.44	0.38	0.30
Sandy loam	0.27	0.24	0.19
Fine sandy loam	0.35	0.30	0.24
Very fine sandy loam	0.47	0.41	0.33
Loam	0.38	0.34	0.29
Silt	0.60	0.52	0.42
Silt loam	0.48	0.42	0.33
Sandy clay loam	0.27	0.25	0.21
Clay loam	0.28	0.25	0.21
Silty clay loam	0.37	0.32	0.26
Sandy clay	0.14	0.13	0.12
Silty clay	0.25	0.23	0.19
Clay	-	0.13-0.29	-

E.2.3.2 Plant Submodel

Forty-three parameters are found in the SPUR2 plant growth model; however, for most situations, the model can be parameterized with adjustments to the following set of parameters. These parameters are located in the submenus of the "Plant Growth Parameters" screen:

1. *Root respiration* — the proportion of root biomass that can be respired on a given day. Increases in this parameter cause decreases in plant biomass, total root-to-shoot translocation, and overall root mortality. Carbon lost through respiration is lost to the organic C pool of the model. In general, grasses have the highest respiratory rates. Shrubs are set slightly lower than grasses, and forbs are given the lowest susceptibility to respiratory loss. Changes to this parameter should be small.

TABLE E.7 *C* Factor for Permanent Pasture, Range, and Idle Land

Vegetation Canopy and Height[a]	Ground Cover (%)[b]	Type[c]	*C* Factor for Percent Ground Cover					
			0%	20%	40%	60%	80%	95%+
No appreciable canopy	-	G	0.45	0.20	0.10	0.042	0.013	0.003
		W	0.45	0.24	0.15	0.091	0.043	0.011
Tall weeds or short brush with average drop fall height of 20 in.	25	G	0.36	0.17	0.09	0.038	0.013	0.003
		W	0.36	0.20	0.13	0.083	0.041	0.011
	50	G	0.26	0.13	0.07	0.035	0.012	0.003
		W	0.26	0.16	0.11	0.076	0.039	0.011
	75	G	0.17	0.10	0.06	0.032	0.011	0.003
		W	0.17	0.12	0.09	0.068	0.038	0.011
Appreciable brush or bushes with average drop fall height of 6.5 ft	25	G	0.40	0.18	0.09	0.040	0.013	0.003
		W	0.40	0.22	0.14	0.087	0.042	0.011
	50	G	0.34	0.16	0.08	0.038	0.012	0.003
		W	0.34	0.19	0.13	0.082	0.041	0.011
	75	G	0.28	0.14	0.08	0.036	0.012	0.003
		W	0.28	0.17	0.12	0.078	0.040	0.011
Trees, but no appreciable low brush; average drop fall height of 13 ft	25	G	0.42	0.19	0.10	0.041	0.013	0.003
		W	0.42	0.23	0.14	0.089	0.042	0.011
	50	G	0.39	0.18	0.09	0.040	0.013	0.003
		W	0.39	0.21	0.14	0.087	0.042	0.011
	75	G	0.36	0.17	0.09	0.039	0.012	0.003
		W	0.36	0.20	0.13	0.084	0.041	0.011

[a]   Canopy height is measured as the average fall height of water drops falling from the canopy to the ground. Canopy effect is inversely proportional to drop fall heights and is negligible if fall height exceeds 33 ft.

[b]   Portion of total area surface that would be hidden from view by canopy in a vertical projection.

[c]   G = Cover surface is grass, grasslike plants, decaying compacted duff, or litter at least 2 in. deep; W = cover at surface is mostly broad-leaved herbaceous plants (as weeds with little lateral-root network near the surface) or undecided residues or both.

TABLE E.8 Slope Length Parameter

Slope (%)	Slope Length Parameter for Slope Length (ft)											
	25	50	75	100	150	200	300	400	500	600	800	1,000
0.5	0.07	0.08	0.09	0.10	0.11	0.12	0.14	0.15	0.16	0.17	0.19	0.20
1	0.09	0.10	0.12	0.13	0.15	0.16	0.18	0.20	0.21	0.22	0.24	0.26
2	0.13	0.16	0.19	0.20	0.23	0.25	0.28	0.31	0.33	0.34	0.38	0.40
3	0.19	0.23	0.26	0.29	0.33	0.35	0.40	0.44	0.47	0.49	0.54	0.57
4	0.23	0.30	0.36	0.40	0.47	0.53	0.62	0.70	0.76	0.82	0.92	1.00
5	0.27	0.38	0.46	0.54	0.66	0.76	0.93	1.10	1.20	1.30	1.50	1.70
6	0.34	0.48	0.58	0.67	0.82	0.95	1.20	1.40	1.50	1.70	1.90	2.10
8	0.50	0.70	0.86	0.99	1.20	1.40	1.70	2.00	2.20	2.40	2.80	3.10
10	0.69	0.97	1.20	1.40	1.70	1.90	2.40	2.70	3.10	3.40	3.90	4.30
12	0.90	1.30	1.60	1.80	2.20	2.60	3.10	3.60	4.00	4.40	5.10	5.70
14	1.20	1.60	2.00	2.30	2.80	3.30	4.00	4.60	5.10	5.60	6.50	7.30
16	1.40	2.00	2.50	2.80	3.50	4.00	4.90	5.70	6.40	7.00	8.00	9.00
18	1.70	2.40	3.00	3.40	4.20	4.90	6.00	6.90	7.70	8.40	9.70	11.00
20	2.00	2.90	3.50	4.10	5.00	5.80	7.10	8.20	9.10	10.00	12.00	13.00
25	3.00	4.20	5.10	5.90	7.20	8.30	10.00	12.00	13.00	14.00	17.00	19.00

2. *Root mortality* — the maximum percentage of root biomass susceptible to mortality per day. Values of 0.5% or less should be used. Changes to this parameter should be small. Grasses are given the most rapid turnover rate, followed by forbs and then shrubs.

3. *Root-to-shoot ratio* — the maximum live-root-to-shoot ratio; for example, a value of 8.0 means that 8 g of live roots is required to support 1 g of green shoots. Increases in this parameter cause a decrease in peak standing crop but an overall increase in total plant biomass. Values for grasses have a much higher root biomass than those for shrubs, and shrub values are higher than those for forbs. Use observed data, if available.

4. Julian dates for beginning of seed production and for the beginning and ending dates for senescence for each functional group are required.

   a. The Julian day that senescence begins is assumed to occur on or around the day of peak standing crop.

   b. The Julian day that senescence ends marks the end of the growing season, although regrowth can occur. This parameter is one of the more sensitive parameters for controlling peak standing crop (MacNeil et al. 1985).

   c. At this time, SPUR2 does not simulate annual plants; therefore, the date for seed production is merely a placeholder. Estimated values are appropriate here.

The following steps are an outline for parameterizing and "fine-tuning" the model to a site:

1. Develop an initial set of conditions as described previously and a useful set of indicator variables to monitor the model's performance.

2. Run SPUR2. Determine if soil organic matter is approximately constant. In a 10-yr simulation, soil organic matter should be approximately the same at the end of the simulation as it was at the beginning.

3. Adjust decomposition rates for soil organic matter.

4. Determine if peak standing crop and the percent composition of the various functional groups are correct. If standing crop and functional group compositions are not in agreement with observed data, adjust the previous parameters. Repeat steps 2 and 3.

5. Study the dynamics of litter and dead roots. If large accumulations or losses occur in either, adjust the decomposition rates accordingly. Go to step 4.

6. If changes have been made to the plant parameters, go to step 2.

7. Begin to fine-tune the model by looking at the various indicator variables. Adjust the parameters so that the dynamics, both in magnitude and timing, agree with current knowledge of the system. Repeat the previous steps as necessary.

E.2.3.3 Forage Interface Model

Parameters for the forage intake model are accessed through the "Animal Growth Parameters" screen under the steer and cow-calf selections.

E.2.3.3.1 Grazing Behavior. Unless the user has empirical data to the contrary, these parameters are to be used as the default.

E.2.3.3.2 Mix Matrix. As explained in Section E.1.4.4, diet is primarily selected on the basis of the quality of the diet; however, the mix matrix is another mechanism for parameterizing the model on the basis of the physical characteristics of the canopy and the animal's preference for a particular functional group. The off-diagonal elements of the matrix determine diet selection on the basis of canopy structure and the animal's ability to differentiate among the various functional groups being simulated; for example, if the animal is selecting the live component of a warm-season grass, the off-diagonal element represents the probability that a bite will also contain the live component of a cool-season grass, assuming that they are present and that the animal could graze both if it so desired. If the structure of the canopy was such that the two functional groups grew so close to one another that the animal could not separate the two, then a value of 1.0 would be entered in off-diagonal element 1,3 of the matrix. If the converse were true (i.e., that the two functional groups were separated spatially such that the animal could not take a bite of each), the value would then be 0.0.

The diagonal elements describe the animal's preference for a particular forage item. These preference values also range from 0.0 to 1.0. Estimates for the diagonal parameters can be obtained from field data on diet selection.

E.2.3.3.3 Site Preference. The values for site preference range from 0.0 to 100.0. Many biotic and abiotic factors determine why animals prefer one site over another. The user must have some idea of where animals prefer to graze when working with a multisite simulation. Although an animal has a preference for a site, whether the animal has physical access to the site is determined by setting the "Grazed" option to "yes" or "no"; for example, an animal may have a strong preference for a particular site, but the site may be in an enclosure. Therefore, the preference may be set to 100, but the grazed option would be set to "no."

    E.2.3.3.4 Grazing Season. The grazing season and the type of grazing management system are controlled from this screen. By definition, more than one site is needed to parameterize the model for any grazing system other than continuous grazing. The following is an example of how to parameterize for a deferred rotational grazing system:

1. Draw a map of the simulation study area that shows how many simulation sites are in the study area, if this has not been done previously. Figure E.3 has five range sites (a-e).

2. Overlay pasture boundaries. Four pastures (1-4) are found on Figure E.3.

3. Label the sites within each pasture. The study area has 11 simulation sites (1-11).

4. Enter the length of the grazing season (Table E.9).

5. Enter the type of grazing management.

6. Enter the sites within each pasture.

7. Enter the dates for grazing in each pasture.

    In the deferred rotational system described previously, pasture 1 is deferred from grazing until year 2. In year 2, pasture 1 will be grazed during the periods 191-210 and 251-270 while pasture 2 is deferred. Pastures 3 and 4 will be grazed during periods 150-170 and 211-230 and periods 171-190 and 231-250, respectively. The pattern would repeat for subsequent years.

    An example of a rotational grazing system is provided in Table E.10. In this system, the period of grazing remains the same for each pasture every year. In other words, for every year of the simulation, pasture 1 will be grazed during the periods 150-158, 186-194, 222-230, and 258-266.

E.2.3.4 Steer (Stocker) Model

    The parameters needed for the steer model are self-explanatory. The user will need to enter the number of animals being simulated, the average genotype being simulated, whether supplemental feed is provided, and the age of the animals when they begin to graze.

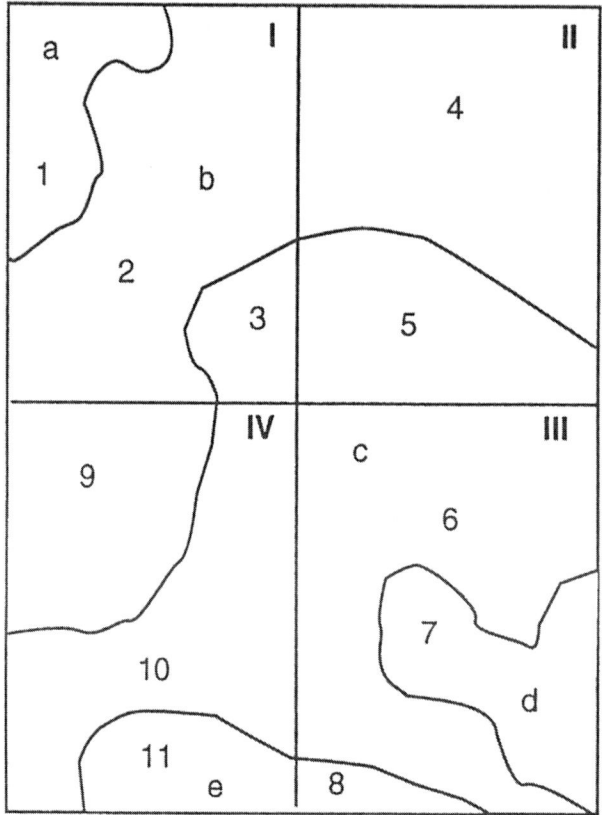

FIGURE E.3 Example of a Four-Pasture Grazing System

TABLE E.9 Sample Menu for a Deferred Rotational Grazing System

Grazing Season 150-270		[ ] Continuous grazing [ ] Rotational grazing [x] Deferred rotational grazing			
		Grazing Intervals			
Pastures	Sites Begin-End	No. 1 Begin-End	No. 2 Begin-End	No. 3 Begin-End	No. 4 Begin-End
1	1-3	0-0	0-0		
2	4-5	150-170	211-230		
3	6-8	171-190	231-250		
4	9-11	191-210	251-270		

TABLE E.10 Sample Menu for a Rotational Grazing System

Grazing Season 150-293		[ ] Continuous grazing [x] Rotational grazing [ ] Deferred rotational grazing			
		Grazing Intervals			
Pastures	Sites Begin-End	No. 1 Begin-End	No. 2 Begin-End	No. 3 Begin-End	No. 4 Begin-End
1	1-3	150-158	186-194	222-230	258-266
2	4-5	159-167	195-203	231-239	267-275
3	6-8	168-176	204-212	240-248	276-284
4	9-11	177-185	213-221	249-257	285-293

E.2.3.5 Cow-Calf (CBCPM) Model

Most of the parameters needed for the cow-calf model are self-explanatory.[5] The user should enter the data in each field that appropriately describe the animal and herd dynamics being simulated. The data files TAPE1.DAT through TAPE8.DAT can be modified by the more experienced user. The CBCPM is a very complex research tool. The intent of this document is to present enough information so that a user may run the model to examine very simple herd simulations.

Note: At this time, the field for entering supplemental feeding for the cow-calf model is not connected. Therefore, the user will have to edit TAPE3.DAT to reflect supplemental feeding. If the default feed groups are not adequate, additional feeding rules may have to be added to the SUBROUTINE in VSRC.FOR.

Feed groups are defined with Boolean logic rules in the subroutine FEED, which is located in the source code VSRC.FOR. These groups correspond to groups of animals that would normally be fed together as a group. The CBCPM is currently dimensioned for up to 20 feed groups. The default is set to 10 groups:

1. Nursing calves,

2. Orphan calves,

3. Starting animals on feed,

4. Animals on feed,

---

[5] Much of the text in this subsection has been taken (all or in part) from AN681 class notes as presented by Dr. Richard M. Bourdon (1992) at Colorado State University.

5.   Animals on pasture (nonreplacement),

6.   Yearling heifer replacements,

7.   Bred heifers,

8.   Two-year-old cows,

9.   Older cows, and

10.   Sires.

The number of feed groups must correspond to the number specified in NOFGPS in TAPE3.DAT.

Seven supplemental feeds are simulated in the model:

- Energy supplement — ESUP;

- Protein supplement — PSUP;

- Creep feed — CREP;

- Rations 1, 2, and 3 — RAT1, RAT2, and RAT3; and

- Harvested forage — HFOR.

RAT1 is considered a finishing ration, RAT3 is a starter or growing ration, and RAT2 is something in between those two rations. The HFOR could be considered to be hay, silage, or whatever harvested forage is fed. The nutritional definitions of the various supplemental feeds are set in TAPE3.DAT.

The CBCPM allows the user to feed fixed levels of supplement, to feed according to body condition (fat-feed), or to feed fixed levels of some of the feeds and variable levels of others. Fat-feeding is generally more appropriate for comparing biological types of animals and to prevent starvation in simulations where management is fixed. *The default is set to fat-feed.*

Target values for empty body fat at month's end are supplied for each feed group and month in the matrix at the bottom of TAPE3.DAT. If fat-feeding is off for a group, these values are ignored.

1. Set the feeding seasons for the supplemental feeds.

2. Select the target fat levels such that supplementation occurs only when needed.

E.2.4 Model Validation

The process of model validation involves testing the model to prove that it is an adequate representation of reality. Because the real system is never known completely, the model can never be an exact representation of the real system. Therefore, validation can only be approached, but never achieved (Neelamkavil 1987). The primary concern of the process of model validation is the degree of fit between the model's output and theoretical or experimental data. Statistical analyses such as analysis of variance, regression, spectral analysis, factor analysis, and chi-square tests are useful for comparing model output with validation data sets.

1. Select indicator variables. Indicator variables are model-derived state or intermediate variables used to test the hypotheses under examination. These variables are determined from scientific precedence, economic importance, or the objectives of the simulation study.

2. Develop the appropriate parameter sets from experimental or theoretical data sets that include representative curves and raw data for the selected indicator variables.

3. Run the model.

4. Conduct a statistical comparison of model output with observed data.

The types of data needed for validation of the plant and animal models are listed in the following subsections.

E.2.4.1 Plant Production

Data needed for validation include the following:

- Percent composition of vegetation for C_3 and C_4 grasses, C_3 and C_4 forbs, and shrubs;

- At minimum, total annual biomass production or total peak standing crop; however, estimates of each functional group's total production or peak standing crop would be useful; and

- Monthly production by functional group through the growing season.

E.2.4.2 Animal Production

Data for validation are as follows:

- Diet composition by functional group through the grazing season,

- Intake of grazed forage through the grazing season,

- Estimate of diet quality through the grazing season,

- Estimate of supplemental forage fed during the year,

- Average daily gain over the grazing period, and

- Estimates of body fat or body condition before and after the grazing season.

E.3 SETTING UP SPUR2 FOR CLIMATE CHANGE SCENARIOS

Data required to conduct the climate change analysis include the following:

- A 30-year historical weather data file for each simulation study area. The format for the data set is described in Section E.2.2.4.1.

- General circulation model (GCM) adjustment statistic for monthly average temperature, precipitation, and solar radiation for each simulation study area. **Note:** At least three equilibrium GCM scenarios for which the CO_2 levels have been doubled (referred to as $2XCO_2$) should be simulated to more adequately describe the sensitivity of the system to climatic perturbations.

E.3.1 Nominal Run

The "Nominal" run is conducted with an unaltered historic climate file:

- Copy all climate files for each study area to the *<device>*:\SPUR2\CLIMATE directory. **Note:** To avoid confusion, name each climate file to reflect the simulation study area.

- Set the length of the simulation to 30 years.

- Enter the appropriate climate file name when prompted by the model.

- Rename output files with a "*.NOM" extension for future use.

E.3.2 $2XCO_2$ Runs

The $2XCO_2$ run is conducted in a manner similar to the "Nominal" run:

- Set the length of the simulation to 30 years.

- Change the value for CO_2 from 330 to 550 ppm in the "Positional Information Screen" in the user interface.

- Enter the appropriate climate file name when prompted by the model.

- Rename output files with a "*.DBL" extension for future use.

E.3.3 General Circulation Model Scenario Runs

Use the following procedure for all GCM scenario runs:

- Set the length of the simulation to 30 years.

- Choose the nearest grid point to a simulation site.

- Change the value for CO_2 from 330 to 550 ppm in the "Positional Information Screen" in the user interface.

- Enter the GCM adjustment statistic for each month and each parameter in the "Climate Adjustment" screen. **Note:** Leave the value for wind set to 1.0 for all runs.

- Run the model.

- Enter the appropriate climate file name when prompted by the model.

- Rename output files with a "*.<GCM *abbreviation*>" extension for future use; for example, a run using the Goddard Institute for Space Studies (GISS) GCM would be "*.GIS," or a run from the Geophysical Fluid Dynamics Laboratory (GFDL) model would be "*.GFD."

E.4 RESULTS GENERATED

To capture the effect of changes in biotic processes due to climatic influences, output data for the indicator variables should be collected daily, weekly, or monthly. Data to be analyzed include monthly averages for the following:

- Peak standing crop (in grams per square meter),

- Transpiration (in millimeters),

- Water use efficiency,

- Potential evapotranspiration (in millimeters),

- Precipitation (in centimeters),

- Soil organic matter (in grams per square meter),

- Carbon-to-nitrogen ratio,

- Digestibility of diet,

- Intake of grazed forage (in kilograms per head),

- Forage-to-supplement ratio,

- Body condition scores,

- Milk production (in kilograms per head), and

- Weight at market (in kilograms per head).

Other specific country or regional data may be needed.

Peak standing crop, transpiration, water use efficiency, potential evapotranspiration, and precipitation are used to examine the potential effects of climatic change on plant and plant-soil moisture interactions. Soil organic matter is used to monitor the status of below-ground nutrient sources. The carbon-to-nitrogen ratio of the aboveground biomass is used to indicate the change in plant tissue quality.

Diet digestibility, intake of grazed forage, forage-to-supplement ratio, and body condition scores are used to evaluated the effect of climatic change on feed intake. Milk production is used to monitor the potential effects of the change in calf performance. Market weights of the animals are used in the economic analysis to determine the economic impact of climatic change.

The modeled values for the indicator variables listed above should be collected for the nominal run, $2XCO_2$ run, and the GCM scenario runs for each site being simulated. The results can be analyzed as described below. These data will be useful as baseline data in the adaptation and mitigation analyses.

E.4.1 Analysis

Multiple comparisons between the various runs, such as Scheffe's multiple comparison procedure, should be conducted to test main effects (year and scenario) on all indicator variables for each simulation site. Data generated from the representative sites should also be analyzed for within-year trends over the simulation period. These data will be used to demonstrate the effect of within-year timing of climate change events, such as earlier or later precipitation or temperature increases.

A geographic information system (GIS) would be useful for demonstrating the spatial component of the impacts (Burke et al. 1991; Baker et al. 1993). If a GIS is available, output data could be interpolated by using kriging, inverse distance weighing, or other appropriate techniques to demonstrate geographic or regional trends.

E.4.2 Economic Analysis

Economic analyses can be conducted to examine the effects of climatic change on the local or producer level, at a more aggregated level to examine the effects on secondary supply and demand within a region or country, and extrapolated even further to explore the effect of world supply and demand of meat animal products produced from grasslands. Several model-generated indicator variables can be used to conduct an economic analysis; however, four variables would have utility for most analyses:

1. Total weight of males sold,

2. Total weight of females sold,

3. Total weight of culled[6] animals sold, and

4. Total weight of supplemental feed used.

Steps in conducting an economic analysis include the following:

- Establish market prices (dollars per kilogram) as live-weight bases for the three classes of animals for all scenarios.

- Multiply market price times the total weight for each class.

- Standardize by dividing by the number of hectares being simulated (dollars per hectare).

- Sum the values for each site.

- Compare results from the climate change simulations to the nominal run. **Note:** The result gives an estimated reduction in the marketed value of rangeland production.

- Enter results in Table E.11.

In some production systems, supplemental feed represents as much as 16% or more of the cash costs for cow-calf production systems. Therefore, changes in the amount of supplemental feed consumed may be an important component in some analyses. Other country- or region-specific economic indicator variables may be needed in the analysis.

---

[6] Culled animals are those animals that are sold because they are too old, have health problems, or are no longer considered to be necessary to maintain the herd.

TABLE E.11 Marketed Output and Values

Site and Item	Males	Females	Cull	Total
Site name				
Nominal run (kg/ha)				
Value; nominal run ($/ha)				
Climate change scenario[a] (kg/ha)				
Value; climate change scenario ($/ha)				
Change in value (%)				

[a] Repeat for as many climate change scenarios as are being simulated.

Note: The figures calculated previously represent the potential impact of climate change on private returns for ranchers in each area. These figures could be aggregated to qualitatively estimate economic impacts of the potential climate change on regional and national livestock production.

E.5 CONDUCTING SIMULATIONS OF ADAPTATIONS TO CLIMATE CHANGE

Specific changes or adaptations will depend on the direction and magnitude of the impact from climate change for the region being simulated. Each option listed subsequently can be simulated with the SPUR2 model by changing parameters in the input data files.

E.5.1 Grazing Season

Changes to the length and dates for the grazing season are made from the "Animal Growth Parameters" set of menus. The method for changing grazing seasons is outlined in Section 2.4.3.4.

E.5.2 Grazing Systems

Changes in the management of grazing systems are made from the "Animal Growth Parameters" set of menus. The methods for changing grazing systems are outlined in Section 2.4.3.4.

E.5.3 Stocking Rates

Two methods are available for changing stocking rates in the SPUR2 model. Either the amount of land that is being simulated can be changed in the "Site Specific" menu, or the number of animals being simulated can be adjusted in the "Animal Growth Parameters" set of menus.

E.5.4 Animal Genotype

For most simulations, animal genotype can be modified from the "Animal Growth Parameters" set of menus. The types of changes that can be made to the steer (stocker) model and the cow-calf (CBCPM) model are outlined in the following subsections.

E.5.4.1 Steer (Stocker) Model

The changes to genotype in this model are limited to structural size:

- Select the "Frame Size" option from the "Steer Parameters" screen.

- Choose a frame size that represents the animal to be simulated.

- Enter the projected 400-day weight of the animal.

- Run the model with the new parameter set.

- Analyze results as outlined previously.

E.5.4.2 Cow-Calf (CBCPM) Model

Because the CBCPM was designed to be a breeding and selection tool, many options are available to the user for changing the genotype of simulated animals. **Note:** Only those changes that are changed from the user interface will be outlined in this document at this time.

- Select the "Genetic Parameters" option from the "Cow-Calf Parameters" screen.

- Enter the appropriate breeding values.

- For climate change simulations, set genetic variation to "Fixed Seed."

- For climate change simulations, set additional animal variation to "Off."

- Run the model with the new parameter set.

- Analyze results as outlined previously.

Additional adaptive strategies to be tested will depend on the current and accepted practices of livestock management within a country or region.

E.6 CAVEATS, LIMITATIONS, AND UNCERTAINTIES

The major disadvantages for using a biophysical simulation approach for this problem are as follows:

- Complete data sets for parameterizing the model rarely exist.

- The use of point models requires simplifying assumptions about spatial heterogeneity when results are aggregated to the regional level.

- Management practices will remain constant over the simulated period.

This type of approach has several advantages:

- Biophysiological simulation models are designed to mechanistically simulate ecological and physiological processes.

- These type of models are useful for integrating the nonlinear effects of climate change.

- Process-driven models can be applied to many different environments.

- The models can also be used to test the sensitivity and stability of the system to a range of changes in climatic conditions.

E.7 APPENDIX E REFERENCES

Arnold, G.W., and M.L. Dudzinski, 1978, *Ethology of Free Ranging Domestic Animals*, Elsevier Scientific Publishing Company, Amsterdam, The Netherlands.

Baker, B.B., et al., 1992, "FORAGE: A Simulation Model of Grazing Behavior for Beef Cattle, *Ecological Modeling* 60:257-279.

Baker, B.B., et al., 1993, "The Potential Effects of Climate Change on Ecosystem Processes and Cattle Production on U.S. Rangelands," *Climatic Change* 25:97-117.

Black, J.L., and P.A. Kenney, 1984, "Factors Affecting Diet Selection of Sheep: II. Height and Density of Pasture," *Australian Journal of Agricultural Research* 35:365-378.

Bourdon, R.M., 1983, *Simulated Effects of Genotype and Management on Beef Production Efficiency*, Ph.D. dissertation, Colorado State University, Fort Collins, Colo.

Bourdon, R.M., and J.S. Brinks, 1987, "Simulated Efficiency of Range Beef Production: I. Growth and Milk Production; II. Fertility Traits; III Culling Strategies and Nontraditional Management Strategies," *Journal of Animal Science* 65:943-969.

Burke, I.C., et al., 1991, "Regional Analysis of the Central Great Plains," *Bioscience* 41:685-692.

Chacon, E., and T.H. Stobbs, 1976, "Influence of Progressive Defoliation of a Grass Sward on the Eating Behavior of Cattle," *Australian Journal of Agricultural Research* 27:709-727.

Chapman, J.A., and G.A. Feldhamer, 1982, *Wild Mammals of North America: Biology, Management, and Economics,* Johns Hopkins University Press, Baltimore, Md.

Cooley, K.R., et al., 1983, "SPUR Hydrology Component: Snowmelt," in J.R. Wight (ed.), *SPUR — Simulation of Production and Utilization of Rangelands: A Rangeland Model for Management and Research,* miscellaneous publication 1431, pp. 45-61, U.S. Department of Agriculture.

Cowan, R.T., 1975, "Grazing Time and Pattern of Grazing of Friesian Cows on a Tropical Grass-Legume Pasture," *Australian Journal of Experimental Agriculture and Animal Husbandry* 15:32-37.

Detling, J.K., et al., 1979, "A Simulation Model of *Bouteloua gracilis* Dynamics on the North American Shortgrass Prairie," *Oecologia* 38:167-191.

Field, L.B., 1987, *Simulation of Beef-Heifer Production on Rangeland,* thesis, Colorado State University, Fort Collins, Colo.

Forbes, T.D.A., and J. Hodgson, 1985, "Comparative Studies of the Influence of Sward Conditions on the Ingestive Behavior of Cows and Sheep," *Grass and Forage Science* 40:69-77.

Hanson, J.D., 1991, "Integration of the Rectangular Hyperbola for Estimates of Daily Net Photosynthesis," *Ecological Modeling* 58:209-216.

Hanson, J.D., et al., 1988, "A Multi-Species Model for Rangeland Plant Communities," *Ecological Modeling* 44:89-123.

Hanson, J.D., et al., 1992, *SPUR2 Documentation and User's Guide,* Great Plains Systems Research Technical Report 1, U.S. Department of Agriculture, Fort Collins, Colo.

Innis, G.S. (ed.), 1978, *Grassland Simulation Model,* Springer-Verlag, New York, N.Y.

Knisel, W.G. (ed.), 1980, *CREAMS: A Field-Scale Model for Chemicals, Runoff, and Erosion from Agricultural Management Systems,* Conservation Research Report 26, U.S. Department of Agriculture.

Lloyd, L.E., et al. (eds.), 1978, *Fundamentals of Nutrition, 2,* W.H. Freeman and Company, San Francisco, Calif.

MacNeil, M.D., et al., 1985, "Sensitivity Analysis of a General Rangeland Model," *Ecological Modeling* 29:57-76.

Minson, D.J., 1990, *Forage and Ruminant Nutrition*, Academic Press, San Diego, Calif.

Monsi, M., and T. Saeki, 1953, "Uber dem Lichtfaktor in den Planzengesellschaften und seine Bedeutung für die Stoffproduktion," *Japanese Journal of Botany* 14:22-52.

Neelamkavil, F., 1987, *Computer Simulation and Modeling*, John Wiley & Sons, New York, N.Y.

Notter, D.R., 1977, *Simulated Efficiency of Beef Production for a Cow-Calf Feedlot Management System*, Ph.D. dissertation, University of Nebraska, Lincoln, Neb.

Renard, K.G., et al., 1987, "Hydrology Component: Upland Phases," in J.R. Wight and J.W. Skiles (eds.), *SPUR: Simulation of Production and Utilization of Rangelands: Documentation and User Guide*, ARS-63, pp. 17-30, U.S. Department of Agriculture, Agricultural Research Service.

Renard, K.G., et al., 1983, "SPUR Hydrology Component: Upland Phases," in J.R. Wight (ed.), *SPUR — Simulation of Production and Utilization of Rangelands: A Rangeland Model for Management and Research*, miscellaneous publication 1431, pp. 17-44, U.S. Department of Agriculture.

Richardson, C.W., et al., 1987, "Climate Generator," in J.R. Wight and J.W. Skiles (eds.), *SPUR: Simulation of Production and Utilization of Rangelands: Documentation and User Guide*, ARS-63, pp. 3-16, U.S. Department of Agriculture, Agricultural Research Service.

Ritchie, J.T., 1972, "Model for Predicting Evaporation from a Row Crop with Incomplete Cover," *Water Resources Research* 8:1204-1213.

Saeki, T., 1960, "Interrelationships between Leaf Amount, Light Distribution, and Total Photosynthesis in a Plant Community," *Bot. Mag. Tokyo* 73:55-63.

Sanders, J.O., and T.C. Cartwright, 1979, "A General Cattle Production Systems Model: I. Description of the Model," *Agricultural Systems* 4:217-227.

Senft, R.L., 1984, "Modeling Dietary Preferences of Range Cattle," *Proceedings of the Western Section of the American Society of Animal Science* 35:192-195.

Skiles, J.W., et al., 1983, "Simulation of Above and Below Ground Carbon and Nitrogen Dynamics of *Bouteloua gracilis* and *Agropyron smithii*," in W.K. Lauenroth et al. (eds.), *Analysis of Ecological Systems: State-of-the-Art in Ecological Modeling*, Elsevier Scientific Publishing Co., New York, N.Y.

Springer, E.P., et al., 1984, "Testing the SPUR Hydrology Component on Rangeland Watersheds in Southwest Idaho," *Transactions of the American Society of Agricultural Engineers* 27:1040-1046, 1054.

Stobbs, T.H., 1973a, "The Effect of Plant Structure on the Intake of Tropical Pastures: I. Variation in Bite Size of Grazing Cattle," *Australian Journal of Agricultural Research* 24:809-819.

Stobbs, T.H., 1973b, "The Effect of Plant Structure on the Intake of Tropical Pastures: II. Differences in Sward Structure, Nutritive Value, and Bite Size of Animals Grazing *Setaria anceps* and *Chloris gayana* at Various Stages of Growth," *Australian Journal of Agricultural Research* 24:821-829.

Stobbs, T.H., 1970, "Automatic Measurement of Grazing Time by Dairy Cows on Tropical Grass and Legume Pastures," *Tropical Grasslands* 4(3):237-244.

Stobbs, T.H., 1974, "Components of Grazing Behavior of Dairy Cows on Some Tropical and Temperate Pastures," *Proceedings of the Society for Animal Production* 10:299-302.

Stout, W.L., et al., 1990, "Use of the SPUR Model for Predicting Animal Gains and Biomass on Eastern Hill Land Pastures," *Agricultural Systems* 34:169-178.

Thornley, J.R.M., 1976,

Urie, S.C., 1990, *The Evaluation of SPUR 1: Early and Late Season Grazing of Crested Wheatgrass in Central Nevada*, M.S. thesis, University of Nevada, Reno, Nev.

USDA, 1976, *Control of Water Pollution from Cropland*, Vol. 1, ARS-H-5-1, U.S. Department of Agriculture, Agricultural Research Service.

Wight, J.R., and J.W. Skiles (eds.), 1987, *SPUR: Simulation of Production and Utilization of Rangelands: Documentation and User Guide*, ARS 63, U.S. Department of Agriculture, Agricultural Research Service.

Williams, J.R., 1975, "Sediment Yield Prediction with Universal Equation Using Runoff Energy Factor," in *Present and Perspective Technology for Predicting Sediment Yield and Sources*, ARS-S-40, pp. 244-252, U.S. Department of Agriculture.

Williams, J.R., et al., 1983, "EPIC — A Model for Assessing the Effects of Erosion on Soil Productivity," in W.K. Lauenroth et al. (eds.), *Analysis of Ecological Systems: State-of-the-Art in Ecological Modeling*, pp. 553-572, Elsevier Scientific Publishing Co., New York, N.Y.

E.8 APPENDIX E BIBLIOGRAPHY

Springer, E.P., and L.J. Lane, 1987, "Hydrology-Component Parameter Estimation," in J.R. Wight and J.W. Skiles (eds.), *SPUR: Simulation of Production and Utilization of Rangelands: Documentation and User Guide*, ARS-63, pp. 260-275, U.S. Department of Agriculture, Agricultural Research Service.

Thornley, J.R.M., 1990, *Mathematical Models in Plant Physiology*, Academic Press, Inc., New York, N.Y.

APPENDIX F:

FOREST IMPACTS

R. Benioff et al. (eds.), Vulnerability and Adaptation Assessments, F-1–F-38.
© 1996 *Kluwer Academic Publishers.*

APPENDIX F NOTATION

The following is a list of the acronyms, initialisms, and abbreviations (including units of measure) used in this appendix. Some acronyms used only in tables are defined in those tables.

ACRONYMS, INITIALISMS, AND ABBREVIATIONS

ABET	actual evapotranspiration
AP	actual precipitation
DBH	diameter
FC	field capacity
GCM	general circulation model
GFDL	Geophysical Fluid Dynamics Laboratory
GIS	geographical information system
GISS	Goddard Institute for Space Studies
IIASA	International Institute for Applied Systems Analysis
OSU	Oregon State University
PET	potential evapotranspiration
UKMO	United Kingdom Meteorological Office
WP	wilting point
$1XCO_2$	single levels of CO_2
$2XCO_2$	doubling of CO_2 levels

CHEMICAL

CO_2	carbon dioxide

UNITS OF MEASURE

°C	degree(s) Celsius
cm	centimeter(s)
ha	hectare(s)
km	kilometer(s)
m	meter(s)
mm	millimeter(s)
yr	year(s)

CONTENTS

TABLES

FIGURES

APPENDIX F:

FOREST IMPACTS

F.1 INTRODUCTION

A brief overview of an approach for evaluating the potential response of forested ecosystems to a global climate change is given in Section 5.3. The purpose of this appendix is to provide a more detailed presentation and discussion of methods, data, models, and procedures associated with the approach outlined in Section 5.3. The approach represents only a subset of the methods currently being applied to evaluate the potential impacts of a global climate change on terrestrial vegetation. It was selected because the models involved in the assessment procedure are relatively simple to parameterize and apply to various forest types and regions. Although many models relate the distribution, abundance, and dynamics of plant species (and communities) to features of the environment, they often require a wealth of very specific data (both species and site descriptions) for application to a given site or region.

This appendix is divided into three main sections: (1) direct analysis of relevant bioclimatic variables, (2) mapping of potential land cover by using bioclimatic models, and (3) forest gap models. Section F.2 discusses climate databases. These databases contain various climatic variables that are important for evaluating the distribution and abundance of forest as a function of climate. The databases include descriptions of both current climate and patterns of climate change as predicted by a number of general circulation models (GCMs).

Section F.3 presents the application of the Holdridge life zone classification model (Holdridge 1967), a climate-vegetation classification model, to evaluate the impacts of a climate change on patterns of land cover. Section F.4 examines the use of forest gap models to predict changes in composition and productivity of forest stands in response to a climate change. Section F.5 discusses the integration of the results obtained from the bioclimatic and forest gap models. Section F.6 identifies the resource requirements needed for use of these models.

F.2 INPUTS TO FOREST MODELS: RELEVANT BIOCLIMATIC VARIABLES

F.2.1 Current Climatic Patterns

F.2.1.1 Monthly Temperature and Precipitation

Variations in temperature and precipitation play a major role in influencing the distribution and productivity of terrestrial vegetation (Walter 1985). Temperature directly affects basic biochemical and physiological processes, whereas precipitation is the major input to soil moisture in most ecosystems. In addition to the direct effects on plants, temperature and soil moisture also directly influence rates of decomposition (Meentemeyer 1987), which, in turn, influence nutrient availability to plants. Because of the importance of these two primary climatic variables to plant processes, these variables are critical inputs to virtually all models of plant distribution and dynamics. The location and number of meteorological stations and the temporal resolution (e.g., hourly, daily, monthly) at which these variables are monitored and reported vary from region to region. As a minimum resolution database for all regions, the climate database of Leemans and Cramer (1991), developed at the International Institute for Applied Systems Analysis (IIASA), is recommended for input to forest models. This database provides global coverage at a spatial resolution of 0.5° × 0.5° longitude and latitude. A full description of the database, including the sources of meteorological station data from which the database was constructed, is in Leemans and Cramer (1991).

F.2.1.2 Estimates of Potential Evapotranspiration

Evapotranspiration is the flux of water from a terrestrial surface resulting from both evaporation and transpiration. Potential evapotranspiration (PET) is the rate of evapotranspiration that would occur at a site as a function of the temperature and irradiance if sufficient (but not excessive) moisture is available. A number of models have been developed to estimate evapotranspiration. In general, these models vary in the degree of complexity with which transpiration is estimated. For this study, the Priestly and Taylor (1972) model was selected.

The Priestley-Taylor model of evapotranspiration differs from more biophysically complex models (e.g., the Penman-Monteith model [Penman 1948; Monteith 1973]) in its simplifying assumptions regarding the process of transpiration. The Priestley-Taylor model is very similar to the Penman-Monteith model, but it uses a simplifying assumption that allows regional scale estimates. The assumption is that, given an adequate supply of soil moisture, atmospheric humidity and evaporation approach an equilibrium over relatively large areas. This equilibrium is independent of wind speed and canopy properties and is controlled primarily by net radiation. The Priestley-Taylor model can be applied on a daily basis by using daily values for temperature, irradiance, and precipitation to provide estimates of PET. Daily values of the climatic variables (precipitation and temperature) can be derived by interpolation of monthly data from the global database.

F.2.1.3 Moisture Deficit

The difference between PET and actual available moisture for a site can be viewed as a measure of moisture deficit. Available moisture for a site can be characterized in a number of ways, the simplest of which is precipitation. Using annual precipitation (AP) as an index of available moisture, moisture deficit can be calculated as (PET − AP).

In most areas, total precipitation is an overestimate of available moisture. Differences in the texture and depth of soil will have a direct influence on the amount of water that can be stored in the soil and the availability of that water to plants. The moisture-holding capacity of a soil (unit of water per volume of soil) is defined by the difference between field capacity (FC) and wilting point (WP), both a function of soil texture. The moisture-holding capacity, together with soil depth, defines the total moisture that can be held by a soil. An estimate of actual evapotranspiration (AET) can be derived by incorporating soil moisture-holding capacity and precipitation as constraints on evapotranspiration in the Priestley-Taylor model. The ratio of AET to PET has been used as an index of moisture deficit (Cramer and Leemans 1992). This ratio correlates with the broad-scale distribution of terrestrial vegetation (Prentice et al. 1992).

Caution should be taken in interpreting the spatial patterns of moisture deficit as defined by either PET – AP or the ratio of AET/PET. Moisture deficit does not necessarily reflect moisture stress. Plants exhibit a wide array of characteristics and adaptations, from stomatal control to patterns of carbon allocation, that enable them to function under conditions of low moisture availability. The large-scale distribution of plants (ecosystems and biomes) reflects these patterns of moisture deficit (Walter 1985), such as the gradient of biomes from desert to rain forest; however, changes in this index within an ecosystem as a function of changes in climate can have a major influence on the distribution and productivity of vegetation.

Another index of moisture availability that has been correlated with species distribution and ecosystem productivity is drought-days (Bassett 1964; Pastor and Post 1985, 1986). Drought-days consist of the sum of the number of days in the growing season that soil moisture is at or below the WP. The index is generally presented as a proportion of the total number of days in the growing season. This index is discussed in Section F.4.2.

F.2.1.4 Growing-Degree Days

Temperature has both a direct and indirect influence on a variety of plant processes. One index of temperature that is related both to plant growth (Bonan and Sirois 1992) and to the large-scale distribution of terrestrial vegetation (Prentice et al. 1992) is growing-degree days. Growing-degree days consist of the sum of daily mean temperatures above some defined threshold (base) temperature over a defined time interval (e.g., growing season, annual); for example, annual growing-degree days for base temperature 0°C would be the yearly sum of daily mean temperatures above 0°C (temperatures below 0°C are not summed). For a base temperature of 5°C, all daily values of d greater than zero would be summed, where $d = t - 5$, and t is the mean daily temperature. Other indices of temperature have been related to plant distribution and have been suggested as critical to understanding the potential response of vegetation to a climate change, such as absolute minimum and maximum temperatures (Woodward 1987).

F.2.1.5 Global Climate Data

To provide a common climate dataset for all investigators, the Leemans and Cramer (1991) global climate database of mean monthly temperature and precipitation is being made available as part of the forest impacts project. The database has a spatial resolution of 0.5° × 0.5°. The temperature data are stored in a compressed form in the file TEMP.ZIP. The dataset can be decompressed using the program PKUNZIP.EXE that is provided on the floppy disk containing the dataset. To decompress the TEMP.ZIP file, copy both the TEMP.ZIP and PKUNZIP.EXE files to a directory on the hard drive. The command "PKUNZIP TEMP.ZIP" will create an ASCII file named TEMP.DAT that contains the temperature dataset. The data are stored as latitude, longitude, and 12 monthly values of mean temperature in degrees celsius. All values are stored as integers with the FORTRAN format (i4,i5,12i4). The values of latitude, longitude, and temperature are reported to one decimal place, and therefore all integer values must be divided by 10.0 to provide real values. Values of latitude in the Northern Hemisphere are reported as positive numbers (0 to 90), latitudes in the Southern Hemisphere as negative numbers (0 to -90). Likewise, values of longitude in the Eastern Hemisphere are reported as positive numbers (0 to 180), values of longitude in the Western Hemisphere as negative values (0 to -180).

The precipitation data are also stored in a compressed form with file name PRECIP.ZIP. To decompress the precipitation dataset, follow the same procedure as with the temperature dataset using the command "PKUNZIP PRECIP.ZIP." The uncompressed file is named PRECIP.DAT. Data are stored as latitude, longitude, and 12 monthly values of precipitation in millimeters. The format of the dataset is (i4,i5,12i4). Latitude and longitude are reported to one decimal place and therefore must be divided by 10.0. Monthly precipitation data are reported in mm and need not be transformed.

In addition to the temperature and precipitation datasets, a global database of PET is being provided. The estimates of PET are derived from the Priestley-Taylor model discussed above using the precipitation and temperature datasets. Estimates of mean monthly cloud cover were used to calculate solar radiation. Cloud cover estimates were from the IIASA global database (Leemans and Cramer 1990). Values of PET are stored in compressed form in the file PET.ZIP. To decompress the file use the command "PKUNZIP PET.ZIP." PKUNZIP will create a file named PET.DAT. Data are stored as latitude, longitude, and 12 monthly values of mean PET in millimeters. The format and units of the PET values are identical to the precipitation data.

The datasets of temperature, precipitation, and PET can be used to calculate indices of GDD and moisture deficit (precipitation – PET) as discussed above.

F.2.2 Climate Change Scenarios and Patterns of Change in Bioclimatic Variables

General circulation model simulations of the current climate (current atmospheric carbon dioxide [CO_2] concentration, single levels of CO_2 [$1XCO_2$]) and climates under a doubling of CO_2 levels ($2XCO_2$) are used to construct climate change scenarios. Climate change scenarios constructed from a number of GCM models have been distributed to all investigators as part of the project (Appendix B). The GCM simulations used were $2XCO_2$ equilibrium model runs. A detailed discussion of the approach used in these GCM simulations is presented in Appendix C.

Changes in mean monthly precipitation and temperature were calculated for each GCM scenario for each computational grid element by taking the difference between simulated $1XCO_2$ and $2XCO_2$ climates. Temperatures were expressed as absolute difference ($2XCO_2 - 1XCO_2$) and precipitation as the ratio of $2XCO_2$ to $1XCO_2$. To provide values for changes in temperature and precipitation at the same spatial resolution as the current climate database provided (i.e., Leemans and Cramer 1991), these data from each GCM need to be interpolated to $0.5° \times 0.5°$ by using the same technique as applied in developing the database for current climate. The technique used was a triangulation of all data points (algorithm developed by Green and Sibson 1978), followed by a smooth surface fitting (Akima 1978).

F.2.2.1 Changes in Monthly Temperature and Precipitation

Predicted mean monthly temperatures for any $0.5°$ grid cell can be constructed by adding the changes in temperature from the GCM to the corresponding monthly values from the current climate database. Changes in precipitation should be multiplied by current monthly values because the changes represent the ratio of $1XCO_2/2XCO_2$ predictions from the GCM. If the GCM data are not interpolated to the grid for the global climate databases, the change values for each GCM grid should be applied to each of the $0.5° \times 0.5°$ grid cells from the global climate database contained within the GCM grid cell. This same procedure should be applied for any dataset used to characterize regional climate patterns (Appendixes B and C give further discussions of scenario construction).

F.2.2.2 Estimates of Potential Evapotranspiration under Climate Change Scenarios

Recalculating estimates of PET (Priestley-Taylor model) on the basis of the changed climates from the GCMs requires the use of the computer code for implementing the Priestley-Taylor model. The model used to calculate the PET estimates under current climate is available from the author upon request. Data requirements include changes in precipitation, temperature, and cloud cover (or solar radiation).

F.2.2.3 Estimates of Moisture Deficit under Climate Change Scenarios

Estimates of moisture deficit under the conditions of the climate change scenarios can be calculated from the estimates of monthly precipitation and PET on the basis of the GCM scenarios.

F.2.2.4 Estimates of Growing-Degree Days under Climate Change Scenarios

Once values of mean monthly temperature have been calculated on the basis of the GCM scenarios, estimates of growing-degree days under the conditions of the climate change scenarios can be calculated by using the same procedures outlined for calculating growing-degree days under current climate conditions.

F.2.2.5 Integrating Land-Cover Mapping and Direct Analysis of Bioclimatic Variables

The climatic indices discussed in the previous sections have all been used to correlate patterns of climate with plant distribution at various levels of plant description (e.g., species, communities, ecosystems, biomes) and spatial scale (e.g., topographic sequence to global). In most applications, these variables are related directly to plant distribution for developing predictive models of the spatial and temporal dynamics of the vegetative pattern; however, the georeferenced databases discussed previously (i.e., latitude, longitude, and climatic variable) can be used directly to evaluate the potential impacts of a climatic change on forest distribution and productivity. By examining the patterns of changes in variables such as moisture deficit and growing-degree days for given forested areas, possible impacts can be interpreted on the basis of known silvicultural characteristics, site indices, or other known patterns of forest productivity and species distribution from the region.

F.3 MODEL IMPLEMENTATION: THE HOLDRIDGE LIFE ZONE CLASSIFICATION MODEL

F.3.1 Description of Model

The Holdridge life zone classification model (Holdridge 1967) (Figure F.1) is a climate classification model that relates the distribution of major ecosystem complexes to the climatic variables of biotemperature, mean annual precipitation, and the ratio of PET to precipitation.

The life zones are depicted by a series of hexagons in a triangular coordinate system. Two climate variables — biotemperature and annual precipitation — determine the classification. Biotemperature is a temperature sum over a year, with the unit temperature values (i.e., average daily, weekly, or monthly temperatures) that are used in computing the index set to 0°C if they are less than or equal to 0°C (i.e., growing-degree days with base temperature of 0°C).

Identical logarithmic (base 2) axes for average annual precipitation form two sides of an equilateral triangle. The PET ratio forms the third side, and an axis for mean annual biotemperature is oriented perpendicular to its base. By striking equal intervals on these logarithmic axes, hexagons are formed that designate the Holdridge life zones.

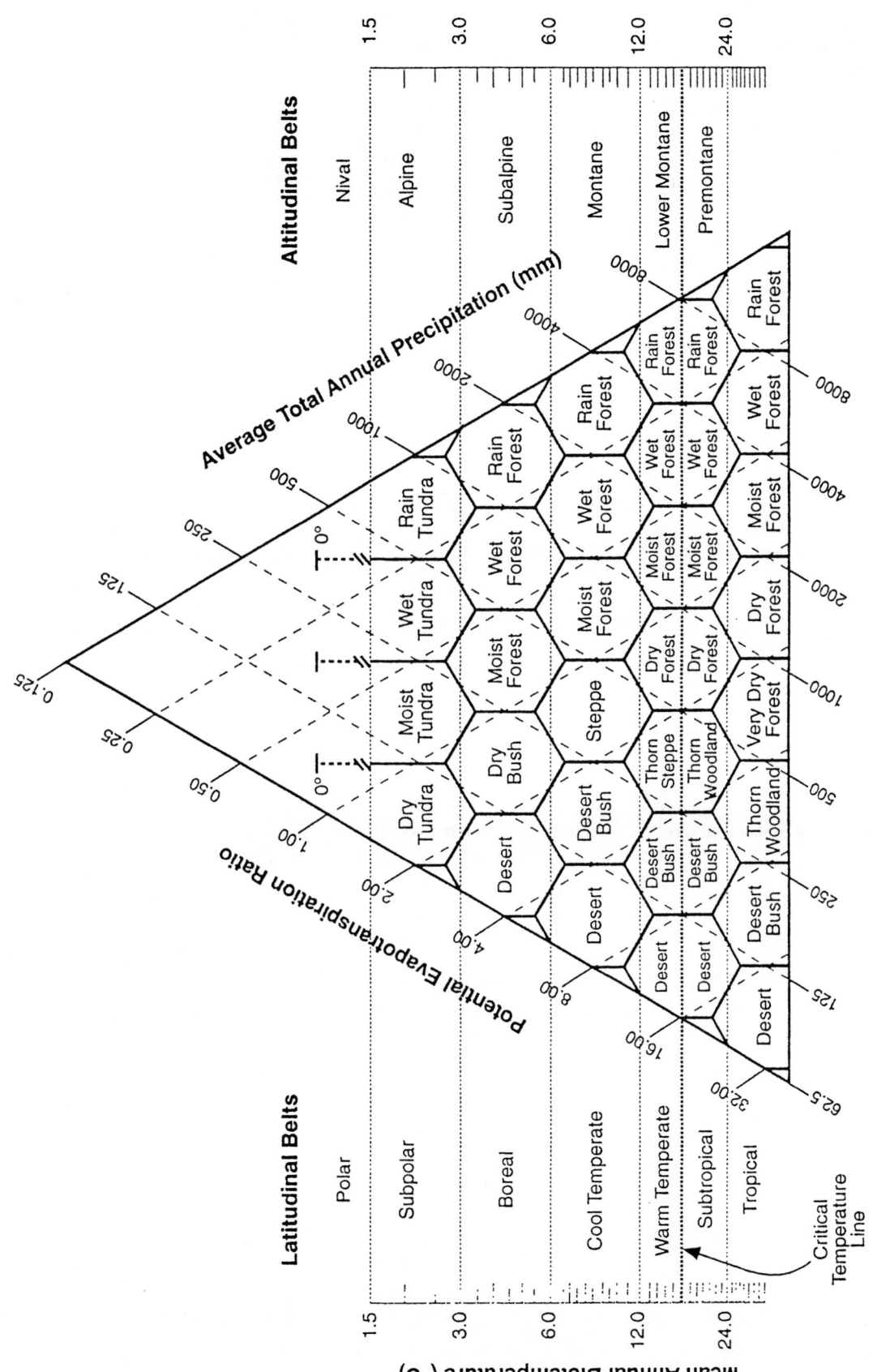

FIGURE F.1 Diagram of the Holdridge Classification Scheme (Source: Holdridge 1967)

The PET ratio is the quotient of PET and average annual precipitation. Holdridge (1959) assumes, on the basis of data from several ecosystem types, that PET is proportional to biotemperature (constant of proportionality = 58.93). The PET ratio in the Holdridge diagram therefore depends on the two primary variables, annual precipitation and biotemperature.

One additional division in the Holdridge classification is based on the occurrence of killing frost. This division is along a critical temperature line that divides hexagons between 12°C and 24°C into warm temperate and subtropical zones. The complete Holdridge classification at this level includes 39 life zones.

Identical axes of biotemperature are shown to the left and right of the triangle. The biotemperature axis to the left of the triangle represents the broad-scale, temperature-defined latitudinal gradient of ecosystems from polar to tropical. The biotemperature axis to the right of the triangle represents the gradient of ecosystems defined by changes in temperature related to elevation (e.g., lowland to montane). The classification of ecosystems by using these two temperature-defined gradients is scale dependent.

At a regional to global scale, variations in temperature are predominantly latitudinal. In contrast, within a given region (or latitudinally defined zone, such as temperate or tropical), the major source of variation in temperature may be topographic variation. In the latter case, the classification of ecosystems on the basis of the latitudinally defined gradient would be more appropriate. Examples will be given in the discussion of regional applications of the model. Application of the Holdridge model to a site requires data on the two primary variables (annual biotemperature and precipitation).

F.3.2 Application: Mapping Potential Land Cover

Although the Holdridge model can be applied at any spatial scale (i.e., site to global) to examine the potential impacts of climate change on forest resources and to subsequently develop strategies of adaptation and mitigation, the model should be used in a spatial context to provide maps of potential land cover for the area (region) of interest. As will be discussed in a later section, developing land-cover maps allows for the direct assessment of impacts for given regions of interest, as well as for the examination of overall changes in land use suitability for the region under the changed environmental conditions (e.g., identification of areas that are currently not suitable for production forestry but that may become suitable under a given climate change scenario). The identification of these areas may assist in developing land use policies or management plans directed at offsetting declines in forest production in other areas resulting from climate change.

The application of the Holdridge model within a spatial context requires a spatially explicit database of temperature and precipitation, from which the two primary variables can be calculated. The spatial resolution of the data should be as high as possible because the resolution at which the land cover is defined depends on the spatial resolution of the climatic data from which it is derived. The spatial resolution is the areal extent (e.g., square meters, hectares, square kilometers) being characterized by a given climatic observation (e.g., annual precipitation); for example, the spatial resolution of the global database of mean monthly temperature and precipitation discussed earlier is 0.5° × 0.5° (approximately 50 × 50 km at the equator). Once the climatic database has been established and the two primary variables have been calculated, each land cell (i.e., area described by a single observation of biotemperature and annual precipitation) can be classified using the Holdridge model, and the results can be mapped.

Figure F.2 shows a map of Holdridge life zones for North America. The map was generated by using the global climate database of mean monthly temperature and precipitation. The map represents potential land cover as predicted by the Holdridge classification. This potential land-cover database should be compared with regional maps of actual vegetation that are based on standard mapping procedures (e.g., ground surveys or areal photography or both) to determine how well the classification-based map matches actual patterns of land cover (i.e., vegetation).

F.3.3 Incorporation of Climate Scenarios

The climate change scenarios derived from the GCMs can be used to evaluate potential changes in land cover by using the Holdridge classification. The altered temperature and precipitation databases corresponding to each of the GCM scenarios are used to calculate biotemperature, annual precipitation, and PET ratio. These new values are then used to reclassify the grid cells (0.5° × 0.5°) using the Holdridge classification. A map of Holdridge life zones for North America under the Goddard Institute of Space Studies (GISS) climate change scenario (Hansen et al. 1988) is presented in Figure F.3.

F.3.4 Interpretation of Results: Identifying Areas of Impact

A direct comparison of the land-cover databases (i.e., latitude, longitude, and life zone) used to generate the maps of North America under current climate (Figure F.2) and the GISS climate change scenario (Figure F.3) can summarize the changes in land cover predicted to occur under the GISS scenario. These comparisons are useful in quantifying overall changes in vegetation, such as the total area of forest cover; however, these estimates represent changes in potential forest cover. Certain land areas that could potentially sustain forest cover are currently in some other form of land use, such as urban development or agriculture. To examine the predicted changes in actual forest cover, the potential land-cover map under current climate (Figure F.2) must be overlaid with a map of existing land cover; for example, areas of natural vegetative cover in North America, as defined by the land-cover map of Olson et al. (1984), are presented in Figure F.4. If this database is overlaid with the predicted changes based on the comparison of Figures F.2 and F.3, the resulting database can be used to examine changes in forested areas. This approach of overlaying predicted changes in land cover for specific categories of land use is important in evaluating the potential impacts of changing climate on particular forest sectors (e.g., conservation or production forestry).

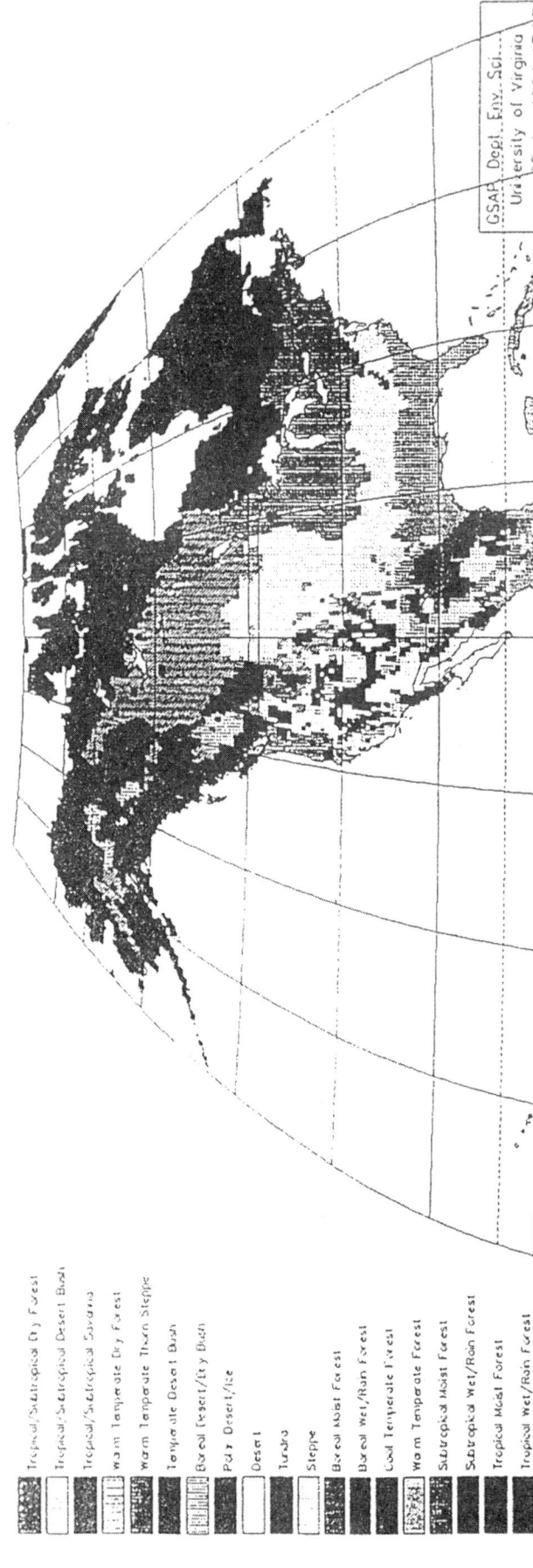

FIGURE F.2 Map of Holdridge Life Zones for North America under Current Climatic Conditions (life zones have been aggregated following the procedure of Smith et al. [1992])

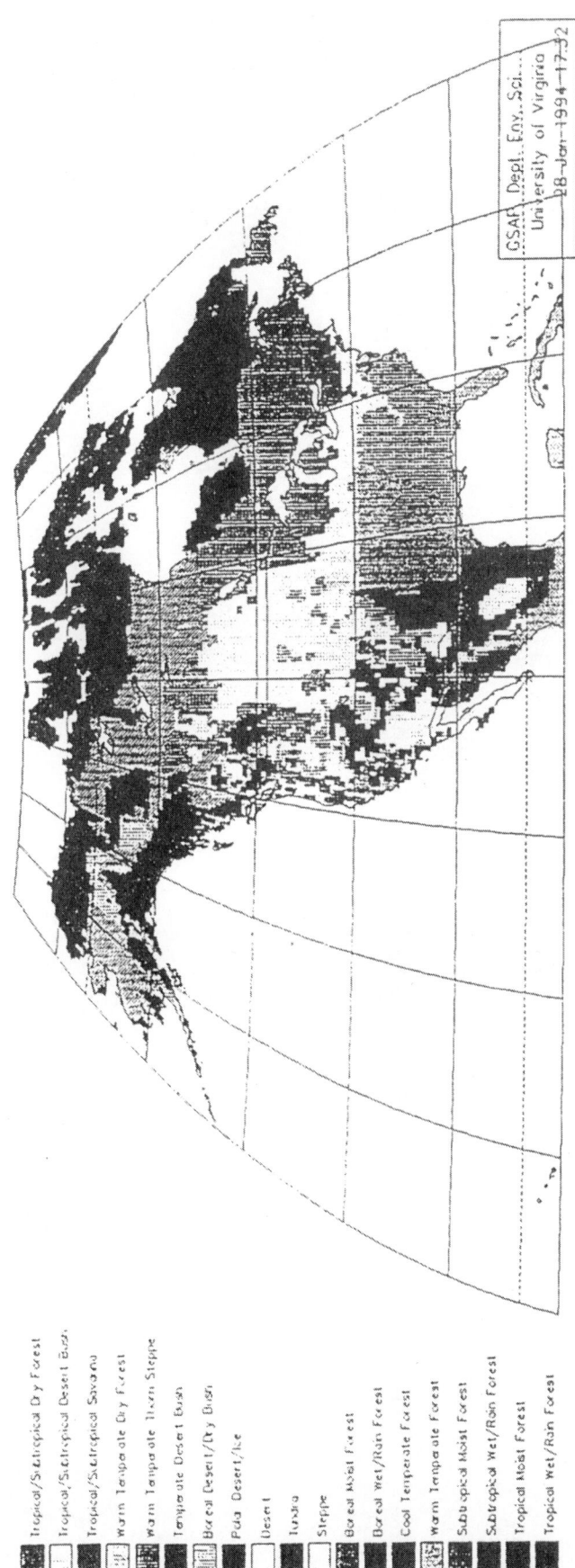

FIGURE F.3 Map of Holdridge Life Zones for North America as Predicted by the GISS-Based Climate Change Scenario (life zones have been aggregated following the procedure of Smith et al. [1992])

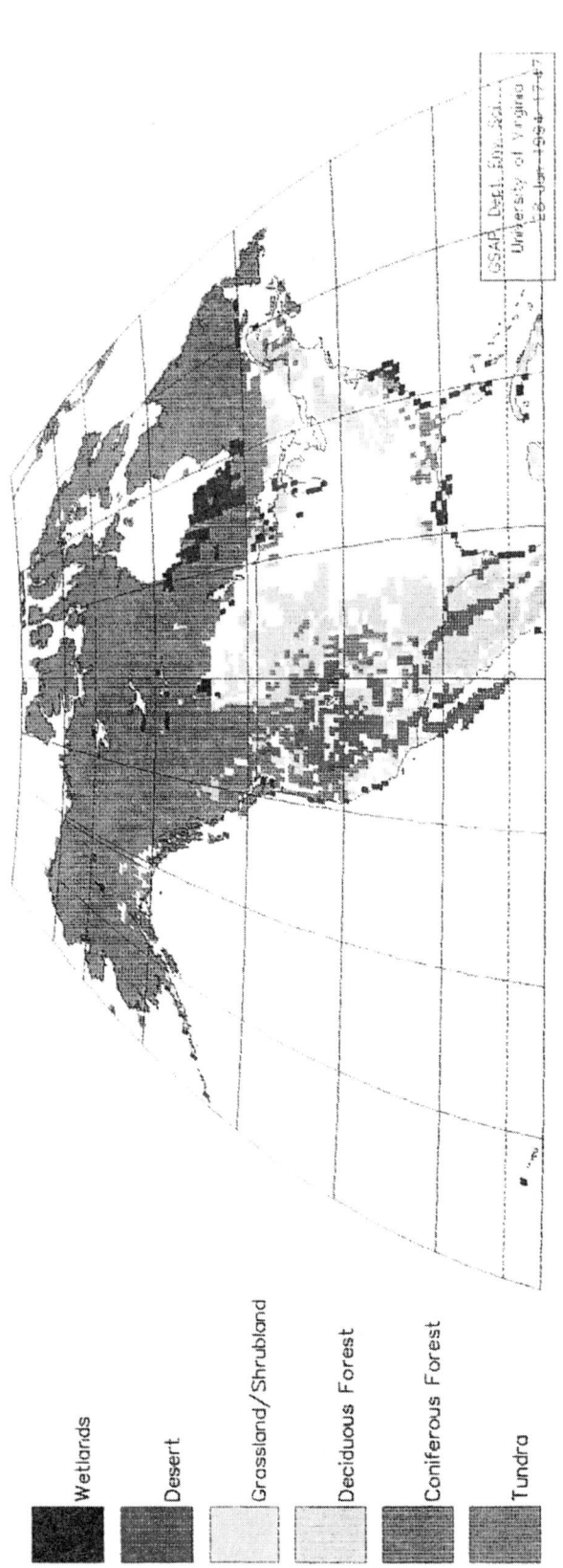

FIGURE F.4 Map of North America Showing Areas of Significant Natural Vegetative Cover (Source: Olson et al. 1984)

A map of forest conservation areas for Costa Rica in Central America) is presented in Figure F.5. To evaluate the potential impacts of a global climate change on these forested areas, a procedure similar to that outlined previously was undertaken (Smith et al. 1994). With the use of a geographical information system (GIS), a 1/250,000-scale Holdridge life zone map of the country (Tosi 1969) was overlaid with a database defining the boundaries of forest conservation areas. The life zone map was then altered by incorporating changes in monthly precipitation and temperature based on two climate change scenarios developed from GCM simulations. Figure F.6 shows areas within the nature reserves that are predicted to undergo a change in forest type. The same type of analysis was undertaken for areas of production forestry to evaluate the impacts of the climate change scenarios on forest productivity.

In addition to evaluating impacts on current forest resources, this procedure can be used to identify areas that are either marginal or unsuitable for production forestry under current climate conditions, but which may become suitable for timber production under the role in developing land-use policies to meet changing environmental conditions (TSC and WRI 1991; Tosi et al. 1992).

FIGURE F.5 Map of Forest Conservation Areas for the Country of Costa Rica

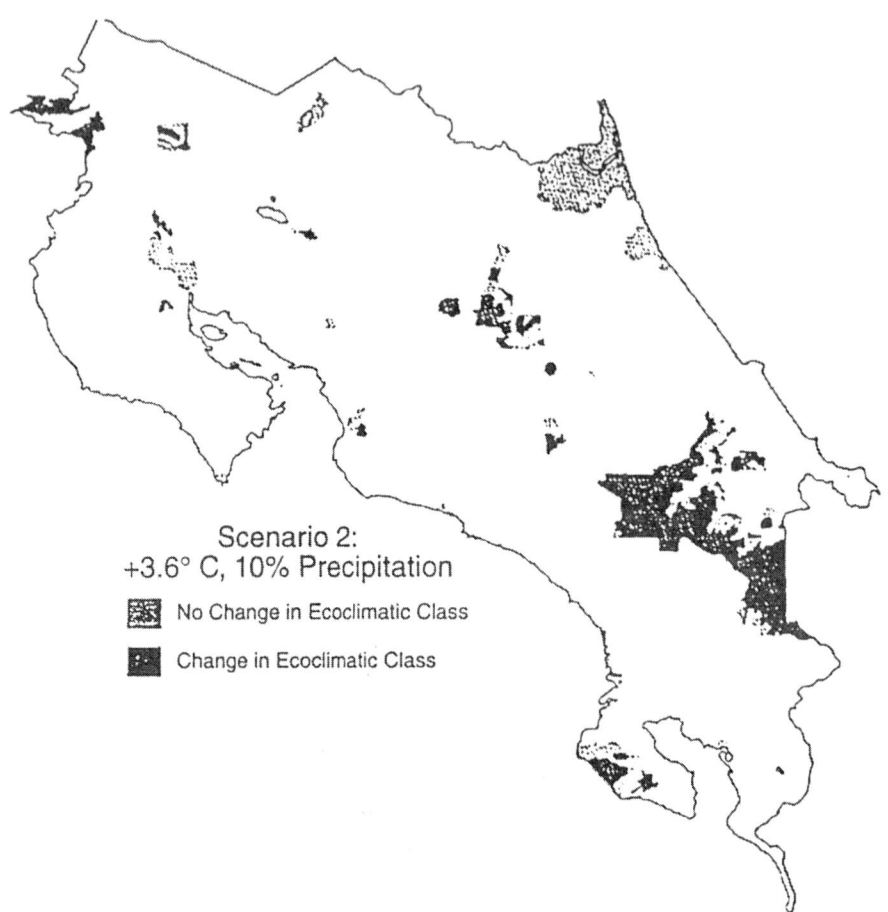

FIGURE F.6   Map of Areas of Impact within the Forest Conservation Regions of Costa Rica as Predicted by the Holdridge Life Zone Classification under a Climate Change Scenario (impact defined as change in life zone [Sources: Halpin et al. 1991; Halpin and Smith 1993])

F.3.5 Computer Program for Holdridge Classification

The computer code for implementing the Holdridge classification model will be distributed to all investigators. The computer program is written in FORTRAN 77 and named HOLDR.FOR. An executable version of the program has been included and is named HOLDR.EXE. The program requires the input variables annual precipitation and average daily biotemperature (Section F.3.1). This versatile program can calculate the life zone (ecosystem type) for either a single site or an array of data. To execute the program, type HOLDR.EXE [enter]. The program will prompt you to define whether you wish to classify a single site [1] or to input an array of sites [2]. If you select option 1 (single site), the program will prompt you for values of annual precipitation and biotemperature; the program will then return a numerical value and the corresponding name of the life zone. The numeric codes and names for the 39 life zone descriptors used in the model are presented in Table F.1. If option 2 is selected (an array of sites), the program will request the name of the input file.

TABLE F.1 Names and Numerical Codes for Holdridge Life Zone Descriptors

Name	Numerical Code
Ice	1
Polar Desert	2
Dry Tundra	3
Moist Tundra	4
Wet Tundra	5
Rain Tundra	6
Boreal Desert	7
Boreal Dry Scrub	8
Boreal Moist Forest	9
Boreal Wet Forest	10
Boreal Rain Forest	11
Cool Temperate Desert	12
Cool Temperate Desert Scrub	13
Cool Temperate Steppe	14
Cool Temperate Moist Forest	15
Cool Temperate Wet Forest	16
Cool Temperate Rain Forest	17
Warm Temperate Desert	18
Warm Temperate Desert Scrub	19
Warm Temperate Thorn Steppe	20
Warm Temperate Dry Forest	21
Warm Temperate Moist Forest	22
Warm Temperate Wet Forest	23
Warm Temperate Rain Forest	24
Subtropical Desert	25
Subtropical Desert Scrub	26
Subtropical Thorn Woodland	27
Subtropical Dry Forest	28
Subtropical Moist Forest	29
Subtropical Wet Forest	30
Subtropical Rain Forest	31
Tropical Desert	32
Tropical Desert Scrub	33
Tropical Thorn Woodland	34
Tropical Very Dry Forest	35
Tropical Dry Forest	36
Tropical Moist Forest	37
Tropical Wet Forest	38
Tropical Rain Forest	39

The input file should include latitude, longitude (for the site or region), annual precipitation, and biotemperature. If the input variables differ from those requested, the program (HOLDR.FOR) will have to be changed and recompiled. The program will read the input file as a free format file, requiring only that there is a space between each variable (i.e., latitude, longitude, annual precipitation, biotemperature). The model will calculate the life zone for each location and output the latitude, longitude, and numeric value for the life zone (Table F.1).

To aid in the initial application of the Holdridge model for any given region of the world, a program has been included that calculates annual precipitation and biotemperature from the global databases provided (Section F.5.2.1.1). The program is named CLIM.FOR. An executable version of the program named CLIM.EXE has also been included. To run the program type CLIM.EXE [enter]. The program requires that the two data files PRECIP.DAT and TEMP.DAT (Section F.5.2.1.1) be in the same directory on your computer as the CLIM.EXE program. The program will then prompt you to input the minimum and maximum latitude and longitude values that define the area you wish to include in the calculations. Remember that latitudes north of the equator are defined as positive values (0 to 90) and southern latitudes as negative values (0 to -90). Likewise, eastern hemisphere longitudes are defined as positive values (0 to 180) and western as negative (0 to -180). For example, a window that includes all of South America could be defined by minimum latitude -60.0, maximum latitude 15.0, minimum longitude -95.0, and maximum longitude -30.0. The program will output a file named CLIM.DAT, which includes latitude, longitude, annual precipitation, and biotemperature. This file is ready to be used as an input file for the HOLDR.EXE program.

F.3.6 Assumptions and Limitations of Method

As with any classification scheme, the Holdridge model is an abstraction of the actual vegetative pattern. First, the Holdridge model is a climate classification, rather than a classification based on actual distribution of vegetation, such as the system developed by Box (1981). Second, the limited number of categories of life zone or biome types results in a coarse resolution of vegetative description. In reality, the patterns of vegetation (i.e., physiognomic structure, species composition, and biomass) vary within any one life zone or classification unit. The approach also assumes that the vegetation (life zone) moves as a fixed unit in time and space. This assumption may not hold, especially under conditions where the changed climate has no current analogue.

The Holdridge classification, like all models for climate-vegetation classification, is correlative and is based on a limited set of variables. Although the bioclimatic indices used in the classification may do a sufficient job of bounding present patterns of vegetation, the actual patterns are a function of additional factors not explicitly considered in the model (e.g., soils) that may vary (both temporally and spatially) under the changed climatic conditions.

Perhaps most important, the approach represents an equilibrium solution for both climate (i.e., $2XCO_2$) and vegetation dynamics. In reality, the vegetation would most likely be unable to track the true transient climate dynamic. Although changes in the climatic pattern, as suggested by the GCM simulations, may occur on a time scale of decades to a century, the response of vegetation and soils to those changes may occur at different and varying time scales.

In areas where biomass values decrease because of moisture stress (i.e., higher PET ratio), the changes may occur quickly as the environmental conditions become such that the present vegetation can no longer be supported (e.g., shift from forest to grassland). In contrast, increases in biomass may require much longer periods of time. In some cases, the present vegetation may show increased growth or recruitment under the more favorable conditions; however, major shifts of forest type (e.g., warm temperate to tropical rain forest, or boreal to cool temperate forest) depend on the movement of species across the landscape and the ability of new species to invade existent communities. These changes in forest type would operate on time scales related to the life cycle or longevity of the component species. In many forest species, this time scale may be on the order of centuries (Davis 1989).

F.4 MODEL IMPLEMENTATION: FOREST GAP MODELS

The use of vegetation-climate classification models to evaluate the response of plants to climate change implicitly assumes a time scale sufficient for migration of vegetation and eventual equilibrium of vegetation to the new "changed" climatic patterns. However, simulating the temporal response of vegetation to changing climatic conditions requires the explicit consideration of plant demographic processes.

Numerous models of vegetation dynamics simulate the demographics of plant populations (Shugart and West 1980). One such class of demographic process models consists of "gap models." Gap models have been developed for a wide range of forest and grassland ecosystems. Although the models differ in their inclusion of processes that may be important in the dynamics of the particular site being simulated (e.g., hurricane disturbance, flooding), all gap models share a common set of characteristics and demographic processes.

In gap models, each individual plant is modeled as a unique entity with respect to the processes of establishment, growth, and mortality. This capability allows the model to track species- and size-specific demographic behaviors. The model structure includes two features important to a dynamic description of vegetation: (1) the response of the individual plant to the prevailing environmental conditions and (2) the way in which an individual modifies those environmental conditions (i.e., the feedback between the structure or composition of vegetation and the environment). Gap models have been applied to examine the response of forested systems to climate changes, both to reconstruct prehistoric Quaternary forests (Solomon et al. 1980, 1981; Solomon and Shugart 1984; Solomon and Webb 1985; Bonan and Hayden 1990; Bonan et al. 1990) and to project possible consequences of future climate change (Solomon et al. 1984; Solomon 1986; Pastor and Post 1988; Urban and Shugart 1989; Bonan et al. 1990; Overpeck et al. 1990; Smith et al. 1992).

In contrast to the Holdridge model, the gap model approach is high resolution in that the model can predict species composition, the structure and associated productivity of vegetation, and standing biomass through time. Although high in resolution, gap models are limited in the spatial extent to which the results can be extrapolated. The reason for this limitation is that the information required to parameterize or initialize a model that can address changes in these features of the vegetation (e.g., species composition and productivity through time) relates to site-specific features such as topographic position, soil characteristics, land-use history, disturbance, and present vegetative structure, all of which may vary over short distances. The application of these models to provide total coverage over broad regions would be virtually impossible because of both computational and data limitations. As an alternative, sampling approaches could provide large-scale coverage over broad environmental gradients (Solomon 1986; Bonan and Hayden 1990; Bonan et al. 1990; Smith et al. 1992).

The discussion of gap models is presented in three parts: (1) model description, (2) requirements for application to a site, and (3) examples of application to issues of climate change.

F.4.1 Model Description

Gap models are individually based in that they simulate establishment, growth, and mortality of each tree on the forest plot. The horizontal position of each tree is not defined; rather, each tree on the plot is assumed to be able to potentially influence (i.e., compete with) any other individual on the plot. This assumption is referred to as "horizontal homogeneity." In contrast to this assumption, the vertical position (i.e., height) of each individual is modeled explicitly. The assumption of horizontal homogeneity makes the size of the plot being simulated a critical parameter. If the plot is too small, the area and resource requirements for an individual to achieve its maximum potential size are not met. Conversely, if the size of the plot is too large, the process of mortality (e.g., death of a canopy-dominant individual) will have a minimal effect on resource availability.

To overcome these constraints, the size of the forest plot must be scaled to an area that can be dominated by a single or a few individuals of maximum size. At this spatial scale, individuals become established and compete for dominance. Eventually, the plot is dominated by a single or a few large trees. With the mortality of the canopy-dominant individuals, the availability of resources at ground level (i.e., light, water, and nutrients) increases, allowing for both the recruitment of new individuals and increased growth of previously suppressed individuals. This process has been described extensively in the ecological literature and referred to as "gap dynamics" (Watt 1947). Gap models derived their name from this view of forest dynamics.

The processes of establishment, growth, and mortality are typically simulated on an annual time step. Calculations defining the environmental conditions on the plot (e.g., moisture availability, and growing-degree days) may occur on daily or monthly time steps, but these subannual measures generally provide an annual index that is related to growth.

Output from the simulations is in the form of individual tree counts by species and size. These data are then used to estimate standing biomass and annual productivity.

The processes of growth, establishment, and mortality are discussed separately. The following materials provide a background to gap models. More detailed descriptions of the models and computer code are available in the references.

F.4.1.1 Growth and Environmental Feedback

The maximum annual growth increment for an individual tree is defined by a species-specific optimal growth curve (Figure F.7). This optimal growth curve can be derived through a number of methods. Botkin et al. (1972) define a method of deriving optimal growth curves as a function of simple silvicultural data on maximum size, height, and age for a species. A detailed description of this method is available in Botkin et al. (1972) or

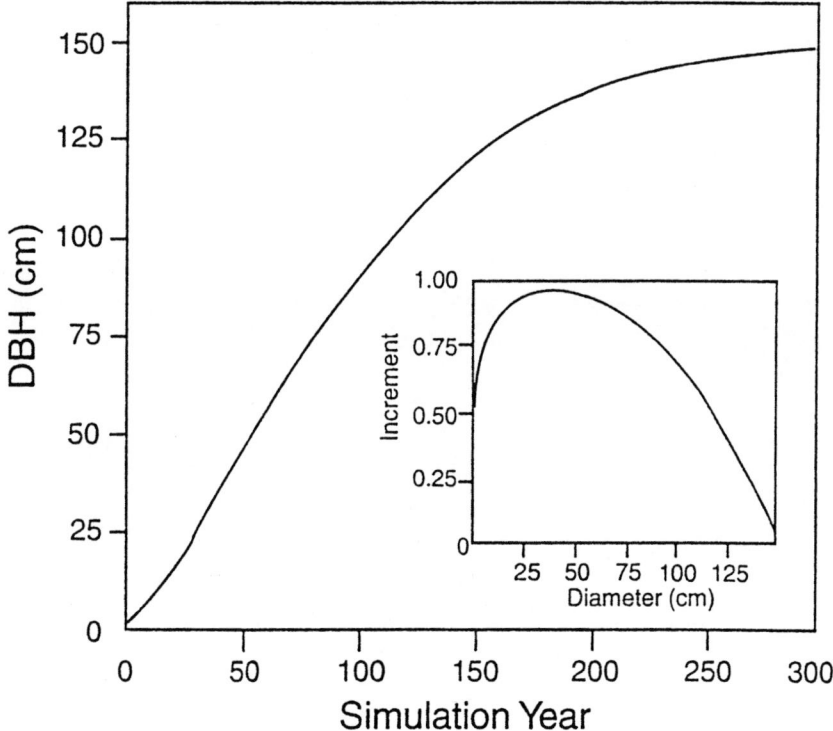

FIGURE F.7 Example of an Optimal Growth Function Used in Forest Gap Models (Growth is defined as diameter [DBH] as a function of time; diameter increment is defined as a function of current diameter [inset].)

Shugart (1984). A second, more data-intensive method is to define the growth curves on the basis of direct observations from diameter remeasurement data or tree ring analysis. In either case, the resulting function relates optimal diameter increment to current diameter. This potential (optimal) growth is then modified by the environmental conditions on the plot.

The response of an individual tree to the environmental conditions on the plot is defined by a number of environmental response functions, generally expressed as a proportion of optimal growth, ranging from 0.0 to 1.0. Examples of these functions are shown in Figure F.8. All gap models consider growth response to light; however, different gap models consider any number of additional environmental variables influencing growth that may be appropriate for a given site or region (e.g., temperature, water, nutrients, permafrost, and herbivory). These environmental response functions have been defined by using various methods, ranging from direct physiological measures to evaluating environmental conditions (e.g., growing-degree days) at the boundaries of the geographic range of the species. A detailed discussion of these methods can be found in Shugart (1984).

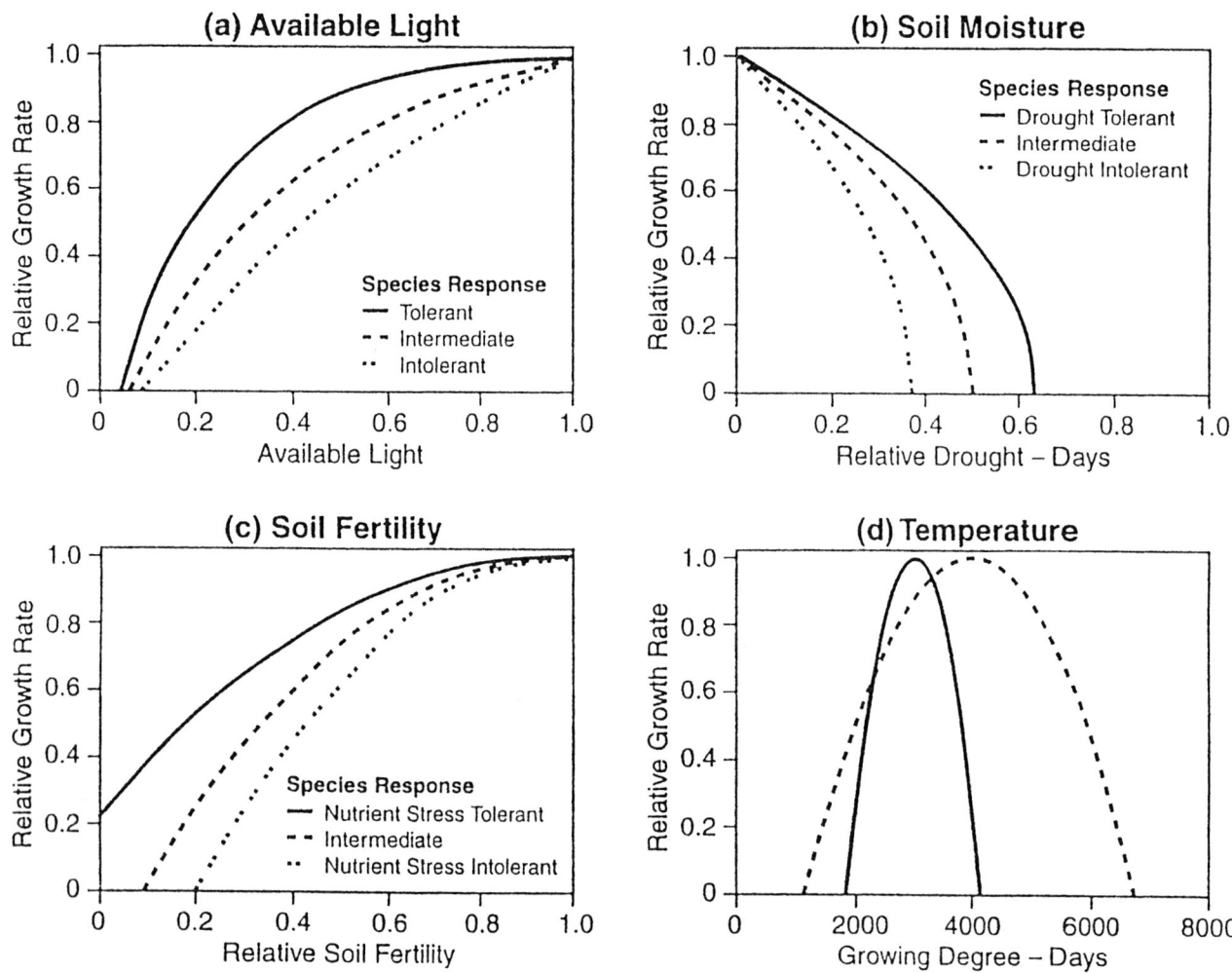

FIGURE F.8 Example of Resource Response Functions Used in the Forest Gap Model (functions define proportion of optimal growth [0.0-1.0] as a function of environmental conditions on the simulated plot)

The height, biomass, and leaf area of individual trees are generally calculated from allometric functions at the species level. By defining height and leaf area for each individual, a vertical profile of leaf area is constructed, and an available light profile is calculated by using a light-extinction equation. This vertical profile of available light is then used to define the available light to each individual tree for determining the growth response.

All gap models simulate the influence of individual trees on the availability of light at height intervals on the plot. Plant influences on other environmental factors are incorporated to varying degrees in different versions of the model; these factors include soil moisture, fertility, and temperature, as well as disturbances, such as fires, and herbivory.

Plant influence on soil moisture has been incorporated by using a variety of models for soil hydrology and evapotranspiration (e.g., the Priestley-Taylor model, the Penman-Monteith model) that allow distribution of roots and the leaf area of the plot to influence the soil moisture profile. In return, soil moisture integrated over the growing season is used to modify plant growth (i.e., biomass and distribution of roots and the leaf area).

The availability of soil nutrients on the plot is influenced by the plants in terms of the quantity and quality of litter input into the soil organic layer for decomposition (Weinstein et al. 1982; Pastor and Post 1985). Species are classified into functional categories on the basis of the characteristics of their leaves (e.g., lignin/nitrogen), which relate to the rates at which they decompose. Nitrogen dynamics (and other mineral nutrients) of the plot are simulated as a function of the prevailing environmental conditions at the surface of the soil and the quantity and quality of litter material.

Soil temperature is influenced by the collective leaf area of individuals on the plot reducing incoming radiation to the soil surface, as well as the insulating effect of litter material on the plot. In turn, soil temperatures are needed as input for calculating rates of decomposition and the development of permafrost. These processes, together with direct effects of temperature on growth, provide feedback to plant processes (Bonan 1988).

In actuality, the modeling framework outlined previously has two components or modules, one involving plant demographics and the other involving the processes, both climatic and biophysical, that define the environment on each plot. An extensive discussion of the biophysical models is beyond the scope of this document, but a complete discussion of these models and their incorporation into the gap model structure can be found in the references cited in the previous text.

In general, these biophysical models require the transfer of certain parameters from the demographic module describing the vegetation of the plot and, in return, define the environmental conditions as a function of the geology and soils, the prevailing climate, and the modifying influence of the plant structure.

F.4.1.2 Establishment

The establishment of individuals on the forest plot is modeled in a number of ways. Original forms of the gap model simulated recruitment as a stochastic process, thereby drawing a fixed number of recruits each year at random from the list of species for the site conditional upon the requirements of the species for germination and establishment (e.g., minerals and soil require fire for germination). In this approach, all species are assumed to have an equal probability of recruitment if environmental conditions are met. Relative rates of establishment of seedlings are not a function of the composition of the canopy.

Alternative approaches that have been adopted include defining relative rates of establishment conditional on the species composition of the canopy (i.e., availability of a seed source) and defining species-specific maximum rates of production of seeds or seedlings. Once individual trees enter the population, annual patterns of growth are determined following the approach outlined previously.

F.4.1.3 Mortality

Mortality of trees is simulated as a stochastic process. Two types (or causes) of mortality are considered: (1) fixed annual probability of mortality and (2) stress-related mortality. Each tree has a fixed probability of mortality each year. This fixed rate is a function of the species-specific maximum longevity (i.e., maximum age) and the assumption that approximately 1% of all individuals achieve the maximum age. In addition to this fixed rate of mortality, the probability of mortality increases as a function of "stress," where stress is defined as failure to achieve some minimum defined proportion of optimal growth.

Mortality is also introduced selectively when forest management (timber harvest) is simulated. An example of an application of a forest gap model to examine implications of forest management practices is presented in Smith et al. (1981).

F.4.1.4 Competition

Individual trees do not interact directly; rather, interaction is through modification of the environment on the plot by the individual trees; for example, through the calculation of the leaf area profiles and available light through the canopy, it is evident that taller individuals shade smaller trees, thus reducing their growth rate. Likewise, individual trees reduce moisture availability, alter the quantity and quality of litter and the subsequent availability of nutrients, and influence the input of solar radiation to the soil surface, thus influencing a range of processes included within the modeling framework (e.g., permafrost dynamics). Under any given set of site conditions, species respond differentially to environmental conditions and influence the availability of resources to other species. Those species that exhibit the highest rates of growth and recruitment under a given set of environmental conditions will reduce the availability of resources to other species and will come to dominate. As environmental conditions change through time on the plot as a function of the vegetation, patterns of recruitment and growth change, and patterns of dominance may shift (i.e., succession).

F.4.2 Application of a Gap Model

A sample of both the tree species and the site data required to parameterize a simple forest gap model for a site in east Tennessee (United States) is presented in Table F.2. The parameters from which the optimal growth function is derived are maximum longevity (AGE_{max}), maximum diameter (DBH_{max}), and maximum height (HGT_{max}). The derivation of the growth function from these parameters is presented by Shugart (1984). The environmental response to light is defined by the parameter L. In this example, species are classified as either shade tolerant ($L = 1$) or shade intolerant ($L = 2$). The functions defining the general forms of the light response curves for these two classes are shown in Figure F.8.

The response to temperature is modeled as a parabola between values of D_{min} and D_{max} (Figure F.8). The values of D_{min} and D_{max} are defined as the maximum and minimum degree-day values associated with the northern and southern boundaries of the geographic distribution of the species (Shugart 1984).

The response to moisture is defined by the parameter DD; DD is the value of drought-days associated with the arid (in this case, western) limit to the geographic distribution of the species. Drought-days as defined here are an index of the proportion of the growing season at which the soil is at or below the WP.

Characteristics of the site include data on climate and soils. Soils are described in terms of WP and FC. These estimates incorporate the depth of the soil. In addition to monthly values of precipitation and temperature, climate parameters include the beginning and end of the growing season (day of year).

In the model presented in this example, soil moisture is simply defined in terms of a single layer of soil with inputs from precipitation and with outputs defined by runoff (precipitation above FC for the soil) and evapotranspiration. Evapotranspiration is calculated on a monthly basis by using the model of Thornthwaite (1948). Other forms of the model use much more complex soil structures and estimates of evapotranspiration.

F.4.3 Incorporation of Climate Change Scenarios

Unlike the equilibrium approach represented by the Holdridge classification, forest gap models track the temporal response of vegetation to changing environmental conditions. Although the changes in climate produced from the $2XCO_2$ equilibrium GCM simulations have been applied to the gap models as a step function (i.e., changes in temperature and

TABLE F.2 Example of Species and Site Parameters for a Forest Gap Model in East Tennessee[a]

Species	DBH_{max} [b] (cm)	HGT_{max} [c] (m)	L [d]	AGE_{max} [e] (yr)	SED_{max} [f]	DD [g]	D_{min} [h]	D_{max} [i]
Acer rubrum	100	30	2	150	75	0.230	1,260	6,600
Acer saccharum	150	30	2	125	20	0.268	1,600	4,700
Carya cordiformis	100	30	1	300	20	0.320	1,910	5,076
Carya glabra	100	30	1	300	20	0.200	1,910	6,960
Carya ovata	100	30	1	275	20	0.200	1,670	5,500
Carya tomentosa	100	28	1	300	20	0.300	1,910	5,993
Castanea dentata	150	35	1	300	20	0.300	1,910	4,571
Cornus florida	25	10	1	100	20	0.250	1,910	5,993
Fagus grandifolia	100	30	1	366	20	0.200	1,326	5,537
Fraxinus americana	100	30	1	300	20	0.280	1,398	5,993
Juglans nigra	150	35	2	250	20	0.300	1,910	4,571
Juniperus virginiana	75	20	2	300	20	0.397	1,721	5,537
Liquidambar styraciflua	125	35	2	250	20	0.300	2,660	5,993
Liriodendron tulipifera	150	35	2	300	400	0.160	2,300	5,993
Nyssa sylvatica	100	30	2	300	20	0.301	1,910	6,960
Pinus echinata	100	30	2	300	20	0.423	2,660	5,076
Pinus strobus	150	35	2	450	140	0.310	1,100	3,165
Pinus virginiana	50	15	2	250	350	0.226	2,660	3,671

TABLE F.2 (Cont.)

Species	DBH_{max} [b] (cm)	HGT_{max} [c] (m)	L [d]	AGE_{max} [e] (yr)	SED_{max} [f]	DD [g]	D_{min} [h]	D_{max} [i]
Prunus serotina	100	30	2	200	20	0.300	2,132	5,993
Quercus alba	100	35	1	400	40	0.330	1,721	5,537
Quercus coccinea	75	25	1	400	20	0.286	2,037	4,571
Quercus falcata	100	35	1	400	20	0.423	2,660	5,993
Quercus prinus	100	30	1	250	20	0.285	1,910	4,110
Quercus rubra	100	30	1	400	40	0.225	1,100	4,571
Quercus stellata	75	25	2	400	20	0.555	2,660	5,993
Quercus velutina	100	30	1	300	20	0.300	1,810	5,076

[a]   Number of plots, 25; years of simulation, 500; number of species; 26.

[b] DBH_{max} = maximum diameter.

[c] HGT_{max} = maximum height.

[d] L = light response category (1 = shade tolerant; 2 = shade intolerant).

[e] Age_{max} = maximum age.

[f] SED_{max} = maximum yearly seedling establishment scaled to plot.

[g] DD = drought-days limit (0.0-1.0).

[h] D_{min} = minimum degree-day limit.

[i] D_{max} = maximum degree-day limit.

Note: Site Variables:

Monthly rainfalls (in centimeters) (mean ± standard deviation) were as follows: 12.2 ± 6.6, 12.1 ± 5.2, 12.8 ± 4.2, 10.0 ± 3.2, 9.4 ± 3.4, 9.6 ± 3.2, 13.5 ± 4.4, 10.4 ± 3.4, 7.4 ± 3.8, 6.8 ± 4.1, 8.4 ± 5.3, and 11.0 ± 4.9.

Monthly temperatures (in degrees Celsius) (mean ± standard deviation) were as follows: 4.6 ± 3.0, 5.5 ± 2.7, 8.9 ± 2.6, 14.4 ± 1.3, 19.2 ± 1.3, 23.5 ± 1.2, 25.0 ± 0.9, 24.5 ± 0.8, 21.4 ± 1.3, 15.3 ± 1.6, 0.86 ± 1.6, and 4.7 ± 2.3.

Soils: FC = 40.0 and WP = 20.0.

precipitation are assumed to occur within a single year), a more realistic means of incorporating the predicted changes is through the development and application of transient climate scenarios. The development of transient scenarios is discussed in Appendix B.

The impacts of four GCM-derived climate change scenarios on forest dynamics for a site in central Alaska are presented in Figure F.9. A transient scenario was constructed from the mean monthly changes in precipitation and temperature by assuming that the changes occurred over a 100-year period. A linear interpolation between current and $2XCO_2$ values was used. Year 0 of the simulation represents the current composition and structure of the forest. The climate change begins at year 0 of the simulation, with full changes in temperature and precipitation being realized by year 100. The climate was then held constant at these new values over the remaining 100 years of the simulation. A control simulation (i.e., no climate change) is also shown in Figure F.9.

F.4.4 Interpretation of Results

Because forest gap models are individual based, the models provide data on the species, age, and size (e.g., diameter, height, leaf area) of individual trees for each simulated plot or forest stand. The data on individual trees can then be aggregated to provide estimates of basal area, biomass, board, feet or any other desired index at a species level. The high resolution of data produced by the models makes the forest gap model approach ideal for commercial forestry applications relating to impacts assessment and the evaluation of adaptation strategies. The results of the simulation of a forest stand on a south-facing slope in central Alaska presented in Figure F.9 provide an example of the application of a forest gap model to evaluate the impacts of a climate change on forest productivity. The results are expressed as changes in species biomass (t/ha) under current climate conditions and climate change as predicted by four GCMs. Year 0 of the simulation represents current stand conditions. At present the stand is dominated by white spruce (*Picea glauca*) and birch (*Betula papyrifera*); aspen (*Populus tremuloides*) is a minor component. If current climate conditions at the site continue (control), the present species composition and patterns of productivity will persist. In contrast, there is a marked decline in stand productivity and biomass under all four climate change scenarios. In the two scenarios showing the greatest degree of warming for the region (Geophysical Fluid Dynamics Laboratory [GFDL] and United Kingdom Meteorological Office [UKMO]) there is a total dieback of the stand by year 50 of the simulation. In the two more moderate scenarios (Goddard Institute for Space Studies [GISS] and Oregon State University [OSU]), the decline occurs more slowly. The decline in productivity and biomass under all four scenarios is a direct function of increased aridity due to both increased temperature and changes in patterns of precipitation. These results are in agreement with the predictions of the Holdridge analysis for the region that predicts a shift from forest to grassland (i.e., steppe).

In the above example, stand structure is expressed in terms of biomass. Estimates of productivity can be derived by examining the annual changes in biomass or basal area for the stand. In addition, size class distributions can also be evaluated. The model can also

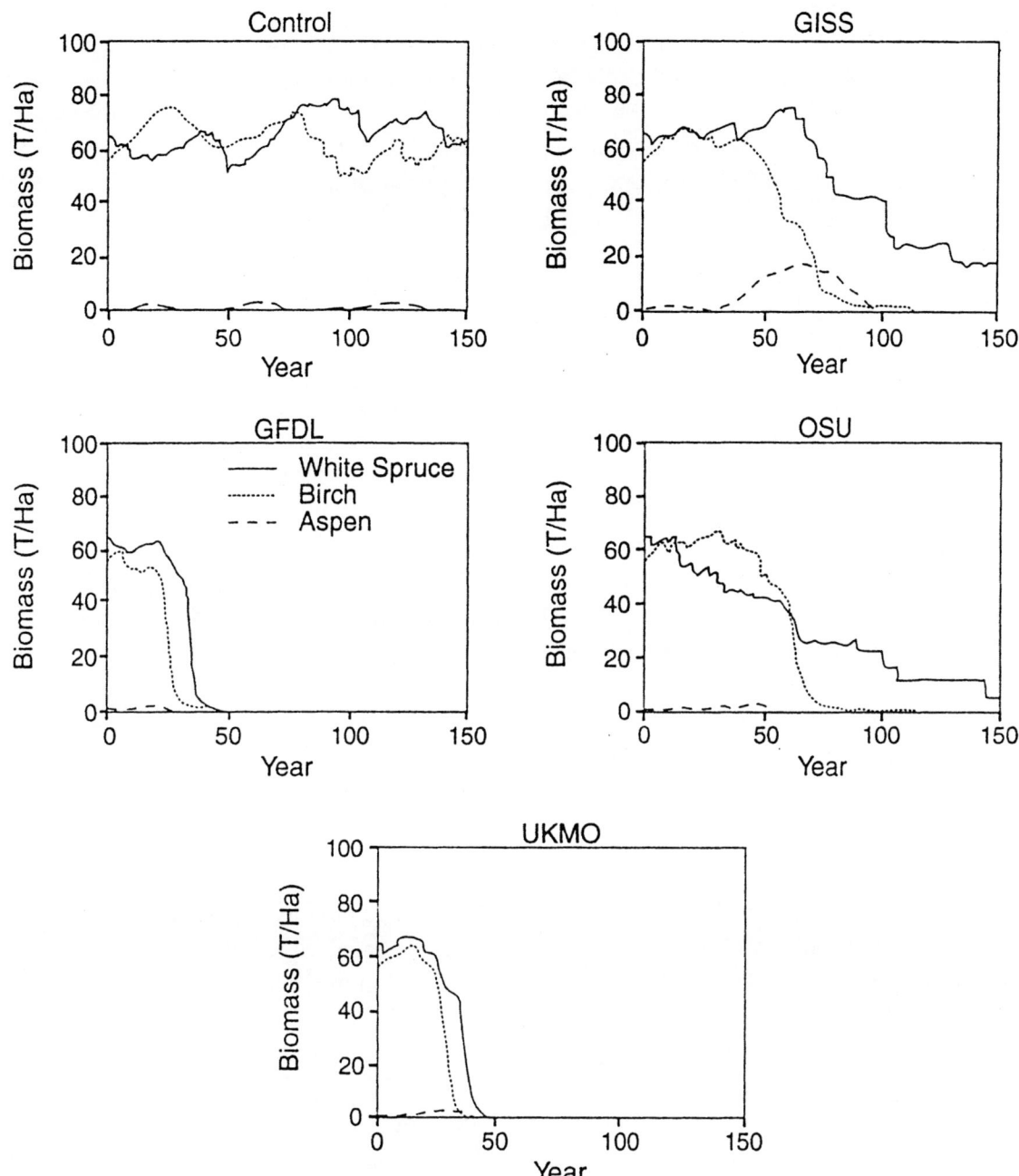

FIGURE F.9 Results from a Forest Gap Model Simulation for a Site in Central Alaska under Current Climate (Control) and Four GCM-Based Scenarios (year 0 of the simulation represents current forest composition and structure; transient climate change scenario is applied for year 0 to year 100; climate is held constant after year 100)

be used to evaluate the possibility of slowing the rate of forest decline through management options such as stand thinning, or screening for tree species that are more suitable for planting (plantation forestry) under the new climate conditions. These new species could be included in the simulations to examine the response of the forest stand with the addition of a new (introduced) species from a warmer, more arid environment.

F.4.5 Identification of Sites for Model Application and Integration of Regional and Site Analyses

The selection of sites for the development and application of forest gap models can be approached in a number of ways. Certain areas that are important for either conservation or forestry can be identified for analysis regardless of the predictions for these areas under the GCMs. Given the current uncertainty in the GCM predictions, understanding the sensitivity of these important forested areas to a climate change is critical. In addition, the regional application of the Holdridge classification for mapping impacts and the direct analyses of the climate indexes (e.g., moisture deficit, growing-degree days) should be used to identify areas where the changes in climate will have significant impacts on the distribution of vegetation (e.g., mesic forest to dry woodland). Sites within these areas can then be located, and the forest gap model(s) can be used to examine the temporal dynamics associated with the predicted shifts in the Holdridge life zone. A further discussion of this procedure is presented in Section F.5.

F.4.6 Assumptions and Limitations of Forest Gap Models

Various assumptions are related to the structure of forest gap models and the methods for deriving site and species parameters. Some of these assumptions have been discussed in Section F.4.1. In addition, a detailed discussion of the assumptions underlying forest gap models can be found in either Botkin et al. (1972) or Shugart (1984). However, a number of assumptions relating to species parameterization and seedling establishment have direct implications on the application of forest gap models for climate change assessment and should be noted. As discussed in Section F.4.1.1, the functions describing plant response to temperature and moisture are often parameterized from data relating environmental conditions (e.g., growing-degree days or drought-days) across a species' range. This form of parameterization assumes that the potential growth for a species at its current distributional boundaries is zero. This assumption may hold for most cases; however, if competition is an important factor limiting the geographic distribution of the species, then this assumption may not hold. Secondly, most current applications of the forest gap models do not consider the direct response of plants to increasing atmospheric concentrations of CO_2. This limitation is due primarily to a lack of species-specific data on plant response to elevated CO_2. Where data are available, the response can be incorporated into the model framework.

F.5 INTEGRATION OF BIOCLIMATIC AND FOREST GAP MODELS

The two primary modeling approaches presented here differ widely in their resolution of vegetation description and the spatial and temporal scales at which they address vegetation pattern and dynamics. The Holdridge model defines broad classes of ecosystem types or life zones, while the forest gap model examines the dynamics of individual species. The Holdridge model has no explicit spatial scale and can be applied at a local, regional, or global scale. The spatial resolution is defined by the scale of the underlying climate database. In contrast, the forest gap model operates at a spatial scale corresponding to a single forest plot (<1 ha). The Holdridge model is an equilibrium model in that the predicted changes in vegetation pattern assume a period of time necessary for a new equilibrium to be achieved between the changed climate conditions and the new distribution of vegetation (Section F.3.6). In contrast, the forest gap model simulates the temporal dynamics of forest structure and composition in response to changing environmental conditions.

Despite the differences between the two models presented, the models are similar in that both involve an explicit link between vegetation distribution and climate. The shifts in ecosystem type or life zone predicted by the Holdridge model imply certain changes in species composition and structure. By combining the two modeling approaches a regional analysis can be undertaken whereby the Holdridge model is used to describe the broadscale shifts in ecosystems, and the gap model(s) can be used to examine both the transient dynamics associated with the predicted shifts, as well as possible changes in composition and structure within a given ecosystem type or life zone. An application of this approach for the North American boreal forest zone is shown in Figure 5.3.9. A forest gap model was used to explore the temporal dynamics of stands of black spruce (*Picea mariana*) for a number of locations for which shifts in forest cover were predicted on the basis of the application of the Holdridge classification for the region.

F.6 RESOURCE REQUIREMENTS

The development and application of a forest gap model can require a significant commitment of time and resources. Fortunately, existing forest gap models represent a variety of forested ecosystems, ranging from boreal to tropical. Before model development is undertaken by any of the investigators involved in the Country Studies Program, a review of the literature should be undertaken in order to take advantage of existing models and research programs.

The computer code and documentation for a generalized forest gap model have been distributed to all investigators under the Country Studies Program. In most cases, the data on species growth rates and environmental responses necessary for the model (Section F.4.2) are available for dominant forest species, especially those of commercial importance. Availability of data can be a problem in species-rich ecosystems such as tropical rainforests.

Simple forest gap models can be compiled and executed on a personal computer; however, a workstation environment, such as the SUN SPARC station 10 with a Unix operating system or equivalent, is preferable.

F.7 OTHER APPROACHES

The two primary modeling approaches presented in the appendix represent only a small subset of the variety of modeling approaches currently being used to assess the possible impacts of a global climate change on terrestrial vegetation. The Holdridge classification is only one of a number of climate-vegetation classifications applied at the regional and global scales. The Box (1981), BIOME (Prentice et al. 1992), and MAPSS (Neilson 1993) models are similar to Holdridge in that they relate the large-scale distribution of vegetation or plant types to climate patterns. However, these models require additional climate variables and calculations of evapotranspiration.

Of the variety of models that simulate the dynamics of stand structure and species composition, the forest gap model is perhaps the most easily adaptable to new sites because parameters can generally be derived from simple silvicultural data and information on species distributions. For this reason, alternative approaches are not examined within the scope of this project.

In addition to the climate-vegetation classification and forest stand models, more physiologically detailed models are available to examine the implications of both climate change and the direct effects of increasing concentrations of atmospheric CO_2 on forests. These models (e.g., Running and Coughlan 1988; Bonan 1992) generally simulate the processes of photosynthesis and transpiration of forest canopies. Although this approach is suitable for examining patterns of CO_2 and water flux on both an intra- and inter-annual basis, typically these models require that vegetation cover be defined as an input. As such, the models are of limited value in examining the potential for changes in forest structure and composition that may occur under changing climate conditions. Recent work has focused on likening these models of canopy processes with individual-based forest gap models (Smith et al. 1992) to examine the potential direct effects of rising CO_2 concentrations on patterns of water use efficiency and net primary productivity under changed climate conditions; however, data requirements for parameterization limit the ease with which this approach can be applied to a variety of forested ecosystems.

Regional to global patterns of net ecosystem productivity and carbon storage can be predicted using a number of ecosystem process (biogeochemistry) models. Two examples of this class of model are TEM (Terrestrial Ecosystem Model; Melillo et al. 1993) and CENTURY (Parton et al. 1988).

F.8 APPENDIX F REFERENCES

Akima, H., 1978, "A Method of Bivariate Interpolation and Smooth Surface Fitting for Irregular Disturbed Datapoints," *ACM Transactions on Mathematical Software* 4:148-159.

Bassett, J.R., 1964, "Tree Growth as Affected by Soil Moisture Availability," *Soil Science Proceedings* 28:436-438.

Bonan, G.B., 1988, "A Computer Model of the Solar Radiation, Soil Moisture, and Soil Thermal Regimes in Boreal Forests," *Ecological Modeling* 45:275-306.

Bonan, G.B., and B.P. Hayden, 1990, "Using a Forest Stand Simulation Model to Examine the Ecological and Climatic Significance of the Late-Quaternary Pine-Spruce Pollen Zone in Eastern Virginia, U.S.A.," *Quaternary Research* 33:204-218.

Bonan, G.B., and L. Sirois, 1992, "Air Temperature, Tree Growth and the Northern and Southern Range Limits to *Picea mariana*," *Journal of Vegetation Science* 3:495-506.

Bonan, G.B., et al., 1990, "The Sensitivity of Some High-Latitude Boreal Forests to Climatic Parameters," *Climatic Change* 16:9-29.

Botkin, D.B., et al., 1972, "Some Ecological Consequences of a Computer Model of Forest Growth," *Journal of Ecology* 60:849-873.

Box, E.O., 1980, *Macroclimate and Plant Forms: An Introduction to Predictive Modeling in Phytogeography*, Junk, the Hague, the Netherlands.

Cramer, W., and R. Leemans, 1992, "Assessing Impacts of Climate Change on Vegetation Using Climate Classification Systems," in A.M. Solomon and H.H. Shugart (eds.), *Vegetation Dynamics and Climate Change*, Chapman and Hall, New York, N.Y.

Davis, M.B., 1989, "Lags in Vegetation Response to Greenhouse Warming," *Climatic Change* 15:75-82.

Green, P.J., and R. Sibson, 1978, "Computing Dirichlet Tessellations in the Plane," *The Computer Journal* 21:168-173.

Halpin, P.N., and T.M. Smith, 1993, "Potential Impacts of Climate Change on Forest Protection in the Humid Tropics: A Case Study in Costa Rica," in *Proceedings of the Symposium on Impacts of Climate Change on Ecosystems and Species*, Amersfoot, the Netherlands, in press.

Halpin, P.N., et al., 1991, *Climate Change and Central American Forest Systems: Costa Rica Pilot Project*, project background report.

Hansen, J., et al., 1988, "Global Climate Changes as Forecast by the GISS-3-D Model," *Journal of Geophysical Research* 93:9341-9364.

Holdridge, L.R., 1959, "Simple Method for Determining Potential Evapotranspiration from Temperature Data," *Science* 130:572.

Holdridge, L.R., 1967, *Life Zone Ecology,* Tropical Science Center, San Jose, Calif.

Leemans, R., and W. Cramer, 1990, *The IIASA Climate Database for Land Area on a Grid of 0.5° Resolution*, WP-41, International Institute for Applied Systems Analysis, Laxenburg, Austria.

Meentemeyer, V., 1978, "Macroclimate and Lignin Control on Decomposition," *Ecology* 59:465-472.

Melillo, J.M., et al., 1993, "Global Climate Change and Terrestrial Net Primary Production," *Nature* 363:234-240.

Monteith, J.L., 1973, *Principles of Environmental Physics*, E. Arnold, London, England.

Neilson, R.P., 1993, "Vegetation Redistribution: A Possible Biosphere Source of CO_2 during Climate Change," *Water, Air and Soil Pollution* 70:659-674.

Olson, J.S., et al., 1983, *Carbon in Live Vegetation of Major World Ecosystems*, ESD 1997, Oak Ridge National Laboratory, Oak Ridge, Tenn.

Overpeck, J.T., et al., 1990, "Climate-Induced Changes in Forest Disturbance and Vegetation," *Nature* 343:51-53.

Parton, W.J., et al., 1988, "Dynamics of C, N, P and S in Grassland Soils: A Model," *Biogeochemistry* 5:109-131.

Pastor, J., and W.M. Post, 1988, "Response of Northern Forests to CO_2-Induced Climate Change," *Nature* 334:55-58.

Pastor, J., and W.M. Post, 1986, "Influences of Climate, Soil Moisture, and Succession on Forest Carbon and Nitrogen Cycles," *Biogeochemistry* 2:3-27.

Pastor, J., and W.M. Post, 1985, *Development of a Linked Forest Productivity-Soil Process Model*, ORNL/TM-9519, Oak Ridge National Laboratory, Oak Ridge, Tenn.

Penman, H.L., 1948, "Natural Evaporation from Open Water, Bare Soil and Grass," *Proceedings of the Royal Society of London*, Series A, 193:120-145.

Prentice, I.C, et al., 1992, "A Global Biome Model Based on Plant Physiology and Dominance, Soil Properties and Climate," *Journal of Biogeography* 19:117-134.

Priestley, C.H.B., and R.J. Taylor, 1972, "On the Assessment of Surface Heat Flux and Evaporation Using Large-Scale Parameters," *Monthly Weather Review* 100:81-92.

Running, S.W., and J.C. Coughlan, 1988, "A General Model of Forest Ecosystem Processes for Regional Applications. I. Hydrological Balance, Canopy Gas Exchange and Primary Production Processes," *Ecological Modeling* 42:125-154.

Shugart, H.H., 1984, *A Theory of Forest Dynamics*, Springer-Verlag, New York, N.Y.

Shugart, H.H., and D.C. West, 1980, "Forest Succession Models," *BioScience* 30:308-313.

Smith, T.M., et al., 1992, "Modeling the Potential Response of Vegetation to Global Climate Change," *Advances in Ecological Research* 22:93-113.

Smith, T. M., et al., 1990, "Global Forests," in *Progress Reports on International Studies of Climate Change Impacts*, U.S. Environmental Protection Agency, Washington, D.C.

Smith, T.M., et al., 1981, "Integrating," in T.M. Smith et al., 1992, "Modeling the Potential Response of Vegetation to Global Climate Change," *Advances in Ecological Research* 22:93-113.

Solomon, A.M., 1986, "Transient Responses of Forests to CO_2-Induced Climate Change: Simulation Modeling Experiments in Eastern North America," *Oecologia* 68:567-569.

Solomon, A.M., and H.H. Shugart, 1984, "Integrating Forest-Stand Simulations with Paleoecological Records to Examine Long-Term Forest Dynamics," in G.I. Agren (ed.), *State and Change of Forest Ecosystems: Indicators in Current Research*, report 13, pp. 333-357, Swedish University of Agricultural Science, Uppsala, Sweden.

Solomon, A.M., and T. Webb, III, 1985, "Computer-Aided Reconstruction of Late-Quaternary Landscape Dynamics," *Annual Review of Ecology and Systematics* 16:63-84.

Solomon, A.M., et al., 1984, *Response of Unmanaged Forests to CO_2-Induced Climate Change: Available Information, Initial Tests and Data Requirements*, TR009, U.S. Department of Energy, Carbon Dioxide Research Division, Washington, D.C.

Solomon, A.M., et al., 1981, "Simulating the Role of Climate Change and Species Immigration in Forest Succession," in D.C. West, H.H. Shugart, and D.B. Botkin (eds.), *Forest Succession*, pp. 154-177, Springer-Verlag, New York, N.Y.

Solomon, A.M., et al., 1980, "Testing a Simulation Model for Reconstruction of Prehistoric Forest-Stand Dynamics," *Quaternary Research* 14:275-293.

Thornthwaite, C.W., 1948, "An Approach Toward a Rational Classification of Climate," *Geogr. Review* 38:55-89.

Tosi, J.A., Jr., 1969, *Mapa Ecologico Republica de Costa Rica*, Tropical Science Center, San José, Costa Rica.

Tosi, J.A., et al., 1992, *Potential Impacts of Climate Change on the Productive Capacity of Costa Rican Forests: A Case Study*, Tropical Science Center, San José, Costa Rica.

Tropical Science Center and World Resources Institute, 1991, *Costa Rica Natural Resources Accounting Study*, San José, Costa Rica.

TSC and WRI: See Tropical Science Center and World Resources Institute.

Urban, D.L., and H.H. Shugart, 1989, "Forest Response to Climate Change: A Simulation Study for Southeastern Forests," in J. Smith and D. Tirpak (eds.), *The Potential Effects of Global Climate Change on the United States,* EPA-230-05-89-054, pp. 3-1–3-45, U.S. Environmental Protection Agency, Washington, D.C.,

Walter, H., 1985, *Vegetation of the Earth and Ecological Systems of the Geo-Biosphere*, 3rd ed., Springer-Verlag, Berlin, Germany.

Watt, A.S., 1947, "Pattern and Process in Plant Communities," *Journal of Ecology* 35:1-22.

Weinstein, D.A., et al., 1982, *The Long-Term Nutrient Retention Properties of Forest Ecosystems: A Simulation Investigation*, ORNL/TM-8472, Oak Ridge National Laboratory, Oak Ridge, Tenn.

Woodward, F.I., 1987, *Climate and Plant Distribution*, Cambridge.

APPENDIX G:

WATER RESOURCE IMPACTS

R. Benioff et al. (eds.), Vulnerability and Adaptation Assessments, G-1–G-64.
© 1996 *Kluwer Academic Publishers.*

APPENDIX G NOTATION

The following is a list of the acronyms, initialisms, and abbreviations (including units of measure) used in this appendix. Some acronyms found only in tables are defined in those tables.

ACRONYMS, INITIALISMS, AND ABBREVIATIONS

AET	actual evapotranspiration
CO_2	carbon dioxide
CPS	consumer-producer surplus
EPA	U.S. Environmental Protection Agency
GCM	general circulation model
GDP	gross domestic product
IIASA	International Institute for Applied Systems Analysis
PET	potential evapotranspiration
SHE	Système Hydrologique Européen
TVA	Tennessee Valley Authority

UNITS OF MEASURE

°C	degree(s) Celsius
cal	calorie(s)
cm	centimeter(s)
d	day(s)
g	gram(s)
h	hour(s)
K	Kelvin(s)
kcal	kilocalorie(s)
kg	kilogram(s)
KJ	kilojoule(s)
km	kilometer(s)
kPa	kilopascal(s)
m	meter(s)
m^3	cubic meter(s)
mbar	millibar(s)
MJ	megajoule(s)
mm	millimeter(s)
m/s	meter(s) per second [flux]
Pa	pascal(s)
s/m	seconds per meter [resistance]
yr	year(s)

CONTENTS

TABLES

FIGURES

APPENDIX G:

WATER RESOURCE IMPACTS

G.1 INTRODUCTION

This appendix provides technical details for participants in the Country Studies Program to conduct an analysis of water resource vulnerability and adaption by using the primary approach outlined in Section 5.4. The appendix is intended for analysts who are familiar with the tools and methods of hydrology, water resource engineering, and economic analysis. It provides extended technical detail and guidance on how to use the standard state-of-the-art water resource planning tools to assess water resource vulnerability and adaptation to climatic change. Therefore, this appendix is not a "handbook" or "how-to book" on the recommended tools: reference documents are listed for persons who require more background material. In addition, the technical approaches for the assessment are based solely on techniques and methodologies found in the references. Each country team should obtain at least one copy of each of the references listed below:

- General reference: *Climate Change & U.S. Water Resources* (Waggoner 1990)

- Water supply: *Handbook of Hydrology* (Maidment 1993)

- Water demand: *Water in Crisis* (Gleick 1994), *Modeling Water Demand* (Kindler and Russell 1984), and *Industrial Water Use and Treatment Practices* (Carmichael and Strzepek 1985)

- Supply-and-demand balance: *World Water Resources and Regional Vulnerability: Impact of Future Changes* (Kulshreshtha 1993)

- Adaptation: *Water Resource System Analysis and Planning* (Loucks et al. 1981) and *Principles of Water Resources Planning* (Goodman 1984)

- Economic impact assessment: *Principles of Water Resources Planning* (Goodman 1984)

The appendix is structured as follows: First, an overview of the steps of the primary approach is given. Background material is then provided on hydroclimatic data and hydrological analyses, including potential evapotranspiration (PET) and snowmelt modeling. Finally, the procedure for water resource impact analyses is discussed in detail. The concluding sections provide information on supplementary approaches to water resource assessment that should be considered.

G.1.1 The Primary Approach

The primary approach to be followed in assessing the vulnerability of a country's water resources to climate change is a comprehensive assessment of the full water resource system, including supply, demand, and management.

Supply is analyzed in two stages. The first stage is to assess river runoff impacts by using one of the methods recommended below. The second stage is to assess the impact of the affected river runoff on the management of the water resource system and the resulting water supply.

Demand is analyzed on an aggregated national basis. Because water demand is closely tied to socioeconomic driving forces, an important part of demand impact assessment depends on the base scenario projection of population, agricultural and industrial production, and energy demand.

Water resource management is assessed by using systems analytic methodologies that include the impacts on supply and demand as well as potential adaptation measures.

G.1.1.1 Supply

A river-basin-scale monthly water balance is recommended as the primary approach for assessing climate change impacts on river runoff. The CLIRUN set of models is the standard water balance tool selected for the Country Studies Program. A monthly mean value assessment is the minimum analytical scale permitted, while a multiple-year time series is highly recommend. Data for the mean values are available from the U.S. Environmental Protection Agency (EPA) or the technical support staff. The Priestley-Taylor method is used to estimate PET. To determine effective precipitation, the temperature index method is used for modeling snowmelt. Once impacts on river runoff have been determined, the performance of the river basin or national water resource system is simulated by applying a water resource systems model.

G.1.1.2 Demand

A simple linear multiplier approach is recommended for forecasting water demand on the basis of estimates of population and economic growth to the base year 2075 based on current climate. Change per unit of water induced by climate change (for example, for agricultural purposes) is estimated by using expert judgment based on the result of the modeling efforts under way in the agricultural analysis component of the Country Studies Program.

G.1.1.3 Vulnerability

The water availability/water use vulnerability criteria presented in Kulshreshtha (1993) are applied for each river basin unit, while national vulnerability is assessed via a gross national supply-and-demand balance or a weighted average of river basin vulnerabilities.

G.1.1.4 Adaptation

National-level adaptations are examined via additional national vulnerability assessments that incorporate the possibility of interbasin transfers of water and shifts in population and economic growth. One or more key or representative river basins are selected to examine alternative management strategies for adaptation to climate change impacts at a basin level. This assessment includes alternative operation of the existing water resource infrastructure as well as the addition of capital investments, such as reservoirs or canal linings. National implications drawn from insights gained from the representative basins study are developed.

G.1.1.5 Economic Analysis

Two approaches are suggested for economic analysis — a micro approach and a macro approach. The micro approach can take place at the firm level or at the sectoral level. The suggested methodology for assessing climate change impact is to estimate the effect climate change has on both the supply and the demand curves and then calculate the changes (positive or negative) in the consumer-producer surplus (CPS). The results can then be scaled to a national level. The macro approach includes climatic effects in a model of the macro economy, such as an input-output model or computable general equilibrium model. These models take into account indirect effects and are national in scale.

Sections G.2-G.6 provide more technical details on the recommended methods for the primary approach as well as a discussion of more detailed hydrological modeling. More detailed hydrological modeling is suggested if the time and resources are available to Country Studies Program teams.

G.1.2 Hydroclimatic Data

Climate evaluations are usually centered in the mean behavior of hydroclimatic variables. Hydrology, on the other hand, is centered in both the mean behavior and the variability of these hydroclimatic variables around that mean behavior.

Hydrological processes are dominated by changes in both temporal and spatial aggregation. The appropriate temporal and spatial scale for hydrological modeling of climate impacts is a difficult question and is discussed below. Niemann et al. (1994) found little difference between monthly time series and monthly means for a water balance model of climate change impacts on runoff. The data needed to perform the primary approach with the recommend methods are listed below:

- Monthly river runoff values for each river basin to be studied (from a local hydro-met service or the Global Runoff Data Centre Koblenz);

- Spatially averaged mean monthly values of temperature and precipitation, covering the watershed of each river basin unit modeled for the time period that coincides with the time period from which the monthly stream-flow values were calculated (from a local hydro-met service or the International Institute for Applied Systems Analysis [IIASA] global climate database);

- Total- and elevation-area values for each river basin unit to be modeled (from local survey or mapping service or IIASA global climate database); and

- Spatially averaged estimates of the changes in mean monthly temperature and precipitation because of a doubling of CO_2 ($2XCO_2$) generated by a general circulation model (GCM) (provided by the Country Studies Program team).

G.1.3 Meteorological Data Requirements

The data requirements for the Priestley-Taylor PET equation are significant. At a minimum, the following meteorological data are required:

- Solar radiation, R_n (langley/d);

- Air temperature, T_a (°C),

- Saturated vapor pressure and actual vapor pressure, e_s and e; and

- Wind speed, u (m/d or km/d).

The following data can be used to estimate R_n:

- Specific humidity, q_v (dimensionless), relative humidity, R_h (%), or dewpoint temperature, T_d (°C); and

- Sunshine hours, n (h), or cloud cover, f.

These data are available globally from the IIASA climate database as mean monthly values. Data on the changes in these values due to climatic change can be taken directly from the GCM, or they can be assumed to remain the same under climate change. Table G.1, adapted from Fennessey and Kirshen (1994), presents the availability of data in four of the most widely used GCMs.

TABLE G.1 Availability of Data for Four GCMs

Hydroclimatic Variable	Model[a]			
	OSU	UKMO	GISS	GFDL
Temperature	*	*	*	*
Precipitation	*	*	*	*
Ratio mixing ratio	*			*
Solar radiation	*	*	*	*
Cloud cover ratio	*		*	*
Wind speed ratio	*	*	*	*

[a] OSU= Oregon State University; UKMO= United Kingdom Meteorological Office; GISS= Goddard Institute for Space Studies; and GFDL= Geophysical Fluid Dynamics Laboratory.

G.1.4 Hydrological Concepts Important in Climate Change Assessment

The PET and snowmelt processes are two key concepts that are directly affected by climate change. The water resources technical team feels that these concepts are so important in understanding the impact of climatic change on hydrology that they are presented in summary here; they are presented in detail in the *Handbook of Hydrology* (Maidment 1993).

G.1.4.1 Potential Evapotranspiration

The evaluation of evaporation and transpiration is important in studies of the impact of climate change on water resources. Numerous models and equations are available for computing evaporation and transpiration as a function of climatological and hydrological data. Unfortunately, these climatological and hydrological data are not available in all parts of the world, and simpler techniques must be used to overcome these limitations. Furthermore, scenarios for climate change generally provide information only on changes in temperature and precipitation. The user is thus forced to assume that the other climatologic variables required to estimate evaporation and transpiration remain at the current levels, or these variables can be scaled by using the information of the GCMs. Lettenmaier et al. (1994) describe some of these adjustments.

This section briefly describes the principal processes that affect evapotranspiration and several methods for estimating evapotranspiration, including energy balance, aerodynamic, combination, and temperature-based methods. The Penman-Monteith method can be used when climatological information is available, or estimated reliably from empirical equations. This equation can be used for daily or longer time steps. The Hargreaves method can be used when climatologic data are not available for time intervals equal to or longer than 1 month. The Priestley-Taylor equation is the method recommended for the primary approach.

G.1.4.1.1 Definitions. The rate of evaporation from an open water surface is primarily based on two factors (Chow et al. 1988):

- Ability to provide latent heat, the energy required for evaporation, because evaporation absorbs heat from its environment; and

- Ability to remove water vapor from the evaporating surface (i.e., wind and the humidity gradient in the air above the evaporating surface).

An additional factor must be considered for the rate of evaporation from a bare soil: the availability of soil moisture to be evaporated.

To facilitate the presentation of the different evaporation and transpiration methods, the following definitions are needed (taken from Shuttleworth [1993]):

- *Evaporation*: rate of liquid water transformation to vapor from open water, bare soil, or vegetation with soil beneath;

- *Transpiration*: part of the total evaporation that enters the atmosphere from the soil through the plants;

- *Evapotranspiration*: combination of evaporation from bare soils and transpiration from plants;

- *Potential evaporation*: quantity of water evaporated per unit area, per unit time, from an idealized, extensive free water surface under existing atmospheric conditions;

- *Reference crop evapotranspiration*: rate of evaporation from an idealized grass crop with a fixed crop height of 0.12 m, an albedo of 0.23, and a surface resistance of 69 s/m; and

- *Actual evapotranspiration*: rate of evaporation when the supply of water is limiting.

The ratio of actual evapotranspiration (AET) to PET is shown, in an idealized form, in Figure G.1. It is assumed that AET achieves the maximum level of evapotranspiration, PET, when the soil moisture content achieves the value q^* that is smaller than the field capacity of the soil. A simpler approximation assumes a linear AET/PET ratio between the permanent wilting point, which is a function of the soil and crop, and the field capacity, which is a function only of the type of soil.

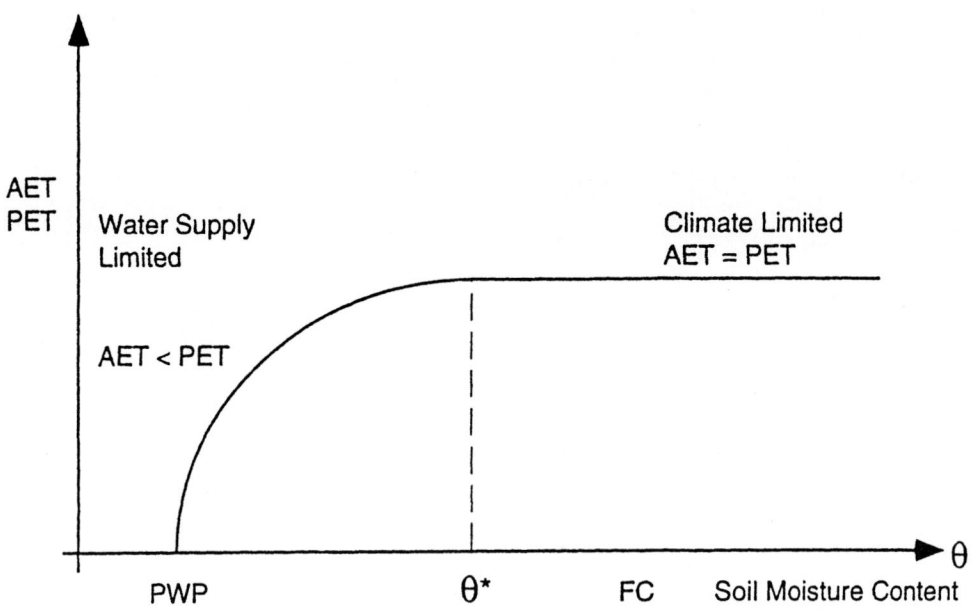

FIGURE G.1 Ratio of Actual Evapotranspiration to Potential Evapotranspiration

G.1.4.1.2 Energy Balance Methods. In the energy balance method, it is assumed that the ability of the system to remove moist air is not limiting to the evaporation process. The energy balance is given by

$$\frac{dH}{dt} = R_n - H_s - G \quad , \tag{G.1}$$

where

dH/dt =   rate of change of storage of energy in the water body;

R_n =   net radiation;

H_s =   sensible heat lost to air; and

G =   sensible heat lost to ground.

After some manipulations, the evaporation estimate, E_r, can be obtained as

$$E_r = \frac{l}{\lambda \rho_w} \left(R_n - H_s - G \right) \quad , \tag{G.2}$$

where

l = latent heat of vaporization (amount of heat needed, in calories or in MJ · K^{-1}, to evaporate 1 gram of water at 1 atmospheric pressure to vapor, >>539 cal/g); and

ρ_w = density of water (kg/m^3).

G.1.4.1.3 Aerodynamic Methods. The second factor that governs evaporation is the ability to transport vapor from the free surface:

• Humidity gradient in the air over the surface, and

• Wind speed across the surface.

The basic equation is as follows:

flux = constant • gradient,

and the flux is either momentum or vapor flux. The evaporation estimate made by assuming unlimited availability of energy for evaporation is

$$E_a = B(e_{as} - e_a) \ , \tag{G.3}$$

where

$$B = \frac{0.622 k^2 \rho_a u_2}{p\rho_\omega \left[\ln\left(\dfrac{z_2}{z_0}\right)\right]^2} \ ; \tag{G.4}$$

and

B = Bowen ratio;

k = von Karman's constant (0.4);

ρ_a = air density;

u_2 = mean wind velocity measured at elevation z_2;

p = atmospheric air pressure;

ρ_w = water density;

z_2 = elevation at which wind measurements are made; and

z_0 = roughness height.

G.1.4.1.4 Combination Methods. In the methods discussed earlier, it was assumed that either the energy available or the ability to remove saturated air was not limiting, but in reality, both factors are limiting in most cases. This condition gives rise to the combined methods, which give a weighted average of the two estimates. The Penman equation is the most widely known combined-method equation.

G.1.4.1.4.1 *Penman Equation.* The general form of the Penman equation (Penman 1948) is

$$E_c = \frac{\Delta}{\Delta + \gamma} E_r + \frac{\gamma}{\Delta + \gamma} E_a \quad , \qquad (G.5)$$

where

E_c = combined evaporation estimate (mm/d);

Δ = slope of the saturated vapor pressure curve; and

γ = psychometric constant = $C_p\, p\, K_h / (0.622\, l_v\, K_w)$.

G.1.4.1.4.2 *Penman-Monteith Equation.* The Penman equation (Penman 1948) has been modified to evaluate the evaporation and transpiration processes in the Penman-Monteith equation (Monteith 1985). This equation has the following form:

$$E = \frac{1}{\lambda} \left[\frac{\Delta A + \rho_a C_p (e_s - e)/r_a}{\Delta + \gamma(1 + r_s/r_a)} \right] \quad , \qquad (G.6)$$

where

A = available energy (MJ) $\gg R_n$ - G;

e_s - e = vapor pressure deficit = D;

r_a = aerodynamic resistance (s/m); and

r_s = surface resistance (s/m).

The other terms have already been defined. The Penman-Monteith equation is one of the most accurate equations for estimating evaporation. It is recommended in the *Handbook of*

Hydrology (Maidment 1993), and it performed well in the comparison study reported by Jensen et al. (1990). However, the data requirements for the Penman-Monteith equation are very large.

The assumptions of the Penman-Monteith method are as follows:

- Steady-state energy flow prevails (no diurnal cycle, $\Delta t > 1$ d), and

- Changes in heat storage are not significant (i.e., not adequate for a lake).

G.1.4.1.4.3 *Priestley-Taylor Equation.* Priestley and Taylor found that for very large areas, the second term of the Penman equation described in Equation (G.6) is approximately 30% of the first term. Thus, an approximation to the Penman equation that is less demanding of data can be written as follows:

$$E_c = \alpha \ \frac{\Delta}{\Delta + \gamma} \ E_r \ ,$$ (G.7)

where α is approximately 1.3 (closer to 1.26 in most references). This approximation performed fairly well in the comparison study reported by Jensen et al. (1990).

G.1.4.1.5 Temperature-Based Equations. Estimates of PET that are mainly based in temperature have been proposed since the 1920s; their main attraction has been the limited data requirements needed to produce the evaporation and evapotranspiration estimates. Of these methods, the most widely used is the Thornwaite method (Thornwaite 1948). It was developed for the east-central United States and meant only to apply to mid-latitude climates similar to those of the east-central United States, but it has been used widely everywhere. Several studies that show that the Thornwaite method usually under-estimates evapotranspiration (see, for example, Jensen et al. [1990]), and it should not be used in this study.

Another commonly used equation is the Blaney-Criddle method (Blaney and Criddle 1962) and its modifications (see, for example, Shuttleworth [1993]; Jensen et al. [1990]). In the original method, the monthly PET estimates for a particular crop are obtained as a function of latitude, mean monthly temperature, and a monthly crop coefficient. Improvements in this original form require additional information and calibration for a particular site.

The method of Hargreaves (Hargreaves 1981; Hargreaves et al. 1985) is the one recommended for this study when monthly time steps are used and the only information available is temperature. The Hargreaves method expresses the reference crop evapotranspiration, E_{rc}, as follows:

$$E_{rc} = 0.0022 \cdot R_A \cdot TD^{0.5} \cdot (T + 17.8) \ ,$$ (G.8)

where

E_{rc} = reference crop evapotranspiration (mm/d) (i.e., evapotranspiration for short grass);

R_A = mean extraterrestrial radiation (mm/d), which is a function of the latitude, f, and can be evaluated by using Equation (G.9);

TD = temperature difference = mean monthly maximum temperature – mean monthly minimum temperature for the month of interest (°C); and

T = mean air temperature (°C).

G.1.4.1.6 Numerical Examples. Table G.2 shows the application of the Penman-Monteith equation with the data supplied by Jensen et al. (1990). The second column shows the computation with nominal and estimated values for some of the coefficients.

Table G.3 shows the steps in a numerical example of the calculation of evapotranspiration estimates by using the modified Penman and Hargreaves methods for Kimberly, Idaho, and data obtained from Jensen et al. (1990).

G.1.4.1.7 Additional Equations. The extraterrestrial radiation, R_A, needed for both the Penman-Monteith and the Hargreaves equations can be estimated by the following (Shuttleworth 1993):

$$R_A = 15.392 \, d_r(w_s \sin f \sin d + \cos f \cos d \sin w_s) \; , \tag{G.9}$$

where

R_A = extraterrestrial radiation (equivalent mm/d) at the site under study;

d_r = relative earth-sun distance, evaluated by

$$d_r = 1 + 0.033 \cos (2pJ/365); \tag{G.10}$$

w_s = sunset hour angle (radians), to be evaluated by

$$w_s = \text{arc cos } (-\tan f \tan d); \tag{G.11}$$

f = latitude of site (plus for Northern Hemisphere, minus for Southern Hemisphere) (radians);

TABLE G.2 Example of Application of Penman-Monteith Method for Kimberly, Idaho, a Sample Site

Data	Term	Value	Value	Unit	Nominal Value
Latitude (+N)	f	42.4	0.740019603	radians	
Elevation	EL	1,195	1,195	m	
Month	J	July	July		
Julian day (mid-July)		195	195		
Air temperature	T_a	20.8	20.8	°C	
Mean monthly maximum temperature	T_M	30		°C	
Mean monthly minimum temperature	T_m	11.7		°C	
Wind speed	u	240	2.78	m/s, km/d	
Relative humidity	R_h	60	60	%	
Sunshine hours		14.8	14.8	h	
Albedo	a	0.05	0.023		0.23
Net radiation observed	R_n	382	15.98861	MJ/m²·d, langley/day	0.75
Mean extraterrestrial radiation	R_a	40.24	16.412	mm/d, MJ/(m²·d)	
Specific heat	C_p	0.24	1.013	kJ/kg·K, kcal/g·°C	
von Karman's constant	K	0.41			
Stefan-Boltzmann constant	s	4.903e-08[a]	4.903e-08[a]	MJ/m²·K 4 d	
Roughness height	z_0	0.25	0.0025	m, cm	
Wind speed measurement elevation	z	2		m	
Computed Parameters					
Latent heat of vaporization	l	584.392	2.4519	MJ/kg, cal/g	
Slope of saturated vapor pressure curve	Δ	1.509	0.1511	kPa/°C	
Atmospheric pressure	p	886.9275	88.6928	kPa, mbar	
Air density	ρ_a	0.00118937	1.0452	kg/m³	
Water density	ρ_w	997	1.000	kg/m³, g/cm³	1,000
Psychrometric constant	γ	0.58560	0.1129	kPa/°C, kg/m³, kPa/°C	

TABLE G.2 (Cont.)

Data	Term	Value	Value	Unit	Nominal Value
Saturated vapor pressure	e_s		24.562	kPa, mb	
Actual vapor pressure	e		14.737	kPa, mb	
Saturated vapor pressure deficit	D		9.825	kPa	
Bowen ratio	B	4.13848e-08	4.07351e-07	m·s/Pa	
Weighting parameter	$D/(D+g)$	0.72	0.719513643		
	$g/(D+g)$	0.28	0.280486357		
Aerodynamic resistance	r_a		74.88	s/m	
Surface resistance	r_s		69	s/m	69
Outgoing heat conduction to soil	G		0	MJ/m². d	
Difference mean monthly maximum/minimum temperature	ΔT		18.3	°C	
Sunset hour angle	w_s		1.944	radians	
Relative sun-earth distance	d_r		0.968		
Solar declination	d		0.380	radians	
Number of daylight hours	N		14.85	h	
Extraterrestrial radiation	S_o	41.08	16.75	mm/d MJ/m²·d	
Net radiation	R_n		30.7378		
Available energy	$A = R_n\text{-}G$		30.7378		
Penman-Monteith estimate	E_{rc}		7.1908	mm/d	

[a] Read as 4.903×10^{-8}.

d =   solar declination (radians), to be evaluated by

$$d = 0.4093 \sin (2pJ/365 - 1.405); \qquad\qquad\text{(G.12)}$$

J =   Julian day.

G.1.4.2 Snowmelt

In many of the high- and mid-latitude, high-elevation watershed basins throughout the world, melting of the snowpack represents the most significant hydrological event. In these regions, runoff from the shallow snow cover often provides 80% or more of the annual surface runoff, replenishes the soil moisture, and recharges the groundwater reservoirs (Maidment 1993). The importance of this meteorological event cannot be neglected in any attempt to model the runoff process by using water balance models. The seasonal, lumped integral water balance models that have been described require the input of effective precipitation; thus, in cold regions, where snow accumulates, it is necessary to derive an "effective precipitation." This term can be defined as follows (Kaczmarek 1991):

$$
\begin{aligned}
P_{eff}(t) = \ &\text{measured precipitation} \times \text{correcting factor} - \text{interception} \\
&+ \text{snow accumulation} - \text{snow melting}
\end{aligned}
\qquad\text{(G.13)}
$$

Kaczmarek (1991) notes that existing precipitation gauges underestimate precipitation because of wind, wetting losses, and gauge evaporation. The correction factor will depend on the type of gauge used, and no attempt is made here to define what this factor might be. The actual precipitation data should be interpreted with the understanding that it might need to be corrected to remove gauge bias. The interception term represents canopy interception and evapotranspiration. No satisfactory theory exists that can determine correct interception factors, as they have been shown to vary widely by region. In most of the seasonal analysis that uses rain gauge data, the correction term is not included, and any

TABLE G.3 Example of Application of Hargreaves Method

Data	Term	Value		Units
Latitude	f	42.4	0.74002	radians
Mean temperature	T	20.8		°C
Difference mean monthly maximum/minimum temperature	ΔT	18.3		°C
Sunset hour angle	w_s	1.944		radians
Relative sun-earth distance	d_r	0.968		
Solar declination	d	0.380		radians
Julian day (mid-July)	J	195		
Extraterrestrial radiation	S_o	16.75	16.412	mm/d
Reference crop evaporation	E_{rc}	6.36	6.23	mm/d
Number of daylight hours	N	14.85		h

error that is introduced in this assumption is assumed to be taken care of in the calibration procedure. The final two terms in the derivation of effective precipitation are snow accumulation and snow melting. These factors play significant roles in many regions and cannot be neglected.

G.1.4.2.1 Snowmelt to Determine Effective Precipitation. Many models are available for forecasting snowmelt runoff. Most use either an energy balance approach or a temperature-index method to compute the melting rate. When water balance models with longer time events, on the order of a month or longer, are used, the temperature index method gives good results for average conditions and is the preferred method in most cases because of its simplicity of application. The energy balance method is described briefly to give the reader a feeling of its complexity.

G.1.4.2.1.1 Energy Balance Method. The energy balance approach applies the law of conservation of energy to a control volume (i.e., a snow pack where the lower boundary is the snow-ground interface and the upper boundary is the snow-air interface). When a control volume is used, energy fluxes are used to express internal energy changes. The balance requires that the sum of the energy fluxes by radiation, convection, conduction, advection, and internal energy change of the volume sum to zero (Gray et al. 1993). As an example of the complexity of the energy balance method, one can take a unit volume of snow and define all the energy transfers that give the amount of energy available for melting snow, Q_m:

$$Q_m = R_n + H + LE + G + A - \Delta U / \Delta t \quad , \tag{G.14}$$

where

R_n = net radiation (radiation exchange);

H = sensible energy (temperature difference at the surface-air interface);

LE = latent energy (vapor movement);

G = ground heat (heat movement to ground by conduction);

A = advective energy (external sources such as rain); and

$\Delta U / \Delta t$ = rate of change of internal energy.

From this expression, it can be easily seen that it would be difficult to identify all these energy elements. Thus, although the energy balance model gives a physical basis for estimating snowmelt, the data required to derive all the elements are too extensive to be immediately practical. However, it has been observed that a temperature-based approach can be a good representation of the physical mechanisms of snowmelt.

G.1.4.2.1.2 *Temperature Index Model.* The temperature index method should be used when analyzing climate sensitivity on a monthly basis with the simplified water balance models proposed in the primary approach. The temperature index method has advantages in that (1) it gives good estimates compared with the energy balance method, and (2) temperature is a measured climatic variable that has a high degree of certainty. No single temperature-based method has been accepted because of the large regional variation. Several temperature-index expressions for calculating daily snowmelt for some North American regions are given in the *Handbook of Hydrology* (Gray 1993). Two simple methods are given below.

The first method uses two temperature threshold values and a linear relationship between these two thresholds to represent the melting process. This relationship gives a simple way of representing effective precipitation and is the recommended method. With this approach, the initial snow accumulation must be known, so it is preferable to begin the data series during the warmer portion of the season (Ozga-Zielinska 1993):

$$Pe_i = \alpha_i (A_{i-1} + Pm_i) \quad , \tag{G.15}$$

where

$$\alpha_i = \begin{cases} 0 & \text{for} \quad T_i \leq T_s \\ 1 & \text{for} \quad T_i \geq T_l \\ \dfrac{(T_i - T_s)}{(T_l - T_s)} & \text{for} \quad T_s < T_i < T_l \quad ; \end{cases} \tag{G.16}$$

$$A_i = (1 - \alpha_i)(A_{i-1} + P_i) \quad ; \tag{G.17}$$

and

Pe_i = effective precipitation at time i (mm);

α_i = accumulation index ($0 \leq \alpha_i \leq 1$);

A_i = accumulation at time step i (mm);

Pm_i = measured precipitation at time i (mm);

T_i = temperature at time interval i (°C);

T_s = solid snow threshold, completely frozen ($\cong$ -3°C); and

T_l = upper temperature threshold, liquid above this value ($\cong$ 3°C).

Table G.4 gives a sample calculation sequence for this temperature method, and Figure G.2 is a plot of these calculations. The initial accumulation is assumed to be 0 mm. The values for the temperature parameters are given as T_l = 3°C and T_s = -3°C.

A second temperature method is given in the *Handbook of Hydrology* (Gray et al. 1993), which gives an expression for the melt water produced during a time interval:

$$M = M_f (T_i - T_b) \quad , \tag{G.18}$$

where

$$M_f = 0.011 \rho_s \quad , \tag{G.19}$$

and

T_i = usually given as the mean temperature in the time interval (°C);

T_b = base temperature ($\cong$ 0°C);

M_f = melt factor (3.5-6 mm/°C d); and

ρ_s = snow density (300-500 kg/m^3).

No values for the melt factor M_f or the snow density ρ_s are universally accepted because each depends on unique basin characteristics and the time of the year, vegetation, topology, and other variables.

G.1.4.2.2 Conclusions on Snowmelt. The importance of the snowmelt process centers around the need to derive an effective precipitation for the rainfall runoff modeling process. The discussion above is not meant to be the definitive guide to modeling the snowmelt process because each region and basin are unique. The recommended reference materials listed in Section G.1 provide a more in-depth discussion of this topic.

TABLE G.4 Example of Temperature-Index Model

Month	Pm	Pe	T	A	a
October	35.60	35.60	8.2	0.00	1.00
November	30.50	29.99	2.9	0.51	0.98
December	59.60	33.06	0.3	27.05	0.55
January	20.60	10.32	-1.7	37.32	0.22
February	30.00	0.00	-12.6	67.32	0.00
March	26.80	21.96	-1.6	72.16	0.23
April	30.70	102.86	5.7	0.00	1.00
May	24.90	24.90	12.6	0.00	1.00
June	80.00	80.00	17.2	0.00	1.00
July	62.80	62.80	17.1	0.00	1.00
August	85.20	85.20	15.3	0.00	1.00
September	39.90	39.90	12.2	0.00	1.00
October	71.10	71.10	7.7	0.00	1.00
November	26.30	7.89	-1.2	18.41	0.30

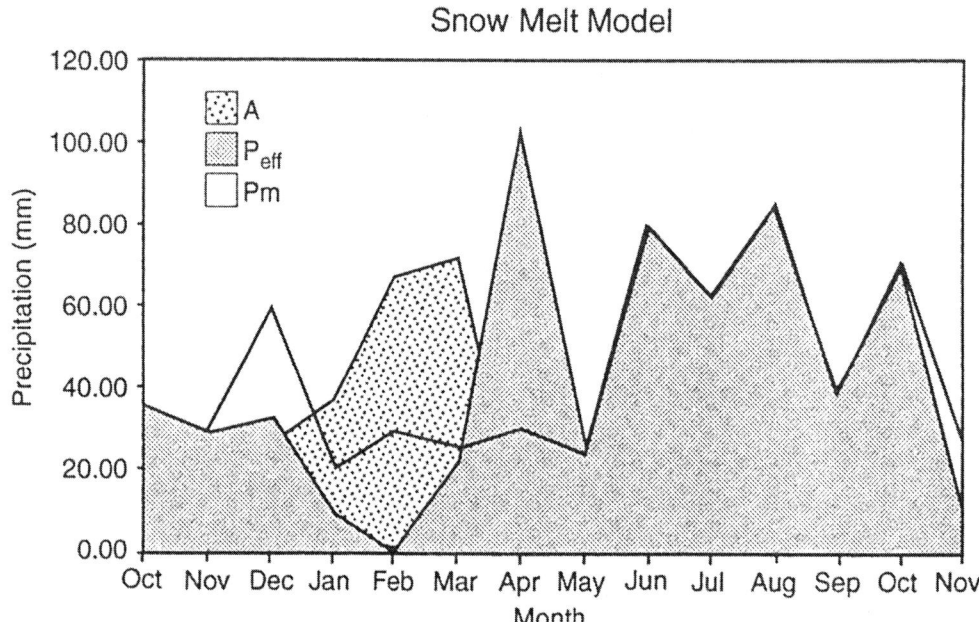

FIGURE G.2 Plot of Measured and Effective Precipitation and Snow Accumulation

G.2 SUPPLY

G.2.1 Regional to National Impacts: Identifying the Water Resource System

The primary approach for assessing hydrological climate impacts is a river basin approach. The relationship of a river basin impact assessment to a national level assessment can take two different forms:

- The river basin covers only part of the total national watershed.

    - The national hydroclimatic characteristics are homogeneous.

    - The nation is made up of two or more distinct hydroclimatic zones.

- The national watershed is completely encompassed in a large international river basin.

    - The country is in an upper basin location.

    - The country is in a lower basin location.

G.2.1.1 The Representative Basin Approach

When the river basin assessed covers only part of the total national watershed, the river basin becomes a representative basin. If the country is located in a single hydroclimatic zone, a single basin can be used. The relative changes in runoff in the representative basin are the basis for a national runoff impact factor. The national water availability is then multiplied by the national runoff impact factor to determine national impacts.

When a nation is made up of two or more distinct hydroclimatic zones, a representative river basin from each zone must be assessed. Then, the relative changes in runoff in the representative basin are the basis for a zonal runoff impact factor. The zonal water availability is then found by multiplying all the streams in the zone by the zonal run-off impact factor to determine zonal impacts. The national impact is determined by an area-weighted summation over all zones.

G.2.1.2 The Subarea Approach

When the national watershed is completely encompassed in a large international river basin, two approaches can be used. If data are not available for the streamflow at the national border, the relative changes in runoff in the entire basin are the basis for a national runoff impact factor. The national water availability is then multiplied by the national runoff impact factor to determine national impacts.

If national data are available, the basin must be separated into a national component and upstream and downstream components. If the country is located in the headwaters of the basin, or in an upper basin location, the single representative basin approach mentioned above is used.

If the country is located in the middle or downstream of the basin, or in a lower basin location, both the upstream portion of the basin and the national portion must be assessed. The multizonal approach discussed above should be used, with the upstream portion of the basin as one zone and the national portion as one or more zones. Kaczmarek (1994) used this approach to assess climate change impacts on the runoff of Poland.

G.2.2 The Hydrological Impact Assessment

The steps involved in assessing a water resource climate change (assuming that the system has been identified) are as follows:

1. Specify the climate change scenario.

2. Implement and validate a hydrological (precipitation) model for historical conditions.

3. Simulate hydrological conditions for the altered climate scenario.

4. Implement and validate a water resources system model for historical conditions.

5. Simulate water resource system performance for altered climate by applying a water resource system model to the results of step 3.

This section describes simple approaches that can be used for each of these steps in the context of the Country Studies Program.

G.2.2.1 The Climate Change Scenario

Among the methods for specifying climate change scenarios are (1) output from GCMs, (2) prescribed climatic change scenarios, and (3) historical and spatial analogues. Given the time and resource limitations of the Country Studies Program, only the first two methods are recommended. They are described briefly below.

G.2.2.1.1 Scenarios Based on GCM Output. General circulation models provide physically based predictions of the way climate might change as a result of increasing concentrations of atmospheric carbon dioxide (CO_2) and other trace gases. The GCMs are mathematical representations of the earth's climate system, and they simulate atmospheric processes at a field of grid points that cover the surface of the earth. The resolution of

GCMs, or grid spacing, has improved as supercomputers have become faster; for long-term climate simulations (runs of multiple decades), most of the major GCM centers are now running at grid resolutions around 5° × 5° (latitude × longitude), and some long-term runs have been made at about half that resolution.

One problem in developing alternative climate scenarios for water resource studies based on GCM simulations is that the coarse spatial resolution of GCMs precludes accurately representing regional or local climatic and hydrological variables, such as temperature, precipitation, evapotranspiration, soil moisture, and runoff, at the spatial scale needed to study climate effects. Because large errors exist in GCM estimates of current climate for many regions of the world (see, for example, Grotch and MacCracken [1990]), direct use of GCM predictions for areas the size of most watersheds is not recommended.

Instead, a commonly used approach is to compute a relative change in climate between a current climate GCM simulation and an altered climate (e.g., $2XCO_2$) simulation. The difference between the two simulations, expressed, for instance, as a ratio or difference of the temporal means of the $2XCO_2$ run and the current ("control") run, is used to adjust an observed series of climate variables. The motivation for this approach is that, although GCMs may not accurately estimate the local statistics of regional climate variables, their internal consistency and strong physical basis may provide plausible estimates of relative changes in climatic variables (Gates 1985). The advantages of this approach are that it is simple and directly uses observed records. The disadvantage is that it retains the historical temporal sequencing of events. For example, the method is unable to reflect changes in the timing of storms (storm interarrival times) that might occur in a warmer climate. Also, there is the problem of determining the appropriate temporal and spatial averaging for the GCM results. If the GCM simulation period is short (e.g., less than 25 years or so), the estimate of the temporal mean will be subject to considerable error; hence, the estimate of the ratio or difference between a $2XCO_2$ and control run may be poor. This problem has been alleviated as faster supercomputers have made GCM runs up to several hundred years long, but it is doubtful that such long runs will be available to the Country Studies Program.

In the ratio and difference methods, the appropriate number of GCM grid cells to average must be selected as well. It is tempting to use the nearest GCM grid node or even to interpolate spatially to the location of interest. However, most GCMs use a spectral representation, the fundamental (Nyquist) frequency of which is represented by two grid nodes in each dimension. Therefore, von Storch et al. (1993) argue that no information at spatial resolutions is higher than twice the distance between grid cells, and that the minimum effective spatial resolution is defined by four or more grid nodes (at least two in each direction). This argument implies that rather than interpolation, a spatial average of at least the four nearest grid nodes should be taken as the basis for computing the ratio or difference of altered and base climate simulations.

G.2.2.1.2 Prescribed Climate Change Scenarios. Prescribed climate change scenarios are based on simple specification (e.g., in a sensitivity analysis context) of plausible changes in surface variables, such as temperature, precipitation, and evapotranspiration, that are of hydrological importance. Prescribed changes can be applied uniformly throughout the year, or they can vary on a monthly or seasonal basis. Appropriate ranges of changes for each climatic variable can be found by searching current climatic change literature, or they can be directed by predictions from GCMs.

Generally, increases in temperature are considered, typically in the range of 2-4°C. Because of the large uncertainty regarding the direction and magnitude of changes in precipitation, both increases and decreases in precipitation, typically in the range of ±20-30%, are often used. Changes in precipitation can be simple changes in monthly or annual precipitation totals, or they can include changes in storm duration, storm intensity, and interstorm period (Wolock and Hornberger 1991). In addition to changes in temperature and precipitation, changes in other variables, such as stomatal resistance to transpiration, may also be prescribed.

G.2.2.2 A Hydrological Model for Historical Conditions

Hydrological models are the tool that allows stream-flow scenarios that would accompany altered surface climate to be predicted. Given the climate scenario, based on either the GCM-based or prescribed approach, modeling produces hydrological scenarios (e.g., stream flow) corresponding to scenarios of surface atmospheric variables.

The selection of an appropriate hydrological modeling strategy for climate effects assessment represents a trade-off between model complexity and data availability. In addition, if the climate scenario is based on GCM simulations, the hydrological model also should account for the need to downscale from a GCM grid cell to a more hydrologically meaningful spatial scale.

The purpose of a hydrological model, in the context of studies on the effects of climate change, is to predict stream flow, given precipitation and other surface meteorological conditions. A key question is the time step of the hydrological model. Even though water resource systems may respond to relatively long time steps, particularly in cases when the reservoir storage is large compared to the mean flow of a river, it is important that the hydrological model operate at a time step short enough to capture the stochastic structure of storm events (wet and dry periods). Generally, this means a time step no longer than 1 d, and quite possibly shorter. As a general rule, the modeling time step should be no longer than several times the time of concentration of the catchment. Usually, the stream flow simulations produced by a hydrological model will have to be aggregated to provide the input to a water resource systems model. For instance, although Lettenmaier and Gan (1990) used a daily time step for their hydrological simulations, the simulated stream flow was aggregated to a weekly time step to simulate a reservoir.

G.2.2.2.1 Water Balance Models to Assess Climate Change: An Introduction to CLIRUN. Most GCMs that assess variations in annual runoff under climate change have

used temperature and precipitation as their key indicators. For a hydrological model to be useful in assessing climate change, it should be physically based in order to capture likely change. Kaczmarek (1993) gives a set of guidelines that a water balance model should follow:

- Its structure should follow physical laws: continuity, and conservation of mass and energy.

- It should be applicable to middle- and large-scale catchments that can be appropriately "lumped" and considered "homogenous."

- It should require relatively simple input data that are compatible with available climate data and climate change data available through GCMs.

- Its input data and model should represent the stochastic nature of runoff through either time series or probability distributions.

A family of water balance models collectively known as "CLIRUNS" is used as part of the primary approach for water resource impact assessment and is presented in detail in the following sections. The CLIRUNS model, developed by Kaczmarek (1993), uses a differential equation to describe the soil moisture, but the basin is modeled as a single unit and is considered homogenous with uniform rainfall. Also, the model represents the water balance components by using continuous function of storage and time. Input data are given as averaged values, but they can be thought of as time-dependent variables. This approach allows the modeler to choose shorter or longer time scales (represented as seasons) (i.e., 10-day, 1-month, or 3-month "seasons").

G.2.2.2.1.1 *Annual Water Balance.* When determining the impact of different processes on runoff, it is necessary to develop functional forms that relate these parameters. Assuming that temperature and precipitation are independent processes, it is possible to develop a functional form that relates temperature and precipitation to runoff:

$$R = f(T,P) \quad ,$$

(G.20)

where

R = runoff,

T = temperature, and

P = precipitation.

The relative importance of runoff processes can be established with the differential described as follows (Kaczmarek 1991):

$$dR = \left(\frac{\partial R}{\partial T}\right) dT + \left(\frac{\partial R}{\partial P}\right) dP + ... \quad , \qquad (G.21)$$

$$R_a = P_a \left[1 - \frac{L_a}{\sqrt{cL_a + P_a^2}} \right] \quad , \qquad (G.22)$$

and

$$L_a = 300 + 25T_a + 0.05T_a^3 \quad , \qquad (G.23)$$

if

$$P_a > (1 - c)^{0.5} L_a \quad . \qquad (G.24)$$

G.2.2.2.1.2   *CLIRUNS — A Conceptual Model.* The CLIRUNS water balance model is based on a simple, conceptual water balance modeling approach (Kaczmarek 1993). In this physically based method to river basin modeling, the basin is conceptualized as a single catchment with lumped parameters. The basin catchment is considered to be a single "bucket" that can be described simply. Monthly data are specified only for precipitation, temperature, PET, and observed runoff for the time series (Figures G.3 and G.4). The objective of modeling the catchment is to determine the relative importance of these parameters in terms of catchment runoff and their potential impact under climate change.

The CLIRUNS model is available as a stand-alone model or as an add-in for users of EXCEL5. The spreadsheet version of CLIRUNS is a more complete modeling tool for assessing climate change because it includes several PET models as well as the snowmelt model. Practically speaking, however, EXCEL5 requires a computer with a great deal of memory and processor capability.

Potential evapotranspiration is a key element in properly modeling catchment runoff. Because of its importance, it is necessary to properly estimate its value before attempting to determine catchment runoff. A significant amount of work has gone into developing functional forms to describe PET (Gray 1993). Potential evapotranspiration varies largely by season because it greatly depends on temperature, wind speed, and vapor pressure.

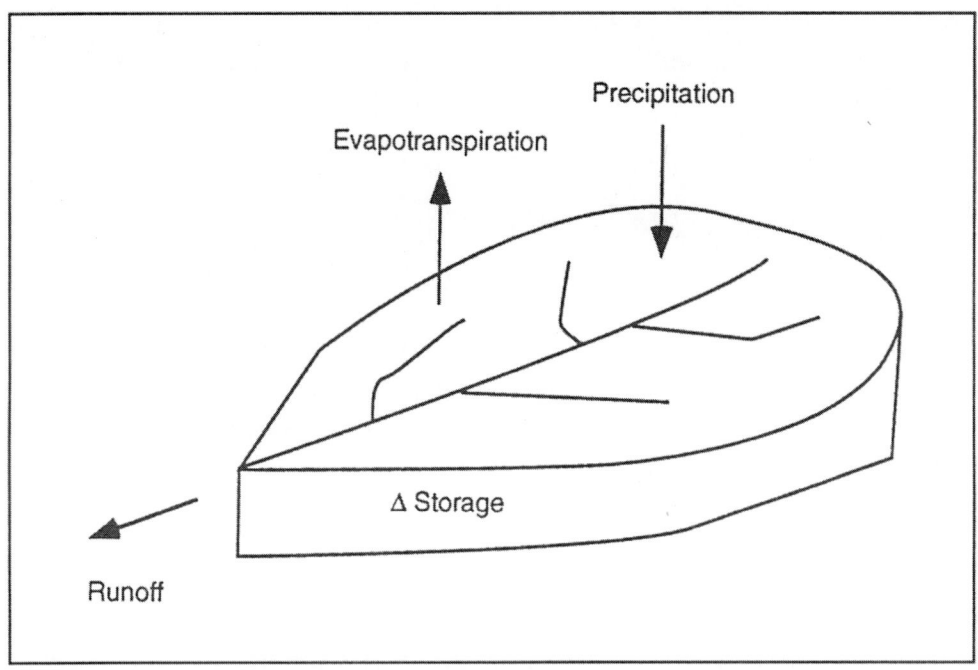

FIGURE G.3 The Water Balance Equation (Kaczmarek 1993)

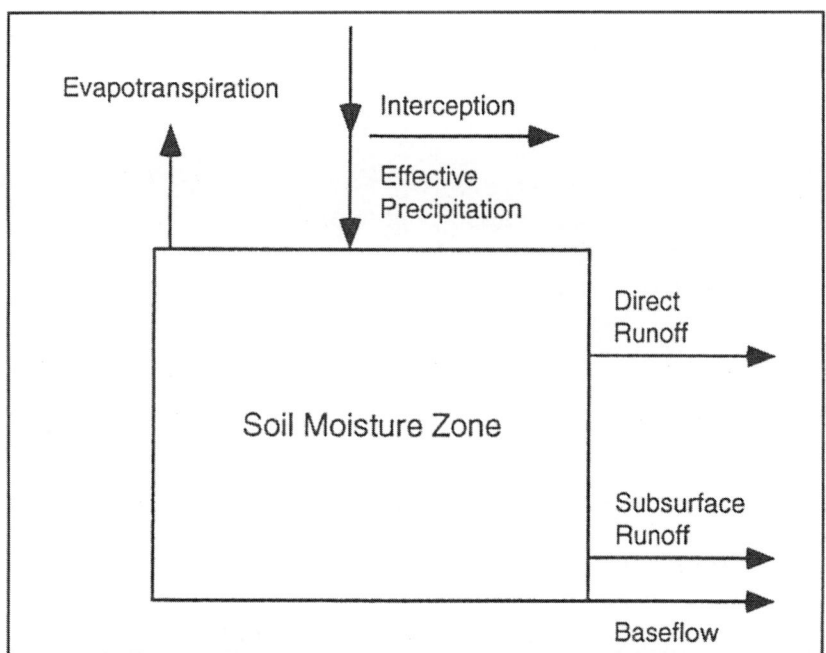

FIGURE G.4 A Conceptual Model of Storage

The recommended PET method for the primary approach is the Priestley-Taylor equation. It has been found that for very large areas, the second term of the Penman equation described in Equation (G.5) is approximately 30% of the first term. Thus, an approximation to the Penman equation that is less demanding of data can be written as follows:

$$E_c = \alpha \; \frac{\Delta}{\Delta + \gamma} \; E_r \;\; , \tag{G.7}$$

where α is approximately 1.3 (closer to 1.26 in most references). This approximation performed fairly well in the comparison study reported by Jensen et al. (1990). Other methods based solely on temperature (Thornwaite 1948; Hargreaves 1981) have given satisfactory results when calibrated (Section G.1.3).

G.2.2.2.1.3 *Catchment Water Balance.* The key to the catchment water balance is the soil moisture accounting, described as follows:

$$\Delta \text{ storage} = \text{inflow} - \text{outflow}. \tag{G.25}$$

Equation G.26 describes the mass balance; for more details regarding the modeling approach, see Kaczmarek (1993):

$$S_{max} \; \frac{dz}{dt} = P(t) - R_s(z, P, t) - R_g(z, t) - E_v(z, \text{PET}, t) - R_b \;\; , \tag{G.26}$$

where

$$S_{max} \quad = \text{ water holding capacity of the catchment (mm);}$$

$$P(t) \quad = \text{ precipitation (mm/d);}$$

$$R_s(z, P, t) \quad = \text{ surface runoff (mm/d);}$$

$$R_g(z, t) \quad = \text{ groundwater runoff (mm/d);}$$

$$R_b \quad = \text{ base flow (mm);}$$

$$E_v(z, \text{PET}, t) \quad = \text{ evapotranspiration (mm/d);}$$

$$Z \quad = \text{ relative storage level } (0 \le Z \le 1); \text{ and}$$

$$t \quad = \text{ time (d).}$$

The continuous functional forms used to express the mass balance are given below. Functional relationships are developed that describe the components of the water balance equation:

- *Evapotranspiration, E_v:* Evapotranspiration is a function of PET and the relative catchment storage state. A number of different expressions have been developed. Both linear and nonlinear expressions are shown. CLIRUN uses the nonlinear expression.

  - A linear expression:

$$E_v(z, \text{PET}, t) = \text{PET}z \quad . \tag{G.27}$$

  - A nonlinear expression:

$$E_v(z, \text{PET}, t) = \text{PET}\left[1 - (1 - z)^{\frac{5}{3}}\right] \text{ or } E_v(z, \text{PET}, t) = \text{PET}\left(\frac{5z - 2z^2}{3}\right) \quad . \tag{G.28}$$

- *Surface runoff, R_s:*

$$R_s(z, P, t) = \frac{\varepsilon}{1 + \varepsilon - z}(P - R_b) \quad . \tag{G.29}$$

  The first parameter of the model, ε, is introduced here in the surface runoff term.

- *Groundwater runoff, R_g:*

$$R_g = \alpha z^2 \quad . \tag{G.30}$$

- Groundwater runoff is assumed to vary as a square of the relative storage state times a coefficient, α. The second model parameter, α, is introduced in the groundwater term.

The third and final model parameter is the maximum catchment holding capacity, S_{max}. The storage variable, Z, is given as the relative storage state: $0 \leq Z \leq 1$. Kaczmarek (1993) gives an analytical solution to the water balance equation that uses the functional forms shown above. The equation can also be solved by using a numerical approach. In the spreadsheet version, a predictor-corrector method was used to solve the differential equation.

Input requirements for the CLIRUN model are as follows:

- Time series: adjusted precipitation, PET, and outflow;

- Initial storage estimated through trial procedure;

- Initial estimates of α, ε, and S_{max}; and

- Number of iterations for numerical procedures.

G.2.2.2.1.4 CLIRUNA — A Stochastic Rainfall-Runoff Model.

The stochastic storage version of CLIRUN, called CLIRUNA, is used when the data are incomplete and only averages are available (along with the coefficient of variation for precipitation). The stochastic model should also be used if mean value data are derived for a remote basin with sparse data. For details of this modeling approach and methodology, see Kaczmarek (1993).

The key concepts in the stochastic storage model include the following:

- Stochastic processes,

- Markov process,

- Storage and transition state,

- Transition probabilities,

- Steady-state probabilities, and

- Probability distribution function (lognormal).

The properties of the model are as follows:

- Stochastic storage theory originally implemented for surface reservoirs to determine probability distributions for model output (storage, runoff, evapotranspiration),

- Same conceptual model developed in the time series approach,

- Discrete time intervals,

- Discrete storage states, and

- Known stochastic properties of input variables.

Definition of the Storage Space. As in the time series model, storage at time i is given as the ratio of actual storage, S_i, to maximum storage, S_{max}, in month i; so, $0 \leq Z_i \leq 1$:

$$Z_i = \frac{S_i}{S_{max}} \quad .$$

          (G.31)

A storage "zone" is conceptualized as a discrete set of storage intervals, r; and the time domain is broken into s equal periods (normally months). An example of the storage space is given in Figure G.5.

Modeling Approach and Objective.

- *Find the steady-state probabilities for each discrete storage state at each time period.* The storage state is defined as a random variable in the m'th period and the i'th zone, given as $\zeta_{i,m}$. This term can be expressed by asking: "What is the probability that the storage in the basin in January will be at half the maximum storage?" It can be written as Prob($\zeta_z = 0.5$, jan).

- *Find transition probabilities.* Transition is defined as the change of the storage level from the initial state $\zeta_{i,m}$ to the final state $\zeta_{i,m+1}$, as seen in Figure G.6.

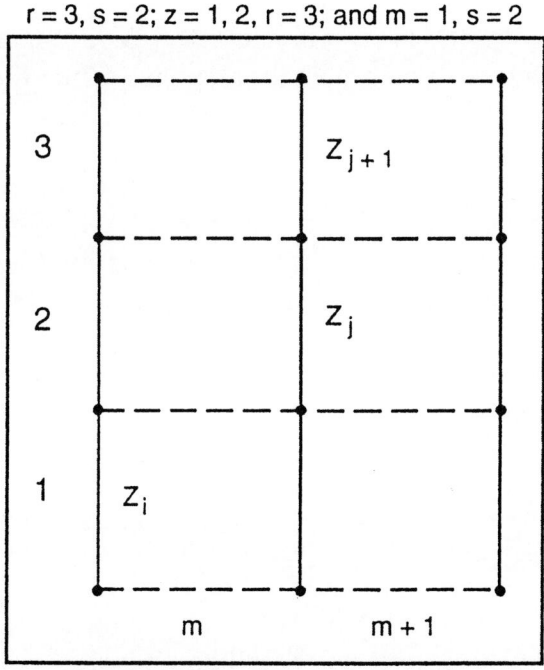

FIGURE G.5 Storage Space in the Stochastic Storage Model

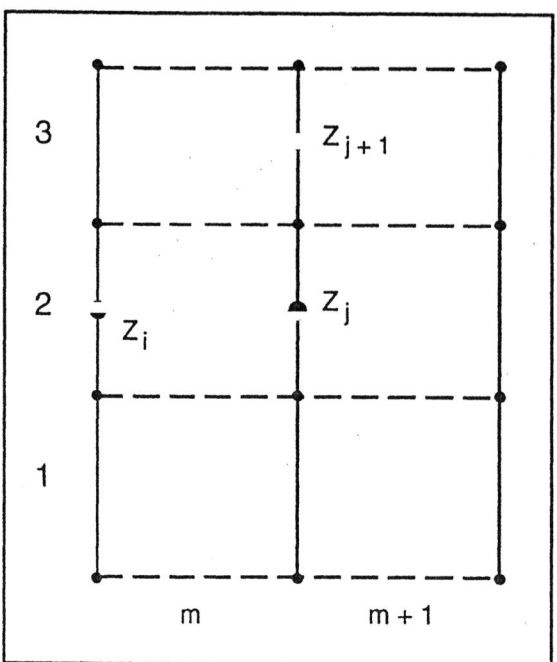

FIGURE G.6 Finding the Transition Probability

For each state and for each period, a transition and a probability can be assigned for moving from state i to state j in period m: $\text{Prob}(\zeta_{i,j,m})$. Transition is defined as the change from the middle of the initial i'th interval to an upper and lower bound of the j'th interval over period m.

The initial storage in period m will meet the requirement $\{(i-1)/r \le Z_m \le i/r\}$, creating the need to define a transition probability matrix. When the upper and lower values z are used, $P_{i,j} = f(Z_i, Z_j)$.

Once the transition probabilities are found, it is possible to define a unique precipitation event that corresponds to the two storage states. The probability distribution function (assumed to be normal for precipitation) is integrated over the lower and upper events to find the transition probability (Figure G.6). Once the transition probabilities are derived, it is then possible to find the steady-state probabilities for each time period (i.e., the probability of the storage state being at level z in time period i). For details about this modeling approach, see Kaczmarek (1991).

G.2.2.2.2 Groundwater Impact. Two types of groundwater resources have been identified: fossil and rechargeable.

G.2.2.2.2.1 *Fossil Groundwater.* Fossil groundwater is water that was deposited in the aquifer over a very long time. The supply of this water no longer exists, or the rate of recharge is so small that it took thousands or tens of thousands of years to provide the water currently in the aquifer. Any human exploitation of this resource would be at a rate so much greater than that of the recharge that the aquifer would have effectively zero recharge over the period of exploitation. In this case, the water is classified as fossil, and its exploitation is considered to be mining of the resource because it will not be replenished in the short to medium term. These types of groundwater resources may play an important short-term role in adapting to climate change impacts, but they are not sustainable long-term solutions. In addition, these resources are not affected by climate change in the short to middle term because the scale of recharge is on the order of thousands of years. Thus, they are affected by a changing climate in the long term, but on a time scale far beyond the scope of this analysis.

Therefore, in the vulnerability and adaptation assessment, the impact of climate change on fossil groundwater resources needs to be addressed; however, these resources can be used as temporary sources of water in the short term, until technology for reducing demand can be developed or surface water resources can be developed.

G.2.2.2.2.2 *Recharges or Tributary Water.*

The other type of groundwater resource is that for which the recharge is on a time scale on the same order of magnitude as the annual hydrological cycle. This water resource is then part of the annual water balance and is often directly linked to stream flow as base flow. The directly linked systems are known as tributary groundwater, and in some U.S. states, they are actually considered to be a surface water resource for water rights appropriation. In the former Soviet Union, most of the cities and towns and most rural settlements use groundwater as their source for water supply (Kovalevsky 1993). In addition, Kovalevsky (1993) reports that groundwater accounts for 10-40% of annual river flow and 80-100% of dry-weather flow. He states that the contribution of precipitation to groundwater recharge is quite variable, depending on annual precipitation level. In years of low precipitation, the contribution is almost 50% of the average year level.

These results point to two important issues in assessing the impacts of climate change on water supply:

- "Shallow, rechargeable, or tributary" groundwater is an important part of the water balance of large-scale river basins.

- The impact on the base flow due to changes in precipitation is not straightforward.

It is necessary to identify ways to assess groundwater resources under climate change and for the approach presented in this document.

In the CLIRUN models, base flow is an input parameter estimated by various different techniques (Maidment 1993). Kovalevsky (1993) suggests that average base flow over a large region can be estimated as a function of precipitation and temperature. By using historical hydroclimatic data, a multivariate statistical relationship can be developed. Thus, when new climate scenarios are analyzed, the base flow that corresponds to new mean temperature and precipitation projections can be used in CLIRUN.

Current, simplified regional hydrological assessments make two key assumptions related to groundwater: (1) all base flow comes from the water balance in the surface water basin; and (2) over the year, no water is lost to deep groundwater percolation. The only outflow of water in the water balance is via evapotranspiration and runoff. This scenario is not always the case, but a more detailed assessment is beyond this analysis.

G.2.2.3 Hydrological Conditions for Altered Climate Scenarios

Once the hydrological model has been implemented, simulation runs are conducted that use the present climate (unadjusted historical climate records) and alternative climate scenarios, at the appropriate time scale, as input. This situation presumes that the spatial scale of the hydrological model is appropriate for this assessment. In previous studies (e.g., Lettenmaier and Sheer [1991], who assessed the Sacramento-San Joaquin River basin in California, and a study by the Tennessee Valley Authority [TVA] of the climate sensitivity of the TVA system), a second step was used to simulate stream flows at nodes, representing relatively large drainage basins and using hydrological model simulations of small "index" catchments. The method used in these studies is based on stochastic stream-flow disaggregation. However, it is anticipated that a more direct approach, based on macroscale hydrological models, will be used for most of the Country Studies sites. The macroscale approach avoids the necessity for the stochastic disaggregation step.

G.2.2.4 From Hydrology to Water Supply

Hydrology is the estimation and description of the natural processes that describe a raw physical resource, that is, water. Water resources or water supply is the human development of this physical resource into a socioeconomic resource by the means of engineered projects or infrastructure development. Water resources sometimes refer to the total potential of the physical resource; this work does not take that perspective. *Water resources or water supply is the amount of the physical resource that has been developed or managed to become socioeconomic resources.* Thus, the amount of water resources available depends on two key factors — hydrology and engineered infrastructure.

While hydrology and runoff are natural spatial processes that can be described easily on an aerial or gridded extent, water supply is an engineered process. Water supply is a function of the seasonal hydrological fluctuation, the storage in the system, and the extent of the water delivery and distribution system. These engineered structures are described as points or vectors and are hard to translate into an area basis. Thus, when assessing the impact of climate change on water supply, the hydrological impacts as well as the impacts on the water resource system must be assessed. Such an assessment is difficult to do on a regional or national basis. It is necessary, therefore, to look at a river basin level.

Some simple, yet useful, techniques can provide insights into the water supply for river basins with no storage or regulation and for basins with a single reservoir. For a river basin with many storage reservoirs and much development, more complex river basin models are needed, as discussed in Section G.2.2.6.

The simplest technique is the flow duration curve, a technique that is most appropriate for unregulated rivers. Curve A in Figure G.7 is a sample flow duration curve for current climate. It provides information on the probability that a given flow will be exceeded. Figure G.7 shows that a flow of 800 will be exceeded 30% of the time. Curve B in Figure G.7 also shows a flow duration curve from a climate change assessment. A yield of 800 is now exceeded only 16% of the time, thus indicating a decrease in water supply reliability.

In a single reservoir system, two graphical methods — the Rippl diagram and the sequent peak method — can be used to determine either the obtainable yield for a given reservoir storage or the storage required to provide a desired yield. As computers are readily available, the modified sequent peak algorithm can be used to determine either of these values. In addition, with the ease and speed of even a small computer, storage-yield relationships can be developed for any reservoir and even provide different levels of reliability (Figure G.8). These curves can then be redeveloped under a new hydrological regime based on a climate change assessment of runoff impacts.

To use these methods, a model of the river basin system must first be developed. Much has been written on river basin modeling, and the reader is referred to Loucks et al. (1981) and Goodman (1984).

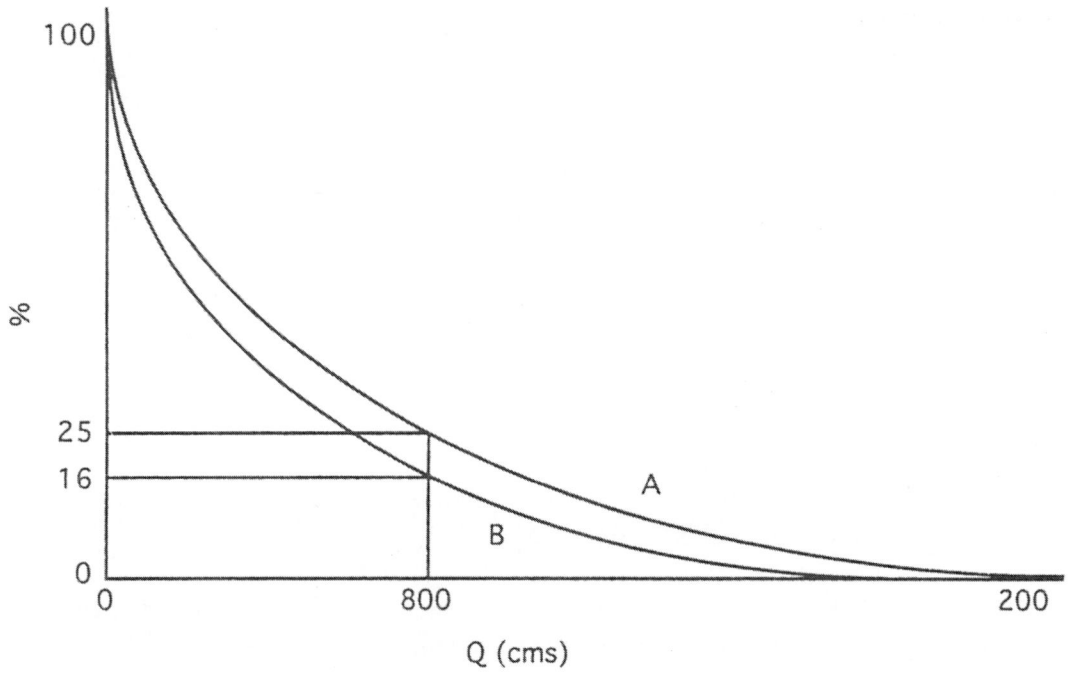

FIGURE G.7 Flow Duration Curves

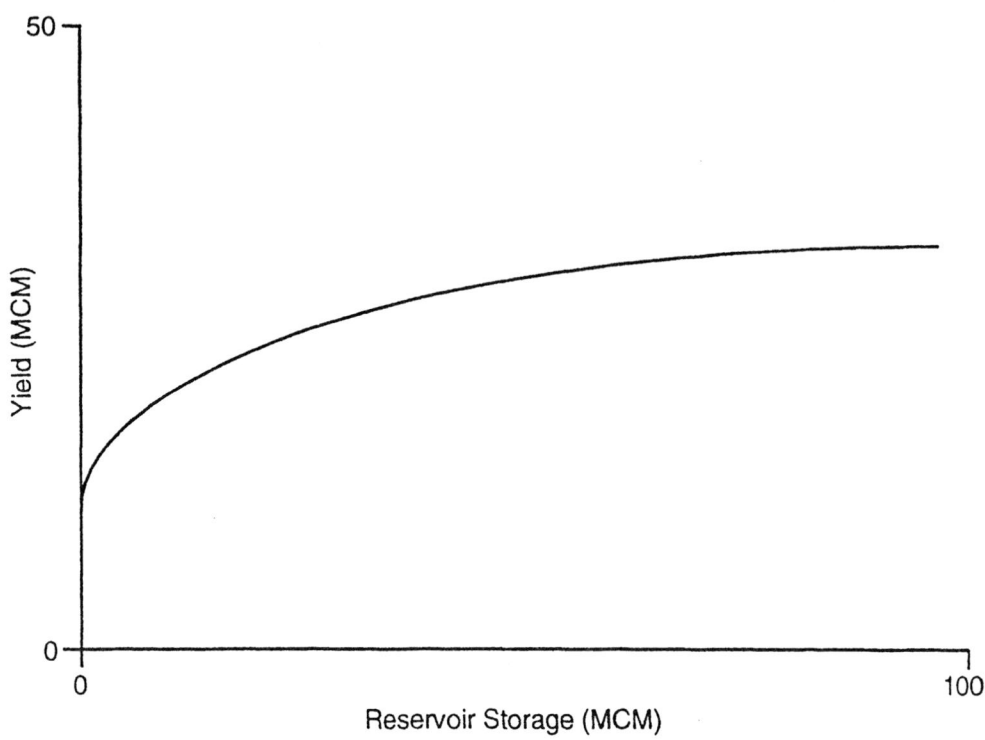

FIGURE G.8 Storage Yield Curve

The issue in river basin modeling for assessing climate change impacts is that the County Studies Program is looking at a base year of 2075. Therefore, the analysis must project the type of development that will be in the basin in that year. Such a projection requires finding long-term development plans and deciding on the most likely development scheme for the current climate and future economic growth.

G.2.2.5 A Water Resource System Model

Most highly developed water resource systems exploit some combination of surface and subsurface storage. A major issue to be addressed in climate sensitivity studies of water resource systems is how well existing water resource systems could buffer the hydrological variability associated with altered climate. It is not surprising, then, that most studies of the effects of climate change on water resources reviewed by the Intergovernmental Panel on Climate Change (Stakhiv et al. 1992) involved some aspect of water resource system simulation or performance assessment. Details of the water resource simulation approach for the Country Studies Program are described in Section G.2.2.2. Note the position of the water resource model in the chain of assessment models: its inputs are provided by the output (aggregated in time and space, if necessary) of the hydrological model.

The final step in a water resource assessment is to simulate water resource system performance. Once steps 1-4 (Section G.2.2) have been completed, it should then be straight-forward to run the hydrological time series corresponding to historical and altered climates through the water resource system model. Two key considerations in performing such simulations are that the length of record be adequate to characterize the natural variability and that known severe droughts, which may have shaped the system operating policy, be included in the period of record. The latter was a major problem in the Sacramento-San Joaquin study (Lettenmaier and Sheer 1991), in which the researchers focused on the 1951-1980 period of record, which was to be standardized among all the projects performed as part of EPA's Reports to Congress. Unfortunately, this period excluded the drought of record, which occurred in the 1930s. The system operating policy must also be determined. Under a changed climate, it may not make sense to operate a water resource system in the same way as was done previously; therefore, direct application of a system simulation model developed for historical conditions may well be inappropriate.

Various means can be used to quantify the effect of climate change on the performance of a water resource system. Depending on the purpose of the system, different performance indices may be appropriate. In water supply systems, such indices might include the magnitude, number, and length of water supply deficits. For flood control systems, flood damage is the obvious performance index. For navigation performance, the length of periods with channel depth below a critical value might be used. For hydropower systems constituting a small part of a combined hydro-thermal system, the total seasonal or annual hydropower avoidance cost (cost that would have been incurred in the absence of hydropower generation) could be used. For base-loaded hydropower systems (such as the Columbia River system), performance measures similar to those used for water supply, such as the length, magnitude, and duration of deficits from "firm power" might be appropriate. Because most systems are operated for multiple purposes, more than one performance measure will often be needed.

A convenient method of comparing system performance for alternative climate scenarios is the empirical probability distribution of the annual values of the performance measure. These values are plotted in the same manner as an empirical probability distribution of an annual flood series. If the sequence $\{Y_j, j = 1, ... N\}$ is the values of a performance index for N years, ranked from smallest to largest, the empirical probability distribution is formed by plotting Y_j against P_j, the empirical cumulative probability. For screening purposes, the Weibull plotting position formula, $P_j = j N + 1$, is usually adequate. The scale for P_j can be distorted for better visual discrimination; a normal probability scale works well for this purpose, even if the data are not normal.

G.2.2.6 Supplementary Modeling Approaches for Climate Change Assessments of Hydrological Resources

Selecting the appropriate hydrological model for climate change assessments is not an easy task. Several issues must be considered:

- Availability of hydroclimatic data in the area of interest to conduct the baseline hydrological assessments;

- Availability of changes in certain hydroclimatic data due to climate change, as obtained in GCM simulations;

- Procedure to disaggregate, in space and in time, those hydroclimatic data to the basin of interest;

- Procedure to determine those hydroclimatic data that are not disaggregated from GCM simulations;

- Other factors, such as

  - Precipitation: interstorm arrival rates, storm types, length of rainy season (e.g., monsoon season), and separation of rain and snow;

  - Overland flow;

  - Evapotranspiration;

  - Groundwater flow: interannual variability;

  - Soil moisture: important for infiltration/overland flow at small temporal/spatial scales; long-term weather forecasts for monthly simulations; droughts;

  - Snowmelt: timing of the hydrograph;

  - ENSO impact on water resources; floods and droughts; and

  - Climate change detection.

G.2.2.6.1 Daily and Event-Scale Rainfall–Runoff Models. For hydrological studies, assessing the impact of climatic change on soil moisture, groundwater, and runoff is of paramount interest. Because the impacts of precipitation are related to the spatial and temporal scale of the rainfall events, models used to predict impacts must be appropriately chosen.

In general, impacts associated with climatic changes occur over a time frame of decades to centuries over very large areas, while small-spatial-scale, short-temporal-scale (hourly to daily) events are important processes for flood prediction, rainwater harvesting, urban drainage, erosion calculations, and generation of single storm hydrographs (Berndtsson and Niemczynowicz 1988). Modeling daily and event-scale processes requires mathematical models appropriate for the scale desired. The remainder of this section discusses models that can be applied to short-temporal and small-scale precipitation events.

G.2.2.6.2 Appropriate Modeling Techniques. A number of techniques have been applied for modeling daily and event-scale rainfall and runoff. These methodologies include, in increasing order of data needs, the following: stochastic models, lumped integral models, distributed integral models, and distributed differential models (Todini 1988). Of these methodologies, purely stochastic models are the simplest to implement and require the least physical information. Because these stochastic models do not assume any direct causality between input and output variables, the results have a high degree of uncertainty and are valid only on average; they are thus best suited for dealing with time increments larger than the actual system dynamics (e.g., monthly time increments). In addition, because stochastic analysis usually assumes a stationary series (without time trends), the underlying algorithms must be adjusted to account for the effects of additional anthropogenic CO_2 in the atmosphere (Linsley et al. 1982).

Lumped integral models use the next most data-intensive methodology. For these models, system dynamics are represented in integral form (e.g., the impulse response and the unit hydrograph), and the catchment is simulated by modeling its overall behavior (Todini 1988). Parameters for these models are usually estimated by statistical optimization techniques. Because the catchment is modeled as a whole, a significant part of it is assumed to be homogeneous and subject to uniform rainfall (Linsley et al. 1982).

The third technique uses distributed integral modeling to represent the catchment response (Todini 1988). This type of modeling is also referred to as conceptual (Linsley et al. 1982). With this methodology, phenomena at a subcatchment scale are represented by either empirical formulas or the impulse response of the subsystem (e.g., surface runoff and infiltration) in integral form. These individual components are then combined by matching their boundary conditions. In general, a large number of parameters are required to implement these models. These parameters are estimated either on physical grounds or on the basis of error minimization. Models in this classification include the Stanford Watershed IV Model (Crawford and Linsley 1966; Anderson 1971); the Sacramento Model developed by the Joint Federal-State River Forecast Center, the U.S. National Weather Service, and the State of California Department of Water Resources (Burnach et al. 1973; Mays and Tung 1992); the TANK Model (Sugarwara et al. 1983); and the Streamflow Synthesis and Reservoir Regulation Model (U.S. Army Corps of Engineers 1972).

In the last technique, the distributed differential model, catchment behavior is represented in terms of all the governing partial differential equations that describe surface flow, interflow, and groundwater flow discretized in time and space. Mass and momentum are balanced for each subsystem, and the subsystems are linked together at each time step by matching their mutual boundary conditions. This methodology requires large computer capabilities as well as large amounts of appropriate, catchment-specific data to solve the coupled, governing equations. One example of this modeling approach is the Système Hydrologique Européen (SHE) discussed by Abbott et al. (1986a,b).

G.2.2.6.3 Model Selection and Data Requirements. To evaluate the impact of changes in climate on the daily or single-event scale, an appropriate model must be selected. A major factor in this selection process is the quantity and quality of data available for implementing the model. If very few physical data are available for the catchment of interest, a stochastic or lumped parameter model should be selected. If there are sufficient quantities of reliable data, and a large computer capability, a distributed differential model, such as SHE, might be preferable. However, because of the uncertainties associated with stochastic models and the inherent complexities involved with distributed differential models, it is recommended that a distributed integral model, such as the Stanford Watershed IV Model or the Sacramento Model, be selected for impact assessment.

G.2.2.6.4 The Stanford Watershed IV Model. Because the Stanford Watershed IV Model is typical of distributed integral models and is one of the first widely accepted models, a brief description of its use and parameters is given below.

The Stanford Watershed IV Model is a sequence of computation routines for each process in the hydrological cycle (interception, infiltration, routing, etc.) (Viessman et al. 1977). The calculations proceed according to the flowchart shown in Figure G.9. All the moisture originally stored in the catchment or input as precipitation during any time period is balanced by the following continuity equation:

$$P = E + R + \Delta S \ , \qquad (G.32)$$

where

P = total precipitation;

E = evapotranspiration;

R = runoff; and

ΔS = total change in storage in upper, lower, and groundwater storage zones.

The change in storage for each zone is calculated as the difference in volume between the inflow and outflow. Before proceeding to a new time step, all hydrological activity is simulated and balanced. Some of the input parameters for this model is listed in Table G.5.

Briefly, the Stanford Watershed IV Model assumes that precipitation falls on an interception storage that must be filled before water is available for other functions (Starosolszky 1987). After interception is satisfied, direct runoff from impervious areas is accounted for, and the remaining moisture is then subject to an infiltration function that determines whether water will move to the stream as overland flow or interflow or infiltrate into the soil. The infiltrating moisture can enter the lower zone storage (subject to evapotranspiration) or the groundwater (subject to drainage as baseflow).

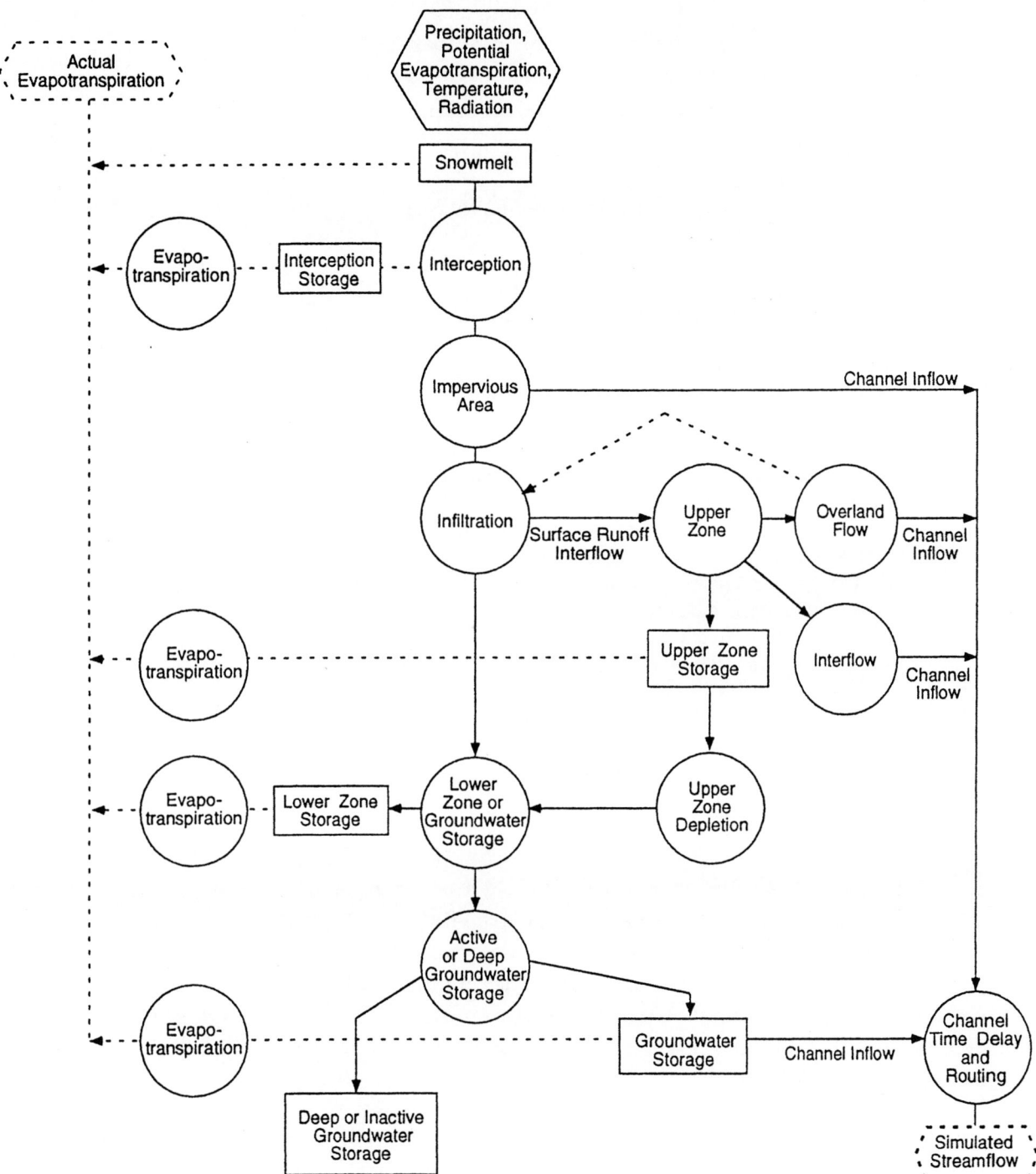

FIGURE G.9 Stanford Watershed IV Model (Source: Crawford and Linsley 1966)

TABLE G.5 Input Parameters for the Stanford Watershed IV Model

---

Time of concentration
Routing interval
Number of elements in the time-area histogram
Time-area histogram ordinates
Watershed drainage area
Impervious area of the watershed surface
Watershed stream and lake surface area fraction
Average length of overland flow
Average overland flow ground slope
Bank-full flow in channel at the gauging station
Daily interflow recession constant
Stream-flow routing parameters for low and flood flow
Manning roughness coefficients for overland flow and impervious areas
Seasonal factor for infiltration
Maximum interception rate for dry watershed
Surface storage capacity index
Soil surface moisture capacity
Soil evaporation parameter
Index of inflow to deep inactive groundwater
Fraction of watershed area in phreatophytes
Infiltration rates factor
Equivalent depth of upper and lower zone storage
Index of antecedent moisture conditions
Volume of water in swamp storage
Monthly evaporation pan coefficients

Once the model has calculated flow to the channel of interest, a hydrological watershed routing technique is used to translate the flow to the watershed outlet. The model views the sum of all channel inflow components as an inflow hyetograph that is then translated in time through the channel to the basin outlet, where it is next routed through an equivalent storage system to account for attenuation caused by storage in the channel. Routing through the linear reservoir is assumed to be directly proportional to the outflow (Dunne and Leopold 1978).

Additional details on the theory and implementation of the Stanford Watershed IV Model can be found in Crawford and Linsley (1966) and Viessman et al. (1977).

G.2.2.6.5 Calibration and Validation. Because the predictive capability of a model is only as good as its input data, calibration, and validation, a great deal of care should be used before applying any model to new conditions. For example, for the Stanford Watershed IV Model, a 3- to 6-yr calibration period is recommended, followed by a period of simulation to validate the calibrated parameters (Viessman et al. 1977). For accurate results, at least hourly records of precipitation and stream discharge are required as well as daily records of evapotranspiration.

Calibration of the model can follow two paths: a statistical, automatic procedure or a procedure based on successive rational attempts (Franchini and Pacciani 1991). The first strategy can often lead to optimized parameters that are not physically meaningful. The second procedure, while more time-consuming, can eliminate that problem by introducing physical understanding and engineering judgment into the process.

In general, the recommended procedure for model calibration involves a visual comparison between computed and historical discharges, aided by an estimate of the determination coefficient, the correlation coefficient, and the explained variance (Franchini and Pacciani 1991). When a best-fit model is developed, it should then be validated against an independent set of data before the model is used for simulation.

G.2.2.6.6 Recommendations. If daily or event-scale rainfall–runoff modeling is desired, a distributed integral model, such as the Stanford Watershed IV Model, or the Sacramento Model is recommended if sufficient data are available for calibration and validation. If few data are available, a lumped parameter model or a stochastic model can be implemented, but the results may have a large degree of uncertainty.

It is imperative that the model chosen be calibrated and validated with as much catchment-specific data as possible to reduce inherent uncertainties in the calculations. It is recommended that the calibration process employ successive rational attempts rather than an automatic statistical technique to avoid unrealistic physical parameters.

G.3 DEMAND

When addressing climate change impacts on water demand, two major questions arise:

- What will the water demand be in the 2075 base scenario?

- What will be the impact of climatic change on the 2075 base scenario?

In many respects, determining the impact of warmer temperatures and changes in precipitation on water demand is more straightforward than "estimating" water demand 80 years into the future, which is a function of population and economic growth and technological change.

This section does not deal with scenario generation, so it is assumed that the Country Studies team has developed a 2075 base scenario for a country on the basis of the guidelines given in Appendix A. Given a population and economic scenario, the next step is estimating the water demand for such a scenario. As stated in Section G.1, this appendix does not provide a handbook (or even a primer) on how to estimate water demand (this information is provided in the references), but it does highlight issues related to estimating water demand 80 years into the future and under a changed climate.

G.3.1 Water Requirements vs. Water Demand

What amount of water is used by an industry, a municipality, or irrigated crop lands? The amount actually used is based on a concept economists call demand. Demand is the willingness of consumers or users to purchase goods, services, or inputs to production processes. This willingness is a function of the price of the thing being purchased, as well as other factors. A demand function is illustrated by curve A is Figure G.10. Water demand is sometimes confused with water requirement. Water requirement is the minimum water needed for social or economic activities. Water requirement does not respond to the price of water supply and is illustrated by line B in Figure G.10 (Kindler and Russell 1984).

Some social or economic activities, such as human daily water consumption, water inputs in the beverage industry, or the low-flow condition for navigation or environmental protection do exhibit "minimum needs" or requirements, but they are usually only a small part of total water use in the activity. For example, human consumption is less than 15% of U.S. domestic water use. Other domestic uses, including law sprinkling sanitation, bathing, and household use, make up the remainder.

G.3.1.1 Price Effects

About 50% of all water withdrawn in the United States in 1975 was for cooling of thermoelectric power stations. While only about 2% of this water is consumed, it is required to be in the stream for withdrawal by less costly once-through cooling systems. Goodman (1984) presents data that show that dry cooling tower technology would require no water for cooling but would mean a 7% reduction in energy generated, while wet towers would require

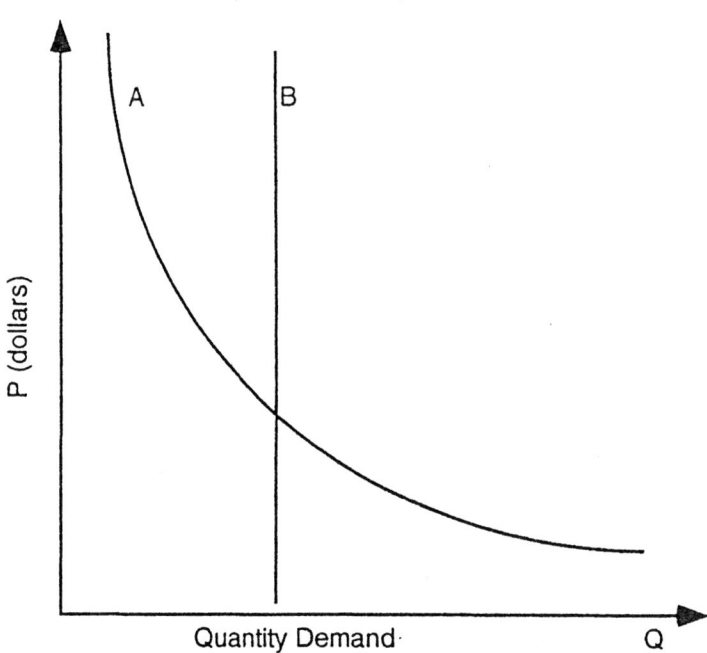

FIGURE G.10 Water Demand versus Water Requirement

only 2% of once-through withdrawals and reduce generation by only 1%. As the price of water increases above the costs of the lost electricity, **alternative** cooling technologies become economic and could free a large amount of instream **water for** other beneficial uses.

G.3.1.2 Institutional Effects

Similar drastic reduction in water use particularly in industry can take place through environmental legislation. Figure G.11 shows forecasted and actual water use in Sweden from 1930 to 1977. The dramatic reduction in water use occurred because it was economical to install water recycling equipment to reduce the costs of meeting stringent water quality legislation. These reductions took place in spite of a substantial increase in industrial production (Falkenmark 1977). Carmichael and Strzepek (1986) report a similar decrease in industrial water use per unit of production in **a number of European** countries over the same period. Thus, a 2075-based water demand **estimate can easily** overestimate industrial water demand.

G.3.1.3 Income Effects

Kindler and Russell (1984) point out that per-capita **domestic or municipal** water use positively correlates to per-capita income and inversely correlates to household size. Most Country Study analyses take place in developing or transitional economies that are likely to

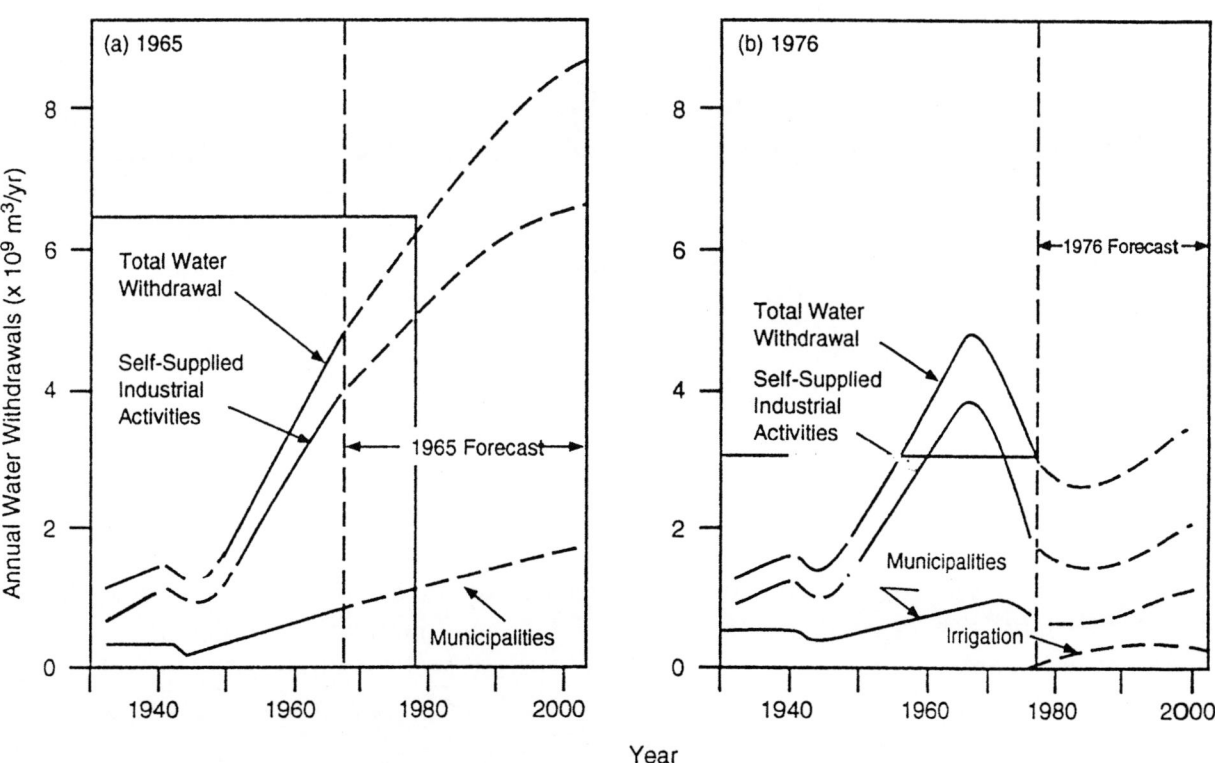

FIGURE G.11 Water Use in Sweden (Falkenmark 1977)

undergo great changes in per-capita income and household sizes over the next 80 years. To assume that per-capita domestic water use will remain at 1990 rates is a rather poor assumption. With increased per-capita income and urbanization will come more water-using appliances, such as flush toilets, washing machines, and dishwashers. Even water-efficient appliances will mean a growth in per-capita water use. These potential changes need to be carefully considered when forecasting so far into the future.

G.3.2 The Forecast

Given all the warnings stated above, it will be very difficult, within the scope of the Country Studies Program, to do more than a simple "statistical" approach to estimate water use:

$$\text{water use}_{2075} = (\text{activity}_{2075}) \times (\text{water use/activity})_{2075} \ . \qquad \text{(G.33)}$$

The activity_{2075} will be provided by those developing the base scenario. However, the water resource analysis team must estimate the $(\text{water use/activity})_{2075}$. The Country Study analyst must be extremely careful when estimating this number and avoid the temptation of simply applying $(\text{water use/activity})_{1990}$ obtained from historical data. It may be that there are no data to justify changing this number. However, given the issues discussed above and the material in the reference documents, a reasonable value should be obtainable. In addition, a sensitivity analysis should be performed on the $(\text{water use/activity})_{2075}$ coefficient to examine the impact on water resource vulnerability.

G.3.3 Climate Change Impacts on Demand

In this effort, climate change is defined as changes in temperature and precipitation. The water resource analysis does not attempt to estimate the impact of climate change on the base socioeconomic activities, such as population, energy-related demand, agricultural demand, or industrial production. The scenario development team should provide these data. The water resource team should estimate how the $(\text{water use/activity})_{2075}$ will be positively or negatively affected by changes in temperature and precipitation. Guidance on how climate change can affect the water use coefficient for each of the four water use sectors is given in the following.

G.3.3.1 Industrial Water Use

In most industries, temperature and precipitation have little direct impact on water use. Reduction of flow in rivers due to climate change may increase the pressure on waste treatment processes, leading to increased water recycling and a decline in industrial water use, as seen in Sweden. However, for the most part, climate change should have little or no effect on industrial water use.

G.3.3.2 Energy-Related Water Use

Energy water use takes two forms — hydroelectric and thermoelectric. Reduced flow will reduce reservoir storage and thus reduce potential energy; warmer temperatures will increase evaporation from hydropower reservoirs. These two impacts will require more stream flow to maintain the same energy production.

Increased river temperatures have little direct effect on thermal efficiency; however, increased river temperatures and reduced flow can result in cooling discharges that violate environmental standards. Reduction in generation or more release of stored water for cooling purpose will then occur. Both are direct effects.

G.3.3.3 Agricultural Water Use

Agricultural water use consists of irrigation and livestock water. The Country Studies Program includes agricultural and livestock assessment teams that will provide the water resource assessment team with the changes in irrigation and livestock water use due to changes in temperature and precipitation.

Should this information not be available, the irrigation water requirement can be estimated by using the appropriate reference crop and PET model (Section G.1.3). Irrigation water use will be the difference between actual evapotranspiration and the new precipitation estimates.

Livestock water use can best be estimated by searching the literature for livestock water use versus climate data. If direct statistical relationships do not exist, use of an analog approach is suggested. The per head water use values for 1990 for a country with 1990 hydroclimatic conditions that are analogous to those forecast for a country of interest under climatic change can be used.

Because base temperature and precipitation as well as changes due to climatic changes can vary widely across a country, the techniques mentioned in Section G.1.10 for going from local/regional to national impacts by using representative crops and regions should be used.

G.3.3.4 Domestic Water Use

Kindler and Russell (1984) observed for residential water use that

- Per residence water use is inversely correlated with rainfall, and

- Per residence water use is positively correlated with average temperature.

Little work has been completed on the impact of long-term climate change on domestic water use. Most analyses to date have drawn information from water users' responses to short-term drought or warm periods. Short-term responses of water use could be much different when users are faced with permanent acceptance of water restrictions and possible lifestyle changes. However, one must be careful when using the analog approach in domestic water use because climate alone is not sufficient to determine an analog. Similar socioeconomic indicators (income and household size) as well as climatic variables must be used in selecting analog countries because domestic water use is correlated with these socioeconomic indicators.

G.3.3.5 Total Water Demand

Total demand is the summation of all sectors of water use. These water uses reflect potential increases or decreases caused by climate change without any adaptation. Possible reduction in water use by adaptations of lifestyle, crop variety, or industrial recycling all come at a cost and should not be analyzed in the impact section but later in the adaptation section.

G.4 VULNERABILITY

In the context of the Country Studies Program, the supply-and-demand balance assessment is done at the national level to identify national vulnerability to climatic change. This type of analysis is what the U.S. Water Resources Council classified as a framework study and assessment. These studies are very general and determine the extent of water and land problems and the needs for a large geographical area over the long term (Goodman 1984).

Section G.2 presents guidelines on how to determine the national level water supply, and Section G.3 presents water demand estimation guidelines. However, these separate analyses cannot tell how vulnerable a nation is to climatic change. Only when the supply and demand are jointly analyzed can vulnerability be assessed.

A number of steps must be carried out to assess the balance of supply and demand:

1. Assess whether a water surplus or shortage exists.

2. Assess the country's water vulnerability index — current, 2075 base, and 2075 under climatic change.

3. Determine the impact of climatic change on vulnerability.

4. If the supply-and-demand balance has been significantly changed, assess adaptations to achieve supply-and-demand equilibrium and the costs of such adaptations.

The analyses suggested in item 4 are discussed in detail in Sections G.5 and G.6.

Looking at the water supply-and-demand balance in a physical sense, following the approach of Kulshreshtha (1993), allows examination of the impact of climate change on a nation's vulnerability by using a variety of alternative vulnerability indexes. This approach is valuable as a "screener" to see whether climate change can cause the nation to make a major shift in water balance and subsequent vulnerability. It can also point out areas of the analysis that need further analysis because the vulnerability index may show a high degree of sensitivity to a single assumption in the supply or demand analysis, such as per-capita water use.

G.5 ADAPTATION

When addressing adaptation in the water resource sector at the national level, a number of issues must be addressed. The first and foremost issue is that most adaptation takes place at the project or regional/river basin scale, not at a national level. The second issue is that adaptations can be divided into two major classes:

- Supply adaptation

    - Construction of new infrastructure,
    - Modification of existing physical infrastructure, and
    - Alternative management of the existing water supply systems

- Demand adaptation

    - Conservation and improved efficiency,
    - Technological change, and
    - Market/price driven transfers to other activities.

This section does not discuss the estimation of the economic impacts of making these adaptations or even how to analyze the "mitigating" effects of such adaptations. Rather, it provides a systematic framework for developing water resource adaptation strategies for water supply and demand for the four main economic sectors.

G.5.1 Water Supply Adaptations

Section G.2.2.4 presents the process for converting hydrological resources into water supply. The key issue is the ability to shift runoff in time and space to be socially and economically beneficial. The conversion process typically takes place via storage reservoirs and distribution systems. Water is stored to provide a water supply with a certain statistical reliability. The system of surface water reservoirs, groundwater wells, and distribution systems is called the water supply infrastructure.

Because a major impact of climatic change is to change the temporal and spatial distribution of precipitation and temperature, the resulting river runoff or hydrological resource may be shifted in time and space. A change in the spatial and temporal distribution of river runoff could greatly affect the efficacy of the water supply infrastructure currently installed. Figure G.12 illustrates a river hydrograph that changes from a nearly uniform distribution with a high to low flow ratio of 3 to a seasonal flow regime with a high to low flow ratio of 10. In this example, the existing storage would be inadequate to meet a consistent within-year demand.

G.5.1.1 Modification of Existing Physical Infrastructure

Many countries have made extensive capital investments in their water supply infrastructure. The water supply infrastructure is designed for a stationary climate. However, because of climate change impacts on runoff, the systems may not perform as designed. Adaptation to climate change may be achieved by modifying the existing system. In some river basins, no suitable projects exist for new development, and thus, an adaptation that uses the existing investment is most economical. A listing of some possible adaptations the Country Study Program analysts should consider follows:

- Decrease flows

    - Change location and/or height of water intakes
    - Canal linings
    - Closed conduits instead of open channels
    - Integration of separate reservoirs into an integrated system
    - Artificial recharge to reduce evaporation

- Increase flows

    - Raising dam height
    - Additional turbines
    - Increasing canal size
    - Sediment removal from reservoirs for more live storage

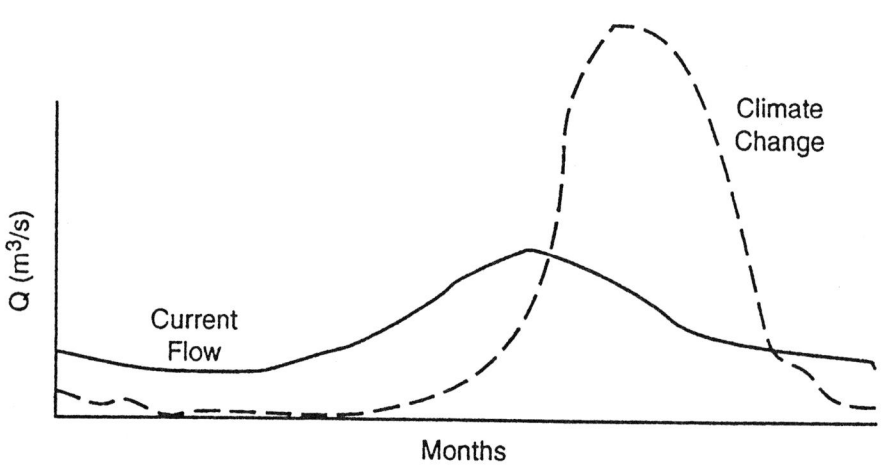

FIGURE G.12 Hydrographic Changes

G.5.1.2 Construction of a New Infrastructure

In river basins where full development has not been realized, it is possible to construct new projects to adapt to the changed runoff and water demand conditions, such as the following:

- Reservoirs,

- Hydroplants,

- Delivery systems,

- Well fields, and

- Interbasin water transfers.

G.5.1.3 Alternative Management of Existing Water Supply Systems

In certain river basins, the nature of the climate change or physical, environmental, or institutional constraints does not warrant or allow new infrastructure projects. Thus, adaptations to the management of the existing system are considered. Possible adaptations are as follows:

- Change in operating rules,

- Conjunctive surface and groundwater supply,

- Change in priority of releases,

- Integrated reservoir operation system, and

- Coordination of supply and demand.

G.5.2 Demand Adaptation

Water demand adaptations take place under two possible regimes:

- *Water is not priced at the margin but is supplied at a fixed price, and it is a limited resource.* Water use is viewed in terms of water requirements. If the climate change impact is a reduction in the amount of water available but the price does not change, supply-and-demand equilibrium can no longer be achieved with a fixed water requirement. The goal of adaptation is to maintain the same economic activity by reducing the water use or requirement via conservation and improved efficiency or technological change, thus reducing the water demand.

Figure G.13 illustrates this adaptation. The initial equilibrium is at point A. The climate change impact is to reduce available supply from Q to Q'. Because the demand is fixed at Q, a supply-and-demand gap of $Q - Q'$ results. Adaptation takes place, and demand is reduced to Q' by shifting the demand curve to D' and establishing a new equilibrium at B. The supply adaptations attempt to increase the supply from Q' to avoid shifting the demand curve or at least reduce the amount that it must move. The objective in the adaptation study is to determine whether it is possible to reduce water use to Q' by changing water use requirements. There is a cost in moving from D to D' as conservation and technological change are introduced. In most cases, the reductions in water use are policy based and are mandated as uniform percentage reductions across all economic sectors or at least across all individual activities within a sector. This solution is very equitable, but not economically efficient.

- *Water supply and demand are price responsive and based on market mechanisms.* The impact of climatic change on water supply drives the market out of supply-and-demand equilibrium. To return to equilibrium, the price of water changes until supply and demand clear. Implicit in the derived demand curve is the economic value of water. Thus, the demand curve represents an economically based collection of conservation and technological changes that are beneficial to the users at the margin. As the price of water increases, individual economic activities pay more for the water until they reach the price where it costs less to reduce water use than to purchase it. They reduce water by introducing conservation, changing technology, or going out of business. The market mechanism provides for the most economically efficient solution, but this solution provides for an equitable sharing of

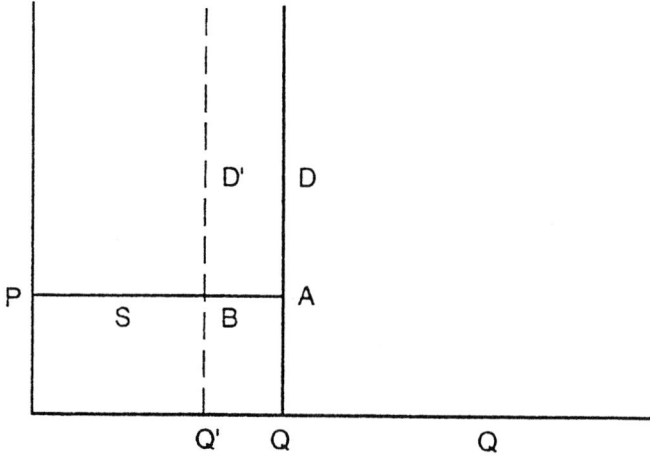

FIGURE G.13 Nonmarket Adaptations

the impact of climatic change. Figure G.14 illustrates this adaptation. The initial equilibrium is at point A. The climate change impact is to reduce available supply from Q to Q' and shift the supply curve from S to S'. Adaptation takes place, demand is reduced to Q' by shifting along the demand curve to P', and a new equilibrium is established at B.

G.5.2.1 Conservation and Improved Efficiency

A variety of measures categorized as conservation and efficiency can be taken to reduce water use. However, these measures vary from sector to sector. Below is a list of potential measures that could be investigated by the Country Studies Team:

- Domestic

  - Reduced toilet flushing
  - Reduced shower or bathing water
  - Reuse of cooking water
  - More efficient appliance use
  - Leak repair
  - Commercial car washing where recycling takes place
  - Rainwater collection for nonpotable uses

- Agricultural

  - Night time irrigation
  - Lining of canals
  - Introduction of closed conduits
  - Improvement in water measurements
  - Drainage reuse
  - Use of wastewater effluent
  - Better control and management of supply network

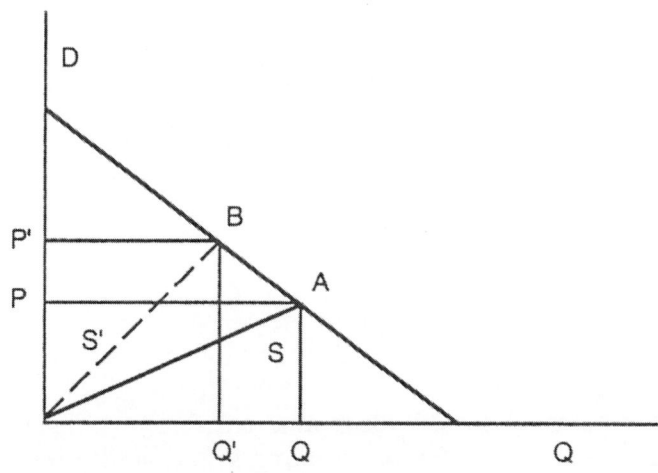

FIGURE G.14 Market Adaptations

- Industrial

    - Reuse of acceptable quality water
    - Recycling

- Energy-Related

    - Keeping reservoirs at lower head to reduce evaporation
    - Changing releases to match other water uses
    - Taking plants off-line in low flow times
    - Cogeneration (beneficial use of waste heat)

G.5.2.2 Technological Change

A variety of changes categorized as technology change can be made in the production process to reduce water use. However, these measures vary from sector to sector. Below is a list of potential measures that could be investigated by the Country Studies Team:

- Domestic

    - Water-efficient toilets
    - Water-efficient appliances
    - Landscape changes
    - Dual supply system — potable and nonpotable
    - Recycled water for nonpotable uses

- Agricultural

    - Introduce low-water-use crops
    - Introduce high value per water use crops
    - Change irrigation systems to drip, micro spray, LEPA
    - Introduce salt-tolerant crops that can use drain water
    - Drainage water mixing stations

- Industrial

    - Introduce "dry" cleaning technologies
    - Introduce closed cycle and/or air cooling
    - Plant design with reuse and recycling of water imbedded
    - Shift products manufactured

- Energy-Related

    - Build additional reservoirs and hydropower stations
    - Introduce low head run of the river hydropower
    - Introduce more efficient hydropower turbines
    - Introduce alternative thermal cooling systems (cooling ponds, wet tower, and dry towers)

G.5.2.3 Market/Price-Driven Transfers to Other Activities

As stated earlier, price can be used to influence water demand. If a derived demand curve for each sector can be constructed, the price change necessary to achieve the desired water use reduction can be determined. The major problem is that in many countries, and even across the economic sectors, it is difficult to construct a derived demand curve for two reasons:

- A curve must be developed for conditions in 2075, and it is very difficult to know the exact technological and economic conditions that will affect demand.

- To construct a demand curve, water use reductions and the costs for all the conservation and efficiency measures plus technological change must be known. The use versus cost information is implicit in the demand curve.

To analyze adaptation via price or market processes, a demand function is needed. Because it is difficult to derive a demand function for 2075, one approach is to estimate a demand curve for current conditions and then to estimate a curve for 2075 on the basis of a number of assumptions.

Following Samuelson and Nordhaus (1989), future economic and population growth should cause an outward shift of the demand curve, as illustrated in Figure G.15. However, the exact distance it shifts and its new slope are very difficult to estimate for 80 years into the future.

G.5.3 Adaptation Summary

Many potential water resource adaptations are available to respond to climatic change impacts on water supply and demand. Some of the more important ones have been highlighted. They have been discussed separately; however, it is most likely that they will occur simultaneously and in the most economic fashion on the basis of the return in investment for each action. The combination of supply-and-demand adaptations will provide for the most efficient adaptation strategy.

Two important questions remain:

- What policy will be implemented to institute adaptation — an equitable adaptation or an economic one?

- What is the economic cost of making these adaptations?

Section G.6 discusses estimating the economic cost or the effect of water resource impact.

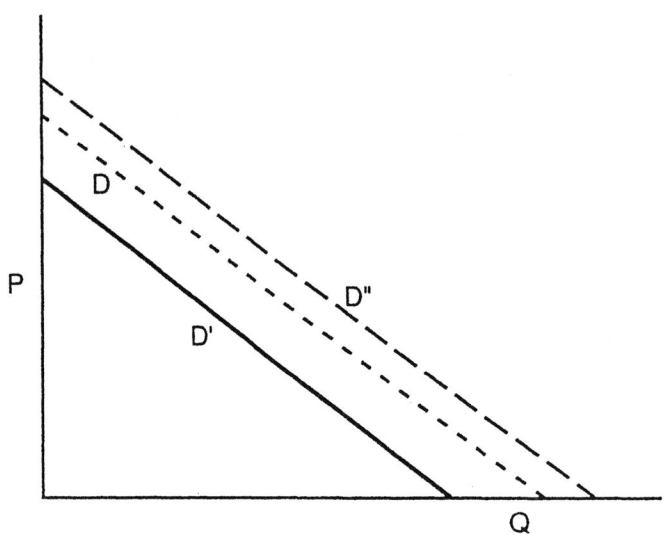

FIGURE G.15 Demand Curve Shift

G.6 ECONOMIC ASSESSMENT

Two approaches to direct measurement of the economic impact of water resources impacts and an approach to measure indirect effects were presented. Within the scope and budget of the Country Studies Program, it will be nearly impossible to fully implement detailed economic modeling unless ongoing research and planning activities are related to the use of these techniques. In lieu of detailed modeling, two simple but insightful approaches are proposed. They are outlined in this appendix, and issues related to their implementation in a climate change assessment are discussed. The reader is referred to the references for this section and to the standard economic literature for the basic procedure for performing the analysis.

G.6.1 Econometric Models

An econometric model is a set of equations that relate an economic indicator (dependent variable) to one or more economic, resource, or climatic parameters (independent variables) by means of regression analysis. The regression analysis provides for a best fit of the equation that relates the independent variables to the dependent variable. Water supply or investment in water resource infrastructure is used frequently as an independent variable in economic development forecasts. While the model is valid only when the underlying economics and infrastructure hold true, it is quite difficult to forecast changes 80 years in the future. As "constant" economic indicators are considered, inflation is not included. If it is felt that the basic engineering/economic processes will change sufficiently to invalidate a model based on historical data, an analog country could be selected. The argument is that the analog's current economy and technology (an econometric model) will better represent the country in 80 years.

An econometric model could be constructed for each of the four proposed economic sectors to relate sector gross domestic product (GDP) as a function of water supply and water resource investment. The economic impact of a climate change scenario and its associated change in national water supply can then be estimated by plugging the new value of water supply into the model and seeing the change in GDP for each sector and then summing across sectors for a total impact. While this method is crude, it is straightforward and will provide some qualitative measure of climate change impact on the economy from the water resources side.

G.6.2 Supply-and-Demand Equilibrium

As discussed in Section G.3, climate change will alter the water supply-and-demand equilibrium if a free market for the resource is in place. Hydrological changes will affect the supply curve, and climatic change will affect the demand curve; in theory, a new equilibrium will be reached. The market provides an efficient mechanism for reallocating the reduced water resource. However, this reallocation results in a cost to the economy and society. In Figure G.16, it is clear that less water is used under climate change; those to the right of the demand curve can no longer afford water. They must find alternative inputs to their productive activity or abandon it. The social cost of this change in productive activity is beyond the scope of this analysis; however, the cost to the economy of the shift in the equilibrium point can be measured.

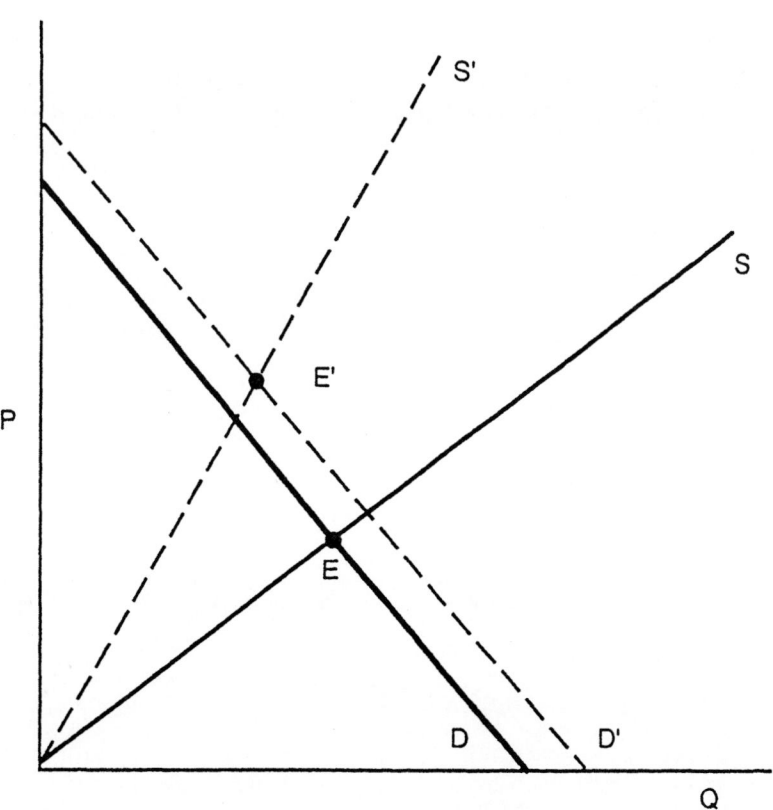

FIGURE G.16 Supply-and Demand Adjustment under Climatic Change

An economic measure of net welfare to society is the consumer-producer surplus (CPS) (Samuelson and Nordhaus 1989). The CPS, a microeconomic concept, is illustrated in Figure G.17. The CPS is used as a measure of economic performance in economic sector models (Hazell and Norton 1986).

In a sectoral economic approach, the supply-and-demand curves for water should be estimated for each of the four sectors (agriculture, industry, domestic, and energy) for the base 2075 scenario.

New supply-and-demand curves for climate change scenarios are then estimated, and a new equilibrium point is found. The CPS for the climate change scenario is calculated and compared with the base CPS. The difference is a measure of the economic impact of climate change on that sector. The impact is then summed over all the sectors for an economywide impact. Figure G.18 illustrates the difference in CPS between the base and a climate change scenario. In this example, the climate change impact is the net change in CPS, equal to the difference between the lost CPS (defined by the triangular area O-A-c) and the increase in CPS (defined by the rectangular area and a-b-c-B).

While it may not be possible to quantitatively estimate the shifts in the supply and demand curves, qualitative estimates of the shifts may provide for insights into the sign (positive or negative) of the impact and a relative indication of the magnitude of the impact. It should be remembered that for some scenarios, water supply may actually increase the shift in supply curve S′ to the right rather than the left, increasing CPS over the base scenario.

The approaches presented here are crude approximations at best, but they can be done within the context of the Countries Studies Program. They may highlight the sectors that need more economic analysis or even those that appear to be qualitatively more affected.

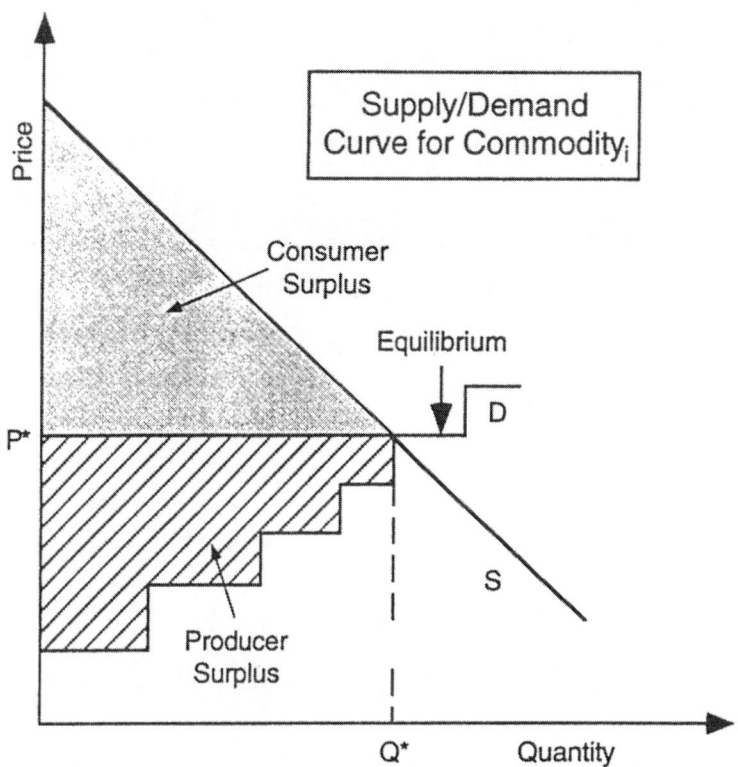

FIGURE G.17 Definition of the Consumer and Producer Surplus

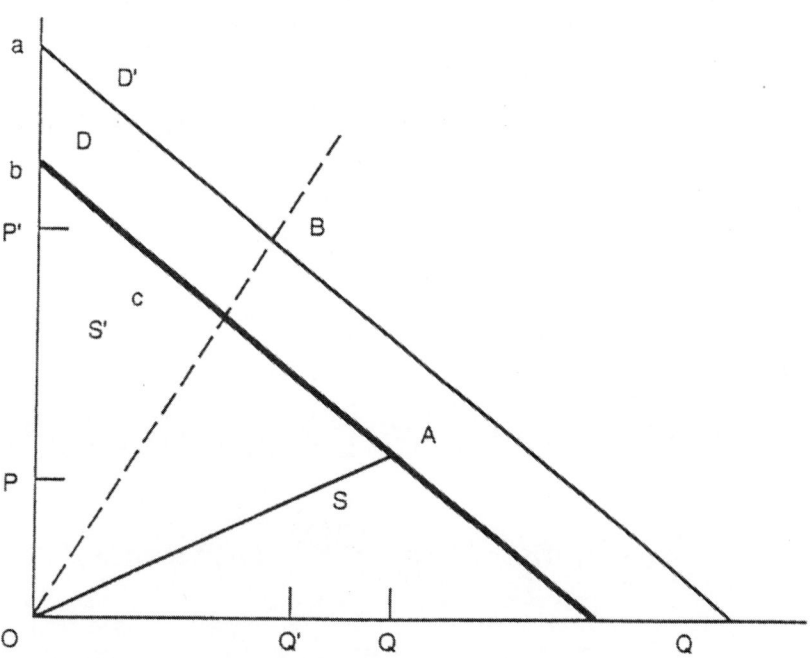

FIGURE G.18 Estimation of Economic Impact via Consumer-Producer Surplus

G.7 APPENDIX G REFERENCES

Abbott, M.B., et al., 1986a, "An Introduction to the European Hydrological System — Système Hydrologique Européen, SHE: 1. History and Philosophy of a Physically-Based, Distribution Modelling System," *Journal of Hydrology* 87:45-59.

Abbott, M.B., et al., 1986b, "An Introduction to the European Hydrological System — Système Hydrologique Européen, SHE: 2 Structure of Physically-Based Distributed Modeling System," *Journal of Hydrology* 87:61-77.

Anderson, E.A., 1971, *FORTRAN-IV Program for the Stanford Watershed Model IV. NQAA,* NWS, Office of Hydrology, Silver Spring, Md.

Berndtsson, R., and J. Niemczynowicz, 1988, "Spatial and Temporal Scales in Rainfall Analysis — Some Aspects and Future Perspectives," *Journal of Hydrology* 100:293-313.

Burnach, J.C., et al., 1973, *A Generalized Streamflow Simulation System: Conceptual Modeling for Digital Computers: Report by the Joint-Federal State River Forecasting Center,* U.S. Department of Commerce, NQAA National Weather Service, and State of California Department of Water Resources, March.

Carmichael, J., and K. Strzepek, 1986, *Industrial Water Use and Treatment Practices,* Tycooly Publishing, Riverton, N.J.

Chow, V.T., et al., 1988, *Applied Hydrology,* McGraw-Hill Book Company, New York, N.Y.

Crawford, N.H., and R.K. Linsley, Jr., 1966, *Digital Simulation in Hydrology: Stanford Watershed Model IV,* Technical Report 39, Department of Civil Engineering, Stanford University, Stanford, Calif., July.

Dunne, T., and L.B. Leopold, 1978, *Water in Environmental Planning,* W.H. Freeman and Company, New York, N.Y.

Falkenmark, M., 1977, "Reduced Water Demand — Result of the Swedish Anti-Pollution Program," *Ambio* 6(2):66.

Franchini, M., and M. Pacciani, 1991, "Comparative Analysis of Several Conceptual RainFall-Runoff Models," *Journal of Hydrology* 122:161-219.

Gates, W.L, 1985, "The Use of General Circulation Models in the Analysis of Ecosystem Impacts of Climatic Change," *Climatic Change* 7:267-84.

Gleick, P.H. (ed.), 1993, *Water in Crisis: A Guide to the World's Freshwater Resources,* Oxford University Press, Inc., New York, N.Y.

Goodman, A., 1984, *Principles of Water Resources Planning,* Prentice-Hall, Englewood Cliffs, N.J.

Gray, D.M., and T.O. Prowse, 1993, "Snow and Floating Ice," in D. Maidment (ed.), *Handbook of Hydrology,* pp. 7.1-7.53, McGraw-Hill Book Company, New York, N.Y.

Grotch, S.L., and M.C. MacCracken, 1990, "The Use of General Circulation Models to Predict Regional Climatic Change," *Journal of Climate* 4:286-303.

Hazel, P., and Notian, 1986, *Mathematical Programming for Economic Analysis in Agriculture,* Macmillan Publishing Company, New York, N.Y.

Jensen, M., et al., 1990, "Evapotranspiration and Irrigation Water Requirements," in *ASCE Manual* 70:132.

Kaczmarek, Z., 1990, *On the Sensitivity of Runoff to Climate Change and Variability,* IIASA Working Paper WP-90-058.

Kaczmarek, Z., 1993, "Water Balance Model for Climate Impact Analysis," *ACTA Geophysica Polonica* 41(4):1-1 6.

Kaczmarek, Z., 1991, *Sensitivity of Water Balance to Climate Change and Variability,* IIASA Working Paper WP-91-047, 7-8.

Kindler, J., and C.S. Russell (eds.), 1984, *Modeling Water Demands,* Academic Press, Inc., San Diego, Calif.

Kovalevsky, V.S., 1993, "The Impact of Predicted Climatic Changes on Groundwater," unpublished manuscript, Water Problems Institute, Russia Academy of Science.

Kulshreshtha, S., 1993, *World Water Resources and Regional Vulnerability: Impact of Future Changes,* RR93-10, IIASA Research Report.

Lettenmaier, D.P., and T.Y. Gan, 1990, "Hydrologic Sensitivities of the Sacramento-San Joaquin River Basin, California, to Global Warming," *Water Resources Research* 26:69-86.

Lettenmaier, D.P., and D.P. Sheer, 1991, "Climatic Sensitivity of California Water Resources," *Journal of Water Resources Planning and Management* 117(1):108-125.

Linsley, R.K., Jr., et al., 1982, *Hydrology for Engineers,* 3rd Ed., McGraw-Hill Book Company, New York, N.Y.

Loucks, D., et al., 1981, *System Analysis and Planning,* Prentice-Hall, Englewood Cliffs, N.J.

Maidment, D. (ed.), 1993, *Handbook of Hydrology,* McGraw-Hill Book Company, New York, N.Y.

Mays, L.W., and Y.K. Tung, 1992, *Hydrosystems Engineering and Management,* McGraw-Hill Book Company, New York, N.Y.

Niemann, J.D., et al., 1994, "Impacts of Spatial and Temporal Data on a Climate Change Assessment of Blue Nile Runoff," draft working paper, IIASA, Laxenburg, Austria.

Qzga-Zielinska, M., and J. Brzenzinski, 1993, *Basin Conceptual Model (BCM): The Simulation Model of the Monthly Runoff,* Institute of Environmental Engineering, Warsaw University of Technology, pp. 3-4.

Samuelson, P., and W. Nordhaus, 1989, *Microeconomics,* McGraw-Hill Book Company, New York, N.Y.

Shaw, E.M., 1983, *Hydrology in Practice,* Van Nostrand Reinhold Company, Ltd., Wokingham, U.K., pp. 260-270.

Shuttleworth, 1993, "Evaporation," in D. Maidment (ed.), *Handbook of Hydrology,* Chap. 4, McGraw-Hill Book Company, New York, N.Y.

Stakhiv, E., et al., 1992, "Hydrology and Water Resources," in W.J.M. Tegart and G.W. Sheldon (eds.), *Climate Change 1992: Supplementary Report to IPCC Impacts Assessment,* Australian Government Publishing Service, Canberra, Australia.

Starosolszky, O., 1987, *Applied Surface Hydrology,* Water Resources Publications, Littleton, Colo.

Sugarwara, M., et al., 1983, *Reference Manual for the TANK Model,* National Research Center for Disaster Prevention, Tokyo, Japan.

Todini, E., 1988, "Rainfall-Runoff Modeling — Past, Present, and Future," *Journal of Hydrology* 100:341-352.

U.S. Army Corps of Engineers, 1972, *Program Description and User Manual for SSARR Model Streamflow Synthesis and Reservoir Regulation: Program 724-K5-G0010,* Dec.

Viessman, W., Jr., et al., 1977, *Introduction to Hydrology,* Harper and Row, New York, N.Y.

Von Storch, H., et al., 1993, "Downscaling of Global Climate Change Estimates of Regional Scales: An Application to Iberian Rainfall in Wintertime," *Journal of Climate* 6:1161-1171.

Waggoner, P.E. (ed.), 1990, *Climate Change & U.S. Water Resources,* John Wiley & Sons, Inc., New York, N.Y.

Wolock, D.M., and G.M. Hornberger, 1991, "Hydrological Effects of Changes in Levels of Atmospheric Carbon Dioxide," *Journal of Forecasting* 10:105-116.

Yates, D. and K.M. Strzepek, 1994, "Comparison of Water Balance Models for Climate Change Assessment Runoff," draft working paper, IIASA, Laxenburg, Austria.

APPENDIX H:

COASTAL RESOURCE IMPACTS AND METHODS OF ADAPTATION ASSESSMENT

R. Benioff et al. (eds.), Vulnerability and Adaptation Assessments, H-1–H-39.
© 1996 *Kluwer Academic Publishers.*

APPENDIX H NOTATION

The following is a list of the acronyms, initialisms, and abbreviations (including units of measure) used in this appendix.

ACRONYMS, INITIALISMS, AND ABBREVIATIONS

AVVA	aerial videotape-assisted vulnerability analysis
DMA	Defense Mapping Agency
EPA	Environmental Protection Agency
GNP	gross national product
GPS	global positioning system
HCOC	high-cost open coast
IPCC	Intergovernmental Panel on Climate Change
LCOC	low-cost open coast
SC	sheltered coast
SLR	sea-level rise
VA	vulnerability analysis

UNITS OF MEASURE

cm	centimeter(s)
ft	foot (feet)
h	hour(s)
km	kilometer(s)
km^2	square kilometer(s)
m	meter(s)
mm	millimeter(s)
min	minute(s)
yr	year(s)

CONTENTS

TABLES

FIGURES

APPENDIX H:

**COASTAL RESOURCE IMPACTS AND METHODS
OF ADAPTATION ASSESSMENT**

H.1 INTRODUCTION

The principal impact of sea-level rise for most coastal areas is land loss through submergence of lowlands and loss of wetlands, as well as erosion of beaches. Coastal environments are incredibly diverse (e.g., coral reef atolls to volcanic islands), so that no single model can be used to determine the impacts of global change. However, all approaches to land loss estimates must use detailed, up-to-date information that provides the following: (1) geomorphic type (e.g., sandy beach and rocky headland); (2) relative elevation (e.g., ability to estimate 1- and 2-m contour locations); (3) land use (e.g., agricultural areas and tourist beaches); and (4) affected populations (e.g., estimated by number of domiciles in impacted areas). Maps never contain the richness and diversity of information; only aerial imagery can provide the needed detail, but conventional vertical aerial photography is often too small scale and too expensive to acquire and analyze. Therefore, other approaches that offer simplicity in acquisition at much lower cost, yet still yield reasonable estimates of coastal land loss, must be used.

In particular, one of the major problems of studying the impacts of a sea-level rise on developing countries is the lack of useful data on coastal elevations. Most maps only have 10-m to 100-m contours, which are virtually useless when analyzing the loss of land from any reasonable scenario for a rise in sea level. Traditional measures for improving this information do not appear to be promising at this time. The method used for determining the topography of the coastline must consider both the lack of existing topographic data and the length of coastline being studied.

Several methods for obtaining the required topographic data for analysis of the impacts of a sea-level rise have been examined. Aerial photography is the traditional method for such studies, but it is prohibitively expensive and time-consuming to analyze. This method also requires a high density of quality benchmark data, which are generally not available in developing countries. Satellite imagery may also be considered, but it is also prohibitively expensive, and, even under ideal conditions, a 10-m contour is the highest accuracy possible. This level is clearly inappropriate for vulnerability and adaptation assessments.

A new technique called aerial videotape-assisted vulnerability analysis (AVVA) is an approach that circumvents these problems and constitutes an important part of the Country Studies primary approach to country vulnerability and adaptation assessment. The AVVA approach is a quick, useful, and cost-effective tool for determining the impacts of sea-level rise in developing countries. The AVVA technique is a natural progression from Intergovernmental Panel on Climatic Change (1990) methods to produce a global analysis of protection costs. The technique uses detailed field data to identify land and infrastructure that are at risk, and determines protection costs for a range of response options. In sum, AVVA serves as a reconnaissance-level survey of the loss of coastal land in response to a rise in sea level. The approach involves:

- Aerial video-recording of the coastline,

- Limited ground-truth information, and

- Archival research in the country.

The AVVA technique involves videotaping the coastline from a small plane at very low elevation (Figure H.1). The video camera captures the aspect of the land relative to the sea. Using both the videotapes and limited ground surveying gives an estimate of contours and, hence, loss of land. This technique is especially attractive because it is the only method that allows individuals to survey large expanses of coastline quickly with a limited budget. The AVVA technique is already used to determine strategies of coastal management in Louisiana, where the coastal zone is retreating so rapidly that a quick and inexpensive method for determining coastal topography is essential.

The video record provides the following information on the coastal zone, including those areas directly affected by a rise in sea level:

- *Index of terrain and changes in relief.* The videotape at the lower elevation will help to determine the relative topography of the coast at the shoreline. It provides a cross-sectional view along the coastline, and at the higher elevation can provide an overview of the topography inland of the shoreline. The videotape also helps to determine if the material that makes up the coastline is hard rock or erodible material — information necessary for determining loss of land. These video data will also point out any errors or omissions such as out-of-date information on the existing maps.

- *Types of coastal environments.* The location and extent of marshes, mangrove swamps, agricultural land, sandy beaches and dunes, deltaic shorelines, and cliffed shorelines can be identified. These data will be very useful in determining if erosion or inundation will be more prevalent with a sea-level rise and also for ascertaining if a dryland area will convert to wetland or directly to water.

- *Land-use practices.* This information will identify if coastal lands are used for agricultural, aquacultural, urban, or other purposes.

- *Infrastructure.* The video record produces a hard record of all infra-structure that will be affected.

- *Population.* The video record shows how much of the population living in the coastal zone will be directly affected by a rise in sea level.

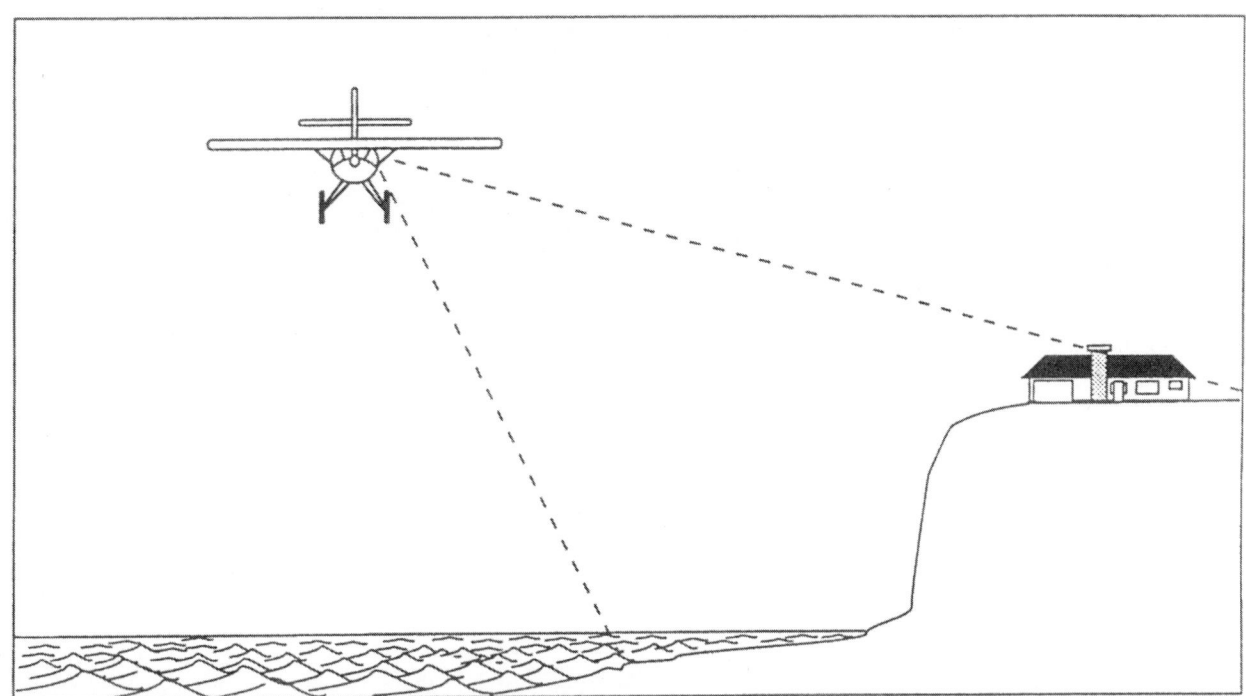

FIGURE H.1 Sketch Showing the Aspect of the Plane and Camcorder to the Land Being Videotaped

The videotape records numerous types of information about the coastal zone, including the relative elevation of the land. When combined with the ground-truth measurements, quantitative elevations can be estimated. Clearly, large errors can be associated with such estimates, but a feasibility study on the Chesapeake Bay demonstrated that such estimates are unbiased (Section H.2). While inappropriate for detailed assessments of sites, this method does provide meaningful results for studies of large areas.

The cost is $60-$70/h to rent a Cessna 172, with pilot, in the United States. This type of plane is available worldwide.

H.2 VERIFICATION OF AVVA METHODOLOGY

To assess errors associated with this method, researchers viodeotaped part of the western shore of the Chesapeake Bay. This coastline is composed of alternating eroding sandy bluffs up to 30 m high and marsh-filled valleys fronted by sandy beaches. In a few areas, the beach is prograding. This diverse coastline is as complex as any encountered in the developing countries and, as such, provides a good test of this method. Only spot heights, which will be available in the countries to be surveyed, were provided in this test. Two operators then *independently* estimated the 3-m contour from the video record.

The results were compared directly at 19 transects with the 3-m (10-ft) contour on the U.S. Geological Survey (USGS) 7.5-min topographic maps. Thus, three data sets were generated, one for each of the two operators and one for the combined data (Table H.1). The agreement was surprisingly good, with a maximum mean error of 6.4 m, which amounted to a 9% error.

TABLE H.1 Errors Associated with Estimating the 3-m (10-ft) Contour[a]

Operator or Data	Mean Error (m)	Standard Error (m)	Confidence Interval (m)
Operator 1	-3.5 (1.6)	7.1	-18.5 (8.1) to 11.5 (5.0)
Operator 2	6.4 (2.8)	4.1	-2.2 (0.9) to 5.0 (6.5)
Combined data	1.4 (0.6)	4.1	-7.0 (3.0) to 9.8 (4.3)

[a] A positive number overestimates land use. The values in parentheses give the error as a percentage of the average distance from the 3-m contour to high water.

In all cases, the mean error was not significantly different from zero, demonstrating that the method is an unbiased estimator of the true contour position (Figure H.2). The 95% confidence intervals show the maximum mean error that might be expected with this method; this error amounts to 18.6 m, which at this site was a 27% error. The combined data show much smaller error bands (up to 9.8 m), so that the method is clearly of sufficient accuracy for a reconnaissance-level estimate of land losses in developing countries. Better results were obtained for the 1-m contour because the surface of the sea provides a level plane of constant elevation (allowing for tidal variation) that is generally close enough horizontally to the 1-m contour to allow comparison. More detailed, time-consuming, and costly surveys that use conventional methods will obviously yield better estimates of land loss than the method outlined here.

A detailed pilot study in Senegal further demonstrated that the method can give useful and cost-effective results compared with alternative techniques. The amount of data collected in a short visit to a country is substantial, and this fact can lead to problems of data analysis.

H.3 PROCEDURE FOR AVVA

The aim of the reconnaissance survey is to produce a quick, but accurate, assessment of the impacts and responses to an accelerated rise in sea level. Data that should be collected and issues to consider are summarized in Table H.2. Past national studies show that AVVA involves spending about three weeks collecting and analyzing data, with about two months of further analysis and compilation, depending on the size of the country. A more detailed analysis of specific areas can be conducted later, depending on the results of AVVA; this analysis will be a more time-consuming effort and may require additional ground-truth measurements.

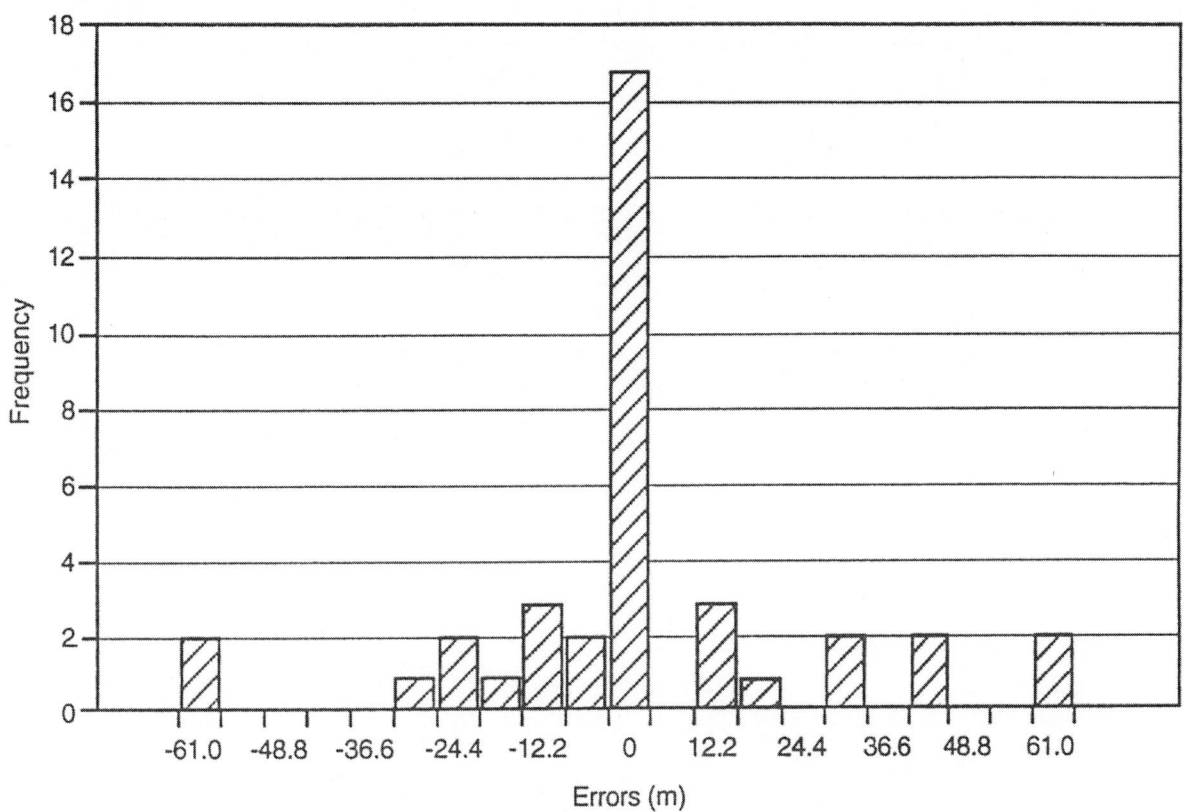

FIGURE H.2 Distribution of Errors for Pilot Study of Calvert County, Maryland, in the Chesapeake Bay

A brief description of the procedure for the fieldwork follows:

- Divide the coast into a working geomorphological classification.

- Videotape the coastline at a low elevation.

- Review the video record, and develop a geomorphological or land-use classification.

- Obtain ground-truth measurements for as many representative coastal types as possible.

- Simultaneously collect topographic and economic data.

Subsequent analysis further develops an inventory of the coastal zone, including coastal geomorphology, coastal land use and development, and estimates of coastal elevation. Thus, the coastline can be classified into sections with similar characteristics that can be considered homogeneous units.

TABLE H.2 Data Requirements and Issues to Consider in Conducting AVVA

A. Collect all existing data that can be used to assess the impacts of a rise in sea level:

1. Maps showing what is located on the coast, within 6, 3, and 1 m of mean sea level

    a. Cities or ports
    b. Resort areas
    c. Tourist beaches
    d. Industrial areas

2. Topographic maps with at least 3-m contour intervals

3. Survey maps

4. Aerial photos

5. Satellite images (useful, but not required)

6. Storm tide tables

7. Population density data

8. Historic data

    a. Flood maps
    b. Photos (aerial or other) taken during floods
    c. Coastline maps from different years to show changes
    d. Photos from different years showing changes caused by erosion

B. Analysis of the effects of a 1-m sea-level rise

1. Increased storm flooding

    a. Description of what is located in flood-prone areas
    b. Historical floods

        (1) Location
        (2) Response of local people
        (3) Response of government (were any policies changed?)

2. Beach or bluff erosion

    a. Description of what is located within 300 m of the ocean coast
    b. Description of beach types
    c. Description of the livelihood of people living in coastal areas

        (1) Fishing
        (2) International tourism

TABLE H.2 (Cont.)

---

    d. Description of beach erosion present, if any
    e. Existing data on erosion

       (1) In country
       (2) British maps
       (3) French maps

    f. Availability of tide gauge data for determining relative sea-level changes
    g. At least a Bruun Rule calculation and preferably a trend analysis of important beach areas

  3. Marsh drowning and loss (the tropical equivalent of temperate-latitude salt-marsh grasses is woody mangroves)

    a. Description of what is located in these areas (do people live here?)
    b. Are mangroves being cut and used?
    c. What is the perception of the need to protect these wetlands?
    d. What extent of wetlands is natural?
    e. Are wetlands viewed as useful for coastal fisheries and hunting or merely thought of as wastelands?

  4. Saltwater intrusion

    a. Is there any problem with the drinking water supply?
    b. Is it likely that salinity will be a problem for surface or subsurface water (or both)?

C. Political ramifications and impacts on future decisions

  1. River dams
  2. New settlements
  3. Beach resorts
  4. Transmigrations

D. Solutions

  1. Once the impacts of a sea-level rise are clearly identified and their magnitude assessed, assess what the country could actually do in response to these problems.
  2. Conduct an informal survey of how other people in influential positions in the country would respond.
  3. Explain laws of property ownership in the country. Can the government tell people that they must move back as the water begins to threaten them? Will or can people stay in place and just build the houses higher for safety and risk reduction?
  4. Describe the history of coastal protection in the country.

Estimates of land loss are made by using appropriate models; hence, economic losses and possible responses are assessed. The video record is still important in this phase of the analysis because (1) losses of infrastructure are estimated directly from the video record, and from any available maps, by simply overlaying the predicted land loss and estimating the number of buildings that would be destroyed; and (2) the lengths of coastline requiring protection are measured from the video record.

The checklist in Table H.2 contains all of the procedures necessary for AVVA. The assumption is that one person trained in the videotaping method will work with another coastal expert throughout the project. This checklist is based on existing experience but should be periodically updated.

The following equipment is essential:

- Two videocameras (one is a backup and is used for copying tapes) with the necessary batteries and rechargers (videocameras with a date stamp included are preferred),

- Two videocassette tapes per 100 km to be taped (one original and one copy),

- One in-plane communication system (usually four-way),

- One handheld global positioning system (GPS) for precise locations (cost, U.S. $1,000 or less),

- One small color television monitor (to view the videotape during the survey mission),

- Transit and rod with leveling books (for ground-truthing), and

- Still cameras (for aerial and ground slides).

H.3.1 Pretrip Preparation

Before the trip, the following preparations are necessary:

- Develop a geomorphological classification of the coastline from all available sources (e.g., Bird and Schwartz [1985]). If the coastal length is too great to videomap entirely, select representative lengths for study.

- Select an appropriate sampling interval for ground-truthing on the basis of natural variability and logistical constraints.

- Obtain all available maps and other relevant information from all possible out-of-country sources (United States, Britain, and France). Defense Mapping Agency (DMA) nautical charts are available in the

United States and provide bathymetric information along the coastlines of the world.

- Obtain all available coastal maps (topographic, bathymetric, city, land use, etc.) and wave data from the appropriate government agencies.

- Obtain tidal predictions of the area of study to aid the planning of the fieldwork.

- Arrange for necessary housing, transportation, and airplane rental (for videotaping) and vehicle rental (for ground-truthing) for the research team.

- Become familiar with the concept of the Bruun Rule (Dean and Maurmeyer [1983]; Hallermeier [1981]).

- Convert the eustatic sea-level scenarios (e.g., 0.5, 1.0, and 2.0 m) to relative sea-level scenarios by estimating uplift or subsidence. To estimate this quantity, obtain the mean sea-level records from the Permanent Service for Mean Sea-Level at Proudman Oceanographic Laboratory, England; assume a eustatic rise of 1.8 mm/yr (Douglas 1991).

H.3.2 Acquisition of Videotape Data

A three-person crew is required, including a pilot, a video-camera operator, and a still-camera operator and coastal expert. The procedure for acquiring the necessary videotape is as follows:

- Videotape the coast, following the shoreline, at two elevations (approximately 30-50 m and 300 m). The video camera is handheld in the backseat of the high-wing plane (e.g., Cessna 172 or 182). The scene should include some sea and sky; do not use the zoom because this feature amplifies any shake from the airplane. As a further guide, view some of the videotapes archived at the University of Maryland's Laboratory for Coastal Research.

- Videotaping from the airplane should include a running commentary by a coastal expert on relevant information, including

  - Place names and locations, with regular positional readings from the handheld GPS unit;

  - Land use and land cover;

  - Existing problems of land loss caused by erosion or submergence;

- Projects for shoreline stabilization; and

- Future plans for development and stabilization.

- To complement the video record, take representative slides of all of the coastal features of the country. Careful records should be kept so that each slide can be accurately located.

- Maintain a written record of the GPS positions.

- It is recommended that the videotapes be backed up (copied) each night to reduce the possibility of losing any data.

H.3.3 Review of Videotape Data

All videotapes should be reviewed nightly. This review should consist of the following tasks:

- Confirm the coastal geomorphological classification used to design the survey (i.e., divide coastlines into units such as barrier beaches, deltas, muddy coasts, coral reefs, hard-rock headlands, and eroding bluffs).

- Sketch the following information on the best available base maps (e.g., large-scale maps, with a scale of 1:200,000 or larger, if feasible) by viewing the aerial videotapes and by using the geomorphology and land-use classification system (Table H.3). Essentially, the coast is partitioned into linear sections on the basis of land use and land cover (Figure H.3).

- Attempt to make a general sketch of the landward extent of low-lying lands that might be impacted by storm flooding, inundation, and sea-level rise. This task also focuses attention on areas where ground-truthing measurements would be most useful.

- Make a preliminary assessment (from videotapes and local expert advice) of areas subject to (1) erosion; (2) inundation, including wetland loss; (3) saltwater intrusion; (4) increased storm flooding and higher water tables; and (5) no impact because land is nonerodible (hard rock).

- Assess the selected sampling interval for ground-truthing measurements.

TABLE H.3 Geomorphology and Land-Use Classification System

I. Coastal geomorphology

   A. Beaches

      1. Barrier beach

         a. Type

            (1) Bay barrier
            (2) Barrier spit
            (3) Barrier island (microtidal or mesotidal)

         b. Morphology

            (1) High (>5 m), continuous foredune
            (2) Extensive dune field
            (3) Low dunes with washovers

      2. Strand plain or headland beach

         a. Type

            (1) Low coastal plain
            (2) Flanked by erodible cliffs
            (3) Flanked by hard rock cliffs

         b. Morphology

            (1) High (>5 m), continuous foredune
            (2) Extensive dune field
            (3) Low dunes with washovers

      3. Pocket beach

         a. Type

            (1) Flanked by erodible headlands
            (2) Flanked by rocky headlands

         b. Morphology

             (1) High (>5 m), continuous foredune
            (2) Extensive dune field
            (3) Low dunes with washovers

TABLE H.3 (Cont.)

B. Wetlands

    1. Estuary

        a. Mangrove
        b. Marsh (grass)
        c. Marsh (scrub shrub)
        d. Marsh (forested)

    2. Delta

        a. Mangrove
        b. Marsh (grass)
        c. Marsh (scrub shrub)
        d. Marsh (forested)

    3. Back-barrier areas

        a. Mangrove
        b. Marsh (grass)
        c. Marsh (scrub shrub)
        d. Marsh (forested)

    4. Tidal flats

C. Cliffs (no beach)

    1. Erodible

        a. Dunes on top of cliff
        b. Flatland on top of cliff
        c. Hilly land on top of cliff
        d. Mountainous land on top of cliff
        e. Height

    2. Rocky

        a. Dunes on top of cliff
        b. Flatland on top of cliff
        c. Hilly land on top of cliff
        d. Mountainous land on top of cliff

TABLE H.3 (Cont.)

    D. Muddy coast (mud beach)

        1. Flatland behind mud beach
        2. Hilly land behind mud beach
        3. Mountainous land behind mud beach
        4. Lake behind mud beach
        5. Lagoon behind mud beach

    E. Hardened (protected) shoreline

        1. Sand dunes behind protection
        2. Flatland behind protection
        3. Hilly land behind protection
        4. Mountainous land behind protection
        5. Wetland behind protection

II. Protection (if present)

    A. Seawall
    B. Bulkhead (note composition: timber, cement, riprap, etc.)
    C. Breakwater
    D. Groins
    E. Jetty
    F. Protected harbor
    G. Beach nourishment

III. Land use

    A. Urban or city
    B. Residential
    C. Industrial
    D. Tourist
    E. Agricultural (crops: note type if known)
    F. Cattle grazing
    G. Sheep grazing
    H. Orchards
    J. Forest
    K. Barren
    L. Shrub lands
    M. Desert
    N. Fishing
    O. Aquaculture

IV. Inland geomorphology

    A. Flatland
    B. Hilly land
    C. Mountainous land
    D. Lake
    E. Wetlands

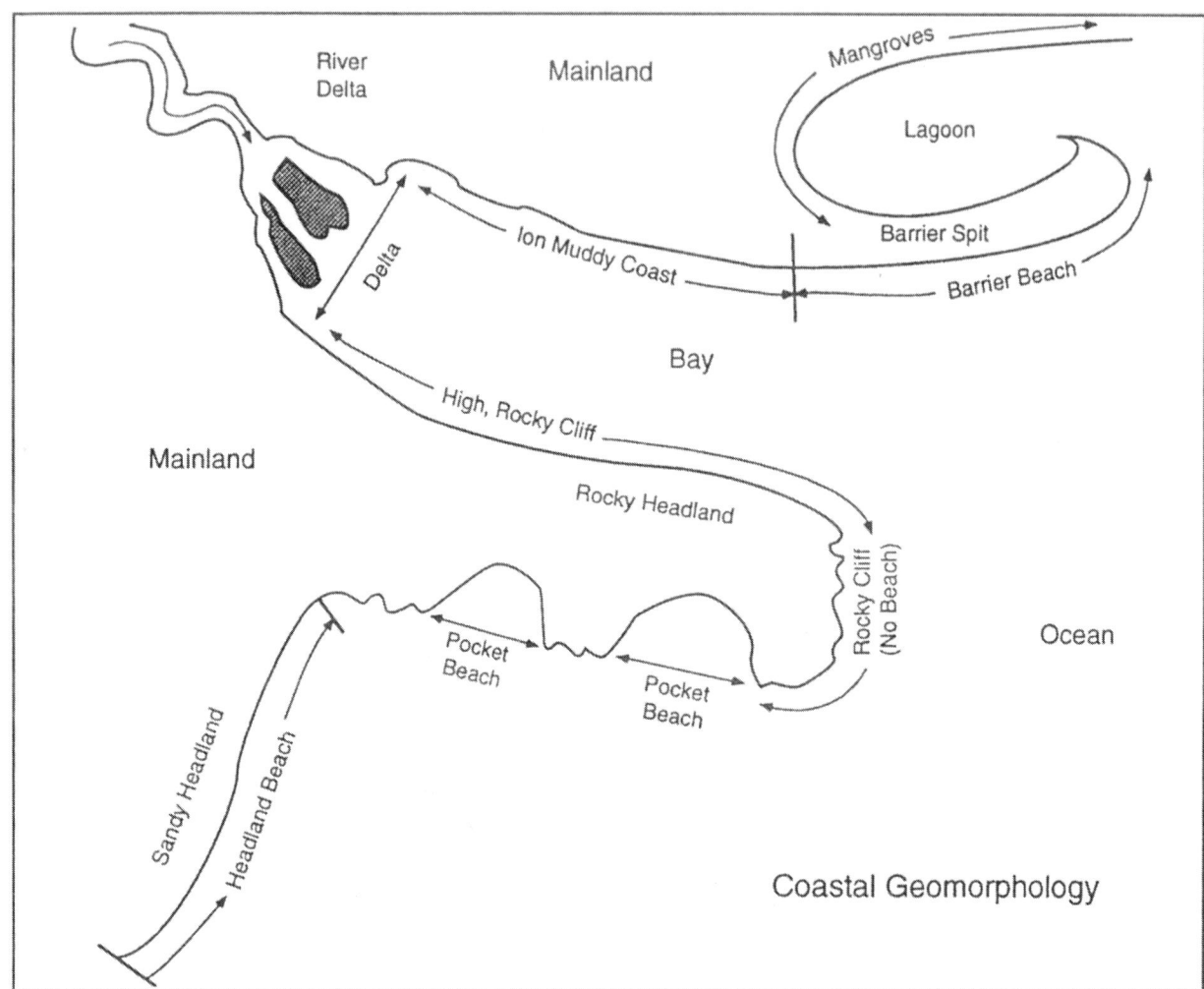

FIGURE H.3 Initial Geomorphological Classification of Coastline

H.3.4 Acquisition of Ground-Truth Data

The acquisition of ground-truth data includes the following tasks:

- Using the refined coastal geomorphological classification, obtain information on spot elevations from available maps or other country surveys.

- If insufficient elevations are available, limited field surveying (from benchmarks or other known elevations) of selected areas should be undertaken.   This process can be extremely time consuming, so surveying is best kept to a minimum; for instance, benchmarks are often located far inland. For areas subject to inundation, approximate spot heights can be established by using high-water marks and tidal predictions. (This procedure should not be used if wave activity exists.) Sampling at regular intervals should be employed, as practical.

Important locations, such as towns, should also be surveyed. When possible, survey at least one beach profile for each type of coastline (if possible, tie these surveys into benchmark information). Remember to note the time and date of the profile. Movement by helicopter will facilitate quick ground surveys where land transportation is difficult. Travel by helicopter is expensive but offers flexibility and rapid movement to remote and inaccessible areas. To date, all of the studies have involved land transport.

- Measure the length of city blocks and individual structures (to aid in quantitative analysis of the video record).

- Obtain long-term development strategies and plans that impact upon coastal areas in terms of future uses.

H.3.5 Collection of Economic Data

- Estimate the value of land. Is land ownership practiced? (In some societies, land cannot be bought or sold.)

- Estimate the value of a city block for erosion or inundation purposes or the value of beachfront houses if destroyed (local real estate agents can provide estimates of average house value, and this number is multiplied by the estimate from the videotape of the number of houses at risk). What are typical weekly and monthly rents?

- For agricultural fields, query farmers on the harvest value of their crops on a per-hectare basis per year.

- For aquacultural operations in intertidal and supratidal areas, obtain harvest estimates and market value on an annual basis.

- Obtain estimates of the value of the coastal infrastructure in conjunction with the responsible government officials (if possible), including (1) buildings, (2) crops, (3) tourist resorts, and (4) commercial establishments (e.g., oil refineries, power plants).

- Obtain estimates of the cost for construction of various types of coastal defenses. Sometimes, finding out the cost of each component or constituent element of a structure is better (e.g., for a dike, find out the cost of building materials by the cubic meter and the cost of moving the material and packing it down; and calculate the total cost yourself by using an estimate of how long, how high, and how wide the dike will have to be to hold back the sea with a predicted sea-level rise). Component or constituent elements include the following: (1) open-coast seawalls (hard stabilization); (2) sheltered-coast bulkheads and dikes (hard stabilization); and (3) beach nourishment (soft stabilization), including capital and operating charges for pumping.

- Obtain information on the country's gross national product (GNP), major sources of income, and the relative importance of coastal activities by category, if possible, on (1) agriculture, (2) aquaculture, (3) commercial facilities, (4) resort areas (tourism), (5) urban areas, and (6) wetlands at risk.

H.3.6 Post-trip Summary

Backup videotapes should always be produced as a precaution against loss or damage. Always work with the first-generation copy, and preserve the original for copying purposes.

- Perform calculations by using the Bruun Rule to estimate land lost on coasts subject to erosion. This estimate requires wave data to calculate a depth of closure and profile information available from the charts and videotapes. The seaward limit of the beach profile is uncertain over the long time scales involved, so scenarios of land loss are constructed. The maximum and minimum scenarios are developed by using the depth definitions d_i and d_1, respectively (Hallermeier 1981), as the seaward limit for calculations.

- Conduct an inundation analysis by using ground-truthing measurements and the videotape.

- Estimate land loss (erosion and inundation).

- Estimate the cost of protection.

- Estimate the economic value of the land and structures that would be lost.

In essence, an index of areas at risk and cost estimates of damage with a 1-m eustatic rise in sea-level or with coast protection should be developed. The other sea-level scenarios are considered similarly, except for the 0.5-m scenario, where linear scaling of the 1-m impact is sometimes necessary.

- *No protection case* (e.g., retreat). Assume that no protection exists (any protection that exists is not taken into consideration).

- *Present case.* Assume that any areas currently protected will remain protected and will be repaired and upgraded as sea-level rises.

- *Important areas protected.* Assume that all important areas will be fully protected (i.e., cities, tourist beaches, and factories), as well as important natural resources (i.e., oil production and coastal fisheries).

- *Total protection.* Assume that areas along the coast with populations greater than 10 inhabitant/km^2 will be protected (IPCC 1990a).

In addition, other questions to be considered discuss (1) dry land converted to wetland, (2) wetland converted to water (including mangroves), and (3) dry land converted to water.

H.4 ESTIMATES OF LAND LOSS

Figure H.4 illustrates an example of land loss estimates using erosion vs. inundation methods.

H.4.1 Erosion

The Bruun Rule is frequently used to calculate erosion caused by a rise in sea level. In the present calculations, the following form is used (Hands 1983):

$$R = G \times S \times [L/(B + h_*)] , \tag{H.1}$$

where

R =   shoreline recession caused by a sea-level rise S;

G =   overfill ratio;

S =   sea-level rise (projected);

L =   active profile width from the dune to the depth of closure;

B =   dune height (note that $H = h_* + B$); and

h_* =   depth of closure.

These parameters can be estimated from field data and available charts. The value of G is assumed to be unity (all eroded material is sand), except where material is not erodible (hard rock), so these estimates of recession are minimum values. The depth of closure h_* (and, hence, the active profile width L) is the variable that is most difficult to estimate. In particular, h_* depends on the time scale; that is, the longer the period of interest, the larger the depth of closure. This fact is important to the analysis because, all things being equal, a larger value of h_* implies more beach erosion (and higher costs for beach nourishment) (Figure H.5). No accepted methods are available for estimating the

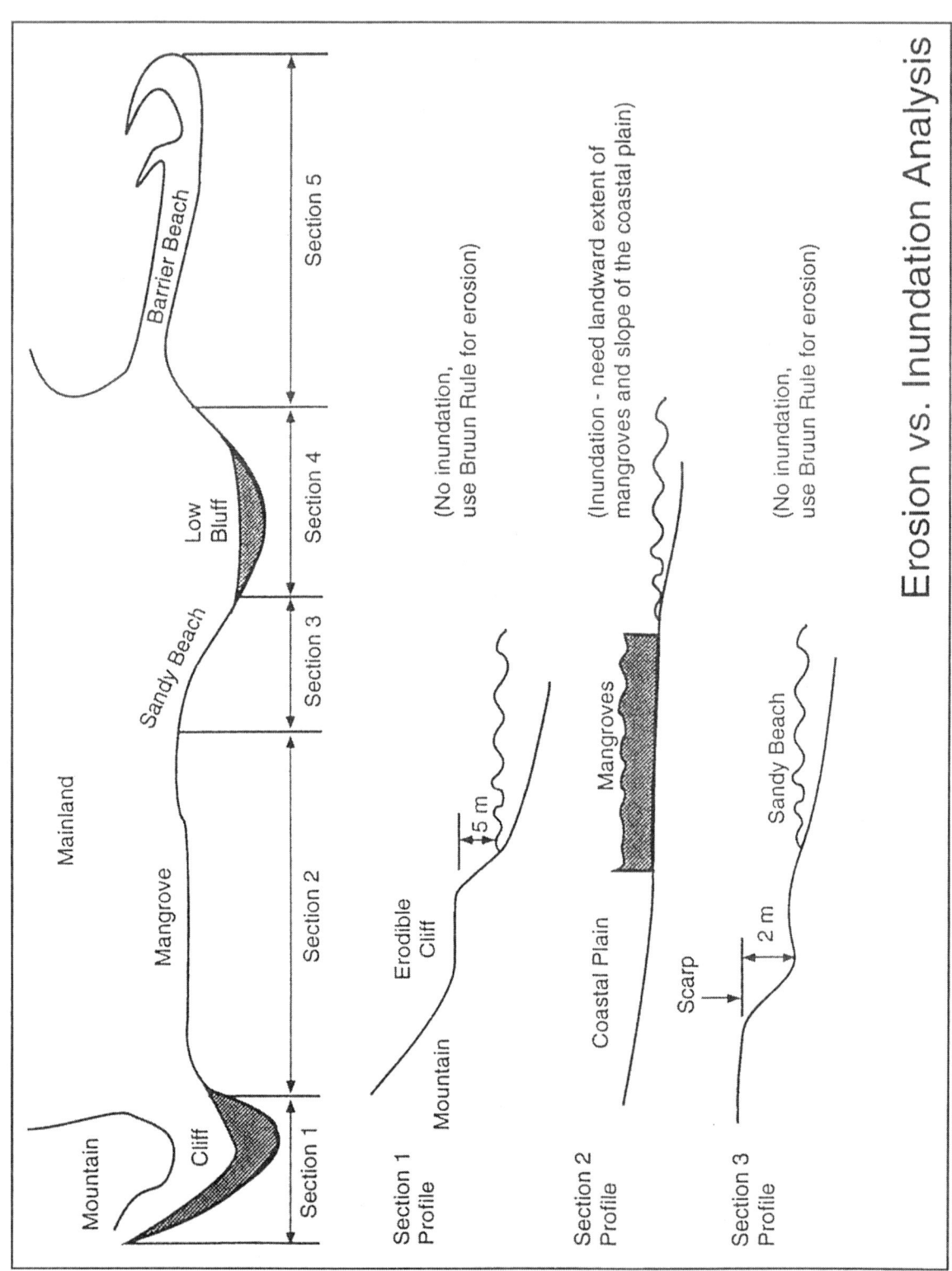

FIGURE H.4 Erosion vs. Inundation Analysis

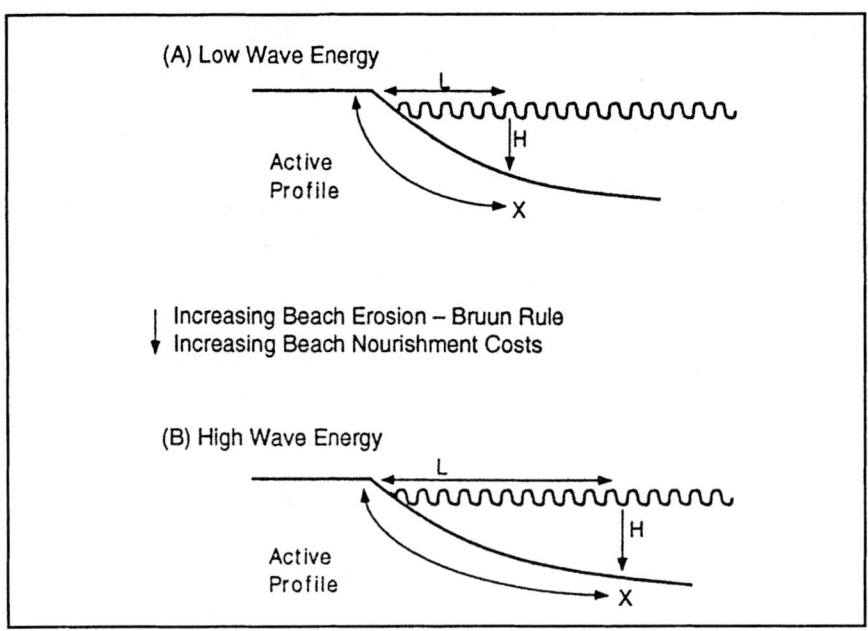

FIGURE H.5 Importance of Active Profile Width

depth of closure over the long time scales of interest. Therefore, for these calculations, low and high estimates of the depth of closure (d_{L1} and d_{L100}, respectively) should be constructed. Here, d_{L1} has the same definition as d_1 of Hallermeier (1981), which is the greatest depth where intense onshore or offshore transport and significant alongshore transport will occur within a typical year. The value of d_{L1} can be approximated as

$$d_{L1} = 2H_s + 11\sigma ,\qquad\qquad\text{(H.2)}$$

where H_s is the mean significant wave height, and σ is the standard deviation of wave height. Experience suggests that d_{L1} is a good estimate for a typical year and provides a robust low estimate. The high estimate is new and is termed d_{L100}. It extends Hallermeier's original concept and has the same definition as d_{L1}, except that the time scale is a century.

On the basis of very limited experience and some new calculations, a simple approximation was developed (Nicholls and Leatherman 1994):

$$d_{L100} = 1.75 \times d_{L_1} .\qquad\qquad\text{(H.3)}$$

Further work to define how h_* varies with time would be useful.

Both values of h_* are referenced to a value 1 m above low water. The corresponding low and high active profile widths can be determined at different locations along the coast by using the best bathymetric charts. The coastline is divided into segments of similar beach width and dune height, and the Bruun Rule is applied.

The availability of data on wave climate can be a major constraint on these analyses. A comprehensive and accessible global database on wave climate is urgently required if the objectives of the IPCC are to be attained.

H.4.2 Inundation

In the case of wetlands and other coastal lowland, a direct inundation or "drowning" concept is applied. Inundation is most significant in deltas and in wetlands around estuaries. For these low-lying areas, the video record is insufficient to define an estimate of elevation, and should be integrated with available maps, plus expert judgment. This integration is not to undervalue the video record because it (1) provided a check on the validity of the maps and (2) helped to define the present extent of wetlands and mangroves.

Two-meter and, where possible, 1-m contours (above high water), plus the area of existing wetlands or mangroves, should be estimated. Sedimentation in the inundated area should be considered as much as possible. Coastal marshes and mangroves can accrete vertically in response to slow rates of sea-level rise and are only inundated above a certain threshold value of sea-level rise. Deltas may or may not be receiving sediment from the river catchment. In most cases, the availability of sediment is poorly understood, and assessment of wetland evolution caused by *relatively* small amounts of sea-level rise (up to 0.5 m) is difficult; often the best estimates of land loss are linear interpolations from the scenarios of higher sea-level rise.

H.5 ECONOMIC ANALYSIS

Completing a thorough examination of the economic cost of sea-level rise is a complicated process that builds on detailed and data-intensive analyses of the values of future economic activities and wealth that might be lost — the sort of analyses that may (or may not) prove to be warranted sometime in the future. Their prospective value can, in fact, be judged only in the context of a preliminary review of the larger "landscape." It is therefore appropriate to begin even the most ambitious assessment with modestly detailed analyses of economic vulnerability — rough and practical analyses designed to produce portraits of how large the sea-level rise problem might be and when it might come to the fore. Properly conducted, these preliminary analyses not only highlight specific geographic regions in which subsequent, more thorough analyses might pay the largest dividends, but also provide order-of-magnitude estimates of potential economic impact — economic bottom line estimates that are essential in judging (1) the relative significance of sea-level rise among a long list of global change impacts *and* (2) the relative significance of global change effects among long lists of national concerns.

Estimates of economic vulnerability to sea-level rise are simple enumerations of the current values of economic assets and activities that might be threatened by inundation. They can be produced by following a straightforward accounting procedure based upon the results of the AVVA impacts analysis described above; and they can be compared with rough estimates of the cost of protection to support reasonable assessments of various adaptive and response strategies. The following pages outline this procedure, which was highlighted in Figure 5.5.4. The procedure is also illustrated in a stand-alone example taken from the U.S. experience.

H.5.1 Defining a Working List of Coastal Units — AVVA and Economic Baselines

The first step in the procedure is one which limits the scope of its coverage. It is not necessary to consider every kilometer of coastline, at least not in the first round of an impacts analysis. Superimposing a list of coastal sites that AVVA surveys identify as vulnerable to inundation from rising seas onto a list of economically important coastal areas should easily provide a (potentially shortened) working list of general coastal areas whose economic vulnerability should be examined.

Each entry on the working list should be divided into as many economically distinguishable sites as necessary to adequately capture the economic, political, or cultural diversity of the identified area. Relatively homogenous areas might be captured adequately without division; identifying only one site would be sufficient in such cases. Adequate coverage of relatively heterogenous areas might, by way of contrast, require long (sub)lists of sites so that the special nature of each (sub)site could be accurately reflected.

Figure H.6 displays a simple worksheet upon which the next steps of the procedure build. One or more of these worksheets should be completed for each coastal (sub)site identified in the first step to be (1) vulnerable to inundation from sea-level rise (from the AVVA survey) and (2) economically (and/or politically and/or culturally) important. The number of worksheets to be completed for site "X" depends on the sensitivity to alternative scenarios.

(1) Year	(2) SLR	(3) Local Subsidence	(4) Total SLR	(5) Lost Land	(6) Value of Lost Land	(7) Value of Structure	(8) Sum of (6) & (7)	(9) Cost of Protection
2000	____	_____	____	_____	_____	_____	_____	_____
2010	____	_____	____	_____	_____	_____	_____	_____
2020	____	_____	____	_____	_____	_____	_____	_____
.	.	.	.	.	.	.	.	.
.	.	.	.	.	.	.	.	.
.	.	.	.	.	.	.	.	.
.	.	.	.	.	.	.	.	.
.	.	.	.	.	.	.	.	.
2100	____	_____	____	_____	_____	_____	_____	_____

FIGURE H.6 Economic Analysis Worksheet

H.5.2 Specifying Sea-Level Rise Scenarios

The worksheet in Figure H.6 records, in Column 1, benchmark years through 2100. Column 1 thereby enumerates the global change planning horizon; it can be lengthened or shortened, and the interval lengths can be adjusted. Table H.4 records six alternative sea-level rise scenarios that might be considered. The first three are drawn from the most recent IPCC scientific assessments: linear (high, medium, and low) trajectories which attribute 33, 67, and 100 cm in sea-level rise through the year 2100 to the threat of "business as usual" global warming. The second three alternatives reach the same endpoints along the quadratic trajectories favored by the U.S. Environmental Protection Agency (EPA) in its impacts assessments. These EPA trajectories record sea-level rise values that are smaller than their linear IPCC counterparts for any year short of 2100, but they exaggerate the rate of later-year sea-level rise.

Continuing with the procedure, choose one or more of these scenarios for site "X" (considering at least a high and low scenario for either the IPCC or the EPA conventions is recommended) and record the corresponding sea-level rise (SLR) values in Column 2 of the worksheet (one for each scenario chosen).

H.5.3 Calibrating the Baseline Sea-Level Rise

Column 3 asks that any local subsidence (positive or negative) that might affect site X be computed and recorded. Since impacts analyses of global change should capture only the additional effects caused by that change, noting any significant, nonzero natural subsidence means that a nonzero benchmark worksheet should also be completed: a "no global change" baseline worksheet with no sea-level rise recorded in Column 2 but with natural subsidence recorded in Column 3.

TABLE H.4 Sea-Level Rise Scenarios

	IPCC			EPA		
Year	33 cm	67 cm	100 cm	33 cm	67 cm	100 cm
2000	3	6	9	0	1	1
2010	6	12	18	1	2	3
2020	9	18	27	2	5	7
2030	12	24	36	4	9	13
2040	15	30	45	7	14	21
2050	18	37	55	10	20	30
2060	21	43	64	13	27	40
2070	24	49	73	17	35	53
2080	27	55	82	22	45	67
2090	30	61	91	27	55	83
2100	33	67	100	33	67	100

H.5.4 Total Sea-Level Rise Scenarios

Column 4 completes the sea-level rise summary by asking for the sum for each year of the sea-level rise values recorded in Columns 2 and 3. Three separate worksheets would be used for site X if two alternative greenhouse-induced sea-level scenarios were chosen and site X were not otherwise stationary.

H.5.5 Land Loss Estimates

Column 5 uses the results of the AVVA survey to identify, on each worksheet, the land in site X that would be lost to rising seas during the specified time intervals. The rate of land lost may not be uniform along the coastline, even within a homogenous and geographically small site. In that case, it is necessary to construct a representative sample across site X: a sample comprising systematically scattered strips of land drawn perpendicular to the coastline (no narrower than the maximum resolution of the AVVA survey, and extending far enough inland to outrun even the highest sea-level scenario). Worksheets for each scenario should then be created for each strip in the sample (for a 10-strip sample across site X two scenarios and a nonstationary baseline case would mean that there are now 30 worksheets for site X).

H.5.6 Estimating the Economic Value of Lost Land

Column 6 is the economic value of the lost land in Column 5. These values can sometimes be extracted on a per-acre (or per-hectare) basis for developed property directly from property tax maps or other public records of market value. If these records do not exist, local real estate agents can often provide some pricing information (of questionable quality unless corroborated). The value of less developed but economically productive land can be deduced by the value of the economic activity that it supports. The present value of per-acre harvest yields, using the current market rate of interest as the discount factor, is a good estimate of the value of an acre of agricultural land. Similarly computed present values of economic return to other activities work as well. Care should be taken to verify that the activity is, indeed, attached to the land. If the economic activity is mobile, over the long term with enough warning it will be moved and not lost to rising seas. There should, finally, be some lower bound land value for undeveloped and currently unused land — a value that represents its "discounted potential use value." Real estate agents can offer some estimates of this as well.

Care need not be taken to capture the location premium associated with some coastal property (high rents for being right on the water). This type of premium will migrate inward as the seas rise, and therefore is not lost; it is simply redistributed from people who used to live right on the water to people who now (at least for a while) live right on the water. Moreover, economic theory suggests that it is the value of interior land that is lost to sea-level rise; but that is sometimes as difficult to capture as the gradient of location premium. In practice, average land values within the various sampling strips must lie somewhere between the high-priced shoreline property and the low-priced interior landscape. They are thus too high for interior land and too low for coastal property, but they are fine middle-of-the-road estimates for a first round vulnerability analysis.

H.5.7 Estimating the Economic Value of Threatened Structure

Column 7 is the value of structures presently located on the land identified in Column 6. These values can be drawn from data, but it is frequently better to simply relate structure value to land value by a simple convention that eliminates the multitude of variables that cause one structure to be worth more or less than another. Tax codes usually record these conventions, or they can be estimated directly from independently collected data sets.

H.5.8 Estimating Total Vulnerability

Column 8 records total vulnerability: the sum of the land and structure vulnerability estimates of Columns 6 and 7. Absent any other change, the value of most structures will be depreciated to zero given enough warning along a gradual sea-level rise trajectory. Column 6 therefore represents a cut at the lower-bound, efficiency-based opportunity cost estimate of vulnerability: the value of land that is indeed lost without consideration of structures whose economic value should be negligible at the time of inundation. Column 8, by way of contrast, represents an upper-bound, no-belief (in either sea-level rise or a policy to retreat from rising seas) estimate of vulnerability. Comparing the two offers a measure of the potential redistribution of wealth, and may indicate the strength of any potential political and/or social pressure that might lead to protecting property that, according to efficiency grounds, should really be abandoned.

The present value of the losses noted in Column 8 (using the private, market rate of interest) is a rough estimate of the present value of protecting the property (within site X or within a sample strip in site X) that is threatened by inundation along the designated sea-level rise scenario. The present value of the losses noted in Column 6 (still using the private, market rate of interest) is a rough estimate of the present value of the cost of abandoning the property (assuming perfect foresight and complete adaptive depreciation of structure) that would be lost to inundation along the designated sea-level rise scenario. The latter is a measure of the benefit of protection under the best of circumstances; the former is a measure of those benefits when people simply do not believe that property will be abandoned.

H.5.9 The Cost of Protection

Column 9 focuses on the cost of protecting the land threatened during each time interval. Cost estimates are best divided into fixed cost (expenditure that is required at the beginning of a protection strategy) and variable cost (subsequent expenditure related to increased sea-level rise and maintenance). Dikes or bulkheads usually involve large initial, fixed cost followed by moderate maintenance and expansion (variable) costs. Beach nourishment or other "as it comes" reactions usually carry small initial investments, but variable costs which expand rapidly as the seas rise. Engineers can provide data for each site (or sample strip) for each type of protection strategy; the data recorded in Column 9 should reflect the least costly alternative (assessed and selected in terms of the minimum present value along the indicated sea-level rise scenario).

The present value of Column 9 is an estimate of the (smallest) cost of protecting site X or a derivative sample strip. It can be compared with the present value of Column 6 to determine whether protection can be justified in terms of economic efficiency, and/or it can be compared with the present value of Column 8 to judge whether or not protection can be justified in terms of maximum economic value and the pain of redistribution.

H.5.10 Aggregate Country-Wide Vulnerability/Protection Estimates

Aggregate vulnerability/protection estimates can now be produced for each sea-level trajectory initially chosen. Suppose that the initial AVVA survey highlighted 10 vulnerable regions and that 6 of them were economically significant. Suppose, further, that 5 of those regions (I, II, III, IV, and V) were sufficiently homogenous and small that sampling was not required. Assume, finally, that a 10-strip sample was created to represent the sixth important site (VI) and that it covered 0.1 ($\times 100\%$) of its coastline. For each sea-level trajectory, there would now be 15 worksheets.

Each worksheet would have a present value of lost land from Column 6 — the efficient cost of abandoning property; call that $PV_{(6)}\{J\}$ for sites I through V and $PV_{(6)}\{VI_k\}$ for sample strip k in site VI. The aggregate cost of abandoning site VI could then be approximated by:

$$PV_{(6)}\{VI\} \approx (1/0.1)[PV_{(6)}\{VI_1\} + \ldots + PV_{(6)}\{VI_{(10)}\}] \tag{H.4}$$

or some more appropriate weighted sum if the sample were not entirely representative. The aggregate cost of abandoning all coastal property would then simply be the sum of $PV_{(6)}\{J\}$ for J = I, ..., VI; and the *extra* cost of abandonment caused by greenhouse warming would be the difference between this sum, computed along some sea-level trajectory, and the sum that emerges in the baseline case when only natural subsidence is allowed.

Summing the present values of Column 8 would produce a different estimate of loss — one which ignored the potential for cost reducing adaptation in advance of abandonment; the difference in the two would be a composite measure of wealth that would be redistributed rather than lost to abandonment.

Applying the same procedure to Column 9 would produce a different estimate — the present value of the aggregate cost of protecting the "important" coastline under the assumption that, all other things being equal, the individual site and/or sample cost projections do not escalate when more than one project is undertaken at the same time. Each worksheet would identify a time trajectory for a protection project; together they can provide some insight into the relative likelihood of coincident investments in protection that might compete for scarce resources.

It should be clear that the worksheets will provide enough information to advance preliminary judgments about whether or not plans should be made to protect each site or sample strip. If the present value of protection (measured more appropriately as the present value of Column 6 unless political, cultural or other imperatives cause Column 8 to be applicable) were greater (less than) the present value of the least-cost means of protection, then the site or sample strip should (should not) be protected. An aggregate bottom line of country-wide exposure to sea-level rise could therefore be created by adding whichever appropriate cost (abandonment or protection) for each site (or sample strip) is smaller into the weighting scheme indicated above. Replicating the procedure for a range of sea-level scenarios would produce a range of cost estimates and, perhaps, an assortment of different protection decisions.

A few caveats should, finally, be offered. If the protection decisions for any site or strip seem to turn on very close comparisons of similar cost and damage estimates, it may pay to do some sensitivity analysis with respect to (1) the interest rate, (2) protection costs, and (3) land and/or structure values. These supplementary "experiments" are the best way to discover when and where more detailed analyses of true economic cost may be warranted. The worksheets will also suggest when various protection strategies should be undertaken. If it appears that they should all begin at roughly the same time, then "bottleneck" problems may arise in both their financing and the delivery of the necessary materials. Some thought should then be given to how best to spread the projects out so that the bottlenecks do not cause costs to rise so much that property that should be protected is abandoned.

H.5.11 Example of the Economic Procedure: Long Beach Island, New Jersey

Figure H.7 records the applicable data for a specific site in the sample that supported vulnerability estimates for the United States (Yohe 1990a). It presents value and cost estimates in 10-yr increments along the quadratic EPA 100-cm sea-level rise scenario for Long Beach Island — a site located midway down the coast of New Jersey with annual natural subsidence of about 3 mm.

Columns 2 through 4 describe the sea-level scenario; notice that natural subsidence adds 34 cm to total sea-level rise through the year 2100. The incremental land loss values portrayed along column 5 were derived from a computerized inundation mapping program devised for the project by Park et. al. (1989); these values are certainly comparable to the types of estimates that would emerge from AVVA.

Columns 6 and 7 were produced by Yohe (1989) from tax maps for the island. The procedure employed to produce them was exactly the procedure that is described in this text.

100 cm EPA Scenario

(1) Year	(2) SLR	(3) Local Subsidence	(4) Total SLR	(5)[a] Lost Land	(6)[b] Value of Lost Land	(7)[b] Value of Structure	(8)[b] Sum of (6) & (7)	(9)[b] Cost of Protection
2000	1	4	5	0.50	17	50	67	14.6
2010	3	7	10	0.25	8	25	33	2.0
2020	7	10	17	0.25	8	25	33	2.2
2030	13	13	26	0.25	8	25	33	2.4
2040	21	16	37	0.25	8	25	33	2.6
2050	30	19	49	1.25	42	134	166	2.9
2060	40	22	62	0.75	25	75	100	3.1
2070	53	25	78	0.50	17	50	67	3.3
2080	67	28	95	0.25	8	25	33	3.5
2090	83	31	114	0.00	0	0	0	3.8
2100	100	34	134	0.25	8	25	33	4.0

[a] Denominated in square kilometers of area.
[b] Current value denominated in millions of 1989 dollars.

FIGURE H.7 Economic Analysis Example: Long Beach Island, New Jersey

They represent the current (in 1989) values of land and structure identified as threatened by the inundation portrait, and they reflect the U.S. convention that roughly 75% of the value of most developed real estate is embodied in structure.

The costs of protection statistics recorded in the last column were derived for a specific protection strategy — raising the island as the seas advance; they were derived from engineering work conducted by Wiegel (1989). Large early costs reflect a considerable investment in preparation for raising houses, roads, and other infrastructure; the extended cost portrait captures the cost of sand computed in terms of a constant supply cost of $6 per cubic yard.

Given a 5% discount rate, the present value of the protection costs is roughly $13.6 million. Compared with a present value of $100.8 million for the total vulnerability of threatened land and structure, protection seems to be the appropriate choice. Even when only land costs are included in the comparison (so that efficient depreciation of structure to zero value in anticipation of potential inundation is assumed), the present value of benefits exceeds the present value of the cost of protection by almost $12 million.

H.6 OPTIONS FOR RESPONSE

Four options for response to sea-level rise should be considered, depending on the socioeconomic value of the resources at risk, and the cost of protection:

- *No protection.*

- *Present protection.* Existing protection is maintained.

- *Protection of developed areas.* Medium to highly developed areas are protected (i.e., cities, tourist beaches, and factories). By using the video record, medium to high development is defined as areas with more than 15% of the ground covered in structures. For tourist areas, beach nourishment is assumed. Elsewhere, seawalls are used.

- *Total protection.* All coastal areas with a population greater than 10/km² are protected (IPCC 1990a). All additional protection above 3. Protection for developed areas is seawalls.

H.6.1 Nourishment and Seawall Designs

The options for protection can be developed as follows by using in-country costs, where possible. For seawalls and groins, use the cost of purchasing, transporting, and placing rock, plus design costs (10%) and maintenance costs (20%), to determine a total cost. The cost of beach fill (sand) can be estimated at U.S. $5/cubic meter, unless alternate cost estimates are available.

H.6.2 Beach Nourishment

The volume of beach fill can be calculated by raising the entire active profile by the sea-level scenario; however, as with beach erosion, the active width is difficult to determine. A logical approach is to use the low and high active widths already defined for the Bruun Rule as the low-cost and high-cost designs, respectively.

An additional cost on long open beaches is alongshore loss of sand out of the nourished area. To stop such losses, large terminal groins at the ends of each nourishment project could be utilized, which, in effect, converts the open beach into a pocket beach. The terminal groins extend from the dune to the seaward limit of the profile, which can be several kilometers offshore. Costs do not rise uniformly with sea-level rise because the groins are a large but declining proportion of the total cost as the sea-level rise scenario is raised. While this procedure raises the costs of beach nourishment significantly, the resulting costs are more realistic than simply considering the cost of a single beach fill. From an environmental perspective, renourishment may be preferred, but the costs will again be much greater than a single beach fill. More work on the optimal application of beach nourishment would be useful.

H.6.3 Seawalls

To calculate the cost of erecting a seawall, three simple seawall designs should be considered, the application of which depends on the wave environment. These designs are a low-cost and high-cost seawall for open (or wave-exposed) coasts (LCOC and HCOC, respectively) and a seawall for sheltered coasts (SCs) (Nicholls and Leatherman 1994) (Figure H.8). All three designs use 1:2 slopes and a 2-m berm. A seawall design for erodible cliffs was also developed.

The LCOC and HCOC designs are necessary because of the uncertain response of the beach in front of the seawall to sea-level rise. In the best case (lowest cost), the beach would not erode at all; in the worst case (highest cost), the beach would be completely lost. Actual behavior will lie somewhere between these extremes, with total beach loss being more likely with scenarios of higher sea-level rise. The LCOC design assumes no beach erosion and only considers the effects of a more severe nearshore wave climate. The HCOC design assumes total beach loss and the consequent need to prevent undermining of the wall. The size and, hence, cost of the HCOC design increases with wave climate. Thus, the LCOC and HCOC designs reflect the two cost extremes. The SC design is less costly than the LCOC design because the sheltered coast would have limited problems with a less severe nearshore wave climate.

The 1:2 slope should be used instead of the 1:4 slope for the cost estimates of the IPCC (1990a) because the 1:2 slope is more economical and is the standard in the United States. The IPCC designs, which use a 1:4 slope and a 4-m berm width, would double the cost of seawall construction; however, the range of costs using the 1:2 slope for wave-exposed coasts embraces the IPCC costs because of the HCOC design.

H.7 EXAMPLE RESULTS FROM PREVIOUS STUDIES

Some results from AVVA for Senegal, Uruguay, and Venezuela are given in Tables H.5, H.6, and H.7. The high and low estimates give a large possible range of land loss and cost, but this range embraces the uncertainty inherent in analyses of this type.

Table H.5 gives land loss with no protection assumed. Note the significant land loss caused by a 0.2-m rise in sea level (the present case). Thus, present rates of sea-level rise are already a significant hazard. Land loss under all scenarios is one or two orders of magnitude smaller in Uruguay than in Senegal and Venezuela because inundation is not an important process in Uruguay. In Senegal and Venezuela, inundation is the dominant process for land loss. Most of the inundated land is wetland or mangroves; however, erosion causes a high proportion of the economic and societal impacts in all three countries.

In Table H.6, the costs of protection are given with the cost estimates of the IPCC (1990a). Uruguay has high protection costs because of the importance of beach-based tourism and, hence, has a large requirement for beach nourishment (over 98% of the costs for protection of developed areas). The IPCC appears to have greatly underestimated the cost

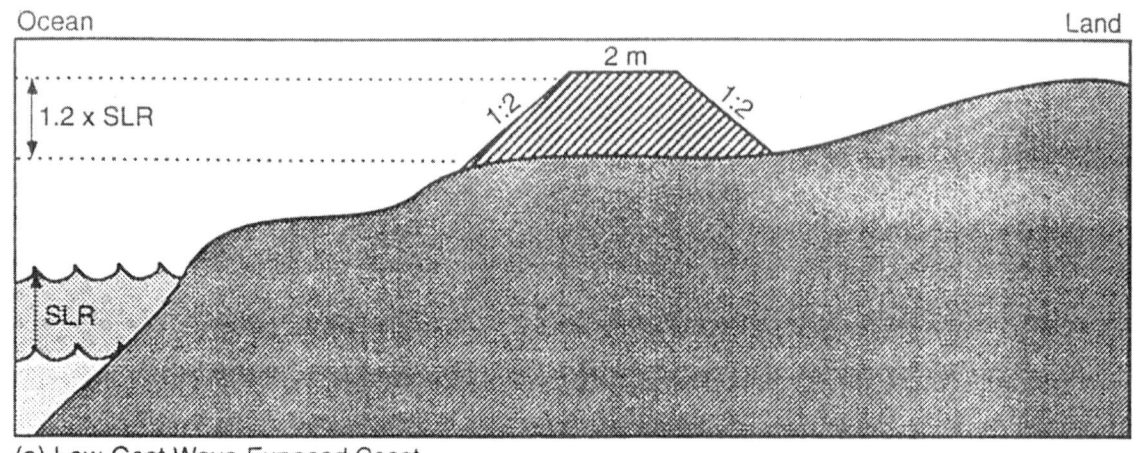

(a) Low-Cost Wave-Exposed Coast

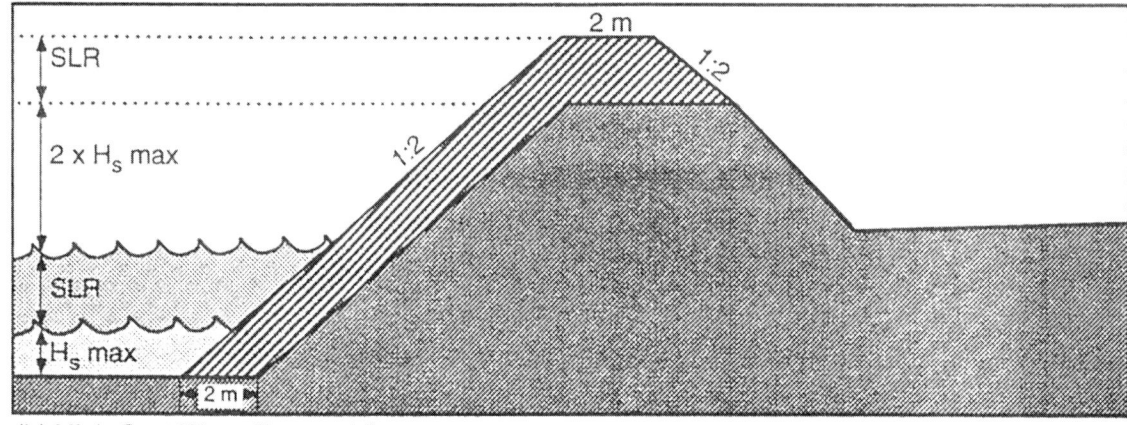

(b) High-Cost Wave-Exposed Coast

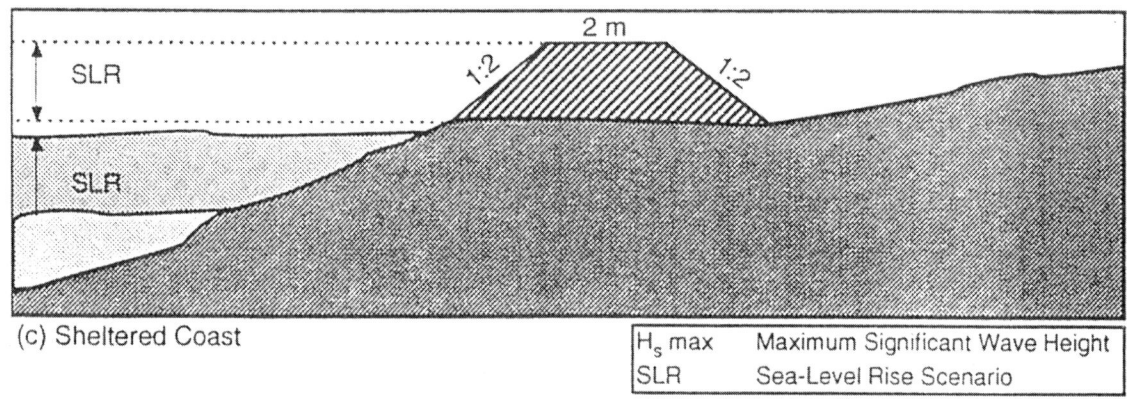

(c) Sheltered Coast

H_s max	Maximum Significant Wave Height
SLR	Sea-Level Rise Scenario

**FIGURE H.8 Seawall Designs for Coasts with Differing Wave Energies
(Source: Nicholls and Leatherman 1994)**

**TABLE H.5 Estimates of Land Loss
with No Protection Assumed**

| Country | Loss of Land (km^2) | | |
	0.2-m Rise	0.5-m Rise	1.0-m Rise
Uruguay	15-20	36-51	72-103
Senegal	349-356	1,947-1,963	6,042-6,073
Venezuela	1,138-1,147	2,844-2,866	5,686-5,730

TABLE H.6 Costs of Protection

| Country | Sea-Level Rise (m) | Cost (10^6 $ U.S.) | | |
		Developed-Areas Protection	Total Protection[a]	Total Protection[b]
Senegal	0.5	146-575	407-1,422	NA[c]
	1.0	255-845	973-2,156	1,596
Uruguay	0.5	2,068-6,871	2,162-7,505	NA
	1.0	2,903-8,578	3,144-9,437	1,805
Venezuela	0.5	454-960	718-1,613	NA
	1.0	999-1,517	1,717-2,634	3,155

[a] Source: Nicholls and Leatherman (1994).

[b] Source: IPCC (1990b).

[c] NA = not available.

of sea-level rise in Uruguay. Senegal has smaller costs for protection in absolute terms, but the costs are still significant. Again, beach nourishment is a large proportion of the costs. A large increase in cost occurs between developed areas and total protection. What benefits Senegal would gain for this increase in expenditure is questionable because much of the additional protection is for low-value desert areas. In general, protection of developed areas, as assessed in these studies, is the most likely option for protection. Venezuela also has large costs for protection; but, given the large length of its coastline, the costs are surprisingly low. This fact is due to the low level of coastal development, occupying only 13% of the coastline in Venezuela. For Venezuela, the IPCC (1990a) cost appears to be a significant overestimation.

**TABLE H.7 Relative Vulnerability
for Developed-Areas Protection
and a 1-m Sea-Level Rise**

Country	Relative Cost (%)[a]	Land Loss (% of Total)
Uruguay	6.5-19.2	3.08
Senegal	0.9-3.0	0.62
Venezuela	0.1-0.2	0.04

[a] Note that relative cost is based on
1987 gross investment and assumes
that the investments occur over
50 years.

In Table H.7, these three countries are ranked by (1) their ability to pay for protection by using gross investment, or (2) land loss (as a percentage of total land area). For the purpose of this analysis, the assumption is made that the investment in protection occurs over 50 years (2050 to 2100). Uruguay has the highest relative costs, consuming a significant proportion of the available investment (at 1987 levels), but has the lowest land loss. Senegal has both high costs and high land loss. These results support the need for a vulnerability profile, as opposed to a vulnerability index, as recommended in the common method.

These studies have shown that the IPCC (1990a) study underestimated the requirement for beach nourishment compared with this analysis. Given the importance and continued growth of coastal tourism, this analysis shows the importance of land-use planning for new tourist developments to reduce future requirements for beach nourishment.

For the 1-m scenario, the cost of protecting all areas in Senegal with a population greater than 10 people per square kilometer (the total-protection case) is between $6,714 million and $8,274 million (Nicholls and Leatherman 1994). These protection costs are much higher than the $1,596 million estimated by the IPCC (1990a) for Senegal using the same protection scenarios. The reasons for this discrepancy are, in increasing importance, as follows:

- Estimate of coastline length to be protected,

- Inclusion of beach nourishment, and

- Cost of building seawalls.

The total length of coastline considered for protection in the IPCC report (1990a) was 1,353 km, compared with 2,065 km in the AVVA study. The IPCC report was a desktop study that determined coastline length by measuring the coastline "as the crow flies" and then applying a multiplier to take into consideration the inland estuary and delta coastlines, irregularities of the coast, and offshore island coastlines. The multiplier used for Senegal was 5. This figure is clearly an underestimation for Senegal because it has two major estuaries and a delta. This difference in the estimate of inland estuary and delta coastlines accounts for most of the discrepancies in the low-lying coastline length between the two studies. Although the coastal length in Senegal was underestimated by IPCC, this fact should not be taken to invalidate their worldwide estimate.

The IPCC report (1990a) does not include beach nourishment because the investigators were unable to identify any recreational beaches in Senegal at the scale of their study; however, Nicholls and Leatherman (1994) show that recreational beaches are very important in Senegal. In 1986, annual tourist arrivals were 235,000 people, and annual tourist receipts were U.S. $118 million (PC Globe, Inc. 1989; Software Toolworks 1990). Therefore, tourist dollars in 1986 represented about 3% of the GNP, and tourism is a growing industry in Senegal. The estimated cost of protecting tourist beaches from a 1-m sea-level rise ranged from $213 million to $1,846 million.

The IPCC report estimates a cost of $0.72 million/km of coastline protected with a seawall, whereas the AVVA study set the cost at $3.3 million/km. In the IPCC report, the cost of seawall construction is based on Dutch unit costs multiplied by a country-specific cost factor. In this study, seawall costs at current rates are based on a study commissioned by the government of Senegal. To take a rise in sea-level into consideration, this study increased the design height by the sea-level rise scenario plus 20% to account for higher wave conditions and increased run-up from the deeper water conditions. This discrepancy in unit costs for seawall construction accounts for most of the difference in the protection costs between the two studies. The estimates of cost in the AVVA study are considered to be much more realistic than those of the IPCC; but with the limited number of seawall construction projects in Senegal, costs may have to be adjusted in the light of more experience.

Nicholls and Leatherman (1994) consider Senegal as of 1990. Population and economic growth can only exacerbate the impacts. Development of a management plan for coastal zones in Senegal would help mitigate the problems of future development and would prove useful with or without an acceleration of sea-level rise.

These results demonstrate the utility of the method described herein. The AVVA approach can provide a better estimate of the potential impacts and cost of accelerated sea-level rise and help to fulfill the IPCC (1990b) aim of quantifying the impacts and possible responses of a 1-m rise in sea level.

H.8 LIMITATIONS OF THE AVVA APPROACH

The AVVA approach is only one method of obtaining the information necessary for a vulnerability analysis (VA). This first-step reconnaissance allows one to focus upon the more critical areas of a coastline. More detailed follow-up studies with the use of more traditional approaches are recommended; however, the videotape record provides much useful information on land use and infrastructure and "brings the coast to your desk." Therefore, the videotape may be useful even if high-quality topographic data are available, as is the case in Louisiana. The AVVA method is least valuable in deltaic areas because of their low topography; however, as already noted, the video record still provides a useful inventory of many coastal characteristics for VA.

As described in this appendix, AVVA has been constructed as a series of modules, whereby individual elements of the overall procedure can be improved, while the remaining elements of the approach remain robust. Therefore, as understanding of the elements improves, VA can be easily and rapidly repeated to generate new and improved estimates of the impacts of sea-level rise.

By the nature of reconnaissance surveys, many possible improvements to the procedures described are possible and would help to more precisely define impacts and appropriate responses to sea-level rise. In particular, improving understanding of beach erosion and, hence, the costs of beach nourishment and seawalls would be useful. In this regard, better wave data are essential. In addition, a more explicit description of the time element of the problem is required, including factors such as economic growth and development.

H.9 SUMMARY

The AVVA is a rapid and low-cost method to conduct reconnaissance VAs on large scales; the AVVA is a lasting source of data, containing almost limitless information on the video images. The AVVA is often the only baseline data available that are up-to-date and contain the detail necessary to define what is at risk. This baseline inventory of resources should be updated by aerial video mapping every decade to determine changes in land use and land cover. The AVVA also provides a focus on which areas can be most usefully studied with more conventional and expensive techniques and, hence, helps to maximize the results from any limited national budget for such an analysis.

Aerial video techniques are useful for more detailed studies when high-quality topographic data already exist; however, other problems remain, such as the appropriate seaward limit of the active beach profile. These problems are independent of the data source used. Procedures have been developed to provide a realistic range of values for these estimates. Further work is required on these problems, including the establishment of observational databases for the world's coastlines, most particularly some meaningful description of wave climate.

H.10 APPENDIX H REFERENCES

Bird, E., and M.L. Schwartz, 1985, *The World's Coastline*, Van Nostrand Reinhold Co., Inc., New York, N.Y.

Dean, R.G., and E.M. Maurmeyer, 1983, "Models for Beach Profile Response," in P.D. Komar (ed.), *Handbook of Coastal Processes and Erosion*, pp. 151-165, CRC Press, Boca Raton, Fla.

Douglas, B.C., 1991, "Global Sea-Level Rise," *Journal of Geophysical Research* 96:6981-6992.

Hallermeier, R.J., 1981, "A Profile Zonation for Seasonal Sand Beaches from Wave Climate," *Coastal Engineering* 4:253-277.

Hands, E.B., 1983, "The Great Lakes as a Test Model for Profile Responses to Sea-Level Changes," in P.D. Komar (ed.), *Handbook of Coastal Processes and Erosion*, CRC Press, Boca Raton, Fla.

IPCC, 1990a, *Strategies for Adaptation to Sea-Level Rise, Appendix D: Report of the Coastal Zone Management Subgroup of the Intergovernmental Panel on Climate Change*, Intergovernmental Panel on Climate Change.

IPCC, 1990b, *Report of Working Group III: Coastal Zone Management Subgroup Report*, Intergovernmental Panel on Climate Change.

Nicholls, R.J., and S.P. Leatherman, 1994, "Sea-Level Rise," in *As Climate Changes: Potential Impacts and Implications*, Cambridge University Press (in press).

P.C. Globe, Inc., 1989, *Computer Software That Compiles Information on the Countries of the World,* P.C. Globe, Inc., Tempe, Ariz.

Park, R., J. Trahan, P. Mausel, and R. Howe, 1989, "The Effects of Global Sea-Level Rise on U.S. Coastal Wetlands," Appendix B to *The Potential Effects of Global Climate Change on the United States*, U.S. Environmental Protection Agency Report to Congress.

Software Toolworks, 1990, *World Atlas* (computer software that compiles information on the countries of the world), Software Toolworks, Chatsworth, Calif.

Wiegel, J., 1989, "The Cost of Defending Developed Shoreline," in Appendix B to *The Potential Effects of Global Climate Change on the United States*, U.S. Environmental Protection Agency Report to Congress.

Yohe, G., 1989, "The Cost of Not Holding Back the Sea — Economic Vulnerability," *Ocean and Shoreline Management* 15:233-255.

Yohe, G., 1990, "The Cost of Not Holding Back the Sea — Toward a National Sample of Economic Vulnerability," *Coastal Management* 18:403-431.

APPENDIX I:

ASSESSMENT OF ADAPTATION OPTIONS

R. Benioff et al. (eds.), Vulnerability and Adaptation Assessments, I-1–I-26.
© *1996 Kluwer Academic Publishers.*

APPENDIX I NOTATION

The following is a list of the acronyms, and abbreviations (including units of measure) used in this appendix. Some abbreviations found only in tables are defined in those tables.

ACRONYMS AND ABBREVIATIONS

DO   dissolved oxygen
IPCC   Intergovernmental Panel on Climate Change
M&I   municipal and industrial
NAS   National Academy of Sciences
OTA   Office of Technology Assessment

UNITS OF MEASURE

d   day(s)
kL   kiloliter(s)
m   meter(s)
ppm   part(s) per million
yr   year(s)

CONTENTS

CONTENTS (Cont.)

TABLES

APPENDIX I:

ASSESSMENT OF ADAPTATION OPTIONS

Section 7 describes why anticipating the effects of climate change on natural resources is important. It also introduces an approach that can be used to examine the need for and the effectiveness of policies to anticipate climate change. That approach uses a matrix to analyze options for anticipatory adaptation. This appendix lists and briefly describes some of these options and provides an example of how the adaptation matrix can be used to examine the relative effectiveness of policies of anticipatory adaptation.

I.1 OPTIONS FOR ANTICIPATORY ADAPTATION

This section describes government policies that could be implemented in anticipation of climate change to reduce its potential effects. The policy options are divided into five climate-sensitive sectors: water resources, sea-level rise, forests, ecosystems, and agriculture. These adaptation options represent the range and type of policies that should be considered; however, the lists are not comprehensive. In conducting an adaptation assessment, additional adaptation options will likely be developed on the basis of specific physical or socioeconomic vulnerabilities and the sensitivity of existing policies to climate change.

This section of the appendix is organized as follows:

- General policy options for adaptation to climate change,

- Policy options for adaptation of water resources,

- Policy options for adaptation to a rise in sea level,

- Policy options for adaptation of forests,

- Policy options for adaptation of ecosystems, and

- Policy options for adaptation of agriculture.

All of the options listed in this appendix can be considered policies of adaption to climate change because the options meet the following criteria:

- *Flexibility.* Each policy enhances the ability to meet stated objectives under a variety of climatic conditions. Each policy is either *robust*, meaning it allows the system to continue functioning under a wider range of conditions, or *resilient*, meaning it allows the system to quickly adapt to changed conditions.

- *Benefits exceed costs.* Each policy produces benefits (measured by the ability to meet objectives) that exceed any increases in costs.[1]

In addition, policy makers should consider all of the adaptation options listed in this appendix in anticipation of climate change, rather than after climate change occurs. Adaptation options should be implemented before climate changes if they fall into one of two categories: (1) they produce benefits independent of climate change, or (2) they will be less effective, or not effective at all, if implemented in reaction to climate change.

- *Benefits independent of climate change.* Adaptation options may bring benefits or reduce costs even if climate change does not occur. Many of these options involve changing plans for responding to particular events or allowing greater flexibility in responses; for example, implementing market-based systems for allocating water would result in more efficient allocation of water under the current climate and would allow for a more rapid and efficient response to climate change than would more rigid schemes for water allocation.

- *High priority.* Adaptation options may warrant consideration in anticipation of climate change because they will be significantly less effective if implemented as reactive policies. Adaptation options that fall into the following categories may be considered high priority:

  - *Irreversible or catastrophic impacts.* These options are policies concerning potentially irreversible or catastrophic impacts of climate change; for example, loss of life or of species, extensive loss of property, or destruction of resources may be irreversible or catastrophic. Such policies warrant consideration because reactive measures will probably be unsuccessful in mitigating the impacts of climate change.

  - *Long-term decisions.* Decisions on many long-term issues, such as the construction of dams, reservoirs, and bridges, have long useful lifetimes and may be affected by climate change. Policies affecting the construction of such structures warrant consideration because the initial costs of making the structures less vulnerable to climate change are likely to be less than the costs of adapting the

---

[1] If the policy will not produce benefits for many decades, the costs of implementation may need to be quite small in comparison with undiscounted future benefits. Suppose an adaptation policy reduces the damages from a catastrophic flood by $1 billion in the year 2075. The benefit in the year 2075 is $1 billion. Assume that it costs $100 million to adopt the policy. At a 3% discount rate, the present value (in 1994) of the $1 billion in damages is $91.2 million. So, the net present value of the benefits of this policy is lower than the costs and, from an economic viewpoint, the investment is not justified.

structures after climate changes (National Academy of Sciences [NAS] 1992).

- *Unfavorable trends.* Certain trends, such as some types of development, may make some types of adaptation more difficult; for example, the fragmentation of habitats is a trend that is unfavorable to wildlife. As climate changes, fragmentation could become an even greater problem as species need to migrate to cooler areas. Policies affecting such trends warrant consideration before climate change, because adaptation may be more difficult in the future or because opportunities to implement low-cost or politically feasible options may be lost.

First, each individual adaptation option is described, and then an indication is given of whether the option has benefits independent of climate change or is of high priority. High-priority options are indicated because they have irreversible or catastrophic impacts, are long-term decisions, or have unfavorable trends.

I.2 GENERAL POLICY OPTIONS FOR ADAPTATION TO CLIMATE CHANGE

I.2.1 Assess Current Practices of Crisis Management

Climate change may exacerbate extreme events with which society must already cope. It is better to prepare for climate change by assessing successes and failures in addressing known extreme events (Johda 1988). *This option is high priority because of irreversible or catastrophic impacts.*

I.2.2 Inventory Existing Practices and Decisions Used to Adapt to Different Climates

An inventory may focus on actual social and economic decisions in light of variable climatic regimes over time or across regions. For example, agricultural practices in arid and semi-arid areas could be used by currently humid areas that may become drier. Compilation of information on various practices may be a cost-effective way to identify feasible options for adaptation. In particular, adaptation options that require long-term decisions should be identified and analyzed for implementation (Johda 1988; Intergovernmental Panel on Climate Change [IPCC] 1990; additional research recommendations in Canadian Climate Program 1993; Office of Technology Assessment [OTA] 1993a,b). *This option is high priority because of long-term decisions, irreversible or catastrophic impacts, and unfavorable trends.*

I.2.3 Promote Awareness of Climatic Variability and Change

The public or decision makers frequently do not understand climatic variability and the potential risks of climate change. Because climatic adaptation will affect the individual, organizational, and policy levels, communication about the human significance of climatic variability is important at all levels in a community. Increasing sensitivity to climatic issues will facilitate adoption of measures to prepare for climatic variability and change (Johda 1988; IPCC 1990; Cooper 1992). *This option is high priority because of irreversible or catastrophic impacts, long-term decisions, and unfavorable trends.*

I.3 POLICY OPTIONS FOR ADAPTATION OF WATER RESOURCES

I.3.1 Use River Basin Planning and Coordination

Comprehensive planning across a river basin may allow for coordinated solutions to problems of water quality and water supply; for example, enhanced coordination of operations of facility systems or expansion of the conjunctive use of groundwater and surface water can improve water yields, which can help to alleviate droughts. Planning can also help to address the effects of population and economic growth and changes in the supply of and demand for water (Gillilan 1992; Schad 1992; Wahl 1992; Warren 1992). *This option receives high-priority consideration because of irreversible or catastrophic impacts and unfavorable trends. Benefits are independent of climate change.*

I.3.2 Adopt Contingency Planning for Drought

Plans for short-term measures to adapt to water shortages could help to mitigate droughts. Planning could be undertaken for droughts of known or greater intensity and duration. The cost of developing contingency plans is relatively small compared with the potential benefits; and plans could be effective in managing current climatic variability, as well as future climate change (Wilhite 1990; Cooper 1992; NAS 1992; Wahl 1992). *This option receives high-priority consideration because of irreversible or catastrophic impacts. Benefits are independent of climate change.*

I.3.3 Make Marginal Changes in Construction of Storage and Distribution Facilities

Marginal changes may be made in the planned construction of storage or distribution facilities to adapt to increased variability in runoff or to the need for greater storage capacity. In planned construction, consider marginal increases in the size of dams or marginal changes in the construction of canals, pipelines, pumping plants, and storm drainages. This change may be much less expensive than adding capacity in the future (NAS 1992; Canadian Climate Program 1993). *This option receives high-priority consideration because of long-term decisions.*

I.3.4 Maintain Options to Develop New Dam Sites

Keep options open to develop new dam sites, as needed. The number of sites that can be used efficiently as reservoirs is limited, and removing structures once an area has been developed may be very costly or politically difficult. Thus, development in potential dam sites should be limited or only allowed under terms that would permit conversion to dam sites (e.g., leases) (Smith and Tirpak 1989). *This option receives high-priority consideration because of unfavorable trends.*

I.3.5 Conserve Water

Reducing demand can increase excess supply, giving a greater margin of safety for future droughts. Demand for water may be reduced through a range of policies that encourage efficient water use, including education, voluntary compliance, price policies, legal restrictions on water use, rationing of water, or the imposition of water conservation standards on technologies (Canadian Climate Program 1993; OTA 1993a). Reduced demand will increase current capacity to cope with drought. *Benefits are independent of climate change.*

I.3.6 Allocate Water Supplies by Using Market-Based Systems

Market-based allocation allows water to be diverted to its most efficient uses. Other mechanisms, such as prior appropriation, may result in inefficient allocation of water supplies. Market-based allocations are able to respond more rapidly to changing conditions of supply and also tend to lower demand, thus conserving water. Thus, market-based allocation increases both the robustness and the resiliency of the water supply system. In addition, it will improve the economic efficiency of the allocation system under current climate (Frederick and Gleick 1989; Wahl 1992). *Benefits are independent of climate change.*

I.3.7 Use Interbasin Transfers

Transfers of water between water basins may result in more efficient water use under current and changed climate. Transfers are often easier to implement than fully operating markets for water allocation. Transfer also can be an effective short-term measure for responding to regional droughts or other problems of water supply (Wahl 1992; Canadian Climate Program 1993). *Benefits are independent of climate change.*

I.3.8 Control Pollution

Polluting water so that it is unfit for drinking or other uses can, in many respects, have an effect similar to reducing water supply. Reducing water pollution effectively increases the supply of water. If this increase results in excess supply over demand, the excess increases the safety margin for maintaining supplies during droughts. In addition, reduced runoff from climate change will most likely increase concentrations of pollutants in the water column. If pollutant loadings are lower, water quality standards are less likely to be violated. These benefits will be realized whether or not climate changes (Schad 1992; Canadian Climate Program 1993). *Benefits are independent of climate change.*

I.4 POLICY OPTIONS FOR ADAPTATION TO SEA-LEVEL RISE

I.4.1 Adopt Coastal Zone Management

Land-use planning in coastal zones, such as the use of setbacks or allocating low-lying vulnerable lands to lower value uses, such as parks rather than housing, will help reduce vulnerability to a rise in sea level. Other land use planning mechanisms, such as construction standards or post-storm reconstruction standards, reduce the risks of living in coastal areas; and financial mechanisms may be created to encourage additional risk-reduction measures. Each of these policies reduces the risks from current climatic variability and protects against potential sea-level rise (Warren 1992; OTA 1993a). *This option receives high-priority consideration because of unfavorable trends. Benefits are independent of climate change.*

I.4.2 Use Presumed Mobility

If a rise in sea level is observed, property owners are required to remove threatened coastal structures. Flexible land use policies, such as presumed mobility, avoid the losses of completely prohibiting development, while allowing markets to decide whether properties are worth developing, given available information about climate change. Variations of this policy include (1) provisions to adjust land ownership on the basis of tidal regimes according to changes in the tides or (2) provisions in coastal land deeds that revert the land to public ownership in so many years if the sea level rises a specified amount (Titus 1991). *This option is high priority because of long-term decisions and unfavorable trends.*

I.4.3 Plan Urban Growth

Redirecting growth from sensitive lands toward less vulnerable areas is an option for reducing the increase in risks from a sea-level rise, as well as reducing the vulnerability to the current severity of coastal storms. This option may be particularly useful in decisions to site large capital facilities or facilities that would pose significant hazards if subject to flooding (Wang et al. 1994). *This option receives high-priority consideration because of unfavorable trends. Benefits are independent of climate change.*

I.4.4 Discourage Permanent Shoreline Stabilization

Permanent shore-hardening structures, such as seawalls and groins, may be banned or discouraged in moderately developed areas. Limiting permanent stabilization of the shoreline will allow a gradual retreat from a sea-level rise and allow the development market to determine whether a property is worth developing, given the risk of erosion or inundation; however, well-developed coastal communities and expensive facilities may represent such a large investment that expansion of coastal barriers to protect the investments from a sea-level rise is warranted (OTA 1993a). *This option receives high-priority consideration because of unfavorable trends and long-term decisions.*

I.4.5 Incorporate Marginal Increases in the Height of Coastal Infrastructure

In building coastal infrastructure, such as bridges or seawalls, marginal increases in the height of the structures may be included to offset a sea-level rise; for example, outflow from sewage treatment plants could be several feet higher. Such additions are less expensive to make while construction is in progress than after the initial work is complete (NAS 1992; Smith and Mueller-Vollmer 1993). *This option is high priority because of long-term decisions.*

I.4.6 Preserve Vulnerable Wetlands

Efforts should be made to maintain wetlands that are more likely to withstand a sea-level rise. Wetlands are valuable natural areas that are difficult to re-create; therefore, current and future efforts are warranted to protect the areas. In setting priorities for protecting wetlands, the likelihood of surviving a sea-level rise or migrating landward should be considered. Protecting wetlands will also improve water quality, flood control, and habitat under current climate (Warren 1992; Canadian Climate Program 1993). *This option receives high-priority consideration because of irreversible or catastrophic impacts. Benefits are independent of climate change.*

I.4.7 Decrease Subsidies to Sensitive Lands

Government subsidies or tax incentives to develop land sensitive to sea-level rise, such as barrier islands, coastal wetlands, estuarine shorelines, and critical wildlife habitat, should be limited. Such policies would include ensuring that any government subsidies for flood insurance reflect the *current* risks of developing in coastal areas and on floodplains and would possibly include prohibiting new insurance policies in risky locations. In addition, private insurance and banking industries may be encouraged to factor risks of climatic variability into investment decisions and, thereby, reduce reliance on government-subsidized insurance and disaster relief. Such policies would allow the markets to reflect the true risks of developing or living in sensitive coastal areas (Warren 1992; Canadian Climate Program

1993; OTA 1993a). *This option receives high-priority consideration because of unfavorable trends. Benefits are independent of climate change.*

I.4.8 Tie Disaster Relief to Hazard-Reduction Programs

Funds for disaster relief could be tied to the implementation of long-term hazard-reduction policies (e.g., dune-protection ordinances). Such requirements may be a valuable investment, particularly if hurricanes and storms increase as a result of climate change (OTA 1993a). This policy would likely have benefits under current climate because current risks from disasters would be reduced. *This option receives high priority because of irreversible or catastrophic impacts and long-term decisions. Benefits are independent of climate change.*

I.4.9 Promote Public Education

The public is often not well informed about the risks associated with living in coastal areas. Timely public education about erosion, sea-level rise, flooding risks, and storm-standard building codes could be a cost-effective means of reducing future expenditures. *This option receives high-priority consideration because of irreversible or catastrophic impacts, long-term decisions, and unfavorable trends. Benefits are independent of climate change.*

I.5 POLICY OPTIONS FOR ADAPTATION OF FORESTS

I.5.1 Encourage Diverse Management Practices

A mix of management practices, strategies, and species may provide a buffer against the uncertainties of climate change. Diverse practices include a mix of planting practices, a greater variety of species, planting of species drawn from warmer climate zones, and an increase in planting densities. One example of how this policy has been implemented is the planting of trees with greater resistance to heat and drought on the southern range of managed forest boundaries if the trees can survive in the current climate. In addition, a mix of different timber-harvesting strategies may be used to promote forest diversity (Smith 1992; OTA 1993b). *This option receives high-priority consideration because of long-term decisions.*

I.5.2 Reduce Habitat Fragmentation and Promote Development of Migration Corridors

Geographic fragmentation may threaten the ability of forests and forest species to migrate or adapt to changing climate. Currently, the health of many forests is stressed by existing fragmentation. This stress will most likely increase under climate change. Forest fragmentation may be reduced through incentive programs for multiple-use management that balances preservation and use within a single parcel or through the negotiation of conservation easements that protect geographically important land parcels from development. Such programs should consider the likelihood that climate change will cause forests to migrate in a poleward or vertical direction. Also, the rate of forest migration may lag far behind the rate of climate change (OTA 1993b). *This option receives high-priority consideration because of irreversible or catastrophic impacts and unfavorable trends. Benefits are independent of climate change.*

I.5.3 Enhance Forest Seed Banks

Maintaining access to a sufficient variety of seeds to allow the original genetic diversity of forests to be rebred assures that the benefits provided by forests are not lost forever. Seed collections should represent the variety of genotypes that exists for each species (OTA 1993b). *This option receives high-priority consideration because of irreversible or catastrophic impacts.*

I.5.4 Establish Flexible Criteria for Intervention

Policies should be in place that establish appropriate criteria for intervention by using existing management practices. The use of management practices such as salvage harvests, silvicultural management, insect and fire control, and restoration activities should be allowed to change as conditions change. Such policies apply to current forest management but should also consider how the structure of the forest might change because of climate change (Cooper 1992; Smith 1992). *This option receives high-priority consideration because of irreversible or catastrophic impacts. Benefits are independent of climate change.*

I.6 POLICY OPTIONS FOR ADAPTATION OF ECOSYSTEMS

I.6.1 Integrated Ecosystem Planning and Management

Integrated planning and management along watershed and ecosystem lines reduce the institutional fragmentation in the management of natural areas and focus on protecting a variety of species and natural systems. Because the impacts of climate change are difficult to predict, the preservation of a variety of species in a healthy ecosystem may be the most effective way to protect those species that will be able to adapt to climate change. In many cases, this preservation will involve international coordination. Such coordination will most likely produce benefits even if climate does not change because coordination will help to address many of the threats to the diversity of species (Cooper 1992; Warren 1992; OTA 1993b). *This option receives high-priority consideration because of irreversible or catastrophic impacts. Benefits are independent of climate change.*

I.6.2 Protect and Enhance Migration Corridors or Buffer Zones

Policies that protect migration corridors or buffer zones help to maintain ecosystems by improving the likelihood of successful adaptation to climate change by animal and plant species. Corridors, where they can be identified, allow species to migrate as climate changes. Buffer zones around current reserve areas that include different altitudes and ecosystems increase the adaptive potential for species within the preserve. A graded system of management may be implemented, where the innermost areas receive the greatest protection and where more uses are allowed in the outer buffer zones or corridors. Agencies that protect natural lands should modify their criteria for acquisition to include the diversity of ecosystems, long-term survivability, climate-sensitive species, and opportunities to connect or enlarge existing protected land parcels. Diverse natural lands and larger land parcels will improve the possibility of species adaptation to climatic variability. The current lack of information on species migration, natural barriers to migration, and the high level of development surrounding many natural areas significantly limits the feasibility of this option (NAS 1992; Lillieholm 1993; OTA 1993b). *This option receives high-priority consideration because of irreversible or catastrophic impacts and unfavorable trends. Benefits are independent of climate change.*

I.6.3 Enhance Methods to Protect Biodiversity Off-site

Threatened or endangered species may be saved off-site, which protects diversity. These methods must be in place before climate changes to avoid the irreversible loss of species extinction. Off-site methods to assist adaptation include gene and seed banks, libraries, gardens, and zoos. Off-site protection may be very important in preventing irreversible loss of biodiversity (NAS 1992). *This option receives high-priority consideration because of irreversible or catastrophic impacts.*

I.7 POLICY OPTIONS FOR ADAPTATION OF AGRICULTURE

I.7.1 Develop New Crop Types and Enhance Seed Banks

Seed banks that maintain a variety of seed types provide an opportunity for farmers to diversify to counter the threat of climate change or to develop a profitable specialization. Development of more and better heat- and drought-resistant crops will help to fulfill current and future world food demand by enabling production in marginal areas to expand. Improvements will be critical because the world population continues to increase, with or without climate change (NAS 1992; OTA 1993a). *This option receives high-priority consideration because of irreversible or catastrophic impacts. Benefits are independent of climate change.*

I.7.2 Liberalize Agricultural Trade

Lowering trade barriers will result in higher levels of global agricultural production under the current climate and under climate change scenarios. Farmers will receive information on changes in global market conditions more quickly than if trade barriers were not lowered (Rosenzweig and Parry 1994). *Benefits are independent of climate change.*

I.7.3 Avoid Tying Subsidies or Taxes to Type of Crop and Acreage

Commodity support programs or tax policies may discourage changing from one cropping system to another that is better suited to a changed climate. Therefore, efforts to stabilize farm supply and to maintain farm incomes should avoid disincentives for farmers to switch crops, rotate crops, and use the full acreage normally planted. This policy approach will increase the efficiency of current farming practices and will also increase the ability of the system to quickly recover from climate change (Lewandrowski and Brazee 1993). *Benefits are independent of climate change.*

I.7.4 Promote Agricultural Drought Management

Encourage management practices that recognize drought as part of a highly variable climate, rather than treating drought as a natural disaster. Farmers can be given information on climatic conditions, incentives can be offered to adopt sound practices of drought management, and farmers can be discouraged from relying on drought relief. This type of policy is particularly useful if farm disaster relief and other government subsidies distort the market and encourage overly risky expansion of farming into marginal lands (OTA 1993a). *Benefits are independent of climate change.*

I.7.5 Increase Efficiency of Irrigation

Many farming technologies, such as efficient irrigation systems, provide opportunities to reduce direct dependence on natural factors such as precipitation and runoff. In evaluating an improvement to irrigation systems, the additional benefit of reducing vulnerability to climatic variations and natural disasters should be considered. Improvements allow greater flexibility by reducing water consumption without reducing crop yields. *Benefits are independent of climate change.*

I.7.6 Disperse Information on Conservation Management Practices

Many practices, such as conservation tilling, furrow diking, terracing, contouring, and planting vegetation to act as windbreaks, will protect fields from water and wind erosion and can help retain moisture by reducing evaporation and increasing water infiltration. Using management practices that reduce dependence on irrigation will reduce water consumption without reducing crop yields and will allow greater resiliency in adapting to future climate changes (Easterling 1993). *Benefits are independent of climate change.*

I.8 USING THE ADAPTATION MATRIX

Section 7 discussed the concept of a matrix to help analyze the benefits and costs of adaptation policy options. This section explains how the matrix can be used by examining a hypothetical example of water resource adaptation policies. All numbers and results reported and discussed here are simply inventions of the authors. The purpose of this section is to explain how one can use the matrix, not to demonstrate that one water policy is superior to another.

I.8.1 Example

For this example, assume that policies are examined for operating a hypothetical reservoir management system for a river basin. Water in the river is used to meet consumptive uses and in-stream uses. The consumptive uses consist of municipal and industrial (M&I) uses and irrigation for agriculture. The in-stream uses consist of water quality and recreation. For simplicity, only these four uses are considered, although many other uses exist for a reservoir system, such as hydropower production and flood control.

The current policy for managing the reservoir system is assumed to be one in which priority is given to irrigation to receive water. The remaining water in the system is used to meet M&I, water quality, and recreational needs.

I.8.2 Matrix Organization

Table I.1 displays the matrix format. The columns list policies, climates, uses and standards, scores, and changes in costs of policies. The rows of values are for current and alternative policies. Within each policy are rows for current and climate change scenarios. The units in the matrix are a measure of success or, in this case, failure. As noted in Section 7, to be able to compare how alternative policies perform in meeting objectives, having quantitative measures of success or failure is crucial. In this case, the number of days on which a certain standard is not met is estimated. Thus, the goal is to minimize failures. The number of days is a realistic standard because a reservoir system operator can easily record it. The measure is not ideal because it does not indicate the degree of damage when a standard is not met, nor does the measure indicate the degree of benefit when a standard is met. The value of just counting the number of days that a standard is violated is, as will be seen below, that days of failure can be combined across different standards.

In M&I uses, assume that the water resource managers attempt to provide 10 million kL/d. Any day in which less than that amount is delivered would not satisfy M&I needs and is recorded as a failure. For irrigation, it is assumed that the farmers demand

TABLE I.1 Water Allocation Policies Matrix

| | | Uses and Standards | | | | Total Weighted Score | Average Total Weighted Score | Change in Costs (10^6/yr) |
| | | Consumptive Uses | | In-Stream Uses | | | | |
Policies	Climate	M&I (No. of days <10 million kL/d) (w = 3)[a]	Irrigation (No. of days <30 million kL/d) (w = 1)	Water Quality (No. of days <5 ppm of DO)[b] (w = 2)	Recreation (No. of days <500-m elevation) (w = 1)			
Current policy	Current							N/A[c]
	Scenario 1							
	Scenario 2							
Enhanced conservation	Current							
	Scenario 1							
	Scenario 2							
Market allocation	Current							
	Scenario 1							
	Scenario 2							

a w = weight.

b DO = dissolved oxygen, measured in parts per million (ppm).

c N/A = not applicable.

30 million kL/d during the growing season. For water quality, the number of days on which a water quality standard for dissolved oxygen (DO) of 5 ppm is not met is measured. If DO levels are below that standard, aquatic life will be harmed, and it is recorded as one day of failure. Finally, it is assumed that recreational needs are satisfied when reservoir levels are maintained at an elevation of 500 m above sea level, and that, at that elevation, activities such as boating and swimming are possible.

To examine how well each policy performs in meeting all of the standards, it is possible to total the number of days of failure across them; however, this approach assumes that the value of meeting each standard is the same. Possibly, more negative consequences exist for failing to meet the M&I standard than for failing to meet the irrigation standard. Weights are determined on the basis of the relative harm from failing to meet a daily standard. They should be determined through consultation with policy makers. If that is not possible, researchers should consult laws, regulations, and other materials that may indicate the relative importance of meeting standards. If these two options do not provide guidance as to setting weights, researchers may use their own judgment to determine weights.

This example assumes that:

- Failing to meet the M&I standard results in three times as much harm as failing to meet the irrigation standard,

- Failing to meet the recreation standard has the same harm as failing to meet the irrigation standard, and

- Failing to meet the water quality standard has twice the harm of failing to meet either the irrigation standard or the recreation standard.

These designations are purely arbitrary and are not meant to mimic reality. A total weighted score for a policy under a particular climate scenario can be derived by simply multiplying the number of days of failure for each use by its weight and then summing across uses. A lower total weighted score implies that the reservoir system is more successful at meeting its standards. A score of zero would mean that all objectives are satisfied.

The average total weighted score indicates how well a policy meets standards under different climate scenarios. This score can be calculated by averaging the total weighted scores for a particular policy under different climate scenarios. These scores are displayed in the next-to-the-last column in the matrix.

The final column in the matrix is change in costs. This column displays the difference in costs between alternative policies and the current policy.

In this example, the current policy and two alternative policies are examined; however, users need not limit themselves to two alternatives. For each policy, how well standards are met under the current climate and climate change scenarios is examined. For the current climate, the climate period of 1951-1980 should be used, as discussed in Section 4.

The matrix has two climate change scenarios, Scenario 1 and Scenario 2. Let us assume that Scenario 1 is a dry climate change scenario and Scenario 2 is a wet climate change scenario. Researchers should use the same scenarios as used in the adaptation assessment. How well standards are met for the current policy under the current climate and climate change is examined first.

I.8.3 Current Policy

I.8.3.1 Current Climate

This climate is the baseline and where the research on adaptation should begin. Table I.2 displays how well the reservoir system meets the standards for the current policy. To repeat, the current policy favors agriculture over other uses. Assume that none of the objectives is fully met under the current climate. On average, under the current climate, the M&I needs are not met on six days per year, the irrigation needs are not met on four days per year, and so on. Using the weights results in a total weighted score of 48. This latter number has no meaning by itself, but is a reference point that can be compared with total weighted scores for other policies.

I.8.3.2 Climate Change

The next step is to examine how the current policy performs in meeting its standards under the climate change scenarios. Use at least one dry scenario and one wet scenario. All of the climate change scenarios will have higher temperatures than those under the current climate. A dry scenario is one in which soil moisture and runoff decrease. A wet scenario is one in which soil moisture and runoff increase. To the extent possible, researchers should use the same scenarios being used in the vulnerability assessment.[2]

Table I.2 displays how the current policy performs in meeting the standards under Scenario 1 and Scenario 2. Scenario 1 assumes that allocations to all uses are reduced. Because the current policy favors deliveries to irrigation over other uses, water deliveries for irrigation are cut back less than deliveries to other uses. This policy translates into a smaller increase in days failing to meet the irrigation standard than the increase in days failing to meet the other standards. The number of days that the agriculture standard is not met doubles. The number of days that the M&I and water quality standards are not met increases by two and a half times, and the number of days on which the recreation standard

---

[2] The methods used to analyze how well current and alternative policies meet objectives are discussed in Section 7. Using the same models or approaches used in the vulnerability assessment is preferable. If that approach is not possible, researchers should use expert judgment to assess how well current and alternative policies will perform.

TABLE I.2 Water Allocation Policies Matrix for Current Policy

Policies	Climate	Consumptive Uses		In-Stream Uses				
		M&I (No. of days <10 million kL/d) (w = 3)[a]	Irrigation (No. of days <30 million kL/d) (w = 1)	Water Quality (No. of days <5 ppm of DO)[b] (w = 2)	Recreation (No. of days <500-m elevation) (w = 1)	Total Weighted Score	Average Total Weighted Score	Change in Costs ($10^6/yr)
Current policy	Current	6	4	8	10	48	68	N/A[c]
	Scenario 1	15	8	20	40	133		
	Scenario 2	3	1	4	5	23		
Enhanced conservation	Current							
	Scenario 1							
	Scenario 2							
Market allocation	Current							
	Scenario 1							
	Scenario 2							

[a] w = weight.

[b] DO = dissolved oxygen, measured in parts per million (ppm).

[c] N/A = not applicable.

is violated increases by four times. The total weighted score for the current policy under Scenario 1 is 133, almost three times higher than the total weighted score under the current climate. A drier climate increases the failure rate significantly over the current climate.

Under Scenario 2, more water is available for consumptive and in-stream uses, so the number of days of failure drops for all categories. Because the current policy is to try to meet agriculture's needs before other uses, the excess water is used to benefit agriculture more than to benefit the other uses. The number of days of failing to meet the agriculture standard is reduced by 75%, while the number of failure days for the other uses is reduced by 50%. The total weighted score for Scenario 2 is 23, a reduction of slightly over 50%.

The column labeled "Average Total Weighted Score" is a simple average of the total weighted scores from the current climate, Scenario 1, and wet climates. If the user decided, he or she could weight the climate scenarios by the probability of occurrence. Because little information is available about the probability of these climate change scenarios occurring, a weighting scheme was not used. For the current policy, the average total weighted score is 68. This score has little meaning by itself but will be useful to compare the effectiveness of current policies with alternative policies. The average total weighted score of 68 is higher than the current climate score of 48. This fact implies that when one factor in climate changes, the expected performance of the current policy declines.

The current policy appears to be vulnerable to a drier climate. The days of failure are significantly increased under Scenario 1. Thus, the analysis of alternative policies under current and changed climates is justified.

I.8.4 Alternative Policies

The hypothetical example examines two alternative policies. The first is a policy of enhancing conservation efforts. It is assumed that the enhanced conservation policy consists of the government subsidizing farmers to use a more efficient irrigation technology and that this technology reduces agriculture's water needs by 20%. The second policy alternative allows the free market to allocate water use. This policy enables users with the highest marginal value for water to pay water users with lower marginal values to use their water. The weighting scheme implies that the highest marginal value uses of water are M&I and then water quality, and that irrigation and recreation have the same lowest marginal value use.

I.8.4.1 Current Climate

The first step is to examine how these alternative policies perform under the current climate. If the alternatives do a better job of meeting objectives than the current policy, the alternatives have "benefits independent of climate change."

The enhanced conservation policy is examined first. Because this policy involves increasing the efficiency of irrigation water use by 20%, agriculture needs less water per day to meet its needs. Rather than needing water at 30 million kL/d, agriculture would now need 24 million kL/d. With agriculture's needs reduced, more water is available in the system to meet all of the needs; however, agriculture still receives the highest priority for allocation of water. Assume that the number of days when agriculture fails to receive its need of 24 million kL/d is reduced from four to one (Table I.3). With excess water in the system, the reservoir operators are able to reduce the number of days of failure for other uses, but by a lower percentage than the reduction for agriculture. The M&I failures are cut by one-third, water quality failures by one-fourth, and recreation failures by one-fifth. The total weighted score for enhanced conservation is 33, which is about 30% lower than the score for the current policy.

Turning to the market allocation policy under the current climate, we assume that higher-marginal-value M&I users bid water away from lower-marginal-value irrigation users, so the M&I number of days of failure drops to three, while the number of days on which the irrigation target is not met increases to five. It is assumed that the trading has no effect on water allocated for in-stream uses, so their days of failure remain unchanged from the current policy. The total weighted score for market allocation is 40, which is less than the total weighted score of 48 for current policy because the three-day reduction for M&I counts as nine (because three days is multiplied by the weighting of three), while the one-day increase for agriculture counts only as one.

Because both alternative policies have total weighted scores lower than the current policy under the current climate, one can conclude that both policies have benefits independent of climate change; however, as seen in the final column of Table I.3 ("Change in Costs"), the alternatives involve additional expenses. The enhanced conservation policy is assumed most expensive because government subsidizes the installation of efficient irrigation technologies. The adoption of market allocation involves additional expenses for administrative costs but is less expensive than enhanced conservation. On the basis of the information here, it cannot be determined whether the benefits of the alternative policies justify their additional costs under the current climate.

I.8.4.2 Climate Change

The way to examine how different policies perform under climate change is to first compare the performance of each policy under the current climate. That comparison helps to demonstrate the difference that a policy change makes in meeting standards. The relative performance of different policies under the same climate change scenario (e.g., How do different policies perform under Scenario 1?) are then compared. Finally, how the different policies perform over a range of climate change scenarios is examined.

The enhanced conservation policy has fewer days of failure in all categories and a lower weighted score under Scenario 1 than the current policy does. That decrease is because, with lower demand for agriculture, more water is available to meet other needs.

TABLE I.3 Water Allocation Policies Matrix for Current Policy and Alternative Policies

| Policies | Climate | Uses and Standards | | | | Total Weighted Score | Average Total Weighted Score | Change in Costs (10^6/yr) |
| | | Consumptive Uses | | In-Stream Uses | | | | |
		M&I (No. of days <10 million kL/d) (w = 3)[a]	Irrigation (No. of days <30 million kL/d) (w = 1)	Water Quality (No. of days <5 ppm of DO)[b] (w = 2)	Recreation (No. of days <500-m elevation) (w = 1)			
Current policy	Current	6	4	8	10	48	68	N/A[c]
	Scenario 1	15	8	20	40	133		
	Scenario 2	3	1	4	5	23		
Enhanced conservation	Current	4	1	6	8	33	40	+20
	Scenario 1	8	2	12	20	70		
	Scenario 2	2	0	3	4	16		
Market allocation	Current	3	5	8	10	40	55	+5
	Scenario 1	6	11	20	40	109		
	Scenario 2	0	2	4	5	15		

[a] w = weight.

[b] DO = dissolved oxygen, measured in parts per million (ppm).

[c] N/A = not applicable.

The enhanced conservation policy also has fewer days of failure and a lower total weighted score under the wet scenario (Scenario 2) than does the current policy. The average total weighted score for the enhanced conservation policy is significantly lower than that for the current policy, so the enhanced conservation policy appears to more flexible (i.e., it performs better under a variety of climate change scenarios than does the current policy).

The story is similar, although less dramatic, for market allocation. In Scenario 1, it is assumed that water use shifts from agriculture to M&I (note that increasing agriculture's failure rate by one day frees 30 million kL, which lowers M&I failure by three days). Because it is assumed that water is only traded between agriculture and M&I, the number of days of failure for in-stream uses does not change. The total weighted score for market allocation in Scenario 1 is 24 less than the total weighted score for the current policy because water has shifted to a higher value use. The market allocation policy has a lower total weighted score in Scenario 2 than does the current policy. The average total weighted score for market allocation is lower than the average total weighted score for the current policy. Like the enhanced conservation policy, the market allocation policy appears to be more flexible than the current policy of water management.

On the whole, both policy alternatives do a better job than the current policy of meeting water policy standards under both the current climate and climate change scenarios. Enhanced conservation does the best at meeting objectives because its average weighted score of 40 is the lowest. The spreadsheet allows ranking of these policies in order of which one does the best job of meeting policy objectives and indicates the relative difference in meeting these objectives. As noted previously, the enhanced conservation policy is also the most expensive policy, with a cost increase of $20 million, followed by market allocation with a cost increase of $5 million, and then the current policy (with no change in cost).

This information on the relative rankings of the benefits and costs of the current policy and alternatives can be presented to policy makers. It is up to them to weigh the pros and cons of these alternatives. They will most likely consider other factors, such as the political feasibility of trying to change current policies. In fact, the results of an analysis are often only one of many factors to be considered by policy makers. Other factors may be more important in making a final decision. The advantage of using the matrix approach is that one can present the benefits and costs of adaptation policies to policy makers in a consistent and systematic manner.

I.9 APPENDIX I REFERENCES

Canadian Climate Program, 1993, *Adaptation to Climatic Variability and Change*, Task Force Report on Climate Adaptation, occasional paper 19, University of Guelph, Guelph, Ontario, Canada.

Cooper, C.F., 1992, *Sensitivities of Western U.S. Ecosystems to Climate Change,* contractor report, prepared for the Office of Technology Assessment, Washington, D.C.

Easterling, W.E., 1993, "Adapting United States Agriculture to Climate Change," in *Preparing for an Uncertain Climate*, Vol. 1, pp. 303-305, OTA-O-567, Office of Technology Assessment, Washington, D.C.

Frederick, K.D., and P.H. Gleick, 1989, "Water Resources and Climate Change," in N. Rosenberg et al. (eds.), *Greenhouse Warming: Abatement and Adaptation,* pp. 133-146, Resources for the Future, Washington, D.C.

Gillilan, D., 1992, *Innovative Approaches to Water Resource Management,* contractor report, prepared by the University of Arizona, Tucson, Ariz., for the Office of Technology Assessment, Washington, D.C.

IPCC, 1991, *The Seven Steps to the Assessment of the Vulnerability of Coastal Areas to Sea Level Rise: A Common Methodology*: *Report of the Advisory Group on Assessing Vulnerability to Sea Level Rise and Coastal Zone Management,* revision 1, WMO/UNEP, Intergovernmental Panel on Climate Change, Response Strategies Working Group.

IPCC, 1990, *Policymakers Summary of the Formulation of Response Strategies*: *Report of Working Group III*, WMO/UNEP, Intergovernmental Panel on Climate Change.

Johda, N.S., 1989, "Potential Strategies for Adapting to Greenhouse Warming: Perspectives from the Developing World," in N. Rosenberg et al. (eds.), *Greenhouse Warming: Abatement and Adaptation*, pp. 147-158, Resources for the Future, Washington, D.C.

Lewandrowski, J.K., and R.J. Brazee, 1993, "Farm Programs and Climate Change," *Climatic Change* 23:1-20.

Lillieholm, R.J., 1993, "Preserves at Risk: An Investigation of Resource Management Strategies, Implications and Opportunities," in *Preparing for an Uncertain Climate*, Vol. 2, pp. 244-250, OTA-O-567, Office of Technology Assessment, Washington, D.C.

National Academy of Sciences, 1992, *Policy Implications of Greenhouse Warming: Mitigation, Adaptation, and the Science Base,* National Academy Press, Washington, D.C.

Office of Technology Assessment, 1993a, *Preparing for an Uncertain Climate*, Vol. 1, OTA-O-567, Office of Technology Assessment, D.C.

Office of Technology Assessment, 1993b, *Preparing for an Uncertain Climate*, Vol. 2, OTA-O-568, Office of Technology Assessment, Washington D.C.

Rosenzweig, C., and M.L. Parry, 1994, "Potential Impact of Climate Change on World Food Supply," *Nature* 367:133-138.

Schad, T.M., 1992, *A New Commission to Study U.S. Water Policy?* draft report prepared for the Office of Technology Assessment (contract I3-5820.0), Washington, D.C.

Smith, J.B., and J. Mueller-Vollmer, 1993, *Setting Priorities for Adapting to Climate Change,* contractor report prepared by RCG/Hagler Bailly, Inc., Arlington, Va., for the Office of Technology Assessment, Washington, D.C.

Smith, J.B., and D. Tirpak (eds.), 1989, *The Potential Effects of Global Climate Change on the United States*, EPA-230-05-89-050, U.S. Environmental Protection Agency, Washington, D.C.

Smith, W., 1992, *Managing Forests under a Changing Climate: Workshop Summary,* presented at a workshop convened by the Office of Technology Assessment, June 18-19, 1992, Washington, D.C.

Titus, J.G., 1991, "Strategies for Adapting to the Greenhouse Effect," *APA Journal* 311:311-323.

Wahl, R., 1992, *The Management of Water Resources in the Western U.S. and Potential Climate Change,* contractor report prepared by the University of Colorado, Boulder, Colo., for the Office of Technology Assessment, Washington, D.C.

Wang, B., et al., 1994, "Potential Impacts of Sea-Level Rise on the Shanghai Area," *Journal of Coastal Research*, special issue 14 (in press).

Warren, R.S., 1992, *Coastal Land Vulnerabilities to Climate Change*, contractor report, prepared for the Office of Technology Assessment, Washington, D.C.

Wilhite, D.A., 1990, *Planning for Drought: A Process for State Government,* IDIC Technical Series 90-1, International Drought Information Center, University of Nebraska, Lincoln, Neb.

APPENDIX J:

DESCRIPTION OF ASSESSMENT METHODS FOR ASSESSING IMPACTS ON FISHERIES

R. Benioff et al. (eds.), Vulnerability and Adaptation Assessments, J-1–J-20.
© *1996 Kluwer Academic Publishers.*

APPENDIX J NOTATION

ACRONYMS, INITIALISMS, AND ABBREVIATIONS

DO	dissolved oxygen
GCM	general circulation model
HSI	habitat suitability index

UNITS OF MEASURE

°C	degree(s) Celsius
gm	gram(s)
ha	hectare(s)
°K	degree(s) Kelvin
kg	kilogram
km	kilometer
m	meter
ppm	parts per million
yr	year(s)

CONTENTS

J.1 INTRODUCTION

The choice of methods for use in evaluating the vulnerability of fisheries resources in any particular country will be dependent on the habitats and fisheries present in that country and on the availability of climate, hydrology, limnology, ecology, and fisheries data, as well as the expected nature of the change in climate. In general, the methods focus on three major climate-related variables; temperature, precipitation, and sea level, and evaluate potential effects at the species or total-fishery level (Fig. J.1).

J.2 LACUSTRINE FISHERIES RESOURCES

The methods for evaluating climate change impacts on lacustrine fisheries include: 1) comparing growth and mortality under historic and predicted temperature regimes; 2) estimating fish yield under historic and predicted temperature regimes; and 3) estimating habitat availability under historic and predicted lake levels and surface water area.

J.2.1   Assessing the Effects of Temperature Changes on Growth and Feeding

Approach: This species-specific method uses a bioenergetics model originally developed by Kitchell et al. (1977) and commercially available from the Wisconsin SeaGrant program as the Bioenergetics Model 2 (Hewett and Johnson 1992). The model processes data on fish physiology, diet composition, energy density, and water temperature and generates consumption and/or growth estimates. The model includes a database containing specific information for 20 taxa (primarily North American), including *Tilapia*. Species not included in this database may be added and modelled using species-specific parameter values. This model has been used by Hill and Magnuson (1990) to evaluate potential climate warming affects on growth of several Great Lakes species. Climate change impacts on fisheries resources are inferred by inputting historic and global circulation model (GCM)-predicted water temperatures and estimating growth and feeding rates under historic and predicted temperature conditions. Because the GCM output provides predicted air temperature values but not water temperatures, multiple regression analysis is used to develop models for predicting water temperatures from air and water temperature (Hill and Magnuson 1990).

Data Requirements:

- Historic air and water temperatures for the habitats of concern.

- Predicted air temperatures.

- Species-specific physiological data for each species of interest, including consumption, respiration, and egestion/excretion data. Bioenergetics 2 model species database can be used to provide surrogate-species data in the absence of species-specific data.

Outcome and Vulnerability Assessment: The Bioenergetics Model 2 will predict growth (gm/gm of body weight/day) and feeding rates (as gm of prey consumed/gm of body weight/day) of individual fish, and the results on individual fish may be extrapolated to the resource as a whole, for historic and predicted water temperatures. The model will also estimate total biomass (gm) for a species under different temperature regimes. Vulnerability is then assessed by evaluating the differences in growth and feeding rates and total biomass production between historic and predicted temperatures.

Limitations: This approach does not include considerations of climate-induced changes to hydrology, limnology, physical habitat availability, water quality, ecosystem structure and function, or changes in fishing pressure, techniques, or success.

J.2.2 Assessing the Effects of Temperature Changes on Mortality

Approach: This species-specific method used the empirical relationships between natural mortality, asymptotic length or weight, K of the von Bertalanffy growth formula[1], and mean annual air temperature developed by Pauly (1980) to estimate the exponential coefficient of natural mortality under historic and GCM-predicted air temperatures. The empirical models were developed using data from 175 fish stocks from 84 species, including marine and freshwater species from tropical to polar habitats. The fish stocks include four cichlid species from Lake Malawi, a single cichlid species from Lake Victoria, and *Stolothrissa tanganicae* from Lake Tanganyika. The empirical models are:

1) $\log M = -0.2107 - 0.0824 \log W_\infty + 0.6757 \log K + 0.4627 \log T$ (r = 0.845)

2) $\log M = -0.0066 - 0.279 \log L_\infty + 0.6543 \log K + 0.4634 \log T$ (r = 0.847)

where M = natural mortality (unitless), W = maximum (asymptotic) live weight (in grams), L_∞ = maximum (asymptotic) total length (in cm), K = species-specific growth coefficient (unitless), and T = mean annual water temperature (°C). Equation 1 is used if the maximum weight, but not total length, is known for the species of concern. Equation 2 is used when the maximum length, but not weight, is known. It is suggested that both equations be used if possible. Spread-sheet versions of both equations should be developed. Multiple regression analysis is used to develop models for predicting mean annual water temperatures from mean annual air temperature.

Data Requirements:

- Maximum weight and length values for the species of interest.
- Growth coefficients for species of interest.
- Historic and predicted air and water temperatures.

Outcome and Vulnerability Assessment: The empirical models will permit estimation of the exponential coefficient of natural mortality for particular fish stocks under historic and GCM-predicted annual temperature conditions. All else being equal, the higher the estimated coefficient of natural mortality, the greater will be the mortality experienced by

---

[1] The von Bertalanffy growth equation for length has the formula: $L_t = L_\infty[1 - e^{-K(t - t0)}]$, where L_t is the length of the fish at time t, L_∞ is the maximum (asymptotic) length, and K is the growth coefficient. An identical growth equation exists for weight, substituting W for L.

a particular fish stock. Vulnerability is assessed by evaluating the difference in natural mortality coefficient between historic and predicted temperature conditions.

Limitations: These models overestimate natural mortality for strongly schooling pelagic fishes such as the herrings (Clupeidae) (Pauly 1983). Pauly (1983) suggests reducing estimates of M for such species by multiplying the estimated values by 0.80. This approach does not include considerations of potential climate-induced changes in physical habitat availability, water quality, or interactions among species. Although this approach does not consider fishing pressure, gear, or success, Pauly (1980) does suggests that the equations can be used to estimate the catchability of a given fishing gear.

J.2.3 Assessing the Effects of Temperature Changes on Fish Yield

Approach: This method uses the empirical model developed by Schlesinger and Regier (1982) which permits estimates of maximum sustainable yield (*MSY*) in lakes from mean annual air temperature and the morphoedaphic index (*MEI*) of lakes. The *MEI* was calculated as the total dissolved solids divided by the mean depth. A similar approach was used by Schlesinger and Regier (1983) to evaluate effects of environmental temperatures on the yields of subarctic and temperate fishes in North America. For intensively fished lakes (n = 43), the models are:

3) $\log_{10} MSY = 0.061 TEMP + 0.043$ $\quad\quad$ ($r^2 = 0.744$)
4) $\log_{10} MSY = 0.050 TEMP + 0.280 \log_{10} MEI + 0.236$ $\quad\quad$ ($r^2 = 0.810$)
5) $\log_{10} MSY = 0.044 TEMP + 0.482 \log_{10} MEI_{25} + 0.021$ $\quad\quad$ ($r^2 = 0.830$)

where *MSY* = maximum sustainable yield (kg/ha/yr), *TEMP* = mean annual air temperature (°C), *MEI* = morphoedaphic index (calculated as total dissolved solids [ppm]/mean depth[m]), and MEI_{25} = *MEI* for a maximum depth of 25 m. Equation 3 should be used if data for estimating the *MEI* are unavailable. Equation 4 should be used for lakes with a maximum mean depth less than 25 m, while equation 5 is used for deep lakes (maximum mean depth > 25 m). The *MEI* for equation 5 is calculated setting the mean depth 25 meters. For each of these equations, *MSY* is assumed to be closely approximated by the reported yield.

For all the lakes (n = 123) evaluated by Schlesinger and Regier (1982), the models are:

6) $\log_{10} yield = 0.051 TEMP + 0.358 EFFORT + 0.161 \log_{10} MEI - 0.383$ $\quad$ ($r^2 = 0.708$)
7) $\log_{10} yield = 0.050 TEMP + 0.349 EFFORT + 0.146 \log_{10} MEI_{25} - 0.367$ $\quad$ ($r^2 = 0.698$)

where *yield* = reported yield (kg/ha/yr) where relationship to *MSY* is unknown, and *EFFORT* = a dummy variable (1 for lakes with light to moderate or unknown fishing intensity or 2 for intensively-fished lakes). Equation 6 is used for lakes where the mean depth is less than 25 m, while equation 7 applies to lakes with maximum mean depths greater than 25 m. Spreadsheet versions of equations 3-7 should be developed.

Data Requirements:

- Historic total dissolved solids (mg/l) for the lake of concern.
- Historic and predicted mean depth and lake levels (meters).

- Historic fish yield.
- Historic and predicted mean annual air temperature.

Outcome and Vulnerability Assessment: Use of the models will permit estimates of *MSY* under different predicted temperature regimes. The effects of precipitation and runoff as they affect lake levels and mean lake depth may also be estimated. Vulnerability is assessed by comparing *MSY* estimates under historic and predicted air temperatures and lake depths.

Limitations: This approach cannot be used to address potential impacts to individual species, but rather addresses all species within the fishery. The models are not suited for use with lakes that: 1) exhibit wide fluctuations in depth (such as lakes Chad and Mweru); 2) do not conform to typical carbonate-bicarbonate chemical type (such as Lake Turkana); and 3) contain a large area of swamp (such as Lake Bangweulu). The use of the *EFFORT* term may not adequately estimate fishing effort in some lakes. This approach does not consider changes in water quality, biotic interactions, or fishing effort, gear, or success, or changes in fishing pressure on various species.

J.2.4 Assessing the Effects of Precipitation Changes on Fish Yields

Approach: This approach employs regression analyses using historic and predicted surface water runoff values from the Water Resources Vulnerability Assessments (see Section 5.4), GCM-predicted precipitation data, and surface water area, to develop lake and reservoir specific models for estimating surface water area under different climate scenarios. The historic and predicted surface water areas are then input into one of three empirical models (Crul 1992) that relate potential yield from lakes and reservoirs in Africa to the surface water area of the water body of concern. The empirical models are:

8) $Catch = 8.32 \, Area^{0.92}$ (n = 71 lakes and reservoirs; $r^2 = 0.93$)
9) $Catch = 8.93 \, Area^{0.92}$ (n = 46 lakes; $r^2 = 0.92$)
10) $Catch = 7.09 \, Area^{0.94}$ (n = 25 reservoirs; $r^2 = 0.94$)

where *Catch* = annual fish yield (metric tons) and *Area* = annual average surface water area (km^2). *Catch* may also be expressed as yield (kg/ha/year) by dividing the estimated catch by the surface area. For the lake or reservoir of concern, historic catch and surface water area data are applied to determine which model is most appropriate for use. Spreadsheet versions of the three models should be developed.

Data Requirements:

- Historic precipitation, surface water runoff, and surface water area data.
- Historic annual fish yields.
- Predicted precipitation, surface water runoff, and surface water area.

Outcome and Vulnerability Assessment: The empirical models permit estimation of *Catch* (annual fish yield) from surface water area, which is a direct function of precipitation and surface water runoff. Vulnerability is assessed by comparing *Catch* estimates under historic and predicted precipitation regimes.

Limitations: This approach will not address impacts to individual species. This approach does not consider potential impacts of temperature or water quality, nor does it consider potential changes in water quality, biotic interactions, or fishing effort, gear, or success.

J.2.5 Assessing the Effects of Precipitation Changes on Habitat Availability

Approach: This approach draws qualitative evaluations of impacts to fisheries (species-specific or overall) from quantitative estimates of changes in habitat availability due to changes in lake levels and inundation. Existing Water Resources Vulnerability Assessments (see Section 5.4) provide data on surface water runoff to and from lakes. Regression analyses using historic precipitation, lake level, surface water area, and surface water runoff data can be employed to develop models for predicting lake level and surface water area from precipitation and runoff data. Bathymetric and topographic maps as well as aerial photographs will also be employed to delineate shorelines and important habitat areas such as spawning grounds and nursery areas. The regression models will be used to predict changes in lake level and surface water area at important fisheries habitats. These predictions will then be applied to the maps and photographs of the fisheries habitats to infer impacts from changes in water depth and inundation. This approach is conceptually similar to the aerial videography approach being employed for the Coastal Impacts Vulnerability Assessment (see Section 5.4), and will require data from the Water Impacts assessments.

Data Requirements:

- Bathymetric and topographic maps and aerial photographs for lake areas with known important fisheries habitats.
- Historic lake level, surface water runoff, and precipitation data.
- Predicted surface water runoff from the Water Resources Assessments.

Outcome and Vulnerability Assessment: The regression models will permit estimation of lake level (water depth) and surface water area for lake areas known to contain important fisheries habitats. Qualitative determinations of impacts will be made by evaluating changes in habitat availability (expressed as area) due to changes in lake level. Vulnerability is assessed comparing habitat availability (area) under historic and predicted precipitation.

Limitations: This approach does not consider habitat suitability, which may be affected by changes in precipitation and surface runoff which bring in nutrients. This approach may also overestimate impacts because it does not address the potential formation of new shallow water habitats as lake levels decline. For example, existing habitats may be lost if nearshore habitats become dewatered as lake levels decline. However, these habitats may be replaced as deeper nearshore areas are brought closer to the lake surface as lake depth decreases.

J.3 RIVERINE FISHERIES RESOURCES

The assessment of potential climate change impacts on fisheries resources is most problematic for riverine-based fisheries. The following methods are focused to assess the impacts of temperature, dissolved oxygen, and precipitation on fish growth, life history, habitat suitability and yield.

J.3.1 Assessing the Effects of Temperature on Productivity

Approach: This approach was developed by Huet (as described in Welcomme 1979) to estimate the annual productivity of a river on the basis of average stream width, the biogenic capacity of the stream, average annual temperature, the acidity or alkalinity of the water, and the type of fish population present in the river. The productivity is calculated using the following formula:

11) $\quad K = B \times L \times (k_1 \times k_2 \times k_3)$

where K = annual productivity (kg/km of river), L = average width of the river (m), B = biogenic capacity, k_1 = annual average temperature, k_2 = acidity or alkalinity of the water, k_3 = the type of fish population present (unitless). Values of B can range from 1-3 for waters with little fish food, 4-6 for average waters, and 7-10 for water rich in fish food. The value for k_3 can be approximated on the basis of the percent of rheophilic (flowing water) and limnophilic (quiet water) species in the fish community using the following equation:

12) $\quad k_3 = (2L + R)/100$

where L = the percentage of the fish community comprised of limnophilic species, and R = the percentage of the fish community comprised of rheophilic species. Spreadsheet versions of equations 11 and 12 should be developed. Water temperature can be estimated from air temperatures using regression analysis to develop a predictive model for water temperature (Stefan and Preud'homme 1993). Mean stream width can be determined from flow and stage data and topographic maps.

Data Requirements:

- Estimates of the food resources available for the river segment of concern - may be assessed using the biomass of zooplankton, benthic invertebrates, or aquatic vegetation.
- Alkalinity data.
- Historic and predicted annual water temperatures.
- Average stream width data.
- Topographic maps.
- Current annual productivity estimates.

Outcome and Vulnerability Assessment: Annual productivity is estimated for current and predicted water temperatures. Vulnerability is assessed by comparing estimated annual productivity under historic and predicted climate scenarios.

Limitations: This approach cannot be used to evaluate potential impacts to individual species. This approach does not consider the effects of changes in water quality or in fishing effort, gear, or success.

J.3.2 Assessing the Effects of Temperature on Growth and Feeding

Approach: This species-specific approach is identical to the approach presented in Appendix J.2.1, and uses the Bioenergetics Model 2 (Hewett and Johnson 1992). Climate change impacts on fisheries resources are inferred comparing Bioenergetics Model 2-predicted growth and feeding rates and biomass production for historic and predicted temperature conditions. Regression analysis (Stefan and Preud'homme 1993) is used to develop models for predicting water temperatures from air temperatures.

Data Requirements:

- Historic air and water temperatures.
- Predicted air temperatures.
- Species-specific physiological data for each species of interest, including consumption, respiration, and egestion/excretion data. Bioenergetics 2 model species database can be used to provide surrogate-species data in the absence of species-specific data.

Outcome and Vulnerability Assessment: The Bioenergetics Model 2 will predict growth and feeding rates to an individual of a species, and biomass production for the entire resource, under historic and predicted water temperatures. Vulnerability is assessed by evaluating the differences in growth and feeding rates and biomass production between historic and predicted temperatures. Results for individuals may be extrapolated to the resource as a whole.

Limitations: This approach does not include considerations of potential climate-induced changes to physical habitat availability, water quality, changes in environmental variables other than temperature, changes in fishing effort, gear, or success.

J.3.3 Assessing the Effects of Temperature on Habitat Suitability - Temperature Tolerance

Approach: This species-specific approach compares predicted maximum water temperatures to species-specific temperature tolerance and preference. Water temperatures will be estimated from predicted air temperatures using the regression approach of Stefan and Preud'homme (1993).

Data Requirements:

- Species upper thermal tolerance and preferred temperature.
- Historic water temperature data.
- Predicted air and water temperatures.

Outcome and Vulnerability Assessment: Monthly or weekly predicted maximum temperature profiles will be developed for selected river segments. A finer time scale (e.g., weekly) will be used if permitted by the GCM predictions. The predicted temperatures will be compared to species-specific preferred temperatures and upper temperature tolerance. This procedure will quantify the amount of riverine habitat that is unsuitable for a particular species on the basis of its thermal biology. Vulnerability will be assessed by comparing amounts of unsuitable habitats for selected species in selected river locations.

Limitations: Thermal data may not be available for some species of concern. This approach does not address changes in water quality parameters other than temperature, biotic interactions, or changes in fishing effort, gear, or success.

J.3.4 Assessing the Effects of Precipitation (as Streamflow) on Life History Parameters

Approach: Stream hydrographs are graphical representations of changes in stream flow over time (e.g., minimum flows in late summer, maximum flows during spring). The shape of the yearly hydrograph and the magnitude, timing, and duration of flood events have been shown to play a very important role in the maintenance of tropical riverine fish stocks (Lowe-McConnell 1987; Welcomme 1976, 1985). In this species-specific approach, predicted changes in the annual hydrograph for a particular river or reach are compared to the historic or current hydrograph, and affects on spawning, growth, and recruitment are qualitatively inferred from the predicted differences in the timing, magnitude, and/or duration of flooding events, as well as differences in minimum and maximum flows. A similar approach was employed by the U.S. Fish and Wildlife Service (Tyus and Karp 1989, 1991) in identifying adverse effects of hydropower facilities on endangered fishes of the upper Colorado River system of North America. Life history information, such as spawning time, nursery period, and flow preference, will be needed for each species of interest. This approach will require the output of the Water Impacts Assessment, and interactions among the fisheries and hydrology technical staff.

Data Requirements:

- Historic hydrograph data identifying stream flow, flood period and duration for area of concern.
- Life history data on spawning period, nursery period, spawning and nursery habitats, and flow preference for species of concern.
- Hydrographs for predicted precipitation scenarios indicating timing, magnitude, and duration of flood periods.

Outcome and Vulnerability Assessment: Historic and predicted hydrographs will be developed for specific river reaches, and important life history parameters (spawning period, nursery period) will be identified on each hydrograph. Vulnerability of the fishery is assessed by comparing changes in major hydrograph components (such as timing, magnitude and duration of flood events) relative to the timing of important life history parameters (spawning, nursery period) under historic and predicted precipitation scenarios. The greater the difference between the historic and predicted hydrographs relative to spawning and nursery periods, the greater the potential for adverse impacts to the fishery.

Limitations: The availability of historic precipitation and flow data or historic hydrographs may be limited. This approach does not consider changes in fishing effort, gear, or success, domestic or industrial pollution, environmental variables other than precipitation, or biotic interactions.

J.3.5 Assessing the Effects of Precipitation on Catch

Approach: Welcomme (1976) and others have reported on the relationships between catch and discharge or floodplain area. Discharge and floodplain area are directly related to precipitation. The approaches described in this section are similar to those used by Welcomme (1976) and others, and use regression analysis to develop empirical models for predicting catch.

Catch vs Floodplain Area: The effects of altered precipitation on catch will be evaluated using historic data to develop catch vs floodplain area relationships. In the absence of sufficient historic catch and floodplain data, either of the following relationships may be used:

14) $C = 2.65 A - 0.98$ $(r^2 = 0.828)$ (Welcomme 1976)

15) $C = 3.83A$ $(r^2 = 0.865)$ (Welcomme 1980)

where C = annual catch (metric tons) per km reach of river and A = floodplain area (km^2) per km reach of river. Historic flood and catch data should be used to determine which equation to use. If adequate historic data are available, river-specific relationships (of the form shown in equations 14 and 15) will be developed. The estimation of floodplain area under predicted precipitation regimes will use the results of the Water Resources impact assessments and will require interaction of the fisheries assessment staff with the hydrology staff performing the Water Resources vulnerability assessment.

Catch vs River Discharge: Fish production has been found to be positively correlated with river discharge, and potential effects of precipitation changes can be evaluated using relationships between river discharge and catch (Welcomme and Hargborg 1977; Sagua 1993). For example, the following relationship was developed between mean total catch (as a 3-yr mean) and mean river discharge (3-yr mean) of the Niger River at Mopti:

16) $Y = 29.338 + 0.075X$ $(r = 0.988)$ (from Sagua 1993)

where Y = total catch (metric tons, averaged over 3 years) and X = mean river discharge (averaged for the same 3–year period). In this present approach, regression analyses will be employed using historic discharge and total annual catch data to develop predictive models of the form described above. The Water Impacts vulnerability assessments (see Section 5.4) will provide predicted discharge values which will then be input to the model to predict catch.

Data Requirements:

- Historic discharge, floodplain area, and catch data.
- Predicted precipitation, surface runoff, flow and discharge data.

Outcome and Vulnerability Assessment: Discharge and floodplain area will be estimated for predicted precipitation regimes. Total annual catch is estimated by inputting the predicted floodplain area or discharge to the appropriate empirical model. Vulnerability is assessed by comparing historic annual catches with predicted catches for the different climate scenario precipitation regimes.

Limitations: The availability of catch data for floodplain areas may be limited. If so, then the relationships developed by Welcomme (1976, 1980) will be used. Neither the floodplain nor the discharge approach consider the potential effects in changes to water quality, fishing effort, gear, or success, or interactions among species.

J.3.6 Assessing the Effects of Precipitation, Temperature, and Dissolved Oxygen on Habitat Suitability Using Habitat Suitability Models

Approach: This species-specific approach includes the development of habitat suitability index (HSI) models for individual species of concern. Habitat suitability modelling was developed by the U.S. Fish and Wildlife Service to aid in impact assessment and habitat management activities, and HSI models have been developed for a variety of North American terrestrial, freshwater, and marine species (see Hays 1987). The models may incorporate environmental variables such as water temperature, current velocity, floodplain inundation duration, dissolved oxygen (DO) concentrations, and substrate composition, and produce an index of habitat suitability between 0 (unsuitable habitat) and 1 (optimally suitable habitat). The specific variables to be included will depend on the known habitats requirements of the target species and the availability of appropriate data. Predicted changes in temperature can be directly input to the models and also be used to predict DO levels (using the approach described in Appendix J.3.3) for model input. Predicted precipitation must be converted to changes in stream flow in conjunction with the Water Resources vulnerability assessments. Development of the habitat models will be aided using Micro-HSI, habitat suitability index modelling software for microcomputers developed by the U.S. Fish and Wildlife Service (Hays 1987). The completed models will be used to estimate habitat suitability under current and predicted climate conditions.

Data Requirements:

- Species-specific habitat and physiology data.
- Habitat characteristics, including but not limited to temperature, DO, substrate, stage, and flow.
- Predicted temperature and precipitation data.

Outcome and Vulnerability Assessment: Following construction of species-specific HSI models, habitat suitability indices will be estimated for specific habitats using historic (or current) climatic, hydrological, and ecological data. Suitability indices will then be calculated for the predicted climatic and hydrological conditions associated with each climate change scenario. Vulnerability will be assessed by comparing historic and predicted HSI estimates.

Limitations: The strength of any HSI model is a direct function of the availability of the life history and physiology data for each species of concern, as well as the availability of habitat data. In the absence of such data, professional judgement may be used to develop some components of the models, although this will lessen the strength of the suitability estimation. This approach does not consider effects of potential changes in fishing effort, gear, or success, or species interactions.

J.4 COASTAL FISHERIES RESOURCES - PENAEID SHRIMP

The assessment of potential climate change impacts on coastal marine fisheries resources focuses exclusively on the penaeid shrimp fishery. However, the assessment approach identified in Appendix J.2.2, which estimates natural mortality from annual water temperature, and the approaches presented in Appendix J.3.6 and J.4.5, which involve the development of habitat suitability models, can also be applied to fish species in coastal marine areas.

J.4.1 Assessing the Effects of Precipitation Changes on Shrimp Yield or Catch

Approach: Studies conducted in the United States, Senegal, the Gulf of Mexico, and Australia have demonstrated the influence of rainfall on shrimp production (see Garcia and Le Reste 1981). Particularly good correlations have been identified between total annual shrimp catch and the previous two years of total annual rainfall, and both positive and negative relationships have been found depending on the specific area or the species. The identified approach uses regression analysis to develop models predicting shrimp yield or catch from total rainfall. Historic yield and catch data and precipitation data are used to develop empirical models for the shrimp fishery of concern. Predicted annual precipitation values are then used to predict shrimp yield for the future climate scenarios.

Data Requirements:

- Historic precipitation data for the area of the shrimp fishery of concern.
- Historic yield data for the fishery area of concern.
- Predicted annual precipitation for the fishery area of concern.

Outcome and Vulnerability Assessment: Using historic data, models are developed to predict shrimp yield or catch from precipitation. GCM precipitation data are then used to predict shrimp yield or catch under different precipitation scenarios. Vulnerability is assessed by comparing predicted annual shrimp yield or catch under historic and predicted precipitation scenarios.

Limitations: This approach only indirectly assesses the effects of continental discharge or the effects of variability in salinity. This approach does not consider the effects of domestic or industrial pollution, environment factors (other than precipitation), sea level rise, or changes in fishing effort, gear, and success. Because many of the relationships that have been developed are not linear (see Garcia and Le Reste 1981), extrapolations of yield estimates should not be made outside the range of the empirical data used to develop the predictive models.

J.4.2 Assessing the Effects of Temperature Changes on Shrimp Yield

Approach: The basis for this approach is the approach for evaluating climate change impacts presented in Regier et al. (1990), where the authors performed regression analyses of empirical data to develop exponential relationships of the Arrhenius form relating ecological rates to absolute temperature. The Arrhenius form of an exponential relationship may be given as:

17) $\log_e k_i = a - b\,(1/T_i)$

where k = rate constant, T = absolute temperature ($^\circ K$), and a (y-intercept) and b (slope) are coefficients estimated by regression analyses. Using empirical data from several sources, Regier et al. (1990) developed the following model for penaeid shrimp yield:

18) $\log_e SCSY = 52.0 - 14312(1/T)$ (r = 0.58)

where $SCSY$ = stabilized commercial shrimp yield (kg/ha of intertidal vegetation) and T = mean annual air temperature ($^\circ K$). In the absence of historic temperature and shrimp yield data, the model developed by Regier et al. (1990) will be used to estimate shrimp yields under different temperature scenarios. A spreadsheet version of this equation should be developed. If historic data are available, regression analyses will be used to develop site-specific Arrhenius models.

Data Requirements:

- Historic data on shrimp yield in kg/ha of intertidal vegetation.
- Historic air mean annual air temperature data.
- Predicted mean annual air temperatures.

Outcome and Vulnerability Assessment: GCM-predicted temperatures will be input to the models to predict shrimp yields under different temperature scenarios. Vulnerability will be assessed by comparing historic yields and predicted yields under each climate scenario.

Limitations: This approach does not consider the effects of domestic or industrial pollution, environment factors other than temperature, sea level rise, or changes in fishing effort, gear, and success. The model developed by Regier et al. (1990) is not based on a particularly strong relationship.

J.4.3 Assessing the Effects of Sea Level Change on Habitat Availability and Shrimp Yield Using Yield and Habitat Data

Approach: This approach uses regression analysis to develop predictive models of the relationship between annual shrimp yield (kg/ha of vegetation) and area of intertidal and estuarine vegetation. The use of this approach is based on the results of studies that have documented positive relationships between penaeid annual yield and estuarine and intertidal vegetation (Turner 1977, 1992; Turner and Boesch 1988). Map and photograph analysis will be performed to quantify the amount of intertidal and estuarine vegetation for each shrimp fishery area. Annual shrimp yield data will also be collected for each fishery area. The loss of intertidal and estuarine vegetation that could result from predicted changes in sea level will be estimated for different climate scenarios using map and photographic analyses and the approach used in the Coastal Impacts vulnerability assessment (see Section 5.4). The predicted changes in vegetation area will then be input to the regression models to predict annual shrimp yield for each climate scenario.

Data Requirements:

- Areal estimates of intertidal and estuarine vegetation and current sea levels for each shrimp fishery area of concern.
- Annual shrimp yield data for each shrimp fishery area.
- Predicted sea level changes for each shrimp fishery area (obtained from the Coastal Resources vulnerability assessment.

Outcome and Vulnerability Assessment: Annual shrimp yields will be estimated for different climate scenarios, using estimated changes in estuarine and intertidal vegetation from sea level change. Vulnerability will be assessed by comparing predicted shrimp yields under each climate scenario to historic and current annual shrimp yields.

Limitations: This approach does not consider the effects of domestic or industrial pollution, environmental factors such as temperature and salinity, or changes in fishing effort, gear, and success.

J.4.4 Assessing the Effects of Sea Level Change on Habitat Availability and Shrimp Yield in the Absence of Yield Data

Approach: In the absence of adequate yield data, it will not be possible to develop predictive models as described in previous sections. This approach allows for qualitative determinations of reduction in shrimp recruitment, growth, and yield on the basis of quantitative estimates of changes in intertidal and estuarine vegetation. The underlying assumption for this approach, based on the work of Turner (1977) and others, is that for a given shrimp fishery area, growth, recruitment, and annual yield are positively related to the areal extent of intertidal and estuarine vegetation. Thus, the loss or gain of vegetation due to changes in sea level will be mirrored by changes in the shrimp resource. Map and photograph analyses, along with the results of the Coastal Impact vulnerability assessment, will be used to estimate changes in intertidal and estuarine vegetation under different climate scenarios.

Data Requirements:

- Bathymetric and topographic maps and aerial photographs of shrimp estuarine and intertidal areas of vegetation.
- Predicted changes in sea level for each area for each climate change scenario (obtained from the Coastal Resource vulnerability assessment).

Outcome and Vulnerability Assessment: The map and photographic analyses will provide quantitative estimates of the area of intertidal and estuarine vegetation under current and predicted sea level conditions of each climate scenario. Vulnerability will be assessed on the basis of changes in the amount (area) of vegetated habitat under current and predicted sea levels.

Limitations: This approach does not provide quantitative predictions of how shrimp yield may be affected by changes in sea level, and does not include considerations of temperature, salinity, or other environmental factors, domestic or industrial pollution, or changes in fishing effort, gear, and success.

J.4.5 Assessing the Effects of Temperature and Sea Level Changes on Habitat Suitability

Approach: This approach, which is identical to that described for riverine fishes in Appendix J.3.6, includes the development of habitat suitability index (HSI) models for the penaeid shrimp. Turner and Brody (1983) developed HSI models for white and brown shrimp inhabiting the northern Gulf of Mexico. The models should incorporate environmental variables similar to those used by Turner and Brody (1983), such as water temperature, salinity, substrate composition, and amount of estuarine and intertidal vegetation, and produce an index of habitat suitability between 0 (unsuitable habitat) and 1 (optimally suitable habitat). Predicted changes in temperature can be directly input to the models, while effects of sea level change will first be related to effects on estuarine and intertidal vegetation following the approach described in Appendix J.4.3 and J.3.4. Development of the HSI models will be aided using Micro-HSI, habitat suitability index modelling software for microcomputers developed by the U.S. Fish and Wildlife Service (Hays 1987).

Data Requirements:

- Delineations of the areal extent of intertidal and estuarine vegetation in known shrimp areas.
- Bathymetric and topographic maps and/or aerial photographs of known shrimp areas.
- Species-specific information such as salinity and temperature preference and tolerance, substrate preferences, and optimal temperatures for growth.
- Current and predicted water temperature and sea level.

In the absence of specific data, professional judgement will be used to develop relationships between individual environmental variables and habitat suitability.

Outcome and Vulnerability Assessment: The outcome of this approach will be an HSI model for predicting habitat suitability for each shrimp species of interest. Vulnerability will be assessed by comparing HSI values for specific shrimp fishery areas using current and predicted water temperatures and areal estimates of intertidal and estuarine vegetation. If the suitability of a particular area is less under predicted climate conditions than under current conditions, then climate change will be considered to pose an adverse threat to that fishery. The magnitude of the difference will also reflect the potential magnitude of the impact to the fisheries.

Limitations: The models will apply only to vegetated areas and not to open bottom areas, and will not consider the effects of changes in water quality, fishing effort, gear, or success, or biotic interactions. The accuracy of the suitability predictions depends on the nature of the environmental and life history data used in the development of the models.

J.6 APPENDIX J REFERENCES

Crul, R.C.M., 1992, *Models for Estimating Potential Fish Yields of African Inland Waters*, CIFA Occasional Paper No. 16, Food and Agriculture Organization of the United Nations, Rome.

Garcia, S. and L. Le Reste, 1981, *Life Cycles, Dynamics, Exploitation, and Management of Coastal Penaeid Shrimp Stocks*, FAO Technical Paper No. 203, Food and Agriculture Organization of the United Nations, Rome.

Hays, R.L., 1987, *A Users Manual for Micro-HSI, Habitat Suitability Index Modelling Software for Microcomputers, Version 2*, National Ecology Center, U.S. Fish and Wildlife Service, Fort Collins, Colorado.

Hewett, S.W. and B.L. Johnson, 1992, *Fish Bioenergetics Model 2*, WIS-SG-91-250, University of Wisconsin Sea Grant Institute, Madison, Wisconsin.

Kitchell, J.F., D.J. Stewart, and D. Weininger, 1977, *Applications of a Bioenergetics Model to Yellow Perch (Perca flavescens) and Walleye (Stizostedion vitreum vitreum)*, J. Fish. Res. Board Can. 34: 1922-1935.

Lowe-McConnell, R.H., 1987, *Ecological Studies in Tropical Fish Communities*, Cambridge University Press, Cambridge, Great Britain.

Pauly, D., 1980, *On the Interrelationships Between Natural Mortality, Growth Parameters, and Mean Environmental Temperature in 175 Fish Stocks*, J. Cons. int. Explor. Mer. 39(2): 175-192.

Pauly, D., 1983, *Some Simple Methods for the Assessment of Tropical Fish Stocks*, FAO Fisheries Technical Paper No. 234, Food and Agriculture Organization of the United Nations, Rome.

Regier, H.A., J.A. Holmes, and D. Pauly, 1990, *Influence of Temperature Changes on Aquatic Ecosystems: An Interpretation of Empirical Data*, Trans. Am. Fish. Soc. 119: 374-389.

Sagua, V.O., 1993, *The Effects of Climate Change on the Fisheries of the Sahel*, Food and Agriculture Organization of the United Nations, Rome.

Schlesinger, D.A. and H.A. Regier, 1982, *Climatic and Morphoedaphic Indices of Fish Yields from Natural Lakes*, Trans. Am. Fish. Soc. 111: 141-150.

Schlesinger, D.A. and H.A. Regier, 1983, *Relationship Between Environmental Temperature and Yields of Subarctic and Temperate Zone Fish Species*, Can. J. Fish. Aquat. Sci. 40: 1829-1837.

Stefan, H.G. and E.B. Preud'homme, 1993, *Stream Temperature Estimation from Air Temperature*, Water Resources Bulletin 29: 27-45.

Turner, R.E., 1977, *Intertidal Vegetation and Commercial Yields of Penaeid Shrimp*, Trans. Am. Fish. Soc. 106(5): 411-416.

Turner, R.E., 1992, *Coastal Wetlands and Penaeid Shrimp Habitat*, pp. 97 - 104 in: *Stemming the Tide of Coastal Fish Habitat Loss*, Marine Recreational Fisheries Report No. 14, R.H. Stroud, ed., National Coalition for Marine Conservation, Inc., Savannah, Georgia.

Turner, R.E. and D.F. Boesch, 1988, *Aquatic Animal Production and Wetland Relationships: Insights Gleaned Following Wetland Loss or Gain*, pp. 25 - 39 in: *The Ecology and Management of Wetlands, Vol. 1, Ecology of Wetlands*, D.D. Hook et al., eds., Timber Press, Portland, Oregon.

Turner, R.E. and M.S. Brody, 1983, *Habitat Suitability Index Models: Northern Gulf of Mexico Brown Shrimp and White Shrimp*, FWS/OBS-82/10.54, U.S. Department of the Interior, Fish and Wildlife Service, Washington, D.C.

Tyus, H.M. and C.A. Karp, 1989, *Habitat Use and Streamflow Needs of Rare and Endangered Fishes, Yampa River, Colorado*, U.S. Fish and Wildlife Service Biological Report 89(14), U.S. Department of the Interior, Fish and Wildlife Service, Washington, D.C.

Tyus, H.M. and C.A. Karp, 1991, *Habitat Use and Streamflow Needs of Rare and Endangered Fishes in the Green River, Utah*, U.S. Fish and Wildlife Service, Flaming Gorge Studies Program, Colorado River Fishes Project, Vernal, Utah.

Welcomme, R.L., 1976, *Some General and Theoretical Considerations on the Fish Yield of African Rivers*, J. Fish Biol. 8: 351-364.

Welcomme, R.L., 1979, *Fishery Management in Large Rivers*, FAO Fisheries Technical Paper No. 194, Food and Agriculture Organization of the United Nations, Rome.

Welcomme, R.L., 1980, *Some Factors Affecting the Catch of Tropical River Fisheries*, pages 268 - 275 in: *Comparative Studies on Freshwater Fisheries*, FAO Fisheries Technical Paper No. 198, Food and Agriculture Organization of the United Nations, Rome.

Welcomme, R.L., 1985, *River Fisheries*, FAO Fisheries Technical Paper No. 262, Food and Agriculture Organization of the United Nations, Rome.

Welcomme, R.L. and D. Hagborg, 1977, *Towards a Model of a Floodplain Fish Population and its Fishery*, Env. Biol. Fish 2(1): 7-24.

Figure J.1 Relationships among the fishery resources, climate change environmental variables, and the potential ecological responses of the fishery resources.

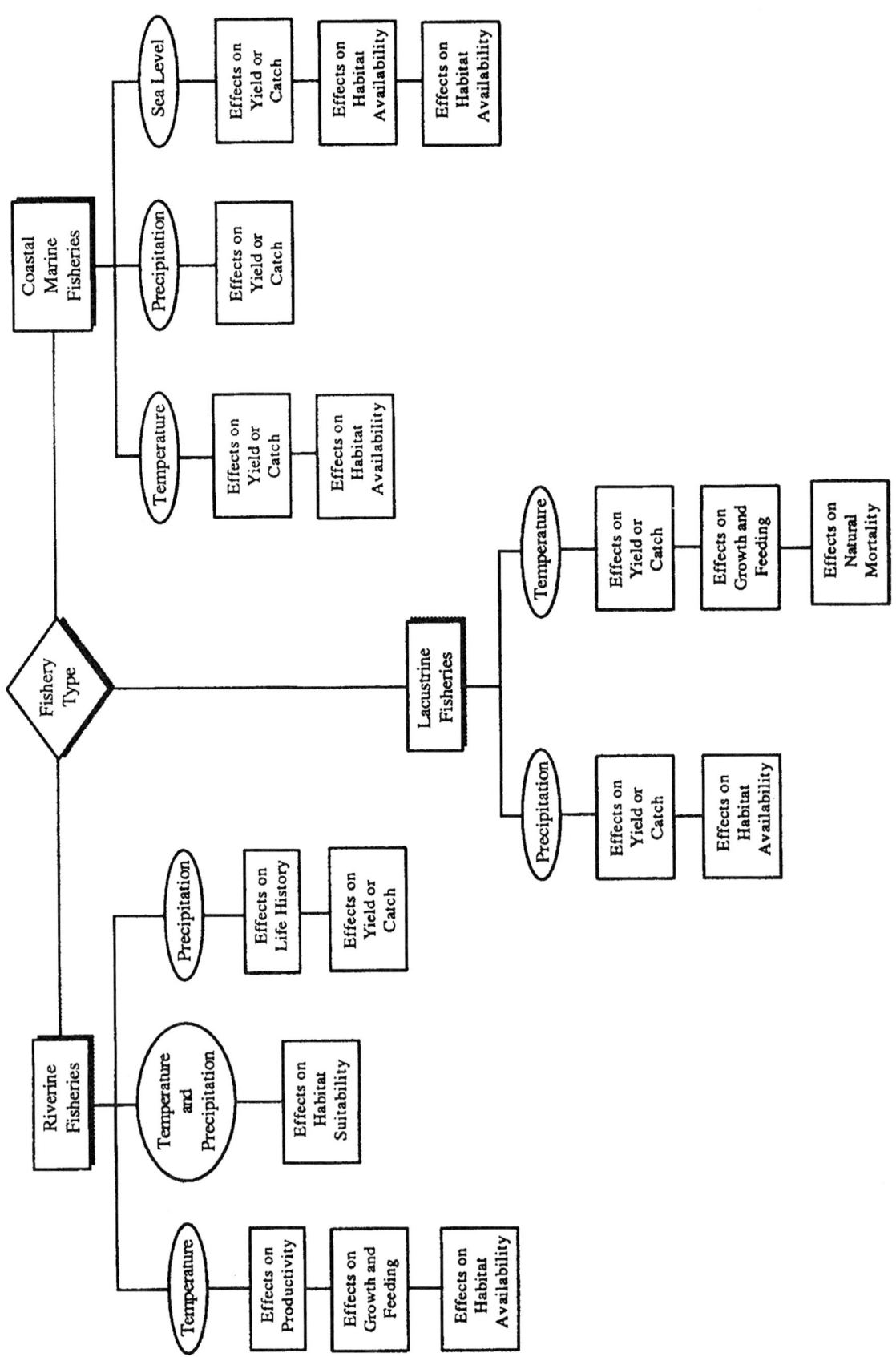

Environmental Science and Technology Library

1. A. Caetano, M.N. De Pinho, E. Drioli and H. Muntau (eds.), *Membrane Technology: Applications to Industrial Wastewater Treatment*. 1995
   ISBN 0-7923-3209-1
2. Z. Zlatev: *Computer Treatment of Large Air Pollution Models*. 1995
   ISBN 0-7923-3328-4
3. J. Lemons and D.A. Brown (eds.): *Sustainable Development: Science, Ethics, and Public Policy*. 1995   ISBN 0-7923-3500-7
4. A.V. Gheorghe and M. Nicolet-Monnier: *Integrated Regional Risk Assessment*. Volume I: Continuous and Non-Point Source Emissions: Air, Water, Soil. 1995   ISBN 0-7923-3717-4
   Volume II: Consequence Assessment of Accidental Releases. 1995
   ISBN 0-7923-3718-2
   Set: ISBN 0-7923-3719-0
5. L. Westra and J. Lemons (eds.): *Perspectives on Ecological Integrity*. 1995
   ISBN 0-7923-3734-4
6. J. Sathaye and S. Meyers: *Greenhouse Gas Mitigation Assessment: A Guidebook*. 1995   ISBN 0-7923-3781-6
7. R. Benioff, S. Guill and J. Lee (eds.): *Vulnerability and Adaptation Assessments*. An International Handbook. 1996   ISBN 0-7923-4140-6
8. J.B. Smith, S. Huq, S. Lenhart, L.J. Mata, I. Nemošová and S. Toure (eds.): *Vulnerability and Adaptation to Climate Change*. Interim Results from the U.S. Country Studies Program. 1996   ISBN 0-7923-4141-4
9. B.V. Braatz, B.P. Jallow, S. Molnár, D. Murdiyarso, M. Perdomo and J.F. Fitzgerald (eds.): *Greenhouse Gas Emission Inventories*. Interim Results from the U.S. Country Studies Program. 1996   ISBN 0-7923-4142-2

KLUWER ACADEMIC PUBLISHERS – DORDRECHT / BOSTON / LONDON